Biological Mechanisms Underlying Physical Fitness and Sports Performance

Biological Mechanisms Underlying Physical Fitness and Sports Performance

Editors

Georgian Badicu
Filipe Manuel Clemente
Eugenia Murawska-Ciałowicz

MDPI • Basel • Beijing • Wuhan • Barcelona • Belgrade • Manchester • Tokyo • Cluj • Tianjin

Editors
Georgian Badicu
Physical Education and Special Motricity
Transilvania University of Brasov
Brasov
Romania

Filipe Manuel Clemente
Polytechnic Institute of Viana do Castelo
School of Sport and Leisure
Melgaco
Portugal

Eugenia Murawska-Ciałowicz
Physiology and Biochemistry Department
Wroclaw University of Health and Sport Sciences
Wroclaw
Poland

Editorial Office
MDPI
St. Alban-Anlage 66
4052 Basel, Switzerland

This is a reprint of articles from the Special Issue published online in the open access journal *Biology* (ISSN 2079-7737) (available at: www.mdpi.com/journal/biology/special_issues/Bio_Fitness_Sports).

For citation purposes, cite each article independently as indicated on the article page online and as indicated below:

LastName, A.A.; LastName, B.B.; LastName, C.C. Article Title. *Journal Name* **Year**, *Volume Number*, Page Range.

ISBN 978-3-0365-5724-3 (Hbk)
ISBN 978-3-0365-5723-6 (PDF)

Contents

About the Editors

Georgian Badicu

Badicu Georgian is an associate professor at Transilvania University of Brasov, Faculty of Physical Education and Mountain Sports, Department of Physical Education and Special Motricity. He has published about 61 papers in JCR journals and is the author or co-author of 47 papers published by MDPI. His main interests are physical activity, fitness, education, obesity, well-being, recreation, football, public health, quality of life, sports activities, physical education, didactics of physical education, and sports.

Filipe Manuel Clemente

Filipe Manuel Batista Clemente has been a university professor since the 2012/2013 academic year and is currently an assistant professor at Escola Superior de Desporto e Lazer de Melgaco (IPVC, Portugal). As scientific merit, Filipe has had 264 articles published and/or accepted by journals indexed with an impact factor (JCR), as well as over 105 scientific articles that have been peer-reviewed indexed in other indexes. In addition to scientific publications in journals and congresses, he is also the author of six international books and seven national books in the areas of sports training and football. He has also edited various special editions subordinate to sports training in football in journals with an impact factor and/or indexed in SCImago. Additionally, he is a frequent reviewer for impact factor journals in quartiles 1 and 2 of the JCR.

Although he started producing research in 2011, he was included in the restricted list of the world's most-cited researchers in the world (where only eight other Portuguese researchers in sports sciences appear), which was published in the journal Plos Biology in 2020. In 2021, the list was updated, with Filipe Manuel Clemente being again included in the top 2% of the world researchers, in which was positioned in the second place in six Portugueses included in the area of sports sciences.

Filipe M. Clemente's SCOPUS h-index is 24 (with a total of 2605 citations), and his Google h-index is 35 (5305 citations). In a list promoted by independent website Expert Escape, he was ranked 40th of 14,875 researchers of football (soccer) in 2020 and in 19th of 15949 in 2021.

Eugenia Murawska-Ciałowicz

Eugenia Murawska-Cialowicz is Editor-in-Chief at the journal *Human Movement* and associate professor at Wroclaw University of Health and Sport Sciences, Physiology and Biochemistry Department, 51-612 Wroclaw, Poland. His main interests are oxidative stress, skeletal muscle exercise physiology, physical activity, exercise, exercise biochemistry, exercise performance, exercise testing, physiology, brain-derived neurotrophic, antioxidants, adipose tissue biology, etc.

biology MDPI

Editorial

Biological Mechanisms Underlying Physical Fitness and Sports Performance: An Editorial

Georgian Badicu [1,*], Filipe Manuel Clemente [2,3,4] and Eugenia Murawska-Cialowicz [5]

1 Department of Physical Education and Special Motricity, Transilvania University of Brasov, 500068 Brasov, Romania
2 Escola Superior Desporto e Lazer, Instituto Politécnico de Viana do Castelo, Rua Escola Industrial e Comercial de Nun'Álvares, 4900-347 Viana do Castelo, Portugal
3 Research Center in Sports Performance, Recreation, Innovation and Technology (SPRINT), 4960-320 Melgaço, Portugal
4 Instituto de Telecomunicacoes, Delegacao da Covilha, 1049-001 Lisboa, Portugal
5 Physiology and Biochemistry Department, Wroclaw University of Health and Sport Sciences, 51-612 Wroclaw, Poland
* Correspondence: georgian.badicu@unitbv.ro

Citation: Badicu, G.; Clemente, F.M.; Murawska-Cialowicz, E. Biological Mechanisms Underlying Physical Fitness and Sports Performance: An Editorial. *Biology* **2022**, *11*, 1425. https://doi.org/10.3390/biology11101425

Received: 27 September 2022
Accepted: 28 September 2022
Published: 29 September 2022

Publisher's Note: MDPI stays neutral with regard to jurisdictional claims in published maps and institutional affiliations.

In general, the concept of a mechanism in biology has three distinct meanings. It may refer to a philosophical thesis about the nature of life and biology, to the internal workings of a machine-like structure, or to the causal explanation of a particular phenomenon [1].

Understanding the biological mechanisms that justify acute and chronic physiological responses to exercise interventions determines the development of training principles and training methods. A strong understanding of the effects of exercise in humans may help researchers to identify what causes specific biological changes and to properly identify the most adequate processes for implementing a training stimulus [1].

Despite the significant body of knowledge regarding the physiological and physical effects of different training methods (based on load dimensions), some biological causes of those changes are still unknown. Additionally, few studies have focused on natural biological variability in humans and how specific human properties may underlie different responses to the same training intervention. Thus, more original research is needed to provide plausible biological mechanisms that may explain the physiological and physical effects of exercise and training in humans.

In this Special Issue, we discuss/demonstrate the biological mechanisms that underlie the beneficial effects of physical fitness and sports performance, as well as their importance and their role in/influences on physical health.

A total of 28 manuscripts are published here, of which 25 are original articles, two are reviews, and one is a systematic review.

Two papers are on neuromuscular training programs (NMTs), training monotony (TM), and training strain (TS) in soccer players [2,3]; five articles provide innovative findings about testosterone and cortisol [4,5], gastrointestinal hormones [6], spirulina [7], and concentrations of erythroferrone (ERFE) [8]; another five papers analyze fitness and its association with other variables [7,9–12]; three papers examine body composition in elite female soccer players [2], adolescents [6], and obese women [7]; five articles examines the effects of high-intensity interval training (HIIT) [7,10,13–15]; one paper examines the acute effects of different levels of hypoxia on maximal strength, muscular endurance, and cognitive function [16]; another article evaluates the efficiency of using vibrating exercise equipment (VEE) compared with using sham-VEE in women with CLBP (chronic low-back pain) [17]; one article compares the effects of different exercise modes on autonomic modulation in patients with T2D (type 2 diabetes mellitus) [14]; and another paper analyzes the changes in ABB (acid–base balance) in the capillaries of kickboxers [18]. Other studies evaluate: the effects of resistance training on oxidative stress and muscle damage in spinal cord-injured rats [19]; the effects of muscle training on core muscle performance in rhythmic gymnasts [20]; the physiological profiles of road cyclist in different age categories [21]; changes in body composition during the COVID-19 [22]; a mathematical model capable of predicting 2000 m rowing performance using a maximum-effort 100 m indoor rowing ergometer [23]; the effects of ibuprofen on performance and oxidative stress [24]; the associations of vitamin D levels with various motor performance tests [12]; the level of knowledge on FM (Fibromyalgia) [25]; and the ability of a specific BIVA (bioelectrical impedance vector analysis) to identify changes in fat mass after a 16-week lifestyle program in former athletes [26]. Finally, one review evaluates evidence from published systematic reviews and meta-analyses about the efficacy of exercise on depressive symptoms in cancer patients [27]; another review presents the current state of knowledge on satellite cell-dependent skeletal muscle regeneration [28]; and a systematic review evaluates the effects of exercise on depressive symptoms among women during the postpartum period [29].

Author Contributions: Conceptualization, G.B., F.M.C. and E.M.-C.; writing—original draft preparation, G.B.; writing—review and editing, F.M.C. and E.M.-C. All authors have read and agreed to the published version of the manuscript.

Funding: This research received no external funding.

Institutional Review Board Statement: Not applicable.

Informed Consent Statement: Not applicable.

Data Availability Statement: Not applicable.

Conflicts of Interest: The authors declare no conflict of interest.

References

1. Nicholson, D.J. The concept of mechanism in biology. *Stud. Hist. Philos. Sci. Part C Stud. Hist. Philos. Biol. Biomed. Sci.* **2012**, *43*, 152–163. [CrossRef] [PubMed]
2. Roso-Moliner, A.; Mainer-Pardos, E.; Arjol-Serrano, J.L.; Cartón-Llorente, A.; Nobari, H.; Lozano, D. Evaluation of 10-Week Neuromuscular Training Program on Body Composition of Elite Female Soccer Players. *Biology* **2022**, *11*, 1062. [CrossRef] [PubMed]
3. Nobari, H.; Gholizadeh, R.; Martins, A.D.; Badicu, G.; Oliveira, R. In-Season Quantification and Relationship of External and Internal Intensity, Sleep Quality, and Psychological or Physical Stressors of Semi-Professional Soccer Players. *Biology* **2022**, *11*, 467. [CrossRef] [PubMed]
4. Zurek, G.; Danek, N.; Żurek, A.; Nowak-Kornicka, J.; Żelaźniewicz, A.; Orzechowski, S.; Stefaniak, T.; Nawrat, M.; Kowal, M. Effects of Dominance and Sprint Interval Exercise on Testosterone and Cortisol Levels in Strength-, Endurance-, and Non-Training Men. *Biology* **2022**, *11*, 961. [CrossRef]
5. Burgos, J.; Viribay, A.; Calleja-González, J.; Fernández-Lázaro, D.; Olasagasti-Ibargoien, J.; Seco-Calvo, J.; Mielgo-Ayuso, J. Long-Term Combined Effects of Citrulline and Nitrate-Rich Beetroot Extract Supplementation on Recovery Status in Trained Male Triathletes: A Randomized, Double-Blind, Placebo-Controlled Trial. *Biology* **2022**, *11*, 75. [CrossRef] [PubMed]
6. Vasto, S.; Amato, A.; Proia, P.; Baldassano, S. Is the Secret in the Gut? SuperJump Activity Improves Bone Remodeling and Glucose Homeostasis by GLP-1 and GIP Peptides in Eumenorrheic Women. *Biology* **2022**, *11*, 296. [CrossRef]
7. Nobari, H.; Gandomani, E.E.; Reisi, J.; Vahabidelshad, R.; Suzuki, K.; Volpe, S.L.; Pérez-Gómez, J. Effects of 8 Weeks of High-Intensity Interval Training and Spirulina Supplementation on Immunoglobin Levels, Cardio-Respiratory Fitness, and Body Composition of Overweight and Obese Women. *Biology* **2022**, *11*, 196. [CrossRef]
8. Dziembowska, I.; Wójcik, M.; Bukowski, J.; Żekanowska, E. Physical Training Increases Erythroferrone Levels in Men. *Biology* **2021**, *10*, 1215. [CrossRef]
9. Jastrzębska, A.D.; Hebisz, R.; Hebisz, P. Temporal Skin Temperature as an Indicator of Cardiorespiratory Fitness Assessed with Selected Methods. *Biology* **2022**, *11*, 948. [CrossRef]
10. Domaradzki, J.; Koźlenia, D.; Popowczak, M. Prevalence of Positive Effects on Body Fat Percentage, Cardiovascular Parameters, and Cardiorespiratory Fitness after 10-Week High-Intensity Interval Training in Adolescents. *Biology* **2022**, *11*, 424. [CrossRef]
11. González-Fernández, F.T.; González-Víllora, S.; Baena-Morales, S.; Pastor-Vicedo, J.C.; Clemente, F.M.; Badicu, G.; Murawska-Ciałowicz, E. Effect of Physical Exercise Program Based on Active Breaks on Physical Fitness and Vigilance Performance. *Biology* **2021**, *10*, 1151. [CrossRef] [PubMed]
12. Gilic, B.; Kosor, J.; Jimenez-Pavon, D.; Markic, J.; Karin, Z.; Domic, D.S.; Sekulic, D. Associations of Vitamin D Levels with Physical Fitness and Motor Performance; A Cross-Sectional Study in Youth Soccer Players from Southern Croatia. *Biology* **2021**, *10*, 751. [CrossRef] [PubMed]

13. Amara, S.; Barbosa, T.M.; Chortane, O.G.; Hammami, R.; Attia, A.; Chortane, S.G.; Tillaar, R.V.D. Effect of Concurrent Resistance Training on Lower Body Strength, Leg Kick Swimming, and Sport-Specific Performance in Competitive Swimmers. *Biology* **2022**, *11*, 299. [CrossRef] [PubMed]
14. Silva, L.R.B.; Gentil, P.; Seguro, C.S.; de Oliveira, J.C.M.; Silva, M.S.; Marques, V.A.; Beltrame, T.; Rebelo, A.C.S. High-Intensity Interval Training Improves Cardiac Autonomic Function in Patients with Type 2 Diabetes: A Randomized Controlled Trial. *Biology* **2022**, *11*, 66. [CrossRef] [PubMed]
15. Arslan, E.; Kilit, B.; Clemente, F.M.; Soylu, Y.; Sögüt, M.; Badicu, G.; Akca, F.; Gokkaya, M.; Murawska-Ciałowicz, E. The Effects of Exercise Order on the Psychophysiological Responses, Physical and Technical Performances of Young Soccer Players: Combined Small-Sided Games and High-Intensity Interval Training. *Biology* **2021**, *10*, 1180. [CrossRef]
16. Karayigit, R.; Eser, M.C.; Sahin, F.N.; Sari, C.; Sanchez-Gomez, A.; Dominguez, R.; Koz, M. The Acute Effects of Normobaric Hypoxia on Strength, Muscular Endurance and Cognitive Function: Influence of Dose and Sex. *Biology* **2022**, *11*, 309. [CrossRef]
17. Zurek, G.; Kasper-Jędrzejewska, M.; Dobrowolska, I.; Mroczek, A.; Delaunay, G.; Ptaszkowski, K.; Halski, T. Vibrating Exercise Equipment in Middle-Age and Older Women with Chronic Low Back Pain and Effects on Bioelectrical Activity, Range of Motion and Pain Intensity: A Randomized, Single-Blinded Sham Intervention Study. *Biology* **2022**, *11*, 268. [CrossRef]
18. Rydzik, Ł.; Mardyła, M.; Obmiński, Z.; Więcek, M.; Maciejczyk, M.; Czarny, W.; Jaszczur-Nowicki, J.; Ambroży, T. Acid–Base Balance, Blood Gases Saturation, and Technical Tactical Skills in Kickboxing Bouts According to K1 Rules. *Biology* **2022**, *11*, 65. [CrossRef]
19. Barros, N.D.A.; Aidar, F.J.; Marçal, A.C.; Santos, J.L.; de Souza, R.F.; Menezes, J.L.; Gomes, M.Z.; de Matos, D.G.; Neves, E.B.; Carneiro, A.L.G.; et al. Effects of Resistance Training on Oxidative Stress Markers and Muscle Damage in Spinal Cord Injured Rats. *Biology* **2022**, *11*, 32. [CrossRef]
20. Esteban-García, P.; Jiménez-Díaz, J.F.; Abián-Vicén, J.; Bravo-Sánchez, A.; Rubio-Arias, J.Á. Effect of 12 Weeks Core Training on Core Muscle Performance in Rhythmic Gymnastics. *Biology* **2021**, *10*, 1210. [CrossRef]
21. Marín-Pagán, C.; Dufour, S.; Freitas, T.T.; Alcaraz, P.E. Performance Profile among Age Categories in Young Cyclists. *Biology* **2021**, *10*, 1196. [CrossRef] [PubMed]
22. Campa, F.; Bongiovanni, T.; Trecroci, A.; Rossi, A.; Greco, G.; Pasta, G.; Coratella, G. Effects of the COVID-19 Lockdown on Body Composition and Bioelectrical Phase Angle in Serie A Soccer Players: A Comparison of Two Consecutive Seasons. *Biology* **2021**, *10*, 1175. [CrossRef] [PubMed]
23. da Silva, L.F.; de Almeida-Neto, P.F.; de Matos, D.G.; Riechman, S.E.; de Queiros, V.; de Jesus, J.B.; Reis, V.M.; Clemente, F.M.; Miarka, B.; Aidar, F.J.; et al. Performance Prediction Equation for 2000 m Youth Indoor Rowing Using a 100 m Maximal Test. *Biology* **2021**, *10*, 1082. [CrossRef] [PubMed]
24. Aidar, F.J.; Fraga, G.S.; Getirana-Mota, M.; Marçal, A.C.; Santos, J.L.; de Souza, R.F.; Ferreira, A.R.P.; Neves, E.B.; Zanona, A.D.F.; Bulhões-Correia, A.; et al. Effects of Ibuprofen Use on Lymphocyte Count and Oxidative Stress in Elite Paralympic Powerlifting. *Biology* **2021**, *10*, 986. [CrossRef]
25. Mendoza-Muñoz, M.; Morenas-Martín, J.; Rodal, M.; García-Matador, J.; García-Gordillo, M.Á.; Calzada-Rodríguez, J.I. Knowledge about Fibromyalgia in Fibromyalgia Patients and Its Relation to HRQoL and Physical Activity. *Biology* **2021**, *10*, 673. [CrossRef]
26. Campa, F.; Matias, C.N.; Nunes, C.L.; Monteiro, C.P.; Francisco, R.; Jesus, F.; Marini, E.; Sardinha, L.B.; Martins, P.; Minderico, C.; et al. Specific Bioelectrical Impedance Vector Analysis Identifies Body Fat Reduction after a Lifestyle Intervention in Former Elite Athletes. *Biology* **2021**, *10*, 524. [CrossRef]
27. Marconcin, P.; Marques, A.; Ferrari, G.; Gouveia, É.R.; Peralta, M.; Ihle, A. Impact of Exercise Training on Depressive Symptoms in Cancer Patients: A Critical Analysis. *Biology* **2022**, *11*, 614. [CrossRef]
28. Kaczmarek, A.; Kaczmarek, M.; Ciałowicz, M.; Clemente, F.M.; Wolański, P.; Badicu, G.; Murawska-Ciałowicz, E. The Role of Satellite Cells in Skeletal Muscle Regeneration—The Effect of Exercise and Age. *Biology* **2021**, *10*, 1056. [CrossRef]
29. Marconcin, P.; Peralta, M.; Gouveia, É.R.; Ferrari, G.; Carraça, E.; Ihle, A.; Marques, A. Effects of Exercise during Pregnancy on Postpartum Depression: A Systematic Review of Meta-Analyses. *Biology* **2021**, *10*, 1331. [CrossRef]

biology

MDPI

Article

Evaluation of 10-Week Neuromuscular Training Program on Body Composition of Elite Female Soccer Players

Alberto Roso-Moliner [1], Elena Mainer-Pardos [1,*], José Luis Arjol-Serrano [1], Antonio Cartón-Llorente [1], Hadi Nobari [2,3,4,5] and Demetrio Lozano [1]

1 Health Sciences Faculty, Universidad San Jorge, Autov A23 km 299, Villanueva de Gállego, 50830 Zaragoza, Spain; aroso@usj.es (A.R.-M.); jlarjol@usj.es (J.L.A.-S.); acarton@usj.es (A.C.-L.); dlozano@usj.es (D.L.)
2 Faculty of Sport Sciences, University of Extremadura, 10003 Cáceres, Spain; hadi.nobari1@gmail.com or nobari.hadi@unitbv.ro
3 Department of Motor Performance, Faculty of Physical Education and Mountain Sports, Transilvania University of Braşov, 500068 Braşov, Romania
4 Sports Scientist, Sepahan Football Club, Isfahan 81887-78473, Iran
5 Department of Exercise Physiology, Faculty of Educational Sciences and Psychology, University of Mohaghegh Ardabili, Ardabil 56199-11367, Iran
* Correspondence: epardos@usj.es; Tel.: +34-976060100

Simple Summary: Soccer performance is complex, requiring mastery of sport specific technical and tactical skills with ideal physical fitness (i.e., includes sprints, hops, accelerations, changes of directions, and so on) and body composition (i.e., increase lean muscle mass and decrease fat mass). In the last decades, performance models have helped to understand the multifactorial mechanisms involved in physical performance in sports. Hence, we tested the hypothesis that a neuromuscular training (NMT) program has an effect on body composition parameters in elite female soccer players. The result showed that implementation of 10-week with thrice-weekly NMT program improves body composition in elite female soccer players.

Abstract: (1) Background: This study was conducted to investigate the effects of a 10-week neuromuscular training program (NMT) on the sum of six skinfolds (Σ6S) and body composition variables in elite female soccer players. (2) Methods: Forty-four Spanish elite female soccer players (age: 24.0 ± 4.2 years; height: 164.3 ± 5.5 cm; body mass: 60.4 ± 5.5 kg; body mass index (BMI): 22.4 ± 2.2 kg/m^2) were randomly assigned to a control group (CG) or to an experimental group (EG). Participants in the EG completed a specific NMT program of 24 min, three times per week, which included exercises from six different categories (mobility, dynamic stability, anterior chain strength, lumbopelvic control, posterior chain strength, and change of direction). The CG followed their normal strength and conditioning program. Pre- and post-intervention assessments included anthropometric measurements (weight, height, limb circumferences, and bone breadths), and subsequently, body composition factors BMI, Σ6S, body mass, muscle mass, and lean body mass were calculated. Nutrition was standardized by a nutritionist and also load monitored. (3) Results: A two-way mixed analysis of variance (group × time) revealed that there was a significant ($p \leq 0.001$) group × time interaction between body mass, fat mass, and Σ6S in favor of NMT. A significant interaction was also observed for body skeletal muscle mass and lean body mass favoring NMT. (4) Conclusions: The application of an NMT program seems to be a useful strategy to improve body composition in elite female soccer players.

Keywords: football; body fat; women; strength training; lean body mass; kinanthropometry

Citation: Roso-Moliner, A.; Mainer-Pardos, E.; Arjol-Serrano, J.L.; Cartón-Llorente, A.; Nobari, H.; Lozano, D. Evaluation of 10-Week Neuromuscular Training Program on Body Composition of Elite Female Soccer Players. *Biology* **2022**, *11*, 1062. https://doi.org/10.3390/biology11071062

Academic Editor: Paulo Gentil

Received: 26 May 2022
Accepted: 14 July 2022
Published: 17 July 2022

1. Introduction

Soccer is arguably the most popular team sport today, and it is played by more than one billion people worldwide [1]. In 2020, the number of federated female soccer players in

Spain reached 77,461 [2]. That year, FIFA's Women's Football Strategy set out its goals of doubling the number of participants by 2026 [3].

Soccer performance is multifactorial and requires mastery of both sport-specific technical–tactical skills and optimal physical fitness. As an intermittent high-intensity sport, soccer involves activities such as sprints, jumps, accelerations, and changes of direction (COD), among others [4,5]. Of note, these high-intensity actions, coupled with the ability to repeat them without fatigue and the somatotype, also account for most actions that cause injury [6]. In addition, the somatotype of the players has shown to be relevant for this purpose [7]. Therefore, the link between anthropometry, muscle performance, and soccer-specific physical performance has been studied extensively (i.e., sprint, repeated sprint ability, vertical jump, etc.). Previous investigations found a strong link between body composition (high levels of lean mass and low levels of fat mass) and vertical jump performance and repeated sprint ability [8–10]. In addition, for leaner body compositions, lower body strength measurements are closely linked to soccer players' acceleration, sprinting, and leaping performance [11,12]. Given these correlations, training methods, such as neuromuscular training (NMT), that increase lower body strength and/or reduce body fat, enhancing the power-to-mass ratio, should result in significant increases in the physical performance parameters of female soccer players [13,14].

In this direction, the relationship between running performance (i.e., aerobic capacity) and body composition has been evaluated in elite males [15], youths [16], and elite females [17], showing high-speed actions and longer distances covered in players with greater lean body mass percentages as a marker of the muscle-to-fat ratio. In addition, previous published kinanthropometric studies in soccer showed different profiles as a function of age, sex, and playing position, and some of them have specifically analyzed the anthropometric profile of female soccer players [18–21].

The research working group on the body composition health and performance of athletes states that low body fat and high lean body mass are strongly correlated with higher levels of performance, especially in weight-sensitive sports such as soccer [22]. However, this relationship should be handled with caution because each sport has its own body composition (i.e., somatotype) that is considered ideal for success [23], and players with low body fat mass do not follow this general rule. A recently published review included kinanthropometric data of elite female soccer players from 2000 to 2020, showing a fat mass percentage between 14.5% to 22% [24]. In addition to the above, it is also important to analyze other performance factors that affect this sport, such as adequate nutritional intake [25] and genetics [26]. A few studies looked at the relationship between different endocrine parameters, such as IGFBP-3, erythropoietin, or estrogen, in female athletes. Additionally, genes related to performance and body composition, such as angiotensin-1 converting enzyme insertion/deletion (ACE I/D) polymorphism or α-actinin-3 (ACTN3) R577X polymorphism, have been studied [11].

In the last two decades, novel training approaches have been developed aiming to improve performance and body composition in female soccer players [17–19,22–24,27,28]. Interestingly, Myer et al. [29] suggested an integrative NMT program including mobility, dynamic stability, core strength, plyometric, agility, and fundamental strength exercises, showing it could improve sport-specific skills and minimize the risk of injuries.

With this rationale, some standardized neuromuscular protocols, such as FIFA11+, Sportsmetrics^{TM}, or Harmoknee, were developed and demonstrated to reduce injury risk [30–35] and improve performance [31,36] in female soccer players. Despite the effects of these training protocols on performance, the effect on body composition remains unknown. Rohmansyah et al. [37] found a reduction in body mass index (BMI), fat mass, and waist circumference in obese young-adult females after a 6-week FIFA 11+ program. Simões et al. analyzed the effects of NMT on body composition in volleyball athletes, finding improvements in body composition [38]. With respect to each of the interventions included in an NMT program, there is some research on the effects of body-weight resistance training [39], eccentric [40], and plyometric-based programs [41] on the body composition of female soccer players. However, the evidence is very scarce regarding the effects of a multicomponent program that combines all of them. Due to this, in the present study we hypothesized that an NMT program can be used to increase the lean body mass and body skeletal muscle mass or reduce the body mass, BMI, fat mass, and skinfold measurements in female soccer players. Thus, the main aim of this study was to evaluate the effects of an NMT program on the body composition of elite female soccer players.

2. Materials and Methods

2.1. Participants

Forty-four Spanish, highly-trained female soccer players voluntarily participated in the study (Table 1).

Table 1. Descriptive data of the participants.

Variable	Control Group (n = 22) Mean ± SD	Experimental Group (n = 18) Mean ± SD	p
Age (years)	24.61± 4.30	23.24 ± 4	0.31
Height (cm)	162.29 ± 5.90	166.09 ± 4.65	0.07
Body Mass (kg)	59.46 ± 6.22	61.61 ± 4.43	0.23
BMI (kg/m^2)	22.46 ± 2.54	22.37 ± 1.85	0.89

SD: standard deviation; BMI: body mass index.

Data collection took place during the competitive period (i.e., seventh month of the season). All the participants played for soccer teams in the Spanish Women's Second Division and completed a similar weekly soccer training regarding volume and methodology (i.e., five 90 min sessions per week and 1 game/week). All the participants met the following inclusion criteria: (i) at least 6 years of experience in soccer training and competition; (ii) participation in regular soccer training and competition for 6 months before data collection; (iii) free from injuries, and iv) refrained from other NMT programs or diets outside this study. Furthermore, participants were excluded if: (i) they missed three or more NMT sessions or (ii) they missed a testing day. The participants were randomly assigned (ABBA distribution) to a control group (CG, n = 22) or to an experimental group (EG, n = 22). Nevertheless, due to NMT attendance and testing days, the final sample was n = 22 for CG and n = 18 for EG (Figure 1). Prior to data collection, written informed consent was obtained from all the participants. The study was developed following the ethical standards of the Declaration of Helsinki and was approved by the Local Ethics Committee of Clinical Research (PI21/011, CEICA, Spain).

2.2. Measurement of Body Composition

The participants were tested according to the guidelines of the International Society of the Advancement of Kinanthropometry (ISAK) at the beginning (i.e., 1 week before) and the end of the intervention (i.e., 1 week after). The instruments were adjusted before their use and data were collected in duplicate [42]. To minimize measurement variation, the same experienced researcher examined all the subjects on the right side of the body during the same time of the day (i.e., 08:00 a.m.–10:00 a.m.). Participants were asked to avoid vigorous activities for at least 48 h before data collection and consumption of large volumes of water 2 h before as well as to follow their ordinary diet. Furthermore, to avoid any possible dietary confounding effects on body-composition assessment, in the pre- and post-test sessions, a 24-h food recall was collected by a registered dietician to check average macronutrient and energy intake (DAPA Measurement Toolkit, Cambridge, UK). After obtaining the players' data, the Spanish Food Composition Database (BEDCA) was used to calculate kilocalories and macronutrient intake. This database includes a compilation of nutritional data from various publications and food composition tables [43]. The results are shown in Table 2.

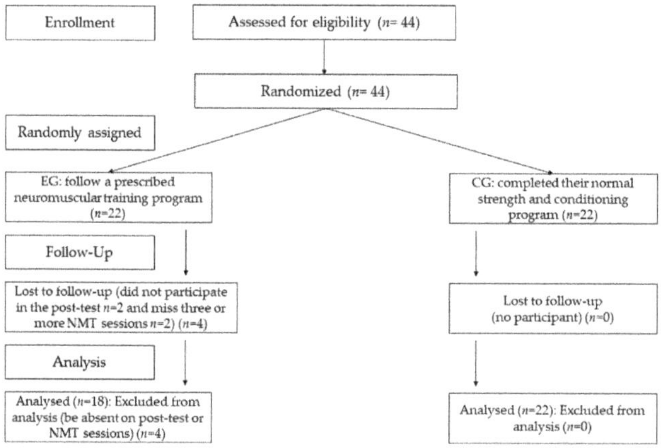

Figure 1. Participant recruitment, allocation, follow-up, and analysis are depicted in a CONSORT diagram. NMT: neuromuscular training; EG: experimental group; CG: control group.

Table 2. Average macronutrient and energy intake.

Variable	Control Group (n = 22)		Experimental Group (n = 18)		p
	1st Registration Mean ± SD	2nd Registration Mean ± SD	1st Registration Mean ± SD	2nd Registration Mean ± SD	
Kilocalories (kcal/day)	2206 ± 377	2222 ± 346	2266 ± 198	2285 ± 189.1	0.48
Carbohydrates (g)	311.9 ± 56.4	315.9 ± 47.9	336.9 ± 29.7	328.8 ± 29.1	0.13
Proteins (g)	92.9 ± 18.2	89.3 ± 13.2	90.8 ± 9.07	90.5 ± 10.6	0.56
Fats (g)	65.1 ± 12.1	66.8 ± 12.8	66.1 ± 6.79	67.8 ± 7.02	0.39

SD: standard deviation; kcal: kilocalories; g: grams.

Anthropometric measurements included: body mass in kilograms (kg) using a digital scale (BC-601, Tanita, IL, USA), height in centimeters (cm) employing a stadiometer (SECA 214, SECA, Hamburg, Germany), limb girths in cm using an anthropometric tape (Lufkin W606PM, Lufkin, NC, USA), bone breadths in cm utilizing a bone caliper (Campbell 10, Rosscraft, CA, USA), and skinfolds in millimeters (mm) using a slim guide skinfold caliper (Harpenden, West Sussex, UK). Specifically, 8 point skinfolds (e.g., triceps, biceps, abdominal, iliac crest, supraspinal, subscapular, front thigh, and medial calf), 4 limb girths (e.g., arm relaxed, arm tensed, mid-thigh, and calf), and 3 bone breadths (e.g., biepicondylar humerus, biepicondylar femur, and bi-styloid diameter of the wrist) were measured. The inter- and intra-observer technical error of measurement was less than 5.5% for skinfolds and less than 1.5% for the other variables.

BMI was calculated as body mass (in kg) divided by height in meters squared (kg/m^2) [44]. The sum of six skinfolds (Σ6S) was obtained as the addition in mm of the standardized 6 skinfolds (triceps, subscapular, supraspinal, abdominal, front thigh, and medial calf) [21]. Body density (BD) was calculated using the equation proposed by Withers et al. (1987) for female athletes [45] (Equation (1)). When BD was calculated, the Siri equation [46] was used to estimate fat mass percentage (Equation (2)). Lean body mass percentage was calculated as the difference between total body mass percentage and fat mass percentage. The body skeletal muscle mass was estimated with the equation of Lee et al. (2000) [47], and once this result was obtained, we converted it to a percentage (Equation (3)).

$$1.17484 - \{0.07229 * [\text{Log}\ (\Sigma 4\text{S.triceps} + \text{subscapular} + \text{supraspinal} + \text{medial calf})]\} \quad (1)$$

$$[(4.95/\text{BD}) - 4.5) * 100] \quad (2)$$

$$\text{Ht} * (0.00744 * \text{CAG2} + 0.00088 * \text{CTG2} + 0.00441 * \text{CCG2}) + (2.4 * \text{sex}) - (0.048 * \text{age}) + \text{race} + 7.8 \quad (3)$$

Ht: height (m); CAG: corrected arm girth (cm); CTG: corrected thigh girth (cm); CCG: corrected calf girth (cm); Sex (1 for male and 0 for female); Race (−2 for Asian, 1.1 for African American and 0 for white or Hispanic)

2.3. Exercise Protocol

Participants in the EG completed a rise, activate, mobilize, and potentiate (RAMP) system warm-up protocol [48] followed by a 24 min NMT program, three times per week, for 10 weeks. Participants in the CG followed the same warm-up protocol. Then, they completed their normal strength and conditioning program for 24 min (Table 3). Both groups were exercised equally and there was no significant difference between groups. Two weeks before the beginning of the intervention, four familiarization sessions were executed in the EG to get to know the exercises included in the NMT program (Figure 2).

Table 3. Training intervention details.

Group	Experimental (n = 18)	Control (n = 22)			Sum
Training Program	NMT (6 Exercises)	Mobility (3 Exercises)	Strength (3 Exercises)	RT (3 Exercises)	9 Exercises
Training Details	Sets: 4 Work: 40 s Rest: 20 s Duration: 24 min	Sets: 2 Work: 30 s Rest: 20 s Duration: 5 min	Sets: 4 Work: 40 s Rest: 20 s Duration: 12 min	Sets: 3 (4 reps) Work: ~10 s Rest: ~20 s Duration: 7 min	Sets: 2–4 Work: 10–40 s Rest: 20 s Duration: 24 min
Work Intensity (RPE)	7.3 ± 0.25		7.26 ± 0.23		p = 0.45

NMT: neuromuscular training; reps: repetitions per set; s: seconds; RPE: rate of perceived exertion (0–10); RT: running technique; SD: standard deviation.

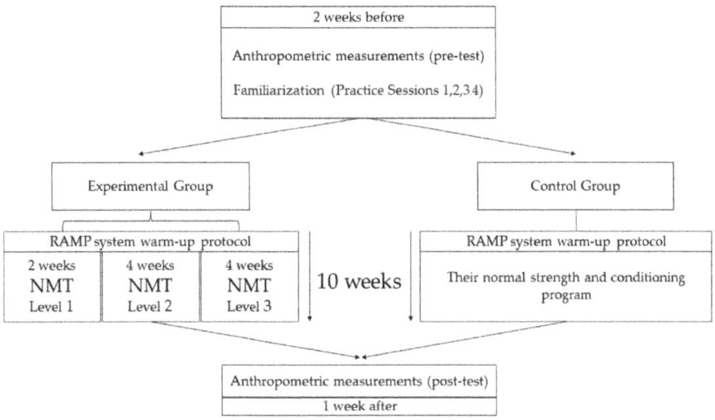

Figure 2. Project design timeline. NMT: neuromuscular training; RAMP: rise, activate, mobilize, and potentiate.

The intensity of the CG and EG training sessions was recorded using the modified Borg scale (0–10 rating), which is valid to control the training intensity and is commonly used by these players in their physical preparation [49,50]. At the end of the physical preparation, the players individually indicated their level of perceived exertion. The average for each of the sessions is shown in Figure 3.

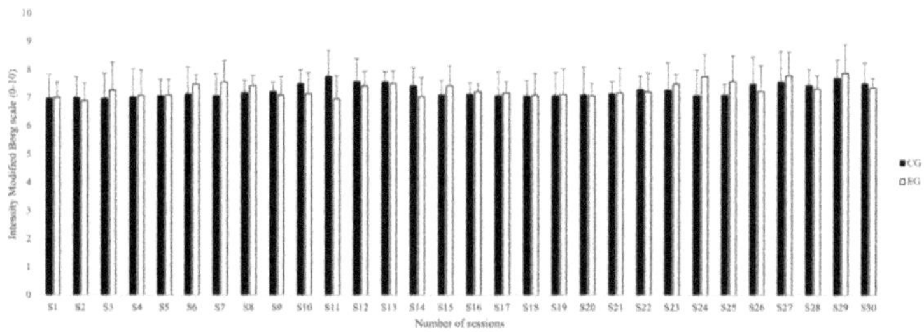

Figure 3. Average intensity (mean and standard deviation) using the modified Borg scale over the thirty training sessions. EG: experimental group; CG: control group.

The NMT program (Figure 4) included exercises from six different categories: (1) mobility, (2) dynamic stability, (3) anterior chain strength, (4) lumbopelvic control, (5) posterior chain strength, and (6) COD and was carried out in 4 sets of the 6-exercise circuit (40 s of work and 20 s of gentle running to change to the next exercise). Level 1 exercises were performed during the first 2 weeks, whereas levels 2 and 3 were performed during weeks 3–6 and 7–10, respectively. For unilateral exercises, the working leg changed between series.

Figure 4. Neuromuscular training protocol.

2.4. *Statistical Analysis*

Data analysis of the present study was carried out as both descriptive and inferential. Normality was inspected for all variables using a Shapiro-Wilk test. Macronutrient and energy intake, descriptive data, and possible differences pre-training were analyzed with independent group *t*-student. Within-group comparisons (Student paired *t*-test) were carried out to detect significant differences between the pre-test and post-test in all variables in both groups. A 2 (group) $\times$ 2 (time) repeated measures ANOVA with Bonferroni post hoc analysis was calculated for each parameter. Hedges' g effect size with a 95% confidence interval was also calculated to determine the magnitude of pairwise comparisons for pre- and post-test and was defined as trivial (<0.2), small (>0.2), moderate (>0.5), and large (>0.8). If the results of the independent sample *t*-test and effect sizes were similar for each group, then the percentage changes were computed and assessed. The significance of statistical analysis was used at the level of $p < 0.05$. All statistical calculations were performed using SPSS (Version 28.0, IBM SPSS Inc., Chicago, IL, USA).

3. Results

The descriptive characteristics of the players of both groups are shown in Table 1. The results of the analysis showed that there were no significant differences between the two groups in these variables. The average intakes of macronutrients and energy are shown in the Table 2. The results of the analysis showed that there were no significant differences in diet during the intervention in either group.

Table 4 shows the mean and SD of the changes in skinfold variables. At the baseline, there were no differences observed between groups in the above variables, except subscapular skinfold (f = 4.91; $p = 0.033$) and sum of six skinfolds (Σ6S) (f = 4.43; $p = 0.04$).

There were significant main effects of time ($p \leq 0.001$, f = 24.52, $\eta_p{}^2 = 0.39$; $p \leq 0.001$, f = 25.46, $\eta_p{}^2 = 0.40$; $p \leq 0.001$, f = 19.81, $\eta_p{}^2 = 0.34$; $p = 0.009$, f = 7.49, $\eta_p{}^2 = 0.16$; $p = 0.007$, f = 8.00, $\eta_p{}^2 = 0.17$; $p \leq 0.001$, f = 24.98, $\eta_p{}^2 = 0.39$) and a group by time interaction ($p \leq 0.001$, f = 29.73, $\eta_p{}^2 = 0.44$; $p \leq 0.001$, f = 47.25, $\eta_p{}^2 = 0.55$; $p \leq 0.001$, f = 41.72, $\eta_p{}^2 = 0.52$; $p \leq 0.001$, f = 51.08, $\eta_p{}^2 = 0.57$; $p \leq 0.001$, f = 25.22, $\eta_p{}^2 = 0.41$; $p \leq 0.001$, f = 68.87, $\eta_p{}^2 = 0.64$) for front thigh, medial calf, subscapular, iliac crest, abdominal, and Σ6S, respectively. The post hoc analysis indicated that front thigh (EG, $p \leq 0.01$, $g = -0.09$), medial calf (EG, $p \leq 0.001$, f = 24.99, $\eta_p{}^2 = 0.40$), subscapular (EG, $p \leq 0.01$, $g = -0.12$), iliac crest (CG, $p = 0.02$, $g = 0.02$, EG, $p \leq 0.01$, $g = -0.04$), abdominal (EG, $p \leq 0.01$, $g = -0.06$) and Σ6S (CG, $p = 0.019$, $g = 0.01$ and EG, $p \leq 0.01$, $g = -0.09$) skinfolds were significantly reduced. Percent changes of skinfold variables between pre- and post-test are shown in Table 4 and Figure 5.

Table 5 shows the mean and standard deviation in body mass, BMI, fat mass, body skeletal muscle mass, and lean body mass. At the baseline, there were no differences observed between groups in the above variables, except the body skeletal muscle mass Lee (f = 16.71; $p \leq 0.001$). There were significant ($p = 0.071$, f = 8.17, $\eta_p{}^2 = 0.18$; $p = 0.006$, f = 8.50, $\eta_p{}^2 = 0.18$; $p \leq 0.001$, f = 16.39, $\eta_p{}^2 = 0.30$; $p \leq 0.001$, f = 32.85, $\eta_p{}^2 = 0.46$; $p \leq 0.001$, f = 16.39, $\eta_p{}^2 = 0.30$) main effects of time and a group by time interaction ($p \leq 0.001$, f = 14.77, $\eta_p{}^2 = 0.28$; $p \leq 0.001$, f = 14.72, $\eta_p{}^2 = 0.28$; $p \leq 0.001$, f = 50.19, $\eta_p{}^2 = 0.57$; $p \leq 0.001$, f = 50.61, $\eta_p{}^2 = 0.57$; $p \leq 0.001$, f = 50.19, $\eta_p{}^2 = 0.56$) for the body mass, BMI, fat mass Withers, body skeletal muscle Lee, and lean body mass, respectively. Post hoc analysis found that the body mass (EG, $p \leq 0.001$, $g = -0.04$), the BMI (EG, $p \leq 0.001$, $g = -0.04$), fat mass Withers (CG, $p = 0.029$, $g = 0.02$, EG, $p \leq 0.01$, $g = -0.10$), the body skeletal muscle mass Lee (EG, $p \leq 0.001$, $g = 0.23$), and lean body mass (CG, $p = 0.03$, $g = -0.02$, EG, $p \leq 0.001$, $g = 0.45$) were significantly reduced post-test vs. pre-test. Percent changes of all body composition variables between pre- and post-test, as shown in Table 5 and Figures 6 and 7.

Table 4. Summary results of skinfold variables within the control group and neuromuscular training group.

Skinfolds (mm)	Control Group (n = 22)					Experimental Group (n = 18)				
	Pre-Test Mean ± SD	Post-Test Mean ± SD	Pre-Post (%)	p	ES (95% CI)	Pre-Test Mean ± SD	Post-Test Mean ± SD	Pre-Post (%)	p	ES (95% CI)
Subscapular	9.62 ± 286	9.66 ± 2.85	0.41	0.143	0.01 (−0.64; 0.66) T	8.64 ± 1.71	8.42± 1.73	−2.54	≤0.001 *	−0.12 (−0.77; 0,53) T
Biceps	4.97 ± 1.40	5.07 ± 1.41	2.01	0.008 *	0.06 (−0.58; 0,72) T	4.27 ± 1.23	4.10 ± 1.21	−3.98	0.001 *	−0.13 (−0.78; 0.52) T
Triceps	11.55 ± 3.42	11.63 ± 3.39	0,69	0.144	0.02 (−0.63; 0.67) T	10.67 ± 3.34	10.45 ± 3.21	−2.06	0.018 *	−0.06 (−0.71; 0.59) T
Supraspinal	8.15 ± 3.24	8.22 ± 3.21	0.85	0.046	0.01 (−0.63; 0.67) T	7.29 ± 2.38	7.16 ± 2.40	−1.78	≤0.001 *	−0.05 (−0.70; 0.60) T
Iliac crest	13.61 ± 4.84	13.69 ± 4.84	0.58	0.002 *	0.02 (−0.64; 0.67) T	13.86 ± 4.39	13.68 ± 4.41	−1.29	≤0.001 *	−0.04 (−0.69; 0.61) T
Abdominal	13.21 ± 4.84	13.27 ± 4.81	0.45	0.110	0.01 (−0.66; 0.67) T	12.77 ± 4.23	12.54 ± 4.25	−1.80	0.001 *	−0.06 (−0.70; 0.60) T
Front thigh	17.76 ± 4.42	17.77 ± 4.42	0.05	0.181	0.01 (−0.65; 0.65) T	17.14 ± 4.48	16.95 ± 4.49	−1.10	≤0.001 *	−0.09 (−0.69; 0.61) T
Medial calf	11.03 ± 3.21	11.05 ± 3.20	0.18	0.137	0.01 (−0.65; 0.65) T	8.61 ± 2.82	8.43 ± 2.82	−2.09	≤0.001 *	−0.06 (−0.71; 0,59) T
Σ6S	71.32 ± 17.14	71.61 ± 16.99	0.40	0.019 *	0.01 (−0.64; 0.67) T	65.12 ± 11.61	63.95 ± 11.60	−1.79	≤0.001 *	−0.09 (0.74; 0.55) T

SD: standard deviation; ES: effect size; CI: confidence interval; T: trivial; Σ6S: sum of six skindfolds; * $p < 0.05$.

Table 5. Summary results of other body composition variables within the control group and neuromuscular training group.

Variable	Control Group (n = 22)					Experimental Group (n = 18)				
	Pre-Test Mean ± SD	Post-Test Mean ± SD	Pre-Post (%)	p	ES (95% CI[4])	Pre-Test Mean ± SD	Post-Test Mean ± SD	Pre-Post (%)	p	ES (95% CI)
Body mass (kg)	59.46 ± 6.22	59.49 ± 6.21	0.05	0.468	0.01 (−0.65; 0.65) T	61.59 ± 4.44	61.37 ± 4.45	−0.35	≤0.001 *	−0.04 (−0.70; 0.60) T
BMI (kg/m^2)	22.46 ± 2.54	22.47 ± 2.53	0.04	0.497	0.01 (−0.65; 0.66) T	22.36 ± 1.85	22.28 ± 1.81	−0.35	≤0.001 *	−0.04 (−0.69; 0,61) T
Fat mass Withers (%)	17.13 ± 3.57	17.21 ± 3.53	0.52	0.029 *	0.02 (−0.64; 0.68) T	15.42 ± 2.68	15.12 ± 2.71	−1.94	≤0.001 *	−0.10 (−0.75; 0.54) T
Body skeletal muscle mass Lee (%)	38.50 ± 4.47	38.45 ± 4.43	−0.10	0.309	0.01 (−0.66; 0.64) T	39.03 ± 1.78	39.46 ±1.75	1.10	≤0.001 *	0.23 (−0.42; 0,88) S
Lean body mass (%)	82.87 ± 3.57	82.79 ± 3.53	−0.10	0.029 *	−0.02 (−0,67; 0.63) T	84.58 ± 2.68	85.88 ± 2.71	1.53	<0.001 *	0.45 (−0.20; 1,12) S

SD: standard deviation; BMI: body mass index; ES: effect size; CI: confidence interval; T: trivial; S: small * $p < 0.05$.

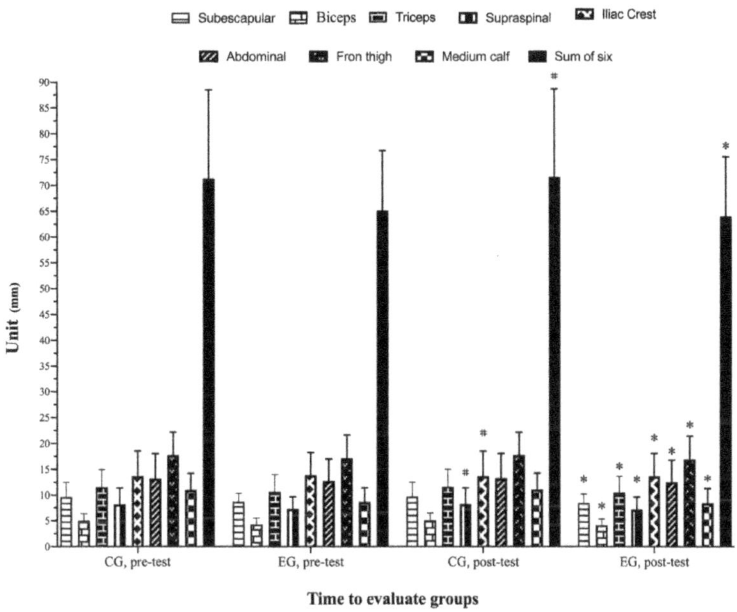

Figure 5. Change in skinfold variables assessment for each group and assessment stage. * Represents a statistically significant difference compared to the pre-test with the superiority of the EG ($p < 0.05$). # Represents a statistically significant difference compared to the pre-test with the superiority of the CG ($p < 0.05$). EG: experimental group; CG: control group.

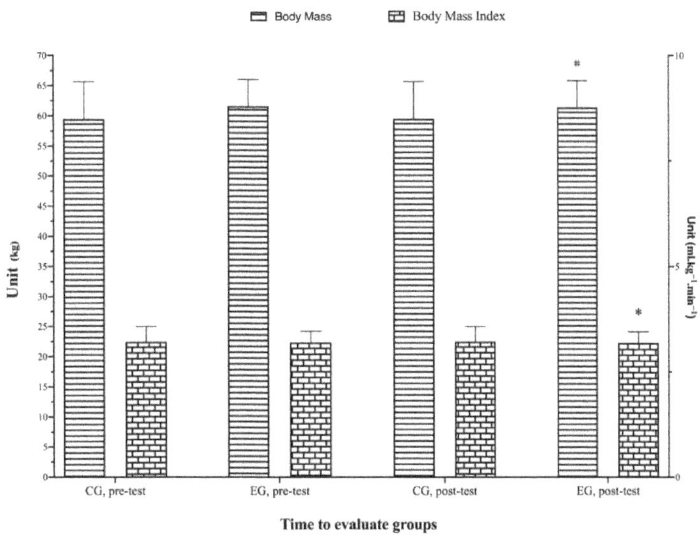

Figure 6. Change in body mass and body mass index variables assessment for each group and assessment stage. * Represents a statistically significant difference compared to the pre-test with the superiority of the EG ($p < 0.05$). # Represents a statistically significant difference compared to the pre-test with the superiority of the CG ($p < 0.05$). EG: experimental group; CG: control group.

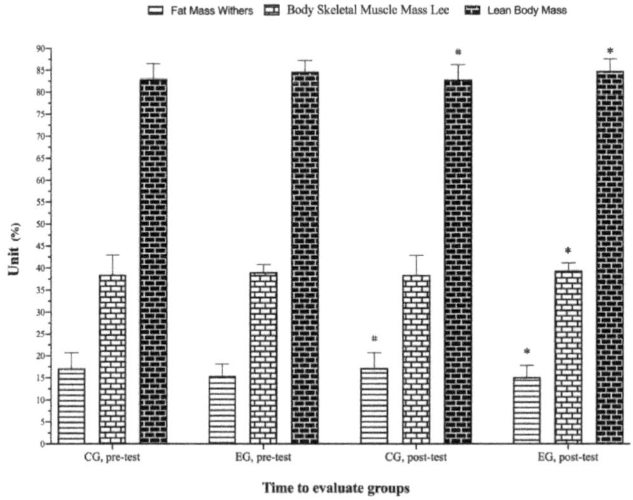

Figure 7. Change in body composition variables assessment for each group and assessment stage. * Represents a statistically significant difference compared to the pre-test with the superiority of the EG ($p < 0.05$). # Represents a statistically significant difference compared to the pre-test with the superiority of the CG ($p < 0.05$). EG: experimental group; CG: control group.

The average intensity registered using the modified Borg's scale (0–10) was recorded over the 30 sessions for both groups (Figure 3). Small magnitudes of differences were found between the average of intensities ($g = 0.20$) between CG and EG.

4. Discussion

The aim of the study was to investigate the effects of a 10-week NMT program on skinfold and body composition variables in highly trained female soccer players. We hypothesized that GE would reduce skinfold values, fat mass, and body mass and increase muscle mass and lean body mass, improving overall body composition.

The main findings of the current work were that 10 weeks of NMT significantly reduced body mass (−0.34%, $g = -0.04$), fat mass (−1.94%, $g = -0.10$), and Σ6S (−1.79%, $g = -0.09$) compared with the CG (0.05%, $g = 0.01$, 0.52%, $g = 0.02$ and 0.4%, $g = 0.01$, respectively). EG and CG were exercised equally, and no significant work intensity was observed between the groups. In addition, body skeletal muscle mass and lean body mass increased in the EG (body skeletal muscle mass: 1.10%, $g = 0.23$; lean body mass: 1.53%, $g = 0.45$) and slightly decreased in the CG (body skeletal muscle mass: −0.10%, $p = 0.3$, lean body mass: −0.10%, $g = 0.02$, respectively).

Previous research in soccer players reported changes in body composition after different resistance training programs [38,40,51–59]. However, no study has been conducted regarding the effects of an NMT program. Of note, the NMT battery applied in the current work includes exercises from six categories: (1) mobility, (2) dynamic stability, (3) anterior chain strength, (4) lumbopelvic control, (5) posterior chain strength, and (6) the ability to COD in this regard, and the effectiveness of each or a combination of these training methods to improve body composition has been considered as a reference for comparing the results of the present study.

Arguably, strength exercises are an effective way to stimulate muscle hypertrophy along with improvements in body composition [60]. Particularly, Falces et al. [55] applied a 16-week strength training program with calisthenics and observed a significant decrease in body mass (ES = −0.08) and fat mass (ES = −0.41) and a significant increase in lean mass (ES = 0.17) in a group of male U17 soccer players. Furthermore, Sánchez-Pérez et al. [54] studied the effects of an 8-week high intensity interval training (i.e., a Tabata workout including calisthenics, plyometrics, and COD ability) in a similar population, showing a reduction in body fat (−1.38%, ES = 0.42) and an increase in lean body mass (1.38%, ES = 0.44), and Suárez-Arrones [40,56] also found differences in the body composition (body fat: ES = −0.99 ± 0.54 and lean body mass percentage: ES = 0.25 ± 0.10) of young male soccer players during a 24-week intervention that included circuit training with some exercises comparable to ours (i.e., posterior chain eccentrics, core stability, and plyometrics). It should be noted that in these last two studies the CG slightly worsened their body composition, just as in the present research.

On the contrary, several studies [58,59,61] assessed training programs that include at least one of the exercise categories applied in the current study in adult soccer players showing no differences in changes in body fat percentage (ES = −0.10) and fat-free mass percentage (ES = 0.09) after 8 weeks or less of intervention. Unfortunately, female soccer players were not included in these works, preventing an accurate comparison with the current data.

Focusing on female soccer players, the study from Polman [52] analyzed the effect of a 12-week physical conditioning program on physical fitness and anthropometric parameters of adult highly trained female soccer players. After the intervention, decreases in body mass (ES = −0.24), BMI (ES = −0.28) and fat mass (ES = −0.16) were found. Although the exercise program in Polman's study is similar to the one included in our study (e.g., balance, jumps, and COD ability), their athletes showed greater improvements than the athletes in the present study. A possible explanation for this little discrepancy could be the longer duration of their intervention and/or the higher body fat percentage of their players at baseline. Remarkably, the mean values for body mass and fat percentage at baseline in the current work fall within the values reported in a review of international female soccer players (56.8–64.9 kg and 14.6–20.1%, respectively) [5], whereas those from the aforementioned study do not.

In contrast, to the best of our knowledge, this is one of the first studies to assess the effects of 12-week plyometric training on body composition, explosive strength, and kicking speed of 20 female soccer players [53]. The results showed an improvement in performance variables but no significant changes in body composition However, changes in muscle strength through plyometric training produce adaptations of the neuromuscular system rather than muscle hypertrophy [62]. Therefore, with unique plyometric training, body composition can be expected to remain unchanged.

Of note, one study [38] analyzed the effects of a 12-week NMT program on the body composition of female volleyball athletes. Though the sports have different metabolic requirements, (football and volleyball), Simões et al. showed an increase in body mass (ES = 0.08) and lean body mass (ES = 0.36) and a reduction in fat mass (ES = −0.50) [38]). In the same direction, the study by Sudha and Dharuman, which evaluated the effects of a 12-week circuit training program combined with different neuromuscular activities in schoolgirls, observed a decrease in BMI (ES = −0.49) [51]. This data, although from a different sample, contribute to reinforcing the results obtained in the present research and highlight that the assessment of body composition is closely related to performance and helps to confirm the training effect [62,63].

Some limitations need to be acknowledged for a correct interpretation of the results. Firstly, it should be mentioned that the sample used is small and the data is limited to a certain group of soccer players, so it would be interesting to carry out further studies to confirm the present results. Female soccer players have characteristics that do not allow us to extrapolate our results directly to other sports. This study did not take into account variables related to the genotype of the female athletes and protein intake above the recommended dietary allowance was not controlled. We recommend that future research examines the relationship between different endocrine parameters (i.e., IGFBP-3, erythropoietin, or estrogen for female athletes) and genes related to performance and body composition, such as angiotensin-1 converting enzyme insertion/deletion (ACE I/D) polymorphism or α-actinin-3 (ACTN3) R577X polymorphism. Future studies should extend these observations to other age groups, competitive levels, and larger samples in order to analyze whether the results are similar. Furthermore, it would be beneficial to observe different intensities and volumes in the NMT program to determine the optimal regimen for this training method as well as observing whether this program can improve body composition in female soccer players.

5. Conclusions

The present study suggests that the implementation of a 10-week NMT program of just 24 min, three times a week improves body composition in highly trained female soccer players compared to a regular physical preparation training. In this regard, the soccer-specific NMT protocol proposed in this study improved female soccer players′ body composition by reducing fat mass and increasing muscle mass. Therefore, female soccer coaches and physical trainers should be aware that combining strength, mobility, lumbopelvic control, dynamic stability, and change of direction exercises based on soccer-specific requirements may also improve the body composition of their female players.

Author Contributions: Conceptualization, A.R.-M., A.C.-L., J.L.A.-S. and D.L.; methodology, A.R.-M., H.N. and A.C.-L.; formal analysis, A.R.-M. and D.L.; investigation, A.R.-M., E.M.-P. and D.L.; data curation, A.R.-M. and D.L.; writing—original draft preparation, A.R.-M., A.C.-L., H.N., E.M.-P. and D.L.; writing—review and editing, D.L., H.N., J.L.A.-S., E.M.-P. and A.C.-L.; supervision, E.M.-P., D.L. and J.L.A.-S.; project administration, A.R.-M. and E.M.-P. All authors have read and agreed to the published version of the manuscript.

Funding: This work was supported by Universidad San Jorge (Internal Research Project 2021–2022) and Departamento de Ciencia, Universidad y Sociedad del Conocimiento from the Gobierno de Aragon (Spain) (Research Group ValorA No. S08_20R).

Institutional Review Board Statement: The study was conducted in accordance with the Declaration of Helsinki, and approved by the local Ethics Committee of CEICA (protocol code PI21/011, 10/02/2021).

Informed Consent Statement: Informed consent was obtained from all subjects involved in the study. Written informed consent has been obtained from the patients to publish this paper.

Data Availability Statement: The data presented in this study are available on reasonable request from the corresponding author. The data are not publicly available due to privacy reasons.

Acknowledgments: The authors thank all the subjects who participated in this study.

Conflicts of Interest: The authors declare no conflict of interest.

References

1. Federation Internationale de Football Association. FIFA Big Count 2006: 270 Million People Active in Football. Available online: https://resources.fifa.com/image/upload/big-count-estadisticas-520058.pdf?cloudid=mzid0qmguixkcmruvema (accessed on 2 March 2022).
2. Consejo Superior de Deportes. Estadística de Deporte Federado 2020. Available online: https://www.culturaydeporte.gob.es/dam/jcr:6b7e9a1a-e3e5-4b45-8ae5-6f187b50235f/estadistica-de-deporte-federado.pdf (accessed on 5 March 2022).
3. Federation Internationale de Football Association. Estrategia de Fútbol Femenino. Available online: https://www.fifa.com/es/womens-football/strategy/strategy-details (accessed on 24 April 2022).
4. Datson, N.; Drust, B.; Weston, M.; Jarman, I.H.; Lisboa, P.J.; Gregson, W. Match Physical Performance of Elite Female Soccer Players during International Competition. *J. Strength Cond. Res.* **2017**, *31*, 2379–2387. [CrossRef]
5. Datson, N.; Hulton, A.; Andersson, H.; Lewis, T.; Weston, M.; Drust, B.; Gregson, W. Applied physiology of female soccer: An update. *Sports Med.* **2014**, *44*, 1225–1240. [CrossRef] [PubMed]
6. Faude, O.; Koch, T.; Meyer, T. Straight sprinting is the most frequent action in goal situations in professional football. *J. Sports Sci.* **2012**, *30*, 625–631. [CrossRef] [PubMed]
7. Salokun, S.O. Minimizing injury rates in soccer through preselection of players by somatotypes. *J. Sports Med. Phys. Fit.* **1994**, *34*, 64–69.
8. Brocherie, F.; Girard, O.; Forchino, F.; Al Haddad, H.; Dos Santos, G.A.; Millet, G.P. Relationships between anthropometric measures and athletic performance, with special reference to repeated-sprint ability, in the Qatar national soccer team. *J. Sports Sci.* **2014**, *32*, 1243–1254. [CrossRef] [PubMed]

9. Rebelo, A.; Brito, J.; Maia, J.; Coelho-e-Silva, M.J.; Figueiredo, A.J.; Bangsbo, J.; Malina, R.M.; Seabra, A. Anthropometric characteristics, physical fitness and technical performance of under-19 soccer players by competitive level and field position. *Int. J. Sports Med.* **2013**, *34*, 312–317. [CrossRef]
10. Nobari, H.; Kargarfard, M.; Minasian, V.; Cholewa, J.M.; Pérez-Gómez, J. The effects of 14-week betaine supplementation on endocrine markers, body composition and anthropometrics in professional youth soccer players: A double blind, randomized, placebo-controlled trial. *J. Int. Soc. Sports Nutr.* **2021**, *18*, 20. [CrossRef]
11. Nobari, H.; Pardos, E.M.; Adsuar, J.C.; García, J.M.F.; Ramos, J.R.; Bolaños, M.A.C.; Alul, L.U.; Gómez, J.P. Association between Endocrine Markers, Accumulated Workload, and Fitness Parameters during a Season in Elite Young Soccer Players. *Front. Psychol.* **2021**, *12*, 702454. [CrossRef]
12. Comfort, P.; Stewart, A.; Bloom, L.; Clarkson, B. Relationships between strength, sprint, and jump performance in well-trained youth soccer players. *J. Strength Cond. Res.* **2014**, *28*, 173–177. [CrossRef]
13. Oliveira, R.; Francisco, R.; Fernandes, R.; Martins, A.; Nobari, H.; Clemente, F.M.; Brito, J.P. In-Season Body Composition Effects in Professional Women Soccer Players. *Int. J. Environ. Res. Public Health* **2021**, *18*, 12023. [CrossRef]
14. Slimani, M.; Nikolaidis, P.T. Anthropometric and physiological characteristics of male soccer players according to their competitive level, playing position and age group: A systematic review. *J. Sports Med. Phys. Fit.* **2019**, *59*, 141–163. [CrossRef] [PubMed]
15. Radzimiński, Ł.; Szwarc, A.; Cabo, A.P.; Jastrzębski, Z. Correlations between body composition, aerobic capacity, speed and distance covered among professional soccer players during official matches. *J. Sports Med. Phys. Fit.* **2020**, *60*, 257–262. [CrossRef]
16. Nobari, H.; Oliveira, R.; Clemente, F.M.; Pérez-Gómez, J.; Pardos-Mainer, E.; Ardigò, L.P. Somatotype, Accumulated Workload, and Fitness Parameters in Elite Youth Players: Associations with Playing Position. *Children* **2021**, *8*, 375. [CrossRef] [PubMed]
17. Gonçalves, L.; Clemente, F.M.; Barrera, J.I.; Sarmento, H.; Fernández, F.T.G.; Vieira, L.H.P.; Figueiredo, A.J.; Clark, C.C.T.; Carral, J.M.C. Relationships between Fitness Status and Match Running Performance in Adult Women Soccer Players: A Cohort Study. *Medicina* **2021**, *57*, 617. [CrossRef] [PubMed]
18. Villa-Vicente, J.G.; García - López, J.; Moreno-Pascual, C. Influence of a Preparation Period in the Kinantropometry Profile of Soccer-Players. *Arch Med Deporte* **2000**, *17*, 9–20.
19. Chamari, K.; Hachana, Y.; Ahmed, Y.B.; Galy, O.; Sghaïer, F.; Chatard, J.C.; Hue, O.; Wisløff, U. Field and laboratory testing in young elite soccer players. *Br. J. Sports Med.* **2004**, *38*, 191–196. [CrossRef]
20. Mujika, I.; Santisteban, J.; Impellizzeri, F.; Castagna, C. Fitness determinants of success in men's and women's football. *J. Sports Sci.* **2009**, *27*, 107–114. [CrossRef]
21. Canda, A.S. *Variables Antropométricas de la Población Deportista Española;* Consejo Superior de Deportes, Servicio de Documentación y Publicaciones: Madrid, Spain, 2012.
22. Meyer, N.L.; Borgen, J.S.; Lohman, T.G.; Ackland, T.R.; Stewart, A.D.; Maughan, R.J.; Smith, S.; Müller, W. Body composition for health and performance: A survey of body composition assessment practice carried out by the Ad Hoc Research Working Group on Body Composition, Health and Performance under the auspices of the IOC Medical Commission. *Br. J. Sports Med.* **2013**, *47*, 1044–1053. [CrossRef]
23. Casajús, J.A.; Aragonés, M.T. Estudio morfológico del futbolista de alto nivel. Composición corporal y somatotipo (Parte 1). *Arch Med Deporte.* **1991**, *8*, 147–151.
24. Randell, R.K.; Clifford, T.; Drust, B.; Moss, S.L.; Unnithan, V.B.; Croix, M.B.A.D.S.; Datson, N.; Martin, D.; Mayho, H.; Carter, J.M.; et al. Physiological Characteristics of Female Soccer Players and Health and Performance Considerations: A Narrative Review. *Sports Med.* **2021**, *51*, 1377–1399. [CrossRef]
25. Steffl, M.; Kinkorova, I.; Kokstejn, J.; Petr, M. Macronutrient Intake in Soccer Players—A Meta-Analysis. *Nutrients* **2019**, *11*, 1305. [CrossRef] [PubMed]
26. Rasulovna, R.M. The Effectiveness of the Program for Determining the Composition of the Body and Neurophysiological Characteristics of Athletes Juniors and Cadets. *Am. J. Med. Med. Sci.* **2021**, *11*, 665–668.
27. Sánchez, M.; Sanchez, J.S.; Nakamura, F.Y.; Clemente, F.M.; Moraleda, B.R.; Campillo, R.R. Effects of Plyometric Jump Training in Female Soccer Player's Physical Fitness: A Systematic Review with Meta-Analysis. *Int. J. Environ. Res. Public Health* **2020**, *17*, 8911. [CrossRef] [PubMed]
28. Mainer, E.P.; Casajús, J.A.; Skok, O.G. Adolescent female soccer players' soccer-specific warm-up effects on performance and inter-limb asymmetries. *Biol. Sport* **2019**, *36*, 199–207. [CrossRef]
29. Myer, G.D.; Faigenbaum, A.D.; Ford, K.R.; Best, T.M.; Bergeron, M.F.; Hewett, T.E. When to initiate integrative neuromuscular training to reduce sports-related injuries and enhance health in youth? *Curr. Sports Med. Rep.* **2011**, *10*, 155–166. [CrossRef] [PubMed]
30. Grandstrand, S.L.; Pfeiffer, R.P.; Sabick, M.B.; DeBeliso, M.; Shea, K.G. The effects of a commercially available warm-up program on landing mechanics in female youth soccer players. *J. Strength Cond. Res.* **2006**, *20*, 331–335. [PubMed]
31. Noyes, F.R.; Westin, S.D.B.; Smith, S.T.T.; Campbell, T. A training program to improve neuromuscular and performance indices in female high school soccer players. *J. Strength Cond. Res.* **2013**, *27*, 340–351. [CrossRef]
32. Gilchrist, J.; Mandelbaum, B.R.; Melancon, H.; Ryan, G.W.; Silvers, H.J.; Griffin, L.Y.; Watanabe, D.S.; Dick, R.W.; Dvorak, J. A randomized controlled trial to prevent noncontact anterior cruciate ligament injury in female collegiate soccer players. *Am. J. Sports Med.* **2008**, *36*, 1476–1483. [CrossRef]

33. Mandelbaum, B.R.; Silvers, H.J.; Watanabe, D.S.; Knarr, J.F.; Thomas, S.D.; Griffin, L.Y.; Kirkendall, D.T.; Garrett, W., Jr. Effectiveness of a neuromuscular and proprioceptive training program in preventing anterior cruciate ligament injuries in female athletes: 2-year follow-up. *Am. J. Sports Med.* **2005**, *33*, 1003–1010. [CrossRef]
34. Bizzini, M.; Dvorak, J. FIFA 11+: An effective programme to prevent football injuries in various player groups worldwide—A narrative review. *Br. J. Sports Med.* **2015**, *49*, 577–579. [CrossRef]
35. Kiani, A.; Hellquist, E.; Ahlqvist, K.; Gedeborg, R.; Michaëlsson, K.; Byberg, L. Prevention of soccer-related knee injuries in teenaged girls. *Arch. Intern. Med.* **2010**, *170*, 43–49. [CrossRef] [PubMed]
36. Liu, R.; Liu, J.; Ma, X.; Li, Q.; An, R. Effect of FIFA 11+ intervention on change of direction performance in soccer and futsal players: A systematic review and meta-analysis. *Int. J. Sports Sci. Coach.* **2021**, *16*, 862–872. [CrossRef]
37. Rohmansyah, N.A.; Hiruntrakul, A.J.S.M. Do Lipid Profiles, Body Composition, and Physical Fitness Improve after a FIFA 11+ Training Programme from Obese Women? *Sport Mont* **2021**, *19*, 51–57.
38. Simões, R.; Salles, G.; Gonelli, P.; Leite, G.; Dias, R.; Cavaglieri, C.; Pellegrinotti, Í.; Borin, J.; Verlengia, R.; Alves, S.; et al. Effects of the Neuromuscular Training in the Cardiorespiratory Fitness and Body Composition of Female Volleyball Athletes. *Rev. Bras. Med. Esporte* **2009**, *15*, 295–298. [CrossRef]
39. Falces-Prieto, M.; de Villarreal-Sáez, E.S.; Raya-González, J.; González-Fernández, F.T.; Clemente, F.M.; Badicu, G.; Murawska-Ciałowicz, E. The Differentiate Effects of Resistance Training with or without External Load on Young Soccer Players' Performance and Body Composition. *Front. Physiol.* **2021**, *12*, 771684. [CrossRef]
40. Suarez-Arrones, L.; de Villarreal, E.S.; Núñez, F.J.; Di Salvo, V.; Petri, C.; Buccolini, A.; Maldonado, R.A.; Torreno, N.; Mendez-Villanueva, A. In-season eccentric-overload training in elite soccer players: Effects on body composition, strength and sprint performance. *PLoS ONE* **2018**, *13*, e0205332. [CrossRef]
41. Peitz, M.; Behringer, M.; Granacher, U. A systematic review on the effects of resistance and plyometric training on physical fitness in youth—What do comparative studies tell us? *PLoS ONE* **2018**, *13*, e0205525.
42. Ros, F.E.; Cristóbal, R.V.; Marfell-Jones, M. *Protocolo Internacional para la Valoración Antropométrica*; Universidad Católica de Murcia Ed.; Univ, U., Ed.; Publicado por Sociedad Internacional para el Avance de la Cineantropometría: Murcia, Spain, 2019.
43. Burgos, M.A.M.; Martínez-Victoria, I.; Milá, R.; Farrán, A.; Farré, R.; Ros, G.; Yago, M.D.; Audi, N.; Santana, C.; Millán, M.B.L.; et al. Building a unified Spanish food database according to EuroFIR specifications. *Food Chem.* **2009**, *113*, 784–788. [CrossRef]
44. Keys, A.; Fidanza, F.; Karvonen, M.J.; Kimura, N.; Taylor, H.L. Indices of relative weight and obesity. *J. Chronic Dis.* **1972**, *25*, 329–343. [CrossRef]
45. Withers, R.T.; Whittingham, N.O.; Norton, K.I.; La Forgia, J.; Ellis, M.W.; Crockett, A. Relative body fat and anthropometric prediction of body density of female athletes. *Eur. J. Appl. Physiol. Occup. Physiol.* **1987**, *56*, 169–180. [CrossRef]
46. Siri, W.E. The gross composition of the body. *Adv. Biol. Med. Phys.* **1956**, *4*, 239–280. [PubMed]
47. Lee, R.C.; Wang, Z.; Heo, M.; Ross, R.; Janssen, I.; Heymsfield, S.B. Total-body skeletal muscle mass: Development and cross-validation of anthropometric prediction models. *Am. J. Clin. Nutr.* **2000**, *72*, 796–803. [CrossRef] [PubMed]
48. Jeffreys, I.J.U.J. Warm up revisited—the 'ramp'method of optimising performance preparation. *UKSCA J.* **2006**, *6*, 15–19.
49. Zamunér, A.R.; Moreno, M.A.; Camargo, T.M.; Graetz, J.P.; Rebelo, A.C.; Tamburús, N.Y.; da Silva, E. Assessment of Subjective Perceived Exertion at the Anaerobic Threshold with the Borg CR-10 Scale. *J. Sports Sci. Med.* **2011**, *10*, 130–136. [PubMed]
50. Haddad, M.; Stylianides, G.; Djaoui, L.; Dellal, A.; Chamari, K. Session-RPE Method for Training Load Monitoring: Validity, Ecological Usefulness, and Influencing Factors. *Front. Neurosci.* **2017**, *11*, 612. [CrossRef]
51. Sudha, K.; Maniazhagu, D. Effects of Circuit Training Combined with Different Neuromuscular Activities on Muscular Endurance and Body Composition of School Girls. *Indian J. Public Health Res. Dev.* **2019**, *10*, 30. [CrossRef]
52. Polman, R.; Walsh, D.; Bloomfield, J.; Nesti, M. Effective conditioning of female soccer players. *J. Sports Sci.* **2004**, *22*, 191–203. [CrossRef]
53. Campo, S.S.; Vaeyens, R.; Philippaerts, R.; Redondo, J.C.; de Benito, A.T.; Cuadrado, G. Effects of Lower-Limb Plyometric Training on Body Composition, Explosive Strength, and Kicking Speed in Female Soccer Players. *J. Strength Cond. Res.* **2009**, *23*, 1714–1722. [CrossRef]
54. Sánchez Pérez, F.J.; Carranque Chaves, G.Á. Efectos del entrenamiento Tabata en la composición corporal del futbolista. *Kronos* **2015**, *14*, 7.
55. Falces, M.; Fernández, F.T.G.; Morales, S.B.; Jiménez, A.B.; Barrero, A.; Conde, L.; Arronez, L.S.; de Villarreal, E.S. Effects of a strength training program with self loading on countermovement jump performance and body composition in young soccer players. *J. Sport Health Res.* **2020**, *1*, 112–125.
56. Suarez-Arrones, L.; Lara-Lopez, P.; Torreño, N.; de Villarreal, E.S.; Di Salvo, V.; Mendez-Villanueva, A. Effects of Strength Training on Body Composition in Young Male Professional Soccer Players. *Sports* **2019**, *7*, 104. [CrossRef] [PubMed]
57. Haghighi, A.; Moghadasi, M.; Nikseresht, A.; Torkfar, A.; Haghighi, M. Effects of plyometric versus resistance training on sprint and skill performance in young soccer players. *Eur. J. Exp. Biol.* **2012**, *2*, 2348–2351.
58. Perez-Gomez, J.; Olmedillas, H.; Delgado-Guerra, S.; Ara, I.; Vicente-Rodríguez, G.; Arteaga-Ortiz, R.; Javier, C.; Calbet, J. Effects of weight lifting training combined with plyometric exercises on physical fitness, body composition, and knee extension velocity during kicking in football. *Appl. Physiol. Nutr. Metab.* **2008**, *33*, 501–510. [CrossRef] [PubMed]
59. Cigerci, A.E.; Genc, H. The Effect of Calisthenics Exercises on Body Composition in Soccer Players. *Prog. Nutr.* **2020**, *22*, 94–102.

60. Staron, R.S.; Leonardi, M.J.; Karapondo, D.L.; Malicky, E.S.; Falkel, J.E.; Hagerman, F.C.; Hikida, R.S. Strength and skeletal muscle adaptations in heavy-resistance-trained women after detraining and retraining. *J. Appl. Physiol.* **1991**, *70*, 631–640. [CrossRef]
61. Sánchez, J.; Pérez, S.; Yagüe, J.M.; Royo, J.M.; Martín, J.L. Aplicación de un programa de entrenamiento de fuerza en futbolistas jóvenes. *Rev. Int. Med. Cienc. Act. Fis. Deporte* **2015**, *15*, 45–59.
62. Häkkinen, K.; Pakarinen, A.; Kallinen, M. Neuromuscular adaptations and serum hormones in women during short-term intensive strength training. *Eur. J. Appl. Physiol. Occup. Physiol.* **1992**, *64*, 106–111. [CrossRef]
63. Stanforth, P.R.; Crim, B.N.; Stanforth, D.; Stults-Kolehmainen, M.A. Body composition changes among female NCAA division 1 athletes across the competitive season and over a multiyear time frame. *J. Strength Cond. Res.* **2014**, *28*, 300–307. [CrossRef]

biology

MDPI

Article

Effects of Dominance and Sprint Interval Exercise on Testosterone and Cortisol Levels in Strength-, Endurance-, and Non-Training Men

Grzegorz Zurek [1], Natalia Danek [2,*], Alina Żurek [3], Judyta Nowak-Kornicka [4], Agnieszka Żelaźniewicz [4], Sylwester Orzechowski [3], Tadeusz Stefaniak [5], Magdalena Nawrat [3] and Marta Kowal [3]

1 Department of Biostructure, Faculty of Physical Education and Sport, Wroclaw University of Health and Sport Sciences, 51-612 Wroclaw, Poland; grzegorz.zurek@awf.wroc.pl
2 Department of Physiology and Biochemistry, Faculty of Physical Education and Sport, Wroclaw University of Health and Sport Sciences, 51-612 Wroclaw, Poland
3 Institute of Psychology, University of Wroclaw, 50-529 Wroclaw, Poland; alina.zurek@uwr.edu.pl (A.Ż.); sylwester.orzechowski@uwr.edu.pl (S.O.); magdalena.nawrat@uwr.edu.pl (M.N.); marta7kowal@gmail.com (M.K.)
4 Department of Human Biology, Faculty of Biological Sciences, University of Wroclaw, 50-137 Wroclaw, Poland; judyta.nowak@uwr.edu.pl (J.N.-K.); agnieszka.zelazniewicz@uwr.edu.pl (A.Ż.)
5 Department of Sport Didactics, Faculty of Physical Education and Sport, Wroclaw University of Health and Sport Sciences, 51-612 Wroclaw, Poland; tadeusz.stefaniak@awf.wroc.pl
* Correspondence: natalia.danek@awf.wroc.pl

Simple Summary: Exercise is a powerful stimulus to the endocrine system, modifying plasma concentrations of many hormones, including testosterone and cortisol, which are often used to describe fatigue in sport. In our investigation, we wanted to explore the hormonal response (testosterone and cortisol) in saliva after acute exercises in men who perform endurance-training and strength-training exercises compared to a non-training group. Participants performed sprint interval exercise. During the whole exercise, the participants' heart rates were measured, and a rating of perceived exertion was assessed immediately after each bout. The study showed that there were no differences in testosterone and cortisol changes in the endurance-training, strength-training, and non-training groups after the sprint interval exercise. We suppose that one session of the sprint interval training should have more volume (more or longer duration of sprints) to provoke testosterone and cortisol reaction in endurance-training and strength-training individuals. However, the heart rates after acute exercise in the endurance-training and strength-training groups were lower than in participants from the non-training group.

Abstract: The aim of the study was to investigate the response of testosterone and cortisol to sprint interval exercises (SIEs) and to determine the role of dominance. The experiment was conducted in a group of 96 men, divided into endurance-training, strength-training, and non-training groups. Participants performed SIEs consisting of 5 × 10-s all-out bouts with a 50-s active recovery. Using the passive drool method, testosterone and cortisol concentrations were measured in saliva samples at rest at 10 min pre and 12 min post exercise. Participants' heart rate (HR) was measured during the whole exercise. Dominance was assessed by the participants before the study; the rating of perceived exertion (RPE) was measured immediately after each bout. The study showed that those who trained in endurance and strength sports had significantly lower mean HRs after five acute 10-s interval bouts than those in the non-training group ($p = 0.006$ and $p = 0.041$, respectively). Dominance has an inverse relation to changes in HR; however, it has no relation to hormone response. No significant differences were observed in testosterone and cortisol changes in the endurance-training, strength-training, and non-training groups after SIE ($p > 0.05$), which may indicate that the exercise volume was too low.

Keywords: acute exercise; hormonal response; saliva

Citation: Zurek, G.; Danek, N.; Żurek, A.; Nowak-Kornicka, J.; Żelaźniewicz, A.; Orzechowski, S.; Stefaniak, T.; Nawrat, M.; Kowal, M. Effects of Dominance and Sprint Interval Exercise on Testosterone and Cortisol Levels in Strength-, Endurance-, and Non-Training Men. *Biology* **2022**, *11*, 961. https://doi.org/10.3390/biology11070961

Academic Editors: Filipe Manuel Clemente and Badicu Georgian

Received: 16 May 2022
Accepted: 21 June 2022
Published: 24 June 2022

1. Introduction

Different exercise modalities and prior training experience (i.e., endurance training, resistance training, interval training), and variables within the modality (i.e., intensity, volume, duration), can result in different hormonal responses [1]. Knowing the hormonal responses after performing a single exercise session can indicate the direction of selecting an appropriate training strategy, which is important because of the great interest in endocrine responses due to the use of training modulation determining the level of changes in testosterone or cortisol concentrations [2]. Previous multifaceted analyses of current responses in a single interval exercise (SIE) session have focused primarily on determining changes in cardiorespiratory parameters that determine physical performance [3]. As reported by Riachy et al. [4], data on the effects of exercise on serum testosterone levels in men show considerable inter-individual and inter-study variability. This variability can be explained by (a) the use of different types of exercise (e.g., endurance, resistance, or interval training), (b) the intensity and duration of the exercise session, and (c) the fitness status of the participants. Testosterone is the main anabolic hormone, important for skeletal muscle growth and maintenance, as well as for neural function [5]. Cortisol, on the other hand, has catabolic effects [6]. Thus, the testosterone/cortisol (T/C) ratio is used as an indicator of the balance between anabolic and catabolic processes [7]. It is identified as an indicator of fatigue (decreased performance, psychological changes, and neuroendocrine disorders after some physical training) [8].

Additionally, changes in testosterone and cortisol concentrations interact to regulate dominance. In a study by Meht and Joseph [9], T was shown to be associated with dominance under conditions of threat or excessive challenge such as extreme physical exertion. Analyses by Batrinos [10] also showed that testosterone levels increase during the aggressive phases of sports games, which is determined by the level of dominance among athletes, while Carré and McCormick [11] indicated that there was no relationship between changes in T concentrations and dominance in the context of competition. Since dominance is associated with acquisition and a high sense of agency, these findings suggest that higher testosterone levels should only support a higher status when cortisol levels are low. When cortisol is high, higher testosterone levels may actually reduce dominance and, in turn, reduce the degree of competition or engagement in extremely hard efforts such as sprint interval exercise (SIE). However, this requires further analysis.

One of the most popular types of interval training is sprint interval training (SIT), which involves performing work at maximal intensity (generating the highest possible power, known as "all-out") [2,12]. To understand the multiple mechanisms that regulate the state of physical adaptation, it is important to determine the hormonal responses to a single SIT session. The single session of SIT (sprint interval exercise—SIE) typically consists of two to six bouts lasting between 10 to 30 s, with recovery of a longer duration (e.g., up to several minutes), and a total duration of one session interval (SIE) of typically 10 to 30 min [13]. However, the 30-s maximal bouts used have been criticized due to negative affective reactions and reluctance to undertake subsequent interval sessions, despite their benefits in overall health improvement [14,15]. Thus, SIT with 10-s repetitions has recently received more interesting consideration given its effectiveness in cardiovascular and respiratory adaptations (increasing maximal VO_2) and in skeletal muscle metabolism [16,17]. As was presented by Islam et al. [18], shorter sprints with more repetitions are perceived as more enjoyable and lead to a greater desire to get involved in SIT. Affect, intention, self-efficacy, pleasure, rating of perceived exertion (RPE), and preference were rated in favour of the shortest 5:40 protocol (24 × 5-s bouts, 40-s rest) versus (A) 30:240 (4 × 30-s bouts, 240-s rest); (B) 15:120 (8 × 15-s bouts, 120-s rest) [18]. However, they did not measure hormonal changes. According to Martinez-Diaz and Carrasco [19], among active male college students, a single interval session composed of 10 × 1-min sets of VO_2 peak power output induced acute changes in mood states that seem to be associated with hypothalamic–pituitary–adrenal axis activation, as the magnitude of the cortisol response in the study reached 37% immediately after the interval session and as high as 77% 30 min after. In contrast, Beaven et al. [20], during maximal single repetition (1-RM) at intensities of 85, 70, 55, and 40% RM, showed no change in testosterone and cortisol.

Given the myriad of possible variants of SIE protocols and the small amount of research on their psychological perception, we chose the 10-s bout protocol and asked whether ratings of perceived exertion would differ depending on the type of activity being practiced (endurance, strength, or no activity in the control group) and questioned the role of domination. We also hypothesized that ratings of perceived exertion after SIE will be greater in the non-training group. In this way, factors capable of producing less-acute SIE protocols could be characterized. Therefore, the present study evaluated preference after an experimental 10:50 (5 × 10-s bouts, 50-s rest) SIE session protocol. The effect of SIE session on T and C levels during the rapid recovery phase in non-training, endurance-training, or strength-training individuals still remains unclear and requires further study. Therefore, the purpose of our study was to determine changes in T and C concentrations after a single SIE session in non-training, endurance-training, or strength-training participants. Analysis of C and T levels can provide valuable information about the physiological stress response and adaptation to one session of sprint interval training. Our initial hypothesis was that the level of T and C will be different depending on the characteristics of the exercise. In the group of non-training participants, the release of C will be higher; however, in the group of endurance- and strength-training participants, higher levels of T will be observed.

2. Materials and Methods

2.1. Participants

Ninety-six healthy men (aged 19–25 years; mean age ($\overline{x}$) = 21.25; SD = 1.79 years) participated in the study. Information about the project was posted on social media, and participants were also recruited through flyers and direct invitations from researchers during physical activity courses. The main criteria for inclusion in the study were generally good physical fitness and a lack of qualification to the risk group of cardiopulmonary and metabolic diseases; a lack of supplementation and/or hormone treatment; no injuries to the mouth and no dental orthodontic treatment during the experiment; and no consumption of alcohol and tobacco or other stimulants 24 h before the experiment. All participants were advised to brush their teeth at least 2 h prior to the study to minimize the impact on the hormonal assessment of saliva [21]. All participants were familiarized with the study procedure and gave written, informed consent to participate in the study. The ethical approval of the study protocol was provided by the Ethical Committee of the Institute of Psychology of the University of Wroclaw (date of the decision: 16 November 2020, decision number: 2020/ARDK), in accordance with the Helsinki Declaration.

None of the participants practiced sports at a professional level, while 30 participants declared the use of regular strength training (bodybuilding, CrossFit, resistant or "resistance" training), and 35 participants declared regular endurance training as a kind of basic training (running, swimming, soccer). Each participant exercised at least 3 times a week. Thirty-one non-training participants declared to perform 2–2.5 h of recreational irregular physical exercise (walking, cycling) twice a week—this made up the control group in our study. The groups were compared in terms of somatic parameters, that is, age ($F(2.90) = 2.34$, $p = 0.103$), body height ($F(2.90) = 2.92$, $p = 0.059$), and body mass ($F(2.90) = 2.19$, $p = 0.117$). Detailed characteristics of the participants are shown in Table 1.

Table 1. Participants' characteristics.

Variables	Endurance Training (N = 35)	Strength Training (N = 30)	Control (N = 31)
Age	20.71 (1.62)	21.53 (1.69)	21.58 (1.96)
Body height (cm)	180.77 (5.55)	183.72 (7.51)	174.45 (5.57)
Body mass (kg)	75.59 (11.34)	80.77 (10.21)	79.81 (9.05)
BMI ($kg \cdot m^{-2}$)	23.11 (3.16)	24.08 (3.75)	24.89 (3.55)
Physical activity (h per week)	6.60 (3.61)	5.25 (2.81)	4.48 (2.29)
HR_{rest} ($beats \cdot min^{-1}$)	68.36 (10.80)	68.03 (11.81)	69.41 (9.43)
HR_{mean} ($beats \cdot min^{-1}$)	152.82 (10.76)	155.83 (13.59)	160.75 (12.12)
Testosterone (pre)	135.46 (62.15)	129.88 (54.21)	135.17 (49.05)
Testosterone change (post–pre)	9.79 (31.07)	20.02 (47.17)	18.37 (50.79)
Cortisol (pre)	7.62 (1.75)	8.07 (1.56)	8.47 (1.72)
Cortisol change (post–pre)	−0.08 (1.33)	0.28 (1.73)	−0.57 (1.60)
T/C (pre)	18.02 (7.46)	16.32 (6.70)	16.37 (6.28)
T/C change (post–pre)	49.14 (118.69)	37.33 (165.49)	−18.58 (126.41)
RPE	5.16 (1.83)	5.60 (2.50)	5.36 (2.19)
Self-reported dominance	3.34 (0.75)	3.31 (0.71)	3.23 (0.63)

Note. Numbers represent means and standard deviations (in brackets). BMI—body mass index (body mass ($kg \cdot m^{-2}$); HR—heart rate. RPE—ratings of perceived exertion (Borg's scale), T/C—ratios of testosterone to cortisol.

2.2. Study Design

The testing session was conducted between 7.00 AM and 11.00 AM in order to avoid daily hormone fluctuations. Participants were instructed to maintain a sleeping pattern and dietary habits and to refrain from undertaking physical activity for 24 h before the testing session in the laboratory in order to reduce any bias in salivary T and C [21]. Before the testing session, participants filled out a questionnaire, reporting their age, how often they exercised, and the type of that activity, as well as the average duration of their trainings. Their body masses (kg) and heights (cm) were measured using a WPT 200 medical scale (RADWAG, Radom, Poland) before the physical exertion. BMI was calculated based on the participants' weights and heights (body mass (kg) (height (m))$^{-2}$). Heart rate (HR) was measured with the Polar S810 sport-tester (Polar Electro, Kempele, Finland) during all exercises. The HR measurement started two minutes before the warm-up and continued until one minute after the SIE. HR_{mean} is the average HR values from the start of the first repetition until the end of the fifth repetition, including recoveries. HR_{rest} is the averaged HR value of one minute of restitution after SIE.

The flowchart and study protocol is presented in Figure 1.

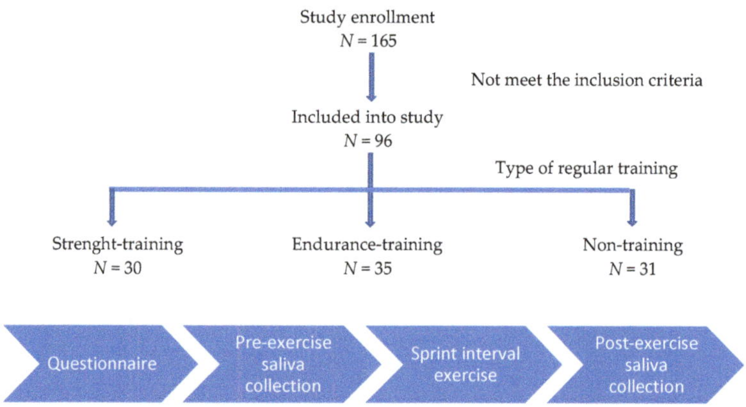

Figure 1. The flowchart and study protocol.

2.3. Salivary Hormone Analysis

Participants provided unstimulated saliva samples at rest 10 min before and 12 min after exercise [22]. T and C concentrations were measured in saliva samples, which were collected by participants using the passive drool method. Participants' identification numbers and the words "pre" or "post" (exercise) were written on every collection tube. The samples were stored using standard procedures (−80 °C) until analysis [23]. Salivary measures of C and T were determined by competitive enzyme-linked immunosorbent assay (ELISA method). Before the analyses, samples were thawed and centrifuged for 10 min at 10,000 RPM. Clear supernatant was used for the quantitative determination of T and C by the commercial ELISA kit (DES6622 and DES6611, DEMEDITEC). The intra- and inter-assay variations for C were: intra: <6.8%; <9.4% with assay sensitivity 0.014 $ng \cdot mL^{-1}$ for C, and for T <9.7%; <9.9% with assay sensitivity 2.2 $pg \cdot mL^{-1}$. The calculation of the results was performed by constructing a standard curve (plotting the absorbance value of the standards (y-axis) against their concentration (x-axis)). Hormone concentrations in saliva samples were calculated in relation to a standard curve and expressed in ng for C and pg/mL for T.

2.4. Sprint Interval Exercise Sessions (SIE)

All participants performed one protocol of sprint interval exercise (Figure 2) on the cycle ergometer (Ergomedic Monark 894, Vansbro, Sweden), which followed the same scientific criteria as tools used in previous studies [24,25]. The physical exertion consisted of a 5 min warm up with a 2 kg load, 1 min of rest, 5 repeated "all-out" bouts of exercise (10 s each) with a Wingtate load—7.5% of the participant's body mass (followed by 50 s of slow-cycling without a load between bouts), and 1 min of slow-cycling without a load at the end. Figure 2 displays the SIE protocol.

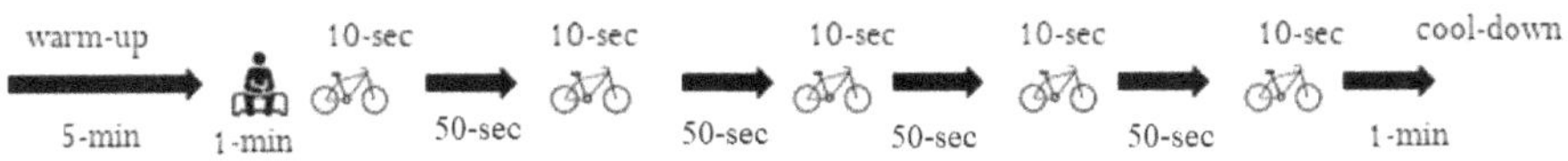

Figure 2. Sprint interval exercise protocol.

2.5. Scales Used

2.5.1. Borg Ratings of Perceived Exertion (RPE) Scale

The RPE is a tool for the subjective assessment of exercise intensity, that is, the degree of fatigue in particular bouts. This scale allows respondents to relate the degree of fatigue during exercise to the fatigue experienced during everyday activities. In general, a score > 18 indicates that a maximal bout was made, and values > 15–16 indicate that the anaerobic threshold was exceeded. Ratings on this scale are related to HR. The principle of the scale is to divide the predicted heart rate for a given exertion by 10; hence, the exertion causing an increase in heart rate to 190 $beats \cdot min^{-1}$ is scored 19 points, and the total rest in which the heart rate oscillates between 60–70 $beats \cdot min^{-1}$ is scored 6–7 points. In the experiment, we used a shortened, 10-point version of the scale, where a score of 9–10 shows the almost maximal or maximal level of exertion [26].

2.5.2. Dominance Scale

Dominance was measured with a 5-item questionnaire previously used in studies by Kowal et al. [27]. This scale has not yet been published. However, the aim of our previous investigations was to construct a reliable and validated scale to measure an individual's dominance. The scale was found to be highly reliable (Cronbach's alpha: 0.78), similar to the present study (Cronbach's alpha: 0.753). The questionnaire included the following items: (1) 'I often persuade others to behave as I suggest'; (2) 'Everything usually turns out to be as I want'; (3) 'I usually make decisions for myself and others'; (4) 'It is rather me who influences others, and not the other way around'; (5) 'I am dominant towards others'. Participants responded to each item on a 5-point Likert scale (range: 1—'I definitely disagree', 2—'I disagree, 3—'I don't have an opinion', 4—'I agree', and 5—'I definitely agree'). In all subsequent analyses, we used a mean value of the dominance scale.

2.6. Statistical Analysis

In the first step, we computed participants' BMI (body mass index (kg)/(height (m))^2), HR_{mean} (a composite score of a mean of five measures of heart rate after five 10-s interval exercises), typical weekly physical activity (the number of trainings in a typical week × a typical length of training), and a mean score of self-reported dominance for each of the participants. Next, we subtracted pre-test testosterone and cortisol from post-test testosterone and cortisol to create testosterone and cortisol change indexes (respectively). In the next step, we calculated the Mahalanobis distance to screen for potential outliers (relying on the usually recommended criteria of $p < 0.001$) [28,29].

We then proceeded with the linear regression models. In the first model, we regressed HR_{mean} on the testosterone change, cortisol change, a mean score of dominance, and a type of physical activity performed by a given participant (i.e., endurance, strength, or other). In the subsequent models, we introduced (2) age, (3) typical weekly physical activity (number of trainings in a typical week × a typical length of training), and (4) BMI, and compared the models' fit. We repeated the above steps, regressing both the cortisol and testosterone change on a type of physical activity performed by a given participant (i.e., endurance, strength, or other) and a mean score of dominance. All analyses were performed in Jamovi (1.8.1) and SPSS (Inc., Chicago, IL, USA).

3. Results

Detailed descriptive characteristics of the participants are shown in Table 1. Analysis of the Mahalanobis distance revealed four potential outliers, which we excluded from all subsequent analyses. When we compared the regression models, the first one showed a superior fit to the other models ($p > 0.05$). Thus, here, we report the results of the first model (see Tables 2–4). Results showed that individuals who trained endurance and strength sports had lower heart rate means following five acute 10-s interval exercises than individuals in the control group (while there were no differences between individuals who trained endurance and strength sports; see Figure 3). Moreover, dominance was negatively related, while the cortisol changes were positively associated with mean heart rates, meaning that those who reported being more dominant had a lower mean heart rate compared to those who reported being less dominant; also, the more acute the cortisol response, the higher the mean heart rate. Furthermore, the control group experienced a larger change in cortisol than the strength-training group, but there were no differences in the cortisol change between the endurance-training and strength-training and control groups (see Figure 4). Dominance was unrelated to the cortisol change, and the type of discipline and dominance were unrelated to the testosterone change.

Table 2. Summary of the linear regression results with the mean heart rate after five bouts as a dependent variable.

Outcome Variable: Mean Heart Rate	**$r^2 = 0.124$, $F_{(5,84)} = 3.513$, $p = 0.006$**			
Predictor	**β**	**95% CI**	**SE**	***p***
Discipline				
Endurance–Control	−0.698	[−1.191, −0.204]	3.127	0.006 **
Strength–Control	−0.531	[−1.041, −0.021]	3.229	0.041 *
Testosterone change	0.075	[−0.128, 0.278]	0.030	0.466
Cortisol change	0.233	[0.023, 0.442]	0.883	0.030 *
Self-reported dominance	−0.221	[−0.420, −0.022]	1.817	0.030 *

Note. * $p < 0.05$, ** $p < 0.01$.

Table 3. Summary of the linear regression results with the cortisol change after five bouts as a dependent variable.

Outcome Variable: Cortisol Change		$r^2 = 0.080$, $F_{(3.86)} = 2.503$, $p = 0.065$		
Predictor	**β**	**95% CI**	**SE**	***p***
Discipline				
Endurance–Control	0.436	[−0.067, 0.939]	0.380	0.089
Strength–Control	0.689	[0.179, 1.199]	0.385	0.009 **
Self-reported dominance	0.028	[−0.178, 0.234]	0.224	0.790

Note. ** $p < 0.01$.

Table 4. Summary of the linear regression results with the testosterone change after five bouts as a dependent variable.

Outcome Variable: Testosterone Change		$r^2 = 0.156$, $F_{(3.86)} = 0.711$, $p = 0.548$		
Predictor	**β**	**95% CI**	**SE**	***p***
Discipline				
Endurance–Control	−0.174	[−0.692, 0.344]	11.281	0.506
Strength–Control	0.065	[−0.461, 0.59]	11.443	0.807
Self-reported dominance	0.120	[−0.092, 0.332]	6.657	0.264

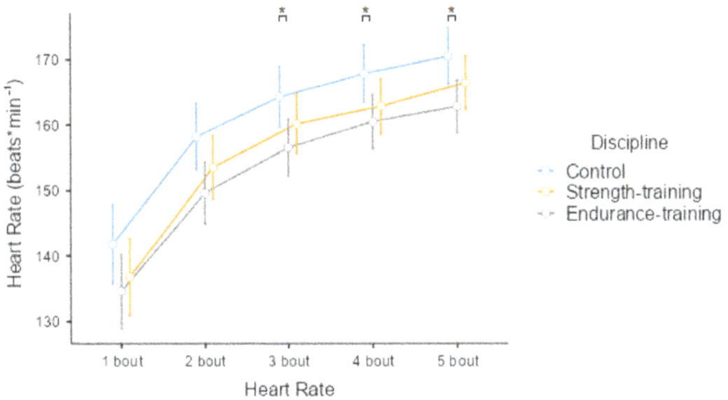

Figure 3. Means and confidence intervals (95%) of the mean heart-rate (beats per minute) measures after five 10-s acute interval exercises in the control, endurance-, and strength-training groups. Asterisks (*) represent significant differences ($p < 0.05$).

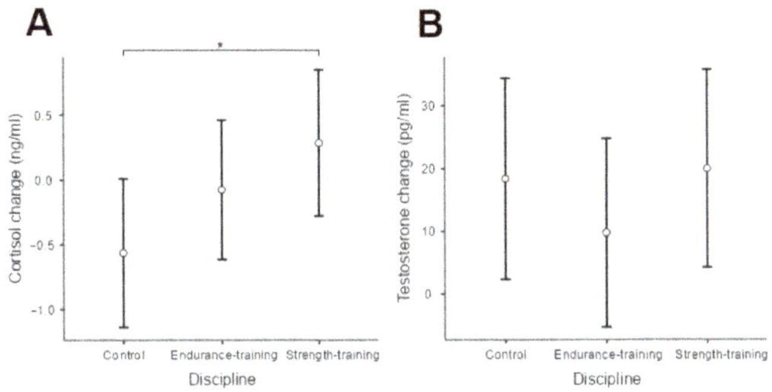

Figure 4. Means and confidence intervals (95%) of the cortisol (**A**) and testosterone (**B**) change after five 10-s acute interval exercises in the control, endurance-, and strength-training groups. An asterisk (*) represents a significant difference ($p < 0.05$).

4. Discussion

The purpose of this study was to determine the effect of a sprint interval exercise on changes in testosterone and cortisol levels and the T/C marker among non-training, endurance-training, and strength-training participants. The main results indicate that the SIE protocol performed by the endurance-training and strength-training participants does not cause significant differences in hormonal conditions compared to the control group, in which a linear relationship was demonstrated between the increase in cortisol concentration and the achievement of higher HR values. We also found that after SIE, there were statistically significant differences in changes of cortisol levels between the strength-training and non-training groups. Dominance has an inverse relation to changes in HR; however, it has no relation to hormone response.

Physical exercise is a particular form of activation of the hypothalamic–pituitary–adrenal (HPA) axis, providing an increase in cortisol levels [30]. The response of the HPA axis to exercise varies with the duration and intensity of exercise [31]. In contrast, the change in cortisol concentration is independent of individuals' fitness status when exercise is performed at similar relative intensities among non-training and trained individuals [32]. Nevertheless, it is also accepted that endurance athletes have a reduced sensitivity to cortisol to protect muscle tissue during and after exercise. Endurance-training individuals show an adaptation of the HPA axis activity to repeated exercise due to reduced tissue sensitivity to glucocorticoids [32]. In their study, Luger et al. [30] indicated that highly trained runners had statistically significantly lower cortisol concentrations compared to sedentary people after prolonged exercise above 60% of maximal oxygen uptake (VO_{2max}). In contrast, as reported by Dote-Montero et al. [4] in their meta-analysis, repeated-sprint training and sprint interval training, despite high intensity, may not be long enough to induce a strong increase in C levels in contrast to interval training bouts ≥ 60 s. This may explain the significant increases in the change of salivary C concentration in all groups after SIE (Table 1). This suggests that the particular exercise was not intense enough to elicit a hormonal response, making it advisable to measure, in future studies, the lactate concentration to determine, among other things, the intensity of the exercise. The value of such a study was shown in an article by Lu et al. [33], who noted that the surge in testosterone immediately following interval exercise is highly correlated with an increase in lactate concentration. Tanner et al. [34], in contrast to our study, showed a significant increase in C levels and obtained HR values close to maximal after a single interval session in trained individuals. However, as indicated by that study's participants themselves, the exercise test was exhausting (six intervals of 3.5 min at a treadmill speed equivalent to 90% VO_{2max}, interspersed by recovery periods of 2 min = at the speed equivalent to 30% VO_{2max}), which could also affect the higher RPE values. In contrast to the results observed in that study, this difference with our findings can be explained by the exercise characteristics and the total duration of the protocol used: 5 × 10-s sprint interval exercise with 50-s recovery versus 6 × 3.5 min with 2-min recovery.

Endurance-training individuals have lower testosterone levels, which may be due to weight loss from this training [35] as opposed to strength-training peoples [36]. Additionally, Kreamer et al. [37] indicated that participants with 2 years of weightlifting training experience showed a significant exercise-induced increase in testosterone, while participants with training experience of less than or equal to 2 years showed no significant differences in testosterone change. Interestingly, Cadore et al. [38] observed lower reactivity of anabolic and catabolic hormonal responses in long-term strength-training men, indicating that higher training volume/intensity is required to induce significant hormonal changes. Statistically significant differences in C change observed between the non-training and strength-training groups can confirm the above-presented research (Figure 4).

The T/C is an appropriate indicator of an organism's anabolic environment [8]. The role of testosterone in the body is to maintain anabolism through the process of protein synthesis. In contrast, cortisol has a catabolic function and is involved in the stress response. Most athletes aim to increase the T/C, thereby enhancing protein synthesis and tissue recovery after physical exercise [37]. However, in this study, we observed no significant differences in T/C after performing SIE in any of the study groups, suggesting that the exercise applied was not sufficient to increase the body's anabolic environment. Similarly, individuals who trained both strength and endurance showed greater resistance to physiological stress by achieving a lower mean heart rate value after five repetitions of 10-s "all-out" exercise (Figure 3). Psychological reinforcement for this thesis is the fact that the dominance scale has an inverse relationship to changes in heart rate—the lower the HR, the higher the level of dominance (Table 1).

An explanation for the results obtained may be found in environmental psychology. Already, Seligman [39] has shown the importance of dominance as a feeling of control related to health and behavior. The adaptation to environmental demands expressed in the lower physiological parameters of our participants favors the increase of the psychological variable of dominance as expressing an internal control of one's own behavior and health. According to Rivers and Josephs [40], dominance is as legitimate an environmental descriptor as pleasure and arousal.

Furthermore, results of Jiménez et al. [41] provided explanations for the lack of SIE effect in T, C, and T/C levels in our participants. After winning a league game, higher testosterone levels were observed in professional soccer players, compared to semi-professional or amateur athletes. In contrast, this temporary hormonal fluctuation was not observed after winning a friendly match or during a normal training day. In the same match, cortisol levels were lower in professional and semi-professional athletes compared to levels in amateur athletes. This means, in soccer players, the increase in testosterone was only noticeable when the team faced the real challenge of a league match. It follows that the desire to achieve a goal (and maintain social status) may be one of the key reasons why testosterone increases rapidly. Conversely, testosterone did not change after friendly games, suggesting that these situations are not true goals in which players do not perceive a real threat (in the sense of dominance) any more than they perceive the preparation for the next game in their daily training, or even in a friendly game. Thus, we speculate that the SIE we conducted did not present a real challenge to our participants in the sense of increased testosterone or dominance, nor did it trigger a stress response in the form of cortisol release. In our study, similar to Jiménez et al. [41], there was an adaptation to the exercise situation in the absence of both T and C output. Results of Jiménez et al. [41] show that cortisol levels were lower in professional and semi-professional athletes compared to those in amateur athletes. Again, similar to Jimenez et al. [41], we observed C changes after SIE between the non-training and strength-training groups. This allows us to suppose that the perception of SIE in groups of professional and amateur athletes depends not only on the type of activity performed, and that the changes in C concentration in amateur athletes are influenced by the intensity and volume of the exercise sessions undertaken.

Finally, some authors [14,15] have shown that SIE can elicit negative affective reactions during exercise, which is responsible for the withdrawal and avoidance of exercise in the future. Most importantly, however, the SIE exercise protocol appears to be perceived as too difficult and demanding in terms of the effort put into it, especially for individuals with sedentary lifestyles [15]. Our experiment did not involve assessing affect, intention, self-efficacy, enjoyment, and preference. However, we used the Borg scale to assess the intensity of rating of perceived exertion after SIE. We found no statistically significant differences in perceived exertion between the endurance, strength, and non-exercise groups (endurance 5.16 RPE; strength 5.60 RPE; control 5.36 RPE); thus, we did not confirm the hypothesis that the non-exercising lifestyle group had a significantly higher perceived exertion. Thus, following Islam et al. [18], a short 10-s sprint with 50-s of rest is perceived as enjoyable and leads to a greater desire to engage in SIE. We can speculate that SIE in our formulation may safeguard the physical activity of a healthy, non-exercising population of young people.

Our results should be interpreted with caution because lactate concentration was not measured in this study, which could enrich the interpretations regarding hormonal changes due to a single interval session. There is also no measurement of peak power output (PPO), work (W), or oxygen uptake (VO_2), which could explain the mechanisms of C changes and the change in the physiological cost of the participants. For a better characterization of the participants, a more detailed analysis of the training experience should be conducted, especially regarding the volume and intensity of physical activity undertaken per week. The dominance scale has been validated, but not published to date, so we see this as a limitation in our study.

5. Conclusions

Sprint interval exercise consisting of 5 × 10-s "all-out" bouts performed by non-training individuals resulted in significant differences in cortisol changes concentrations compared to strength-training individuals. The lack of significant changes in T and C hormone concentrations among strength-training and endurance-training participants may indicate that the exercise volume was too low.

Non-training participants had higher HR after SIE than those from endurance- and strength-training groups; however, there were no differences in the ratings of perceived exertion. This suggests that it may be used by non-professional individuals.

Author Contributions: Conceptualization: G.Z., N.D., A.Ż. (Alina Żurek) and M.K.; methodology: N.D., A.Ż. (Alina Żurek), J.N.-K. and A.Ż. (Agnieszka Żelaźniewicz); validation: G.Z., N.D. and M.K.; formal analysis: G.Z., N.D., A.Ż. (Alina Żurek), J.N.-K., A.Ż. (Agnieszka Żelaźniewicz) and M.K.; investigation: G.Z., A.Ż. (Alina Żurek), S.O., T.S., M.N. and M.K.; resources: G.Z., N.D., A.Ż. (Alina Żurek), J.N-K., A.Ż. (Agnieszka Żelaźniewicz), S.O., T.S., M.N. and M.K; writing—original draft preparation: G.Z., N.D. and A.Ż. (Alina Żurek); writing—review and editing: G.Z., N.D., J.N.-K., A.Ż. (Agnieszka Żelaźniewicz) and M.K.; visualization: N.D., S.O. and M.K.; supervision: G.Z., A.Ż. (Alina Żurek) and M.K.; project administration: M.K. All authors have read and agreed to the published version of the manuscript.

Funding: This research received no external funding.

Institutional Review Board Statement: The study was conducted according to the guidelines of the Declaration of Helsinki, and approved by the Ethics Committee of the Institute of Psychology at the University of Wroclaw, Poland.

Informed Consent Statement: Informed consent was obtained from all participants involved in the study.

Data Availability Statement: The datasets used and/or analyzed during this study are available from the corresponding author on reasonable request.

Acknowledgments: We would like to express our great gratitude to the participants and our respect and appreciation for their unbroken attitude.

Conflicts of Interest: The authors declare no conflict of interest.

References

1. Hayes, L.D.; Elliott, B.T. Short-Term Exercise Training Inconsistently Influences Basal Testosterone in Older Men: A Systematic Review and Meta-Analysis. *Front. Physiol.* **2019**, *9*, 1878. [CrossRef] [PubMed]
2. Dote-Montero, M.; Carneiro-Barrera, A.; Martinez-Vizcaino, V.; Ruiz, J.R.; Amaro-Gahete, F.J. Acute effect of HIIT on testosterone and cortisol levels in healthy individuals: A systematic review and meta-analysis. *Scand. J. Med. Sci. Sports* **2021**, *31*, 1722–1744. [CrossRef] [PubMed]
3. Danek, N.; Smolarek, M.; Michalik, K.; Zatoń, M. Comparison of Acute Responses to Two Different Cycling Sprint Interval Exercise Protocols with Different Recovery Durations. *Int. J. Environ. Res. Public Health* **2020**, *17*, 1026. [CrossRef]
4. Riachy, R.; McKinney, K.; Tuvdendorj, D. Various Factors May Modulate the Effect of Exercise on Testosterone Levels in Men. *J. Funct. Morphol. Kinesiol.* **2020**, *5*, 81. [CrossRef] [PubMed]
5. Kraemer, W.J.; Ratamess, N.A.; Hymer, W.C.; Nindl, B.C.; Fragala, M.S. Growth Hormone(s), Testosterone, Insulin-Like Growth Factors, and Cortisol: Roles and Integration for Cellular Development and Growth with Exercise. *Front. Endocrinol.* **2020**, *11*, 33. [CrossRef] [PubMed]

6. Keizer, H.; Janssen, G.M.E.; Menheere, P.; Kranenburg, G. Changes in Basal Plasma Testosterone, Cortisol, and Dehydroepiandrosterone Sulfate in Previously Untrained Males and Females Preparing for a Marathon. *Int. J. Sports Med.* **1989**, *10*, S139–S145. [CrossRef]
7. Hayes, L.D.; Bickerstaff, G.F.; Baker, J.S. Interactions of cortisol, testosterone, and resistance training: Influence of circadian rhythms. *Chronobiol. Int.* **2010**, *27*, 675–705. [CrossRef]
8. Moreira, A.; Aoki, M.S.; Franchini, E.; Machado, D.; Paludo, A.; Okano, A.H. Mental fatigue impairs technical performance and alters neuroendocrine and autonomic responses in elite young basketball players. *Physiol. Behav.* **2018**, *196*, 112–118. [CrossRef]
9. Mehta, P.H.; Josephs, R.A. Testosterone and cortisol jointly regulate dominance: Evidence for a dual-hormone hypothesis. *Horm. Behav.* **2010**, *58*, 898–906. [CrossRef]
10. Batrinos, M.L. Testosterone and aggressive behavior in man. *Int. J. Endocrinol. Metab.* **2012**, *10*, 563–568. [CrossRef]
11. Carré, J.M.; McCormick, C.M. Aggressive behavior and change in salivary testosterone concentrations predict willingness to engage in a competitive task. *Horm. Behav.* **2008**, *54*, 403–409. [CrossRef] [PubMed]
12. Gibala, M.J.; Hawley, J.A. Sprinting Toward Fitness. *Cell Metab.* **2017**, *25*, 988–990. [CrossRef] [PubMed]
13. Vollaard, N.; Metcalfe, R. *Research into the Health Benefits of Sprint Interval Training Should Focus on Protocols with Fewer and Shorter Sprints (Forthcoming/Available Online)*; Springer: Berlin/Heidelberg, Germany, 2017.
14. Hardcastle, S.J.; Eray, H.; Beale, L.; Hagger, M. Why sprint interval training is inappropriate for a largely sedentary population. *Front. Psychol.* **2014**, *5*, 1505. [CrossRef] [PubMed]
15. Biddle, S.J.; Batterham, A.M. High-intensity interval exercise training for public health: A big HIT or shall we HIT it on the head? *Int. J. Behav. Nutr. Phys. Act.* **2015**, *12*, 95. [CrossRef]
16. Kujach, S.; Byun, K.; Hyodo, K.; Suwabe, K.; Fukuie, T.; Laskowski, R.; Dan, I.; Soya, H. A transferable high-intensity intermittent exercise improves executive performance in association with dorsolateral prefrontal activation in young adults. *NeuroImage* **2018**, *169*, 117–125. [CrossRef]
17. Kujach, S.; Olek, R.A.; Byun, K.; Suwabe, K.; Sitek, E.; Ziemann, E.; Laskowski, R.; Soya, H. Acute Sprint Interval Exercise Increases Both Cognitive Functions and Peripheral Neurotrophic Factors in Humans: The Possible Involvement of Lactate. *Front. Neurosci.* **2020**, *13*, 1455. [CrossRef]
18. Islam, H.; Townsend, L.K.; Hazell, T.J. Modified sprint interval training protocols. Part I. Physiological responses. *Appl. Physiol. Nutr. Metab.* **2017**, *42*, 339–346. [CrossRef]
19. Martínez-Díaz, I.C.; Carrasco, L. Neurophysiological Stress Response and Mood Changes Induced by High-Intensity Interval Training: A Pilot Study. *Int. J. Environ. Res. Public Health* **2021**, *18*, 7320. [CrossRef]
20. Beaven, C.M.; Gill, N.D.; Cook, C.J. Salivary Testosterone and Cortisol Responses in Professional Rugby Players After Four Resistance Exercise Protocols. *J. Strength Cond. Res.* **2008**, *22*, 426–432. [CrossRef]
21. Granger, D.A.; Shirtcliff, E.A.; Booth, A.; Kivlighan, K.T.; Schwartz, E.B. The "trouble" with salivary testosterone. *Psychoneuroendocrinology* **2004**, *29*, 1229–1240. [CrossRef]
22. Crewther, B.T.; Lowe, T.E.; Ingram, J.; Weatherby, R.P. Validating the salivary testosterone and cortisol concentration measures in response to short high-intensity exercise. *J. Sports Med. Phys. Fit.* **2010**, *50*, 85–92.
23. Allgrove, J.E.; Gomes, E.; Hough, J.; Gleeson, M. Effects of exercise intensity on salivary antimicrobial proteins and markers of stress in active men. *J. Sports Sci.* **2008**, *26*, 653–661. [CrossRef] [PubMed]
24. Kowal, M.; Sorokowski, P.; Żelaźniewicz, A.; Nowak, J.; Orzechowski, S.; Żurek, G.; Żurek, A.; Juszkiewicz, A.; Wojtycka, L.; Sieniuć, W.; et al. No relationship between the digit ratios (2D:4D) and salivary testosterone change: Study on men under an acute exercise. *Sci. Rep.* **2020**, *10*, 10068. [CrossRef] [PubMed]
25. Kowal, M.; Sorokowski, P.; Żelaźniewicz, A.; Nowak, J.; Orzechowski, S.W.; Żurek, A.; Żurek, G. A positive relationship between body height and the testosterone response to physical exercise. *Evol. Hum. Behav.* **2021**, *42*, 179–185. [CrossRef]
26. Williams, N. The Borg Rating of Perceived Exertion (RPE) scale. *Occup. Med.* **2017**, *67*, 404–405. [CrossRef]
27. Kowal, M.; Sorokowski, P.; Żelaźniewicz, A.; Nowak, J.; Orzechowski, S.; Żurek, G.; Żurek, A.; Nawrat, M. Are Beards Honest Signals of Male Dominance and Testosterone? *Arch. Sex. Behav.* **2021**, *50*, 3703–3710. [CrossRef]
28. Mahalanobis, P.C. A method of fractile graphical analysis. *Économ. J. Économ. Soc.* **1960**, *28*, 325–351. [CrossRef]
29. Penny, K.I. Appropriate Critical Values When Testing for a Single Multivariate Outlier by Using the Mahalanobis Distance. *J. R. Stat. Soc. Ser. C* **1996**, *45*, 73. [CrossRef]
30. Luger, A.; Deuster, P.A.; Kyle, S.B.; Gallucci, W.T.; Montgomery, L.C.; Gold, P.W.; Chrousos, G.P. Acute hypothalamic–pituitary–adrenal responses to the stress of treadmill exercise. *N. Eng. J. Med.* **1987**, *316*, 1309–1315. [CrossRef]
31. Spanoudaki, S.; Maridaki, M.; Tsironi, M.; Baltopoulos, G.; Bogdanis, G.; Baltopoulos, P. Immune responses during and after exercise of constant and alternating intensity above the lactate threshold. *J. Sports Med. Phys. Fit.* **2010**, *50*, 363.
32. Lanfranco, F.; Strasburger, C.J. (Eds.) *Sports Endocrinology*; Karger Medical and Scientific Publishers: Basel, Switzerland, 2016.
33. Lu, S.S.; Lau, C.P.; Tung, Y.F.; Huang, S.W.; Chen, Y.H.; Shih, H.C.; Wang, P.S. Lactate and the effects of exercise on testosterone secretion: Evidence for the involvement of a cAMP-mediated mechanism. *Med. Sci. Sports Exerc.* **1997**, *29*, 1048–1054. [CrossRef] [PubMed]
34. Tanner, A.V.; Nielsen, B.V.; Allgrove, J. Salivary and plasma cortisol and testosterone responses to interval and tempo runs and a bodyweight-only circuit session in endurance-trained men. *J. Sports Sci.* **2013**, *32*, 680–689. [CrossRef] [PubMed]

35. Wheeler, G.D.; Singh, M.; Pierce, W.D.; Epling, W.F.; Cumming, D.C. Endurance Training Decreases Serum Testosterone Levels in Men without Change in Luteinizing Hormone Pulsatile Release. *J. Clin. Endocrinol. Metab.* **1991**, *72*, 422–425. [CrossRef] [PubMed]
36. Vingren, J.L.; Kraemer, W.J.; Ratamess, N.A.; Anderson, J.M.; Volek, J.S.; Maresh, C.M. Testosterone Physiology in Resistance Exercise and Training. *Sports Med.* **2010**, *40*, 1037–1053. [CrossRef]
37. Kraemer, W.J.; Ratamess, N.A. Hormonal Responses and Adaptations to Resistance Exercise and Training. *Sports Med.* **2005**, *35*, 339–361. [CrossRef]
38. Cadore, E.; Lhullier, F.; Brentano, M.; Silva, E.; Ambrosini, M.; Spinelli, R.; Silva, R.; Kruel, L. Correlations between serum and salivary hormonal concentrations in response to resistance exercise. *J. Sports Sci.* **2008**, *26*, 1067–1072. [CrossRef]
39. Seligman, M.E. *Helplessness: On Depression, Development and Death*; Scribner: New York, NY, USA, 1975; pp. 35–42.
40. Rivers, J.J.; Josephs, R.A. The social psychology of power. *Domin. Health* **2010**, 87–112.
41. Jiménez, M.; Alvero-Cruz, J.R.; Solla, J.; García-Bastida, J.; García-Coll, V.; Rivilla, I.; Ruiz, E.; García-Romero, J.; Carnero, E.A.; Clemente-Suárez, V.J. Competition Seriousness and Competition Level Modulate Testosterone and Cortisol Responses in Soccer Players. *Int. J. Environ. Res. Public Health* **2020**, *17*, 350. [CrossRef]

biology

MDPI

Article

Temporal Skin Temperature as an Indicator of Cardiorespiratory Fitness Assessed with Selected Methods

Agnieszka Danuta Jastrzębska *, Rafal Hebisz and Paulina Hebisz

Department of Physiology and Biochemistry, Wroclaw University of Health and Sport Sciences, 51-612 Wroclaw, Poland; rafal.hebisz@awf.wroc.pl (R.H.); paulina.hebisz@awf.wroc.pl (P.H.)
* Correspondence: agnieszka.jastrzebska@awf.wroc.pl

Simple Summary: The aim of this study was to investigate whether it is possible to use infrared thermography to assess cardiovascular fitness and aerobic capacity. Changes in temporal temperature during and after a single bout of high-intensity exercise were measured from subjects with varying levels of physical activity. Significant correlation between the temporal temperature measured during recovery time with cardiovascular fitness parameters (HRR and HRV) and maximum oxygen consumption confirm the usefulness of thermal imagining in aerobic capacity evaluation. These results could foster the employment of infrared thermography to monitor athletic/athletes' performance.

Abstract: The aim of this study was to determine whether there are associations between cardiovascular fitness (and aerobic capacity) and changes in temporal skin temperature during and after a single bout of high-intensity exercise. Twenty-three men with varying levels of physical activity (VO_2max: 59.03 ± 11.19 (mL/kg/min), body mass 71.5 ± 10.4 (kg), body height 179 ± 8 (cm)) participated in the study. Each subject performed an incremental test and, after a 48-h interval, a 110%Pmax power test combined with an analysis of the thermal parameters, heart rate recovery and heart rate variability. Thermal radiation density from the body surface (temple) was measured using a Sonel KT384 thermal imaging camera immediately after warm-up (Tb), immediately after exercise (Te) and 120 sec after the end of exercise (Tr). The differences between measurements were then calculated. The correlation analysis between the thermal and cardiovascular function parameters during the recovery period showed strong positive associations between the Tr-Te difference and measures of cardiovascular fitness ($50 < r < 69$, $p < 0.05$). For example, the correlation coefficient between Tr-Te and VO_2max reached 0.55 and between Tr-Te and Pmax reached 0.68. The results obtained indicate that the measurement of temporal temperature during and after an intense 3-min bout of exercise can be used to assess aerobic physical capacity and cardiovascular fitness.

Keywords: cardiovascular fitness; aerobic capacity; skin surface temperature; high-intensity exercise; thermal imaging; recovery

Citation: Jastrzębska, A.D.; Hebisz, R.; Hebisz, P. Temporal Skin Temperature as an Indicator of Cardiorespiratory Fitness Assessed with Selected Methods. *Biology* **2022**, *11*, 948. https://doi.org/10.3390/biology11070948

Academic Editors: Filipe Manuel Clemente, Georgian Badicu and Eugenia Murawska-Ciałowicz

Received: 24 April 2022
Accepted: 20 June 2022
Published: 21 June 2022

1. Introduction

The evaluation of cardiovascular fitness during exercise has been the subject of numerous research papers [1–9]. In exercise testing, the measurement of maximal oxygen uptake (VO_2max) [1–4], assessment of heart rate recovery (HRR) [5–7] and heart rate variability (HRV) [8,9] are popular methods used in its assessment.

The efficiency of the cardiovascular system (especially the cardiac minute volume) and the blood volume [10] influence the regulation of blood distribution to the muscles and skin during exercise [11]. Regulation of blood flow is largely an adrenergic response. During intense exercise, the concentration of norepinephrine in blood increases [12,13]. Norepinephrine acts on the postsynaptic receptors (alpha1 and alpha2) of the sympathetic adrenergic system, causing vasoconstriction of the cutaneous circulation vessels [11,14]. As a result, at the onset of intense exercise, there are changes in blood flow, consisting of an increase in flow in the limbs loaded with exercise and a decrease in flow to inactive limbs and skin microcirculation [14–16]. Through the described reactions, the muscles are better supplied with oxygen during intense work and the convection of flowing blood is more effective [16–21]. However, when exercise is performed over a long period of time, the internal body temperature increases and the heat removal mechanisms must be activated [22]. These include the vasodilation of the skin vessels and an increase in the intensity of cutaneous blood flow, intended to increase the heat release to the atmosphere [22]. The magnitude of the increase in cutaneous blood flow during exercise depends on aerobic capacity [11] and increased adaptation to training and heat stress [23]. This adaptation lowers the internal temperature threshold at which vasodilation of the skin vessels occurs [23]. Exercise-induced changes in blood flow affect body surface temperature [16,24].

The previous study by Hebisz et al. (2019) [25] showed that the level of maximal oxygen uptake correlates strongly with the change in the temporal temperature observed during recovery. At the same time, no similar correlations (with maximal oxygen uptake) by measuring exercise and recovery changes in arm temperature were established. It is surmised that those effects may be related to the degree of the vascularisation of the inspected skin region, which is large in the temporal area (compared with the vascularisation of the arm), due to a branch of the external carotid artery (superficial temporal artery) [26].

The exercise protocol used in the aforementioned study consisted of 4 sprints of 30 s each, separated by 90 s rest periods [25]. Repeated sprint efforts (similar to Wingate tests) lead to a severe disturbance of acid-base homeostasis [27] and high levels of subjective fatigue [28]. For these reasons, there are people who avoid such efforts [28]. An alternative protocol for testing thermal parameters on the temporal surface could be based on a single effort of a few minutes, typical for HIIT-type training or tests aimed at verifying maximal oxygen uptake. Such efforts are also very intensive as they achieve an oxygen uptake close to the maximum [29], but at the same time, the level of subjective fatigue may be lower afterwards [28].

The aim of this study was to determine whether there are associations between cardiovascular fitness (and aerobic capacity) and changes in temporal skin temperature during and after a single bout of high-intensity exercise. It was assumed that, among other things, the rate of maximal oxygen uptake (as well as the rate of heart rate recovery) would positively correlate with the change in temporal temperature post-exercise.

2. Material and Methods

2.1. Participants

A group of 23 men with varying levels of physical activity participated in the study. The participants were free of any known neuromusculoskeletal, cardiovascular and respiratory systems impairment. A total of 6 participants led a sedentary lifestyle, 10 were classified as a physically active (exercise duration 3 to 5 h per week: 5—runners, 3—cross-fit, 2—swimming, 1—racket games;) and 7 were classified as athletes (regularly participating in sport competitions; exercise duration 9 to 15 h per week: 3—runners, 2—cross-country skiing, 2—team games). Table 1 shows the values of parameters characterising the physical capacity and physique of the study group.

Table 1. Physiological and anthropometric characteristics of the group.

VO_2max	Pmax	Age (y)	BM (kg)	LBM (kg)	BH (cm)
59.03 ± 11.19	332.5 ± 48.4	22.8 ± 5.4	71.5 ± 10.4	64.3 ± 8.6	179 ± 8.1

VO_2max—maximum oxygen consumption; Pmax—maximum power in graded exercise test; BM—body mass; LBM—lean body mass; BH—body height.

The study was approved by the University Ethics Committee and conducted in accordance with the ethical standards established by the Declaration of Helsinki. Participants were made aware of the experimental protocol and gave written consent to participate prior to the study.

2.2. Study Design

The subjects had not performed heavy physical exercise in the 48 h prior to the exercise tests. Each subject performed an incremental test for cardiovascular fitness and aerobic efficiency assessment as well to determine the power of the verification test equal to 110% of Pmax obtained in GXT. After a 48-h interval, a 110%Pmax power test combined with an analysis of the thermal parameters, heart rate recovery and heart rate variability were carried out (Figure 1).

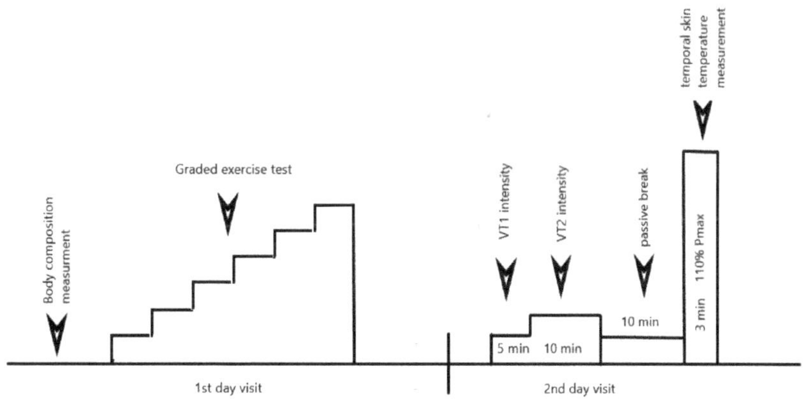

Figure 1. Scheme of visit in laboratory.

2.2.1. Body Composition

Before the test, body composition was measured using a near-infrared device NIR (6100/XL, Futrex, Hagerstown, MD, USA). The device measures the optical density of body tissues on the *biceps brachii* of the dominant hand to estimate the body fat expressed as kilograms and percentages of body weight, lean body mass (LBM) and body water content in liters and as a percentage of the body mass. It is a commonly used method for body composition measurements [30,31].

2.2.2. Graded Exercise Test (GXT)

A graded exercise test (GXT) was performed to determine aerobic power [25,32]. The test was conducted on a Lode Excalibur Sport cycloergometer (Lode B.V., Groningen, The Netherlands), which was calibrated before the start of the study. The seat height was adjusted individually so that the angle of deflection was no greater than 5° when the leg was fully extended. The effort started at a load of 50 W; every 3 min, the load was increased by 50 W until the subject refused to continue. If during the last load of the test the subject failed to exert for 3 min, then for each missing second, 0.28 W was subtracted from the current power value [33]. In this way, the maximal aerobic power (Pmax) value was obtained.

During testing, subjects breathed through a mask as their expired air was sampled breath by breath and analysed by a Quark CPET gas analyser (Cosmed, Milan, Italy). The apparatus was calibrated with atmospheric air and a gas mixture composed of the following elements: CO_2—5%, O_2—16% and N_2—79%. The respiratory parameters (oxygen uptake (VO_2), exhaled carbon dioxide (VCO_2), and minute pulmonary ventilation (VE)) were measured. The analysis of the data was carried out with the results averaged every 30 s. The highest VO_2 recorded in GXT defines VO_2peak1. Heart rate (HR) was recorded with the V800 cardiofrequencimeter (Polar, Oy, Kempele, Finland).

The first ventilatory threshold (VT1), at the point before the first non-linear increase in $VE \cdot VO_2^{-1}$ equivalent without a concomitant increase in $VE \cdot VCO_2^{-1}$, was determined from the recorded respiratory data from the GXT test; the second ventilation threshold (VT2) was determined at the point preceding the second non-linear increase in $VE \cdot VO_2^{-1}$ or $VE \cdot VCO_2^{-1}$ equivalent, following the methodology described by Beaver et al. (1986) [34] and Davis et al. (1980) [35].

2.2.3. Test at 110% of Pmax

A Lode Excalibur Sport cycloergometer and a Quark ergospirometer were used to perform the test. The test was preceded by a 15-min warm-up, consisting of exercising for 5 min at an intensity corresponding to the power achieved at the VT1, followed by 10 min at a power corresponding to halfway between the VT1 and the VT2. The warm-up was followed by a 10-min passive break. The verification test lasted 3 min and was performed at an intensity of 110% of the Pmax achieved in the GXT test performed two days earlier. Recording of the respiratory parameters began 1 min before the verification test and ended 5 min after the test. The analysis of the data was carried out using the results averaged every 30 s. The highest recorded oxygen uptake value (from 30 sec averaging) was considered to be the peak oxygen uptake in a verification test performed on a separate day (VO_2peak2). Maximal oxygen uptake (VO_2max) was considered to be the higher than VO_2peak1 and VO_2peak2, as in earlier work [32]. The power and oxygen uptake parameters analysed in this study were recalculated in relation to body mass (VO_2peak1, VO_2peak2 and VO_2max) and lean body mass ($Pmax \cdot LBM^{-1}$, $VO_2max \cdot LBM^{-1}$). Each exercise test was performed under controlled thermal conditions. The air temperature was 21 °C and the humidity was between 40–45%.

During the test at power 110%Pmax, the density of thermal radiation from the body surface was measured using a thermal imaging camera Sonel KT384 (Sonel S.A., Świdnica, Poland). The thermal imaging camera was used in accordance with the manufacturer's guidelines. The camera features IR resolution of 384 × 288; spectral range of 8–14 µm; thermal sensitivity of 0.08 °C. The software provided by the manufacturer (Sonel Thermo-Analyze ver. 1.0) converted radiation density into body surface temperature expressed in °C. During playback of the recorded video, individual frames were analysed and the mean temperature was recorded within a square box (with a side length of 10 pixels), individually marked on the temple (Figure 2), as in an earlier publication [25]. Baseline temperature was determined after warm-up (T_b), immediately before the start of the exercise at 110%Pmax. Further temperature determinations were made immediately after exercise (T_e), and 120 s after exercise (T_r). The difference between T_e and T_b ($T1 = T_e - T_b$), the difference between T_r and T_b ($T2 = T_r - T_b$) and the difference between T_r and T_e ($T3 = T_r - T_e$) were then calculated.

During the warm-up and a 10-min passive break before the 110%Pmax power test, the time interval between heartbeats (RR) was recorded using a V800 cardiofrequency meter (Polar, Oy, Kempele, Finland). Heart rate values were averaged for 59–61″ (HRR1′), 119–121″ (HRR2′), 179–181″ (HRR3′), 239–241″ (HRR4′) and 299–301″ (HRR5′) recovery after warm-up, as recovery RR interval measurements have a high variability. The RR intervals allowed for the analysis of consciously selected sections for HRR analysis. A simpler method, such as HR recording, would result in a random entry segment for the HRR analysis, as the cardiofrequencymeter's software calculates the HR based on the averaging data at 3 s intervals. The conversion of RR to HR was performed automatically by PolarFlow (www.flow.polar.com accessed on 22 October 2021) in a precisely marked part of the saved data. The changes in heart rate during recovery after warm-up were then calculated as the difference between the heart rate measured at the end of warm-up and HRR1′ (ΔHRR1′), HHR2′ (ΔHRR2′), HRR3′ (ΔHRR3′), HRR4 (ΔHRR4′) and HRR5′ (ΔHRR5′), respectively, similar to Suzic Lazic et al. (2017) [6].

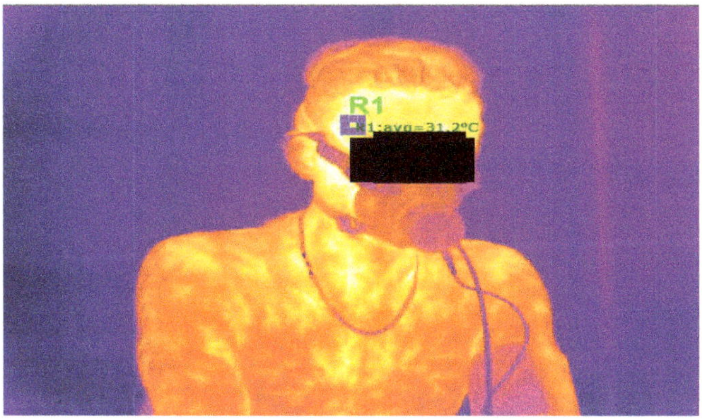

Figure 2. Single frame of film recorded during a test at 110%Pmax in which the field (R1) for temporal surface temperature analysis is marked.

For the temporal HRV parameters, the square-root of the mean squared difference between successive normal-to-normal RR intervals ($RMSSD_{3–5'}$) and standard deviation of normal-to-normal RR intervals ($SDNN_{3–5'}$) were calculated from the data recorded by the cardiofrequencimeter during recovery after the warm-up. For the frequency domain, a spectral analysis was performed using fast Fourier transformation to obtain low-frequency spectral power ($LFP_{3–5'}$) and high-frequency spectral power ($HFP_{3–5'}$). For this purpose, the Kubios HRV Standard software (KubiosOy, Kuopio, Finland) was used. HRV parameters were calculated for the portion of the recording covering the third, fourth and fifth minutes of recovery, similar to the methodology previously used by Buchheit et al. (2009) [36]. Medium threshold data filtering was applied in the calculation of the above variables.

For the analyses of HRV, HRR and ΔHRR, we used data collected in post-warm-up recovery as these data allow analyses to be performed after standardised moderate-intensity exercise, similar to the procedures used by other authors [36,37].

2.3. Statistical Analysis

The Shapiro–Wilk test was used to assess the distribution of the parameters studied. Analysis of variance with repeated measurements was used to compare the parameters that were measured several times during the study. Pearson's simple correlation was used to determine the strength of the relationships between the changes in temporal temperature and measurements of cardiovascular fitness (and physical capacity). The scale modified by Hopkins et al. (2009) [38] was used to interpret the correlation coefficient, which is as follows: 0.1–0.29 = trivial, 0.30–0.49 = moderate, 0.50–0.69 = strong, 0.70 to 0.89 = very strong, 0.90–0.99 = nearly perfect, and 1 = perfect.

Using the formula for the critical value of the correlation coefficient, the minimum group size was calculated assuming that the acceptable level of statistical significance (alpha) is 0.05 and that the strength of the correlation should be high, $r > 0.5$. The following formula was used:

$$r = \sqrt{\frac{t_\alpha^2}{n - 2 + t_\alpha^2}} \quad (1)$$

On this basis, we determined that the minimum number of study participants was 16 at $t \approx 2.15$.

3. Results

Using the analysis of variance, a statistically significant effect of the repeated measurements was demonstrated for $Temp_{b,e,r}$ ($F = 39.95$; $p = 0.000$; $\eta^2 = 0.645$). Using a post-hoc test, significant differences were shown between $Temp_e$ and $Temp_b$ and between $Temp_r$ and $Temp_e$ (Table 2).

Table 2. Thermal parameters and baseline measurements of the physical capacity of the study participants.

Items	X ± SD	Lower 95%CI	Upper 95%CI
$Temp_b$ (°C)	33.0 ± 2.5	31.9	34.0
$Temp_e$ (°C)	30.8 ± 2.5 *	29.7	31.9
$Temp_r$ (°C)	32.6 ± 2.6 **	31.5	33.7
T1 (°C)	−2.2 ± 1.1	−2.7	−1.7
T2 (°C)	−0.6 ± 1.3	−1.2	0.0
T3 (°C)	1.7 ± 1.0	1.3	2.1
Pmax (W)	332.5 ± 48.4	311.6	353.5
$Pmax{\cdot}LBM^{-1}$ ($W{\cdot}kg^{-1}$)	5.27 ± 1.04	4.82	5.72
VO_2peak1 ($mL{\cdot}min^{-1}{\cdot}kg^{-1}$)	55.56 ± 13.02	49.93	61.19
VO_2peak2 ($mL{\cdot}min^{-1}{\cdot}kg^{-1}$)	58.25 ± 10.82	53.57	62.93
VO_2max ($mL{\cdot}min^{-1}{\cdot}kg^{-1}$)	59.03 ± 11.19	54.19	63.87
$VO_2max{\cdot}LBM^{-1}$ ($mL{\cdot}min^{-1}{\cdot}kg^{-1}$)	65.35 ± 11.21	60.50	70.20

$Temp_b$—baseline temporal temperature; $Temp_e$—temporal temperature measured immediately after the exercise at 110%Pmax; $Temp_r$—temporal temperature measured during recovery—120 sec after the exercise at 110%Pmax; T1—difference between $Temp_e$ and $Temp_b$; T2—difference between $Temp_r$ and $Temp_b$; T3—difference between $Temp_r$ and $Temp_e$; Pmax—maximum aerobic power; LBM—lean body mass; VO_2peak1—peak oxygen uptake determined in GXT test; VO_2peak2—peak oxygen uptake determined in exercise with 110%Pmax; VO_2max—maximum oxygen uptake; CI—confidence interval; *—$p < 0.000$ vs. $Temp_b$; **—$p < 0.000$ vs. $Temp_e$.

The simple correlation coefficient indicated moderate statistically significant relationships of T1 with VO_2peak2, VO_2max, $VO_2max{\cdot}LBM^{-1}$ ($0.30 < r < 0.49$) and strong relationships of T3 with Pmax, $Pmax{\cdot}LBM^{-1}$, VO_2peak1, VO_2peak2, VO_2max, $VO_2max{\cdot}LBM^{-1}$ (Table 3) ($0.50 < r < 0.69$).

Table 3. Pearson correlation between thermal parameters and baseline measurements of physical capacity.

Items	T1 (°C)	T2 (°C)	T3 (°C)
Pmax (W)	−0.33	0.11	0.63 *
$Pmax{\cdot}LBM^{-1}$ ($W{\cdot}kg^{-1}$)	−0.32	0.06	0.68 *
VO_2peak1 ($mL{\cdot}min^{-1}{\cdot}kg^{-1}$)	−0.37	−0.07	0.46 *
VO_2peak2 ($mL{\cdot}min^{-1}{\cdot}kg^{-1}$)	−0.42 *	−0.01	0.51 *
VO_2max ($mL{\cdot}min^{-1}{\cdot}kg^{-1}$)	−0.43 *	−0.05	0.51 *
$VO_2max{\cdot}LBM^{-1}$ ($mL{\cdot}min^{-1}{\cdot}kg^{-1}$)	−0.42 *	−0.02	0.55 *

T1—difference between $Temp_e$ and $Temp_b$; T2—difference between $Temp_r$ and $Temp_b$; T3—difference between $Temp_r$ and $Temp_e$; Pmax—maximum aerobic power; LBM—lean body mass; VO_2peak1—peak oxygen uptake determined in GXT test; VO_2peak2—peak oxygen uptake determined in exercise with 110%Pmax; VO_2max—maximum oxygen uptake; *—$p < 0.05$.

Using the analysis of variance, statistically significant effects of the repeated measurements were demonstrated for HRR (F = 15.24; $p = 0.000$; $\eta^2 = 0.409$) and ΔHRR (F = 15.57; $p = 0.000$; $\eta^2 = 0.414$). Using post-hoc tests, it was shown that HRR and ΔHRR measurements performed in the first minute of recovery were significantly different from those performed in the second, third, fourth and fifth minutes of recovery (Table 4).

Using the simple correlation coefficient, strong ($0.50 < r < 0.69$), statistically significant relationships of T1 with ΔHRR2′, ΔHRR3′ and T3 with HRR1′, HRR2′ HRR4′, HRR5′, ΔHRR1′ ΔHRR2′ ΔHRR4′ ΔHRR5′ were demonstrated (Table 5).

Using the simple correlation coefficient, moderate ($0.30 < r < 0.49$), statistically significant associations ($p < 0.05$) of T2 with SDNN1 and strong ($0.50 < r < 0.69$) of T3 with all HRV parameters analysed in recovery were demonstrated (Table 6).

Table 4. Recovery heart rate and recovery heart rate variability among study participants.

Items	1′	2′	3′	4′	5′
HRR (bpm)	99.2 ± 17.1	90.5 ± 18.6 *	88.4 ± 16.8 *	86.7 ± 14.8 *	86.2 ± 12.5 *
95%CI L-U	91.8–106.6	43.7–54.6	81.2–95.7	80.3–93.1	80.8–91.6
ΔHRR (bpm)	49.1 ± 12.6	58.1 ± 13.9 *	60.2 ± 13.3 *	62.0 ± 11.2 *	62.4 ± 11.0 *
95%CI L-U	43.7–54.6	52.1–64.2	54.4–65.9	57.1–66.8	57.7–67.2
$SDNN_{3–5′}$ (ms)	—	—		43.10 ± 22.46	
95%CI L-U				33.39–52.81	
$RMSSD_{3–5′}$ (ms)	—	—		31.17 ± 19.72	
95%CI L-U				22.65–39.70	
$HFP_{3–5′}$ (ms^2)	—	—		657.1 ± 1005.7	
95%CI L-U				222.2–1092.0	
$LFP_{3–5′}$ (ms^2)	—	—		1488.2 ± 1541.9	
95%CI L-U				821.5–2155.0	

HRR—heart rate recovery; ΔHRR—difference between heart rate measured at the end of the warm-up and the heart rate measured at the end if 1′, 2′, 3′, 4′, 5′ minute of recovery; $SDNN_{3–5′}$—the standard deviation of NN intervals; $RMSSD_{3–5′}$—the root mean square of successive differences between normal heart beats $HFP_{3–5′}$—high-frequency power; $LFP_{3–5′}$—low-frequency power; CI L-U—confidence interval for lower - upper value; 1′, 2′ 3′, 4′, 5′—1′, 2′ 3′, 4′, 5′—measurements taken in the next minutes of recovery after the warm-up; *—$p < 0.05$ vs. 1′.

Table 5. Pearson correlation between thermal parameters and recovery heart rate.

Items	T1 (°C)	T2 (°C)	T3 (°C)
HRR1′ (bpm)	0.09	−0.28	−0.53 *
ΔHRR1′ (bpm)	−0.34	0.11	0.63 *
HRR2′ (bpm)	0.20	−0.18	−0.50 *
ΔHRR2′ (bpm)	−0.48 *	−0.02	0.57 *
HRR3′ (bpm)	0.17	−0.16	−0.36
ΔHRR3′ (bpm)	−0.44 *	−0.07	0.35
HRR4′ (bpm)	0.10	−0.3	−0.58 *
ΔHRR4′ (bpm)	−0.40	0.07	0.64 *
HRR5′ (bpm)	0.12	−0.48 *	−0.62 *
ΔHRR5′ (bpm)	−0.41	0.22	0.57 *

T1—difference between $Temp_e$ and $Temp_b$; T2—difference $Temp_r$ and $Temp_b$; T3—difference $Temp_r$ and $Temp_e$; HRR—heart rate recovery; ΔHRR—difference between heart rate measured at the end of the warm-up and the heart rate measured at the end if 1′, 2′, 3′, 4′, 5′ minute of recovery; 1′, 2′, 3′, 4′, 5′—measurements taken in the next minutes of recovery after the warm-up; *—$p < 0.05$.

Table 6. Pearson correlation between thermal parameters and recovery heart rate variability parameters.

Items	T1 (°C)	T2 (°C)	T3 (°C)
$SDNN_{3–5′}$	−0.18	0.43 *	0.65 *
$RMSSD_{3–5′}$	−0.29	0.36	0.67 *
$HFP_{3–5′}$	−0.22	0.29	0.53 *
$LFP_{3–5′}$	−0.17	0.31	0.50 *

T1—difference between $Temp_e$ and $Temp_b$; T2—difference between $Temp_r$ and $Temp_b$; T3—difference between $Temp_r$ and $Temp_e$; SDNN—the standard deviation of NN intervals; RMSSD—the root mean square of successive differences between normal heart beats HFP—high-frequency power; LFP—low-frequency power; 1′, 2′ 3′, 4′, 5′—measurements taken in the next minutes of recovery after the warm-up; *—$p < 0.05$.

4. Discussion

The results of the study presented in this paper confirm the previous findings that during a few minutes of intense physical exercise, the body surface temperature decreases [24,39,40], which is linked to vasoconstriction of the vessels of cutaneous circulation and consequently to a decrease in blood flow in cutaneous circulation [24,39]. Furthermore, the results presented herein indicate that the level of cardiovascular fitness is related to the magnitude of the decrease in temporal temperature during an effort of 110%Pmax lasting a few minutes. This is evidenced by the results of the Pearson analyses, which showed a negative correlation of moderate strength between T1 and VO_2max and ΔHRR2′ and ΔHRR3′. The presence of the above correlations may be due to the fact that individuals with higher VO_2max [41] and greater HRR [42] are able to achieve greater power output in incremental tests and, therefore, also in the 110%Pmax intensity test. During the initial phase of high-power exercise, oxygen deficit and muscle oxygen demand increase rapidly [43], resulting in vasoconstriction in the inactive tissues (including the skin) and redirection of blood towards the active tissues [21]. Hence, the magnitude of the decrease in exercise body surface temperature is dependent on the intensity of the exercise [24]. The mechanism described above may also explain the correlations that relate T1 and VO_2max, as well as T1 (Table 3) and ΔHRR, in our study (Table 2). Thus, the magnitude of the drop in temporal temperature can be used as an indicator of the level of cardiovascular fitness alongside generally accepted indicators such as maximal oxygen uptake [44] and heart rate recovery [7–9].

The mechanical efficiency of muscles is only 20–30% [45,46]. When exercise is continued over a long period of time, the internal body temperature increases [47] because the energy created by the skeletal muscle metabolism is largely converted into thermal energy [48]. Maintaining a balance between heat production and heat removal during exercise ensures that you can continue exercising, preventing the development of thermal shock [49]. Thermal energy is most effectively removed from the muscles by blood convection [50]. At the same time, vasodilatation of the vessels of cutaneous circulation takes place, which enables an increase in cutaneous blood flow, an increase in body surface temperature and the release of thermal energy through radiation and sweat evaporation [21]. In the process of training, vascular endothelial function improves [16]. This has been demonstrated by administering acetylcholine and sodium nitroprusside by iontophoresis [51,52]. In conjunction with improved vascular endothelial function, it was observed that endurance-trained individuals had higher cutaneous blood flow during exercise at intensities up to 90% VO_2max, compared to non-trained individuals [53]. In addition, endurance-trained athletes have a higher blood volume and a higher cardiac stroke volume than non-trained athletes [54,55]. High cutaneous blood flow, high cardiac stroke volume and high plasma volume are factors that determine VO_2max levels [2,55] and enable efficient thermal energy removal during exercise [56,57]. Therefore, it can be assumed that high cardiovascular fitness should favour a higher increase in body surface temperature, which is confirmed by the strong correlation between T3 and VO_2max ($0.50 < r < 0.69$) (Table 3). The direction of the correlation of T3 with VO_2max is opposite to the correlation between T1 and VO_2max probably because for short high-intensity efforts, the increase in body surface temperature occurs only after the end of the effort [58].

The relationship between T3 and VO_2max (Table 3) that we observed in recovery is similar to the observations described in our earlier publication [25], in which we showed a correlation between VO_2max and recovery temporal temperature change after sprint interval training (4 all-out sprints of 30 s each, interspersed with breaks of 90 s). However, the exercise test protocol used in previous studies [25] is not commonly used in the diagnosis of performance capacity. Efforts lasting several minutes at 110% Pmax are more commonly performed in tests to verify VO_2max [32,59–61]. Similar efforts are also used during training [29]. For this reason, the correlation described in this study, with an exercise protocol lasting 3 min at 110%Pmax, seems to be of greater applicative importance.

In the current study, HRV recovery and ΔHRR were measured after a moderate-intensity warm-up, and T3 was measured during the high-intensity exercise that followed the warm-up. Nevertheless, our results show strong correlations of T3 with ΔHRR and with HRV (Table 6) parameters in each of the five recovery minutes analysed (Table 5). As mentioned above, HRR is a measure of cardiovascular fitness [8,9] as well as fitness level in endurance disciplines [6]. During recovery, the heart rate is reduced by a decrease in sympathetic, and an increase in parasympathetic, nervous system activity [6,62]. HRV parameters measured during recovery are also a measure of sympathovagal balance [8]. This balance depends, among other things, on the level of heat stress [63,64]. It is, therefore, possible that T3 correlates with ΔHRR and HRV recovery because it is a measure of the ability to reduce heat stress. The significance of this relationship appears to be high, as the coefficient of determination (R^2) between T3 and recovery HRV parameters reached 45% in our analyses (T3 and $RMSSD_{3–5'}$).

Maximal oxygen uptake is the parameter that determines the possible amount of energy obtained through aerobic metabolism [65]. In incremental tests, aerobic metabolism is responsible for the vast majority of the work carried out and even in the final phase of these tests, it is the dominant energy source [66]. Moreover, the correlation between maximal power in incremental tests and VO_2max level is very strong [67]. In view of this, the relationship of T3 and VO_2max should allow a correlation of T3 with Pmax to exist and the current results shows this correlation, which was strongest (r = 0.68) when Pmax was expressed as $Pmax·LBM^{-1}$ (Table 3). This is a significant outcome of the current study, as it allows us to conclude that physical capacity (ability to perform intense aerobic work) can be assessed on the basis of changes in post-exercise body surface temperature.

The results of the research presented in this paper indicate that the temperature of the temples' surface increases in recovery after intense aerobic exercise, especially in people with high physical capacity. Such results confirm the thesis from the studies by Hebisz et al. [25] that blood flow increases in the temporal region as a result of intensive work. At the same time, already at an intensity above 60% VO_2max, the cerebral blood flow decreases, which protects the brain from excessive thermal stress [68]. It is possible that the reduction in cerebral blood flow additionally increases the blood flow in the branches of the external carotid artery. Thus, a decrease in cerebral blood flow may result in a greater increase in the temples' surface temperature following exercise. However, this statement requires empirical confirmation.

5. Conclusions

As hypothesised, the recovery changes in temporal temperature (after a 3-min bout of intense exercise) were positively correlated with the measurements of cardiovascular fitness applied. It is also possible to assess the level of aerobic capacity (ability to perform intense aerobic work) on the basis of recovery changes in temporal temperature. Furthermore, exercise-induced changes in temporal temperature correlated with the measurements of cardiovascular fitness, although in this case, the relationship was negative. The results obtained indicate that the measurements of temporal temperature during and after an intense 3-min bout of exercise can be used to assess aerobic physical capacity and cardiovascular fitness.

Author Contributions: Conceptualization A.D.J., R.H. and P.H.; methodology R.H. and P.H.; software R.H.; validation R.H. and A.D.J.; formal analysis R.H. and P.H.; investigation A.D.J., R.H. and P.H.; resource R.H.; data curation A.D.J., P.H.; writing-original draft preparation R.H.; writing-review and editing A.D.J.; visualization A.D.J.; supervision A.D.J., R.H. and P.H.; project administration R.H.; funding acquisition R.H. All authors have read and agreed to the published version of the manuscript.

Funding: This work was supported by the University School of Physical Education in Wrocław, under grant number PN/BK/2020/07.

Institutional Review Board Statement: The study was approved by the University Ethics Committee and conducted in accordance with the ethical standards established by the Declaration of Helsinki, No. 39/2019.

Informed Consent Statement: Informed consent was obtained from all subjects involved in the study.

Data Availability Statement: The data are available on request from the corresponding author.

Conflicts of Interest: The authors declare no conflict of interest.

References

1. Clausen, J.S.R.; Marott, J.L.; Holtermann, A.; Gyntelberg, F.; Jensen, M.T. Midlife cardiorespiratory fitness and the long-term risk of mortality: 46 years of follow-up. *J. Am. Coll. Cardiol.* **2018**, *72*, 987–995. [CrossRef] [PubMed]
2. Levine, B.D. VO_{2max}: What do we know, and what do we still need to know? *J. Physiol.* **2008**, *586*, 25–34. [CrossRef] [PubMed]
3. De Lorenzo, A.; Da Silva, C.L.; Castro Souza, F.C.; De Souza, L.; Lima, R. Value of the oxygen pulse curve for the diagnosis of coronary artery disease. *Physiol. Res.* **2018**, *67*, 679–686. [CrossRef] [PubMed]
4. Lopes, A.J.; Vigário, P.S.; Hora, A.L.; Deus, C.A.L.; Soares, M.S.; Guimaraes, F.S.; Ferreira, A.S. Ventilation distribution, pulmonary diffusion and peripheral muscle endurance as determinants of exercise intolerance in elderly patients with chronic obstructive pulmonary disease. *Physiol. Res.* **2018**, *67*, 863–874. [CrossRef] [PubMed]
5. Daanen, H.A.; Lamberts, R.P.; Kallen, V.L.; Jin, A.; Van Meeteren, N.L. A systematic review on heart-rate recovery to monitor changes in training status in athletes. *Int. J. Sports Physiol. Perform.* **2012**, *7*, 251–260. [CrossRef] [PubMed]
6. Suzic Lazic, J.; Dekleva, M.; Soldatovic, I.; Leischik, R.; Suzic, S.; Radovanovic, D.; Djuric, B.; Nesic, D.; Lazic, M.; Mazic, S. Heart rate recovery in elite athletes: The impact of age and exercise capacity. *Clin. Physiol. Funct. Imaging* **2017**, *37*, 117–123. [CrossRef] [PubMed]
7. Watson, A.M.; Brickson, S.L.; Prawda, E.R.; Sanfilippo, J.L. Shortterm heart rate recovery is related to aerobic fitness in elite intermittent sport athletes. *J. Strength Cond. Res.* **2017**, *31*, 1055–1061. [CrossRef]
8. Facioli, T.P.; Philbois, S.V.; Gastaldi, A.C.; Almeida, D.S.; Maida, K.D.; Rodrigues, J.A.L.; Sanchez-Delgado, J.C.; Souza, H.C.D. Study of heart rate recovery and cardiovascular autonomic modulation in healthy participants after submaximal exercise. *Sci. Rep.* **2021**, *11*, 3620. [CrossRef]
9. Arena, R.; Arrowood, J.A.; Fei, D.; Shelar, S.; Helm, S.; Kraft, K.A. The influence of sex on the relationship between heart rate recovery and other cardiovascular risk factors in apparently healthy subjects. *Scand. J. Med. Sci. Sports* **2010**, *20*, 291–297. [CrossRef]
10. Sawka, M.N.; Convertino, V.A.; Eichner, E.R.; Schnieder, S.M.; Young, A.J. Blood volume: Importance and adaptations to exercise training, environmental stresses, and trauma/sickness. *Med. Sci. Sports Exerc.* **2000**, *32*, 332–348. [CrossRef]
11. Johnson, J.M. Physical training and the control of skin blood flow. *Med. Sci. Sports Exerc.* **1998**, *30*, 382–386. [CrossRef] [PubMed]
12. Fisher, J.P.; Young, C.N.; Fadel, P.J. Autonomic adjustments to exercise in human. *Compr. Physiol.* **2015**, *5*, 475–512. [CrossRef] [PubMed]
13. Berkin, K.E.; Walker, G.; Inglis, G.C.; Ball, S.G.; Thomson, N.C. Circulating adrenaline and noradrenaline concentrations during exercise in patients with exercise induced asthma and normal subjects. *Thorax* **1988**, *43*, 295–299. [CrossRef] [PubMed]
14. Johnson, J.M. Exercise and the cutaneous circulation. *Exerc. Sport Sci. Rev.* **1992**, *20*, 59–97.
15. Kamon, E.; Belding, H.S. Dermal blood flow in the resting arm during prolonged leg exercise. *J. Appl. Physiol.* **1969**, *26*, 317–320. [CrossRef]
16. Simmons, G.H.; Padilla, J.; Young, C.N.; Wong, B.J.; Lang, J.A.; Davis, M.J.; Laughlin, M.H.; Fadel, P.J. Increased brachial artery retrograde shear rate at exercise onset is abolished during prolonged cycling: Role of thermoregulatory vasodilation. *J. Appl. Physiol.* **2011**, *110*, 389–397. [CrossRef]
17. Ichinose, M.; Saito, M.; Fujii, N.; Ogawa, T.; Hayashi, K.; Kondo, N.; Nishiyasu, T. Modulation of the control of muscle sympathetic nerve activity during incremental leg cycling. *J. Physiol.* **2008**, *586*, 2753–2766. [CrossRef]
18. Katayama, K.; Ishida, K.; Saito, M.; Koiko, T.; Hirasawa, A.; Ogoh, S. Enhanced muscle pump during mild dynamic leg exercise inhibits sympathetic vasomotor outflow. *Physiol. Rep.* **2014**, *2*, e12070. [CrossRef]
19. Katayama, K.; Kaur, J.; Young, B.E.; Barbosa, T.C.; Ogoh, S.; Fadel, P.J. High-intensity muscle metaboreflex activation attenuates cardiopulmonary baroreflex-mediated inhibition of muscle sympathetic nerve activity. *J. Appl. Physiol.* **2018**, *125*, 812–819. [CrossRef]
20. McNamara, T.C.; Keen, J.T.; Simmons, G.H.; Alexander, L.M.; Wong, B.J. Endothelial nitric oxide synthase mediates the nitric oxide component of reflex cutaneous vasodilatation during dynamic exercise in humans. *J. Physiol.* **2014**, *592*, 5317–5326. [CrossRef]
21. Kenney, W.L.; Johnson, J.M. Control of skin blood flow during exercise. *Med. Sci. Sports Exerc.* **1992**, *24*, 303–312. [CrossRef]
22. Smith, C.J.; Johnson, J.M. Responses to hyperthermia. Optimizing heat dissipation by convection and evaporation: Neural control of skin blood flow and sweating in humans. *Auton. Neurosci.* **2016**, *196*, 25–36. [CrossRef] [PubMed]
23. Roberts, M.F.; Wenger, C.B.; Stolwijk, J.A.; Nadel, E.R. Skin blood flow and sweating changes following exercise training and heat acclimation. *J. Appl. Physiol. Respir. Environ. Exerc. Physiol.* **1977**, *43*, 133–137. [CrossRef] [PubMed]
24. Torii, M.; Yamasaki, M.; Sasaki, T.; Nakayama, H. Fall in skin temperature of exercising man. *Br. J. Sports Med.* **1992**, *26*, 29–32. [CrossRef] [PubMed]
25. Hebisz, R.; Hebisz, P.; Borkowski, J.; Wierzbicka-Damska, I.; Zatoń, M. Relationship between the skin surface temperature changes during sprint interval testing protocol and the aerobic capacity in well-trained cyclists. *Physiol. Res.* **2019**, *68*, 981–989. [CrossRef] [PubMed]
26. Gofur, E.M.; Bordoni, B. *Anatomy, Head and Neck, Cerebral Blood Flow*; StatPearls Publishing: Treasure Island, FL, USA, 2019.
27. Hebisz, R.; Hebisz, P.; Borkowski, J.; Zatoń, M. Differences in physiological responses to interval training in cyclists with and without interval training experience. *J. Hum. Kinet.* **2016**, *50*, 93–101. [CrossRef] [PubMed]

28. Follador, L.; Alves, R.C.; Ferreira, S.D.S.; Buzzachera, C.F.; Andrade, V.F.D.S.; Garcia, E.D.S.A.; Osiecki, R.; Barbosa, S.C.; de Oliveira, L.M.; da Silva, S.G. Physiological, perceptual, and affective responses to six high-intensity interval training protocols. *Percept. Mot. Skills* **2018**, *125*, 329–350. [CrossRef]
29. Buchheit, M.; Laursen, P.B. High-intensity interval training, solutions to the programming puzzle: Part I: Cardiopulmonary emphasis. *Sports Med.* **2013**, *43*, 313–338. [CrossRef]
30. Murawska-Cialowicz, E.; Wojna, J.; Zuwala-Jagiello, J. Crossfit training changes brain-derived neurotrophic factor and irisin levels at rest, after Wingate and progressive tests, and improves aerobic capacity and body composition of young physically active men and women. *J. Physiol. Pharmacol.* **2015**, *66*, 811–821.
31. Fukada, D.H.; Wray, M.E.; Kendall, K.L.; Smith-Ryan, A.E.; Stout, J.R. Validity of near-infrared interactance (FUTREX 6100/XL) for estimating body fat percentage in elite rowers. *Clin. Physiol. Funct. Imaging* **2017**, *37*, 456–458.
32. Hebisz, P.; Jastrzębska, A.D.; Hebisz, R. Real assessment of maximum oxygen uptake as a verification after an incremental test versus without a test. *Front. Physiol.* **2021**, *12*, e739745. [CrossRef] [PubMed]
33. Jeukendrup, A.; Saris, W.H.; Brouns, F.; Kester, A.D. A new validated endurance performance test. *Med. Sci. Sports Exerc.* **1996**, *28*, 266–270. [CrossRef] [PubMed]
34. Beaver, W.L.; Wasserman, K.; Whipp, B.J. A new method for detecting anaerobic threshold by gas exchange. *J. Appl. Physiol.* **1986**, *60*, 2020–2027. [CrossRef] [PubMed]
35. Davis, J.A.; Whipp, B.J.; Wasserman, K. The relation of ventilation to metabolic rate during moderate exercise in man. *Eur. J. Appl. Physiol. Occup. Physiol.* **1980**, *44*, 97–108. [CrossRef]
36. Buchheit, M.; Al Haddad, H.; Laursen, P.B.; Ahmaidi, S. Effect of body posture on postexercise parasympathetic reactivation in men. *Exp. Physiol.* **2009**, *94*, 795–804. [CrossRef]
37. Peçanha, T.; de Paula-Ribeiro, M.; Nasario-Junior, O.; de Lima, J.R. Post-exercise heart rate variability recovery: A time-frequency analysis. *Acta Cardiol.* **2013**, *68*, 607–613. [CrossRef]
38. Hopkins, W.G.; Marshall, S.W.; Batterham, A.M.; Hanin, J. Progressive statistics for studies in sports medicine and exercise science. *Med. Sci. Sports Exerc.* **2009**, *41*, 3–13. [CrossRef]
39. Tanda, G. The use of infrared thermography to detect the skin temperature response to physical activity. *J. Phys. Conf. Ser.* **2015**, *655*, 012062. [CrossRef]
40. Knyszyńska, A.; Radecka, A.; Lubkowska, A. Thermal imaging of exercise-associated skin temperature changes in swimmers subjected to 2-min intensive exercise on a VASA swim bench ergometer. *Int. J. Environ. Res. Public Health* **2021**, *18*, 6493. [CrossRef]
41. Hawley, J.A.; Noakes, T.D. Peak power output predicts maximal oxygen uptake and performance time in trained cyclists. *Eur. J. Appl. Physiol. Occup. Physiol.* **1992**, *65*, 79–83. [CrossRef]
42. Wernhart, S.; Guazzi, M.; Halle, M. Correlation of heart rate recovery, aerobic physical activity and performance. A sub-analysis of the EURO-Ex trial. *Dtsch. Z. Sportmed.* **2020**, *71*, 19–24. [CrossRef]
43. Gore, C.J.; Withers, R.T. The effect of exercise intensity and duration on the oxygen deficit and excess post-exercise oxygen consumption. *Eur. J. Appl. Physiol. Occup. Physiol.* **1990**, *60*, 169–174. [CrossRef] [PubMed]
44. Kang, S.J.; Ko, K.J. Association between resting heart rate, VO_2max and carotid intima-media thickness in middle-aged men. *Int. J. Cardiol. Heart Vasc.* **2019**, *23*, 100347. [CrossRef] [PubMed]
45. Jabbour, G.; Majed, L. Mechanical efficiency at different exercise intensities among adolescent boys with different body fat levels. *Front. Physiol.* **2019**, *10*, 265. [CrossRef]
46. Kenny, G.P.; Jay, O. Thermometry, calorimetry, and mean body temperature during heat stress. *Compr. Physiol.* **2013**, 3, 1689–1719.
47. Kenny, G.P.; McGinn, R. Restoration of thermoregulation after exercise. *J. Appl. Physiol.* **2017**, *122*, 933–944. [CrossRef]
48. Hargreaves, M.; Spriet, L.L. Skeletal muscle energy metabolism during exercise. *Nat. Metab.* **2020**, *2*, 817–828. [CrossRef]
49. Sawka, M.N.; Wenger, C.B. Physiological responses to acute exercise-heat stress. In *Human Performance Physiology and Environmental Medicine at Terrestrial Extremes*; Pandolf, K.B., Sawka, M.N., Gonzalez, R.R., Eds.; Benchmark: Indianapolis, IN, USA, 1988; pp. 97–151.
50. Kenny, G.P.; Reardon, F.D.; Zaleski, W.; Reardon, M.L.; Haman, F.; Ducharme, M.B. Muscle temperature transients before, during, and after exercise measured using an intramuscular multisensor probe. *J. Appl. Physiol.* **2003**, *94*, 2350–2357. [CrossRef]
51. Wang, J.S. Effects of exercise training and detraining on cutaneous microvascular function in man: The regulatory role of endothelium-dependent dilation in skin vasculature. *Eur. J. Appl. Physiol.* **2005**, *93*, 429–434. [CrossRef]
52. Boegli, Y.; Gremion, G.; Golay, S.; Kubli, S.; Liaudet, L.; Leyvraz, P.F.; Waeber, B.; Feihl, F. Endurance training enhances vasodilation induced by nitric oxide in human skin. *J. Investig. Dermatol.* **2003**, *121*, 1197–1204. [CrossRef]
53. Fritzsche, R.G.; Coyle, E.F. Cutaneous blood flow during exercise is higher in endurance-trained humans. *J. Appl. Physiol.* **2000**, *88*, 738–744. [CrossRef] [PubMed]
54. Hagberg, J.M.; Goldberg, A.P.; Lakatta, L.; O'Connor, F.C.; Becker, L.C.; Lakatta, E.G.; Fleg, J.L. Expanded blood volumes contribute to the increased cardiovascular performance of endurance-trained older men. *J. Appl. Physiol.* **1998**, *85*, 484–489. [CrossRef] [PubMed]
55. Dawson, E.A.; Green, D.J.; Gable, N.T.; Thijssen, D.H. Effects of acute exercise on flow mediated ditatation in healthy humans. *J. Appl. Physiol.* **2013**, *115*, 1589–1598. [CrossRef] [PubMed]
56. Périard, J.D.; Racinais, S.; Sawka, M.N. Adaptations and mechanisms of human heat acclimation: Applications for competitive athletes and sports. *Scand. J. Med. Sci. Sports* **2015**, *25* (Suppl. 1), 20–38. [CrossRef]

57. Wendt, D.; van Loon, L.J.; Lichtenbelt, W.D. Thermoregulation during exercise in the heat: Strategies for maintaining health and performance. *Sports Med.* **2007**, *37*, 669–682. [CrossRef]
58. Neves, E.B.; Vilaca-Alves, J.; Antunes, N.; Felisberto, I.M.; Rosa, C.; Reis, V.M. Different responses of the skin temperature to physical exercise: Systematic review. In Proceedings of the 2015 37th Annual International Conference of the IEEE Engineering in Medicine and Biology Society (EMBC), Milan, Italy, 25–29 August 2015; pp. 1307–1310. [CrossRef]
59. Possamai, L.T.; Campos, F.S.; Salvador, P.C.D.N.; de Aguiar, R.A.; Guglielmo, L.G.A.; de Lucas, R.D.; Caputo, F.; Turnes, T. Similar maximal oxygen uptake assessment from a step cycling incremental test and verification tests on the same or different day. *Appl. Physiol. Nutr. Metab.* **2020**, *45*, 357–361. [CrossRef]
60. Scharhag-Rosenberger, F.; Carlsohn, A.; Cassel, M.; Mayer, F.; Scharhag, J. How to test maximal oxygen uptake: A study on timing and testing procedure of a supramaximal verification test. *Appl. Physiol. Nutr. Metab.* **2011**, *36*, 153–160. [CrossRef]
61. Astorino, T.A.; DeRevere, J. Efficacy of constant load verification testing to confirm VO_2 max attainment. *Clin. Physiol. Funct. Imaging* **2018**, *38*, 703–709. [CrossRef]
62. Michael, S.; Graham, K.S.; Davis, G.M. Cardiac autonomic responses during exercise and post-exercise recovery using heart rate variability and systolic time Intervals—A review. *Front. Physiol.* **2017**, *8*, 301. [CrossRef]
63. Flouris, A.D.; Bravi, A.; Wright-Beatty, H.E.; Green, G.; Seely, A.J.; Kenny, G.P. Heart rate variability during exertional heat stress: Effects of heat production and treatment. *Eur. J. Appl. Physiol.* **2014**, *114*, 785–792. [CrossRef]
64. Peçanha, T.; Forjaz, C.L.M.; Low, D.A. Passive heating attenuates post-exercise cardiac autonomic recovery in healthy young males. *Front. Neurosci.* **2017**, *11*, 727. [CrossRef]
65. Green, H.J.; Patla, A.E. Maximal aerobic power: Neuromuscular and metabolic considerations. *Med. Sci. Sports Exerc.* **1992**, *24*, 38–46. [CrossRef] [PubMed]
66. Bertuzzi, R.; Nascimento, E.M.; Urso, R.P.; Damasceno, M.; Lima-Silva, A.E. Energy system contributions during incremental exercise test. *J. Sports Sci. Med.* **2013**, *12*, 454–460.
67. Denham, J.; Scott-Hamilton, J.; Hagstrom, A.D.; Gray, A.J. Cycling power outputs predict functional threshold power and maximum oxygen uptake. *J. Strength Cond. Res.* **2020**, *34*, 3489–3497. [CrossRef] [PubMed]
68. Ogoh, S.; Ainslie, P.N. Cerebral blood flow during exercise: Mechanisms of regulation. *J. Appl. Physiol.* **2009**, *107*, 1370–1380. [CrossRef] [PubMed]

biology

MDPI

Review

Impact of Exercise Training on Depressive Symptoms in Cancer Patients: A Critical Analysis

Priscila Marconcin [1,2,*], Adilson Marques [1,3], Gerson Ferrari [4], Élvio R. Gouveia [5,6,7], Miguel Peralta [1,3] and Andreas Ihle [7,8,9]

1 Centro Interdisciplinar de Estudo da Performance Humana, Faculdade de Motricidade Humana, Universidade de Lisboa, 1495-751 Lisbon, Portugal; amarques@fmh.ulisboa.pt (A.M.); mperalta@fmh.ulisboa.pt (M.P.)
2 KinesioLab, Research Unit in Human Movement Analysis, Piaget Institute, 2805-059 Almada, Portugal
3 ISAMB, Universidade de Lisboa, 1649-020 Lisbon, Portugal
4 Laboratorio de Rendimiento Humano, Grupo de Estudio en Educación, Actividad Física y Salud (GEEAFyS), Universidad Católica del Maule, Talca 3460000, Chile; gerson.demoraes@usach.cl
5 Department of Physical Education and Sport, University of Madeira, 9020-105 Funchal, Portugal; erubiog@uma.pt
6 Laboratory of Robotics and Systems in Engineering and Science (LARSyS), Interactive Technologies Institute, 9020-105 Funchal, Portugal
7 Center for the Interdisciplinary Study of Gerontology and Vulnerability, University of Geneva, 1205 Geneva, Switzerland; andreas.ihle@unige.ch
8 Department of Psychology, University of Geneva, 1205 Geneva, Switzerland
9 Swiss National Centre of Competence in Research LIVES—Overcoming Vulnerability: Life Course Perspectives, 1015 Lausanne, Switzerland
* Correspondence: priscilamarconcin@fmh.ulisboa.pt; Tel.: +351-21-414-9100

Simple Summary: Cancer patients need to overcome several issues, leaving them more vulnerable to depressive symptoms. Exercise is recognised as a practice that helps to deal with depressive symptoms. This study is an umbrella review of meta-analyses about the effect of exercise on depressive symptoms among cancer patients. Six studies were included. A significant reduction in depressive symptoms was observed because of exercise. However, the studies varied in methodological terms, making a broad generalisation difficult. We can conclude that exercise is a good alternative to deal with depressive symptoms among cancer patients. Still, more studies are needed to clarify some aspects that are not answered yet.

Abstract: Background: Cancer patients must deal with several health challenges, including emotional distress and depressive symptoms. This study aimed to evaluate evidence from published systematic reviews and meta-analyses about the efficacy of exercise on depressive symptoms in cancer patients. Methods: We searched for previous meta-analyses of randomised controlled trials on PubMed, Web of Science and Scopus, with data inception to 30 December 2021. Two independent researchers assessed the methodological quality using the Assessment of Multiple Systematic Reviews 2 (AMSTAR2) instrument. Six meta-analyses were integrated. All included middle-aged and older adults. Five presented moderate quality, and one presented low quality. Results: Overall, a significant reduction in depressive symptoms was observed among the included studies. However, the heterogeneity between studies was high, and high-quality evidence for the efficacy of exercise on depressive symptoms was limited. Conclusions: Exercise could be a possibility in the treatment of depressive symptoms in cancer patients, especially when supervised and outside the home. The better dose of exercise needs to be clarified. More high-quality evidence is needed to better prescribe exercise to this vulnerable population.

Keywords: tumour; exercise; depression; mental health; cancer survivorship

Citation: Marconcin, P.; Marques, A.; Ferrari, G.; Gouveia, É.R.; Peralta, M.; Ihle, A. Impact of Exercise Training on Depressive Symptoms in Cancer Patients: A Critical Analysis. *Biology* **2022**, *11*, 614. https://doi.org/10.3390/biology11040614

Academic Editor: Redha Taiar

Received: 1 March 2022
Accepted: 14 April 2022
Published: 18 April 2022

1. Introduction

Cancer is a global public health issue, with 19.3 million new cases of cancer diagnosed in 2020 and 10 million individuals dying from the disease [1]. Cancer occurs mostly with older age and in the United States of America, and 90% of cancers are diagnosed in those aged >50 years [2]. Female breast cancer is the most commonly diagnosed cancer (11.7%), followed by lung (11.4%), colorectal (10.0%), prostate (7.3%), and stomach (5.6%) cancers [1]. Despite the lethality of different types of cancer, many cancer patients survive. However, cancer patients are in a vulnerable situation since they go through several health challenges, as cancer diagnosis and treatment have a serious impact on their physical and mental well-being [3]. Cancer patients experience several emotional disruptions, such as fear of death, interruption of life plans, decreased body image and self-esteem, and changes in social role and lifestyle [4]. One of the most common impacts is depression, which affects up to 20% of patients with cancer [5], however, the prevalence rate of depression among cancer patients is heterogeneous, according to clinical setting [6], the stage of the disease [5,7] and type of cancer [8], ranging between 5% and 49% [9]. Aggravating this issue, depression in cancer patients is associated with low chemotherapy compliance [10] and an increased risk of death [11]. Therefore, the treatment of depression among cancer patients should be a priority. However, there is still the notion that depression is inevitable and untreatable [12]. In addition, there is limited trial data on depressive symptoms' treatment efficacy in cancer patients [13]. Pharmacological therapy, consisting of antidepressant medications, is usually considered for the treatment of moderate to severe major depression; also, a combined modality approach, including psychosocial and pharmacologic interventions, is a feasible alternative [14].

Alongside pharmacological and psychosocial therapy, exercise can have a positive impact on depressive symptoms [15]. Several mechanisms are involved in the association between exercise and depression, from neurobiological to behavioural mechanisms [16]. One is the inflammation-related factors (IRFs) [17], where studies have shown an association between inflammatory markers and depressive symptoms, including fatigue, impaired sleep and cognitive dysfunction [18,19]. Exercise could create an anti-inflammatory environment and reduce the serum level of leptin and fibroblast growth factors (FGF) [20]. IL-10, produced by exercise, acted as an anti-inflammatory cytokine and is stimulated by the release of adrenaline and cortisol from the adrenal gland, which reduces the release of pro-inflammatory cytokines in the hippocampus [21]. Regarding behavioural mechanisms, exercise can promote several behavioural changes. Engagement in exercise programs and learning new movements skills or completing physically challenging exercises may lead to gaining a sense of mastery [22]. The activity-based perception of physical strength and flexibility is associated with increased physical self-esteem and consequently, an increase in global self-esteem [23].

Regular exercise after diagnosis increases survivorship by 50–60%, with strong evidence for breast and colorectal cancers [15]. In addition to improving depressive symptoms, exercise positively impacts other depression- and cancer-related outcomes, such as anxiety, fatigue, physical functioning, and health-related quality of life [3]. Although the efficacy of exercise interventions in reducing depressive symptoms among cancer patients was already established by previous systematic reviews and meta-analyses [24–29], previous studies substantially vary in scope, quality and methodology, which can cause considerable confusion and misdirect efforts in the implementation of exercise interventions. An umbrella review of previous research is warranted to better inform future trials needs, as well as establish a consistent message for health policies targeting this vulnerable population. The specific questions that we should answer with this study are: (1) regarding some aspects of exercise intervention, such as the type of exercise, the dose of exercise, the difference between home-based exercise and other locations, which are the most effective to deal with depressive symptoms? (2) Regarding the difference between the type of cancer, the moment of the exercise intervention, before, during or after cancer treatment, are there any differences? Therefore, this study aimed to present an umbrella review of an exercise intervention on depressive symptoms among cancer patients, appraising hints of uncertainty and bias in the body of literature and providing recommendations for future research.

2. Materials and Methods

2.1. Literature Search

The protocol of this umbrella review was registered under PROSPERO (CRD42021254843) and followed the Preferred Reporting Items for Systematic reviews and Meta-Analyses (PRISMA) 2020 guidelines [30]. Two researchers performed the literature search in PubMed, Web of Science, and Scopus, focusing on meta-analyses published until 30 December 2021 to investigate the efficacy of exercise in reducing depressive symptoms among cancer patients. In cases of disagreement, a third researcher was asked to arbitrate. The search terms were: ("physical activ*" OR "physical inactiv" OR exercise OR training OR sport* OR fitness OR "movement behavio*" OR walking OR running OR yoga OR jogging OR swimming OR cycl*) AND (depress* OR "mental health" OR mood OR "psychological health" OR "psychological function*" OR "mental function*" OR worries OR worry OR "depressive disorder*" OR "baby blues") AND (cancer OR neoplas* OR tumor OR chemo* OR radiat* OR malign* OR carciniom*) NOT Rats. No language limitation was established. Records previously known to the authors were also identified.

2.2. Eligibility Criteria

Included articles in the systematic review met the PICOS (participants, intervention, comparison, outcome, study design) criteria [31]. The criteria included characteristics of participants (cancer patients); intervention (any type of exercise); comparison: regular care or physical activity; outcome (depressive symptoms diagnosed using a structured clinical interview, screened for probable depression using a validated assessment, or diagnosed according to the judgement of a health professional); study design (meta-analyses of parallel designs, controlled trials. Meta-analyses were excluded if the studies involved animals.

2.3. Quality Assessment

Two authors assessed the methodological quality of the included meta-analyses using the Assessment of Multiple Systematic Reviews (AMSTAR 2) checklist. Scores range from 0 to 11, with higher scores indicating greater quality [25]. The AMSTAR checklist involves the dichotomous scoring (0 or 1) of 11 items related to the rigour of systematic reviews and meta-analyses (e.g., comprehensive search strategy, publication bias assessment). AMSTAR scores are graded as high (8–11), medium (4–7), and low (0–3) quality [32]. The authors discussed grading discrepancies and reached a consensus.

2.4. Data Extraction

Study characteristics were extracted from full texts, including the number of randomised controlled trials (RCTs) and participants; participants' characteristics; exercise intervention's characteristics; comparisons; and outcomes measures. Data on the standardised mean difference (SMD) and heterogeneity (I2 statistic) in meta-analytic comparisons were also extracted. The SMD was classified as trivial (<0.20), small (0.20 to 0.49), medium (0.50 to 0.79) or large ($\geq$0.80) [33]. I2 statistic values were considered to be representative of low (0 to 25%), moderate (25 to 50%), large (50 to 75%) or a very large (>75%) inconsistency [34].

3. Results

3.1. Literature Search

The study selection process is summarised in Figure 1. A total of 54 records were identified in the literature search, 53 from the databases and 1 from other sources, i.e., previously known to the authors. After removing the duplicates (n = 32), two researchers reviewed the remaining 22 records' titles and abstracts. Ten records were excluded at this stage. The remaining 12 records' full text were assessed for eligibility. From this analysis, six records were excluded for the following reasons: another type of intervention (n = 1); without meta-analysis (n = 3); without data on depressive symptoms (n = 2). Therefore, six records were included in this study [24–29].

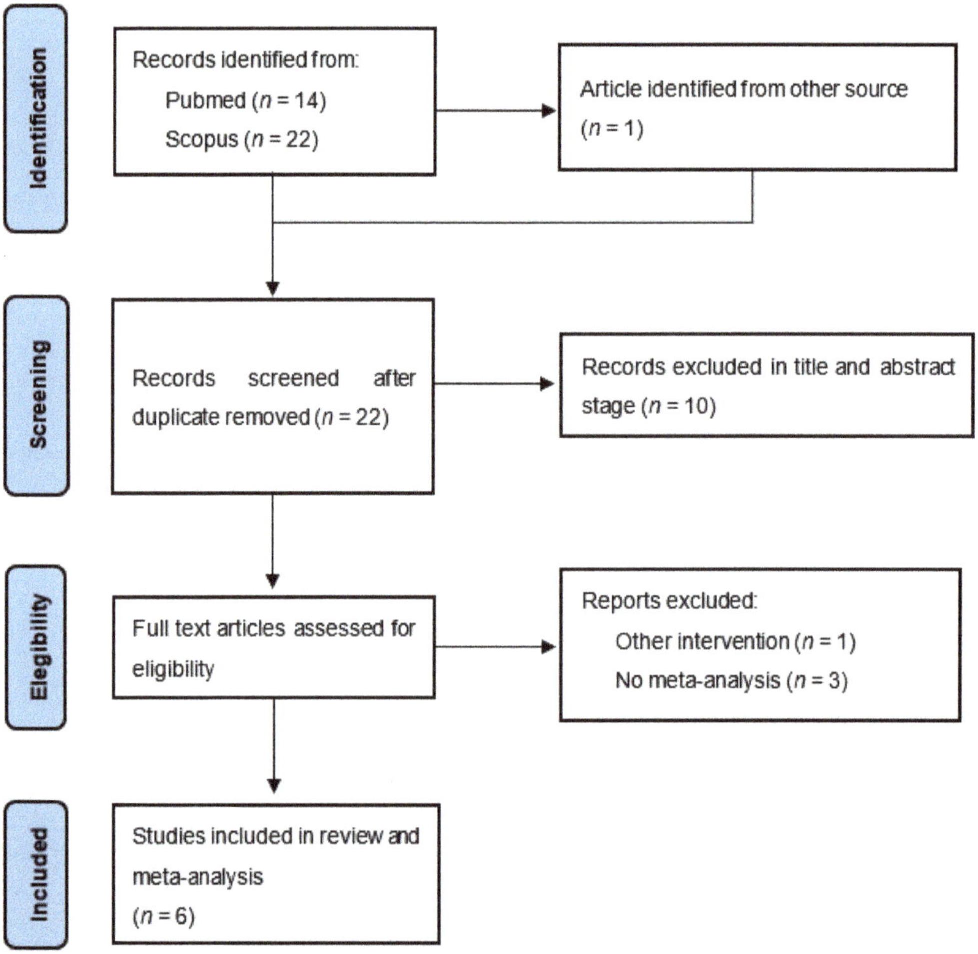

Figure 1. PRISMA flow diagram of study selection.

3.2. Study Characteristics

The characteristics of the meta-analyses included in this umbrella review are presented in Table 1.

Table 1. Characteristics of the meta-analyses included in the study.

Reference	No. of RCTs and Participants	Participants Characteristics	Exercise Intervention's Characteristics	Comparison	Outcomes Measures
Brown et al. [24]	37 RCTs; 2929 participants.	Age: mean 51.3 years (range: 39–70); Gender: 87% women; Cancer: breast cancer (24 studies), other types of cancer (13 studies).	Type: walking (16 studies), stationary cycling (5 studies), resistance machines (2 studies), resistance bands (3 studies), yoga (8 studies); Duration/frequency: mean of 13.2 ± 11.7 weeks with 3.0 ± 2.5 sessions/week lasting 49.1 ± 27.1 min/session.	Usual care.	Depressive symptoms (CES-D, POMS, BDI, HADS) and Symptom Assessment Scale.
Craft et al. [25]	15 RCTs; 1371 participants.	Age: mean 51.6 years; Gender: no information about gender; Cancer: breast cancer (60% of the included studies).	Type: aerobic (10 studies), aerobic and resistance (5 studies); Duration/frequency: ranged from 4 to 14 weeks; Supervised, facility-based programs (3 studies), unsupervised home-based programs (6 studies), some exercise programs supervised (4 studies).	Usual care (12 studies); Educational print material (3 studies).	Depression inventory and clinician interview.
Gonzalez et al. [26] *	26 RCTs; 1486 participants.	Age: mean 54.4 years (range 44–68.7 years); Gender: 86.1% women; Cancer: breast cancer (18 studies), mixed cancers (2 studies), other types of cancers (6 studies).	Type: hatha yoga (11 studies), other types of yoga (15 studies); Duration/frequency: mean of 9.3, with 1 to 3 sessions/week lasting 45–120 min/session.	Usual care (19 studies); Psychosocial or educational interventions (6 studies); Other physical activity interventions (2 studies).	Depressive symptoms (HADS, BDI-II, CES-D, POMS, PHQ-2, PHQ-9).
Patsou et al. [27]	14 RCTs; 1701 participants.	Age: mean 52 years; Gender: only women; Cancer: only breast cancer.	Type: aerobic, resistance, aerobic and resistance, yoga; Duration/frequency: no information;	Usual care; Health education intervention; Waitlist; Relaxation and stretching.	Depressive symptoms (POMS, HADS, CES-D).
Vashistha et al. [28] **	3 RCTs; 192 participants.	Age: mean between 67 and 73 years; Gender: only men; Cancer: only prostate cancer.	Type: qigong (1 study), aerobic and resistance (1 study), aerobic and light resistance (1 study); Duration/frequency: no information;	Usual care; Stretching.	Depressive symptoms (BSI-18, CES-D).
Yi et al. [29] ***	6 RCTs; 446 participants.	Age: mean between 45 and 60 years; Gender: only women; Cancer: only breast cancer.	Type: only yoga; Duration/frequency: no information;	Usual care.	Depressive symptoms (BDI-II, POMS, HADS, CES-D, SDS).

Abbreviations: BDI, Beck Depression Inventory; BDI-II, Beck Depression Inventory-II; BSI-18, Brief Symptom Inventory; CES-D, Center for Epidemiological Studies—Depression; HADS, Hospital Anxiety Depression Scale; Patient Health Questionnaire-2; PHQ-2, Patient Health Questionnaire-9; PHQ-9; POMS, Profile of Mood States; SDS, Self-rating Depression scale. * Gonzalez et al. [26] analysed 26 RCTs, but 1 was not included in the meta-analysis. ** Vashistha et al. [28] analysed 13 RCTs, but only 3 included measures of depressive symptoms. *** Yi et al. [29] analysed 8 RCTs, but only data regarding the 6 that analysed depressive symptoms is presented.

3.3. Number of RCTs and Participants

The number of participants included in each meta-analysis varied according to the number of included RCTs. The Brown et al. research had the largest sample, including 37 RCTs and 2929 participants [24], while Vashistha et al. presented the smallest sample, with three RCTs and 196 participants [28]. Considering all the included meta-analyses, this study undertook 100 RCTs and 8125 participants. The overlap of single studies within the six included meta-analyses was low (27%), leading to a final number of 79 RCTs.

3.4. Participants' Characteristics

The mean age of the participants ranged between 45 [29] and 73 years [28]. The mean age of five [24–27,29] out of the six included studies is above 50 years, that is, patients of older age. Two meta-analyses only included women [27,29], the other two meta-analyses included men and women [24,26], and one meta-analysis only included men [28]. One meta-analysis did not present participants' gender information [25]. Two meta-analyses were focused on breast cancer [27,29]. In three other meta-analyses, most RCTs were focused on breast cancer (24 out of 37 [24], 60% of participants [25], and 18 out of 26 [26]). One meta-analysis was focused on prostate cancer [28].

3.5. Exercise Intervention Characteristics

Different types of exercise interventions were included, such as aerobic training (e.g., walking, cycling) [24,25,27,28], resistance training (e.g., weight machines, resistance bands) [24,25,27,28], yoga [24,26,27] and qigong [28]. The duration and session frequency were also different for each intervention. For instance, the mean duration of the intervention in the three meta-analyses that reported this information was 13 weeks [24], 4 to 14 weeks [25], and 9 weeks [26].

3.6. Comparison of Experimental Conditions

Exercise interventions were compared with different control conditions, including: no exercise program [24], usual care [25–29], educational print material [25], psychosocial or educational interventions [26], and stretching [27,28].

3.7. Outcome Measures

Depressive symptoms were assessed by different instruments, such as the Center for Epidemiological Studies—Depression [24,26–29], the Profile of Mood States [24,26,29], the Beck Depression Inventory [24,26,29], the Hospital Anxiety Depression Scale [24,26,27,29], the Symptoms Assessment Scale [24], the Patient Health Questionnaire [26], the Brief Symptom Inventory [28], and the Self-rating Depression Scale [29].

3.8. Quality Assessment of Studies

All included meta-analyses conducted a risk of bias analysis regarding single studies. Three of them used the PEDro scale. The mean PEDro score was 7.0 ± 1.0 in Brown's study [24], representing high quality. In the Craft's study, all with the exception of three studies attained high quality [25]. Additionally, in Patsou's study, the mean PEDro score indicated high quality (6.1 ± 2.0) [27]. The other three studies used the Cochrane risk of bias tool, assessing six aspects of the trial methodology. Under each domain, studies were classified as low, high or unclear risk of bias. More details about each domain for each study analysed can be seen in the original paper [26,28,29].

3.9. Quality Assessment

Table 2 presents the results obtained with the AMSTAR 2 checklist regarding the methodological quality of the meta-analyses. All meta-analyses, except one, presented a moderate-quality review. Vashistha et al. [28] had a low-quality review, mostly because it did not account for the risk of bias in individual studies when interpreting the review results.

Table 2. Quality of the meta-analyses included in the study according to AMSTAR 2 criteria.

AMSTAR 2 Criteria	**Brown et al. [24]**	**Craft et al. [25]**	**Gonzalez et al. [26]**	**Patsou et al. [27]**	**Vashistha et al. [28]**	**Yi et al. [29]**
1. Did the research questions and inclusion criteria include the components of PICO?	V	V	V	V	V	V
2. Did the review report contain a statement that the review methods were established before the conduct of the review, and did the report justify significant deviations from the protocol?	X	X	V	X	V	X
3. Did the review authors explain their selection of the study designs for inclusion in the review?	V	V	V	V	V	V
4. Did the review authors use a comprehensive literature search strategy?	V	V	V	V	V	V
5. Did the review authors perform study selection in duplicate?	X	V	V	X	V	V
6. Did the review authors perform data extraction in duplicate?	V	V	V	X	V	V
7. Did the review authors provide a list of excluded studies and justify the exclusions?	V	V	V	V	V	V
8. Did the review authors describe the included studies in adequate detail?	V	V	V	V	V	V
9. Did the review authors use a satisfactory technique for assessing the risk of bias (RoB) in individual studies included in the review?	V	V	V	V	V	V
10. Did the review authors report funding sources for the studies included?	X	X	X	X	X	X
11. If meta-analysis was performed, did the review authors use appropriate methods for the statistical combination of results?	V	V	V	V	V	V
12. If meta-analysis was performed, did the review authors assess the potential impact of RoB in individual studies on the results of the meta-analysis or other evidence synthesis?	V	X	V	V	V	V
13. Did the review authors account for RoB in individual studies when interpreting/discussing the review results?	V	V	V	V	X	V
14. Did the review authors provide a satisfactory explanation for, and discussion of, any heterogeneity observed in the review results?	V	V	V	V	V	V
15. If they performed quantitative synthesis, did the review authors carry out an adequate investigation of publication bias (small study bias) and discuss its likely impact on the review results?	V	V	V	V	V	X
16. Did the review authors report any potential sources of conflict of interest, including any funding they received for conducting the review?	V	V	V	V	V	V
	Moderate	Moderate	Moderate	Moderate	Low	Moderate

Note: V when it fulfills the evaluation criterion and X when it does not fulfill the evaluation criterion.

3.10. Synthesis of Results

3.10.1. Main Results

The main results of each included meta-analysis are summarised in Table 3. Different methods were used to present aggregate effects, including Cohen's d, Hedges' g statistic, and the standardised mean difference (SMD) using a random-effects model. In four out of the six included meta-analyses, the authors observed a significant reduction in depressive symptoms favouring the exercise group. In studies from Patsou et al. [20] and Vashistha et al. [21], no statistically significant decrease in depressive symptoms for the exercise group was observed. Three meta-analyses, Brown et al. [17], Craft et al. [18] and Patsou et al. [20] observed small effect sizes, whereas two meta-analyses, Gonzalez et al. [19] and Yi et al. [22], reported moderate effect size. All the included meta-analyses presented a large or very large heterogeneity (I2 from 55% to 84%).

Table 3. Results of the meta-analyses included in the study.

Reference	Effect on Depressive Symptoms (95% CI)	I2 (%)	Conclusions
Brown et al. [24]	d = −0.13 (−0.26, −0.01)	55%	Significant small reduction in depressive symptoms compared to usual care among all types of cancer.
Craft et al. [25]	d = −0.22 (−0.43, −0.009)	The test for heterogeneity was significant ($p < 0.001$).	Significant small reduction in depressive symptoms when comparing exercise interventions to control groups.
Gonzalez et al. [26]	g = −0.55 (−0.78, −0.32)	77%	Significant medium effect size in favour of yoga interventions for reducing depression symptoms in comparison to control conditions.
Patsou et al. [27]	g = −0.38 (−0.89, 0.13)	77%	Non-significant reduction in depressive symptoms for the exercise group.
Vashistha et al. [28]	SMD = −3.02 (−7.83, 1.79)	78%	Non-significant reduction in depressive symptoms for the exercise group.
Yi et al. [29]	SMD = −0.56 (−1.05, −0.07)	84%	Significant improvement in depressive symptoms for yoga interventions.

Abbreviations: d, mean change scores (Cohen's d); g, Hedges' g statistic to estimated effect size; I2, I-squared statistic for heterogeneity; SMD, standardised mean difference.

3.10.2. Sensitivity and Subgroup Analyses

Four of the six included meta-analyses presented sensitivity or subgroup analyses. Regarding the type of cancer, the Brown et al. [17] subgroup analysis revealed significant reductions in depressive symptoms among breast cancer patients (d= −0.17; 95% CI: −0.32, −0.02), but non-significant differences for prostate, leukaemia, lymphoma and colorectal cancer patients. Gonzalez et al. [19] proceeded with a subgroup analysis of only breast cancer patients and presented a significant moderate reduction in depressive symptoms favouring exercise (g = −0.41; 95% CI: −0.59, −0.23).

A subgroup analysis compared supervised vs. non-supervised exercise. Brown et al. [17] showed that supervised exercise was the most effective in reducing depressive symptoms (ß = −0.26, $p = 0.01$). Moreover Craft et al. [18] founded that supervised exercise presented a greater reduction in depressive symptoms (ES = −0.67; 95% CI: −1.11, −0.23) than non-supervised exercise (ES = 0.25; 95% CI: −0.01, 0.50).

Another aspect observed in the subgroup analyses was the exercise dose. Craft et al. [18] founded that exercise bout durations >30 min had larger effects (ES = −0.57; 95% CI: −0.91, −0.23) on depression than exercise bouts ≤30 min (ES = 0.01; 95% CI: −0.20, 0.22). Lastly, Patsou et al. [20] demonstrated that exercising ≤135 min/week yielded a moderate to a large effect (g = −0.82; 95% CI: −1.54, −0.10; I2 = 35%) and exercising ≥135 min/week presented no significant effect. Moreover, exercise up to 12 weeks yielded a moderate to a large effect (g = −1.69; 95% CI: −2.66, −0.73; I2 = 32%), while exercise duration over 12 weeks presented no significant effect.

Regarding participants' age, only Brown et al. [17] explored this subgroup analysis and showed that exercise was the most effective when cancer patients were between 47 and 62 years (ß = −0.27, p = 0.01). Among older adults, the effect of exercise on depressive symptoms was not significant.

Craft et al. [18] also analysed potential moderators of effect, including exercise location, observing that home-based exercise was associated with increased depressive symptoms (ES = 0.16; 95% CI: −0.15, 0.47), while other exercise locations presented a reduction in depressive symptoms (ES = −0.45; 95% CI: −0.77, −0.14).

The Gonzalez et al. [19] sensitivity analysis showed that removing one study greatly reduced heterogeneity (I2 = 36.9%) but also reduced the effect size to the small-medium range (g = −0.41; 95% CI: −0.55, −0.28). Additionally, comparing studies that used active and inactive control interventions did not find differences; both had a significant, medium to small effect size post-intervention.

Lastly, Patsou et al. [20] performed several subgroup analyses showing that: aerobic exercise interventions yield a large and significant effect on depression (g = −1.23; 95% CI: −1.97, −0.49; I2 = 0%), no significant effect was found regarding resistance exercise interventions, combined aerobic and resistance exercise and Yoga interventions. Exercise during treatment yielded a moderate effect (g = −0.54; 95% CI: −1.16, 0.08; I2 = 25%), while exercise post-treatment yielded no significant effect.

4. Discussion

This umbrella review included six meta-analyses that comprised 100 individual studies with little overlap that investigated the effect of exercise on depressive symptoms among cancer survivors. Overall, a small significant reduction in depressive symptoms in this vulnerable population was observed in the studies. However, high-quality evidence for the efficacy of exercise on depressive symptoms is limited. For a more detailed analysis, some points need to be considered, such as the type of cancer, the specificity of exercise prescription, the time of interventions, and during or after cancer.

In our umbrella review, participants had mainly breast cancer in the included meta-analysis and were mostly women. Only one study did not include breast cancer [27] and was with prostate cancer patients, and it was the one that did not observe a significant effect of exercise on depressive symptoms. In a subgroup analysis, Brown et al. found a significant reduction in depressive symptoms among breast cancer survivors but did not find the same in prostate, leukaemia, lymphoma and colorectal cancer [17]. The prevalence of depression among breast cancer survivors is higher than in other cancers and can achieve 32.8% [35]. Moreover, depression is more prevalent in women than men [36], and breast cancer is prevalent in women. Evidence suggests that depression in breast cancer patients decreases over time and is more common throughout the disease and in the recurrent phase of breast cancer [37]. The occurrence of depression among patients with breast cancer is due to several factors, such as treatment-related distress, worries regarding fear of death and disease recurrence, and altered body image, sexuality and attractiveness [38–40]. In addition, a study exposes the association between depression and tumour levels of estrogen receptors and progesterone receptors [41]. A study found that fatigue and pain are significant risk factors for developing depression among breast cancer survivors [37]. Fatigue is also a recognised barrier to exercise [42]; however, exercise can reduce fatigue among women with breast cancer [43]. The benefits of exercise can be extended to improve physical functioning and multiple aspects of quality of life among cancer patients [44]. Moreover, exercise is a feasible alternative to control symptoms burden and improve well-being among breast cancer patients [39].

Another sample characteristic that must be highlighted is that most patients were older adults (>50 years old). In the general population, the prevalence of depression symptoms rises with increasing age, 10% to 15% of older adults have clinically significant depressive symptoms [45]. Older patients with cancer often experience depression, fatigue, pain, and sleep disturbance [46]. Only one included meta-analysis directly explored the role of age in the effectiveness of exercise on depression symptoms and found that the efficacy seems to disappear among old age patients [24]. However, an RCT with older cancer patients receiving chemotherapy found that after the six-week structure exercise program, participants' anxiety and mood improved [47]. Besides the effects of exercise on mental health, physically active old age patients improve general health, such as physical fitness outcomes, quality of life and increased life expectancy [48].

When considering the effects of exercise on depressive symptoms, it is necessary to consider the characteristics of the exercise we are referring to. Many dimensions of exercise exist, which are captured in part by the principle (frequency, intensity, time and type of exercise), as well as the way of practising, whether accompanied or not and if exercise occurs indoors or outdoors. However, the included systematic meta-analyses showed great variance concerning exercise. Except for Gonzalez et al. and Yi et al. [26,29], which analysed the effects of yoga intervention, the others included meta-analyses that examined a variety of exercises, such as aerobic (e.g., walking, cycling), resistance (e.g., weight machine, resistance bands) and qigong. Only the Patsou et al. study explored the difference between the types of exercise and found that aerobic intervention yields a large significant effect on depressive symptoms. At the same time, resistance training presents a small significant effect, and combined aerobic and resistance training yielded a moderate effect [27]. This statement is in accordance with the American College of Sports Medicine (ACSM), which describes that resistance training alone does not seem effective for depression [3]. Aerobic activities are cost-effective and should be popularised in clinical practice.

Regarding yoga, both included meta-analyses that analysed only intervention found significant and medium effects on depressive symptoms [26,29]. However, in the Patsou et al. study, which included aerobic exercise, resistance exercise and yoga intervention, when a subgroup analysis proceeded and considered only yoga intervention, no significant difference in depression symptoms was observed [27]. The contradictory results found in the three studies can be explained by the fact that yoga combines breathing (pranayama) and meditative techniques during a series of postures (asanas), but different types of yoga were being practised, which made it difficult to understand the effects of this practice [49].

Two included meta-analyses found that supervised exercise is more efficient than non-supervised exercise [32,38], which also appears in the ACSM recommendation [3]. Craft et al.'s study [25] explored the effects of exercise session durations and found that more than 30 min had larger effects compared with less than 30 min of the exercise session. In Patsou et al. [27] $\leq$135 min/week yielded a moderate to large effect and no effect with $\geq$135 min/week of exercise. The ACSM describes that aerobic training performed three times per week and for at least 12 weeks or twice weekly with combined aerobic plus resistance training lasting 6 to 12 weeks, can significantly reduce depressive symptoms in cancer survivors during and after treatment. However, the exact exercise duration per week has not yet been established by the ACSM. Gonzalez et al.'s study [26] explored the frequency and found no differences between one class per week and two or more classes per week. In contrast, Patsou et al. [27] explored the exercise intervention program duration and found that exercise for up to 12 weeks yielded a moderate to large effect compared with a small effect of over 12 weeks. Aside from the efficacy of depressive symptoms, exercise has other benefits in health outcomes among cancer survivors that must be considered such as improving cardiorespiratory fitness [50], and muscle strength [51,52].

Another important aspect of the efficacy of exercise on depressive symptoms among cancer survivors is the time of intervention, before the diagnosis, during treatment or in a recovering phase. This aspect was explored, and no difference was found between patients receiving cancer treatment, following treatment or mixed treatment status [26]. On the other hand, patients under treatment yielded a moderate effect, and patients post-treatment yielded a small effect [27]. Exercise increases the chemotherapy completion rate during treatment without causing lymphedema or significant adverse events [53]. In addition, exercise appears to reduce chemotherapy-induced peripheral neuropathy symptoms in patients receiving taxane-, platinum-, or vinca alkaloid-based chemotherapy [54].

Concerning the methodological quality, the AMSTAR 2 scores show that the majority were of moderate methodological quality. Nevertheless, as for item #10, "Did the review authors report on the sources of funding for the studies included in the review", no study reported the source of funding which can entail a risk of bias.

Strengths and Limitation

The strength of our study is that we included two recent meta-analyses, one from 2020 [19] and one from 2021 [22], and compiled current data regarding the effectiveness of exercise in depressive symptoms among cancer patients. However, some limitations should be exposed. The prevalence of breast cancer was substantial, which prevented us from generalising the findings to other types of cancer. Two studies did not perform subgroup analysis, and those who performed subgroup analysis did not analyse the same constructs, specifically in relation to the FITT principles of exercise. Future research is required to clarify the effectiveness of different intervention modalities (frequency, intensity, time and type) for depressive symptoms among cancer survivors to allow determination of exercise dose-response. Moreover, additional studies are needed to evaluate the cost-analysis and adverse events of exercise, and there is an urgent need for innovative methods to generate high-quality evidence.

5. Conclusions

Our critical review contributed to the evidence of the effects of exercise on depressive symptoms among cancer patients. Four in six studies found a significant effect of exercise. Some aspects should be highlighted and should be used for future interventions. Supervised exercise is better than non-supervised. The dose of exercise seems to be important, however, the finding did not present a specific dose-response relation. Exercise outside of the home is better than home-based exercise. Aerobic exercise is the most effective type of exercise. The effect of exercise on depression seems to be more effective among breast cancer patients. Future studies should explore other types of cancer.

Author Contributions: P.M.: conceptualisation, formal analysis, writing—original draft, writing review and editing; A.M.: supervision, formal analysis, writing review and editing; G.F.: writing review and editing; É.R.G.; writing review and editing; M.P.: formal analysis, writing review and editing; A.I.: supervision, writing review and editing. All authors have read and agreed to the published version of the manuscript.

Funding: This work was supported by the Swiss National Centre of Competence in Research LIVES—Overcoming vulnerability: life course perspectives, granted by the Swiss National Science Foundation (grant number: 51NF40-185901). AI acknowledges support from the Swiss National Science Foundation (grant number: 10001C_189407).

Institutional Review Board Statement: Not applicable.

Informed Consent Statement: Not applicable.

Data Availability Statement: Not applicable.

Conflicts of Interest: The authors declare no conflict of interest.

References

1. Sung, H.; Ferlay, J.; Siegel, R.L.; Laversanne, M.; Soerjomataram, I.; Jemal, A. Global Cancer Statistics 2020: GLOBOCAN estimates of incidence and mortality worldwide for 36 cancers in 185 countries. *CA Cancer J. Clin.* **2021**, *71*, 209–249. [CrossRef] [PubMed]
2. Siegel, R.L.; Miller, K.D.; Jemal, A. Cancer statistics, 2018. *CA Cancer J. Clin.* **2018**, *68*, 7–30. [CrossRef] [PubMed]
3. Campbell, K.L.; Winters-Stone, K.M.; Wiskemann, J.; May, A.M.; Schwartz, A.L.; Courneya, K.S. Exercise guidelines for cancer survivors: Consensus statement from international multidisciplinary roundtable. *Med. Sci. Sports Exerc.* **2019**, *51*, 2375–2390. [CrossRef] [PubMed]
4. Pasquini, M.; Biondi, M. Depression in cancer patients: A critical review. *Clin. Pract. Epidemiol. Ment. Health* **2007**, *3*, 2. [CrossRef]

5. Mitchell, A.J.; Chan, M.; Bhatti, H.; Halton, M.; Grassi, L.; Johansen, C. Prevalence of depression, anxiety, and adjustment disorder in oncological, haematological, and palliative-care settings: A meta-analysis of 94 interview-based studies. *Lancet Oncol.* **2011**, *12*, 160–174. [CrossRef]
6. Naser, A.Y.; Hameed, A.N.; Mustafa, N.; Alwafi, H.; Dahmash, E.Z.; Alyami, H.S. Depression and anxiety in patients with cancer: A cross-sectional study. *Front. Psychol.* **2021**, *12*, 585534. [CrossRef]
7. Caruso, R.; Nanni, M.G.; Riba, M.; Sabato, S.; Mitchell, A.J.; Croce, E.; Grassi, L. Depressive spectrum disorders in cancer: Prevalence, risk factors and screening for depression: A critical review. *Acta Oncol.* **2017**, *56*, 146–155. [CrossRef]
8. Pilevarzadeh, M.; Amirshahi, M.; Afsargharehbagh, R.; Rafiemanesh, H.; Hashemi, S.-M.; Balouchi, A. Global prevalence of depression among breast cancer patients: A systematic review and meta-analysis. *Breast Cancer Res. Treat.* **2019**, *176*, 519–533. [CrossRef]
9. Walker, J.; Holm Hansen, C.; Martin, P.; Sawhney, A.; Thekkumpurath, P.; Beale, C. Prevalence of depression in adults with cancer: A systematic review. *Ann. Oncol.* **2013**, *24*, 895–900. [CrossRef]
10. Ayres, A.; Hoon, P.W.; Franzoni, J.B.; Matheny, K.B.; Cotanch, P.H.; Takayanagi, S. Influence of mood and adjustment to cancer on compliance with chemotherapy among breast cancer patients. *J. Psychosom. Res.* **1994**, *38*, 393–402. [CrossRef]
11. Onitilo, A.A.; Nietert, P.J.; Egede, L.E. Effect of depression on all-cause mortality in adults with cancer and differential effects by cancer site. *Gen. Hosp. Psychiatry* **2006**, *28*, 396–402. [CrossRef] [PubMed]
12. Dwight-Johnson, M.; Ell, K.; Lee, P.J. Can collaborative care address the needs of low-income Latinas with comorbid depression and cancer? Results from a randomised pilot study. *Psychosomatics* **2005**, *46*, 224–232. [CrossRef]
13. Laoutidis, Z.G.; Mathiak, K. Antidepressants in the treatment of depression/depressive symptoms in cancer patients: A systematic review and meta-analysis. *BMC Psychiatry* **2013**, *13*, 140. [CrossRef]
14. Rodin, G.; Katz, M.; Lloyd, N.; Green, E.; Mackay, J.A.; Wong, R.K.S. The treatment of depression in cancer patients: A systematic review. *Supportive Care Cancer* **2007**, *15*, 123–136. [CrossRef] [PubMed]
15. Newton, R.U.; Galvão, D.A. Exercise in prevention and management of cancer. *Curr. Treat. Options Oncol.* **2008**, *9*, 135–146. [CrossRef] [PubMed]
16. Smith, P.J.; Merwin, R.M. The role of exercise in management of mental health disorders: An integrative review. *Annu. Rev. Med.* **2021**, *72*, 45–62. [CrossRef] [PubMed]
17. Yuan, N.; Chen, Y.; Xia, Y.; Dai, J.; Liu, C. Inflammation-related biomarkers in major psychiatric disorders: A cross-disorder assessment of reproducibility and specificity in 43 meta-analyses. *Transl. Psychiatry* **2019**, *9*, 233. [CrossRef]
18. Bower, J.E.; Huedo-Medina, T.B.; Pescatello, L.S.; Ryan, S.M.; Pescatello, S.M.; Moker, E. Fatigue and proinflammatory cytokine activity in breast cancer survivors. *Psychosom. Med.* **2002**, *64*, 604–611. [CrossRef]
19. Motivala, S.J.; Sarfatti, A.; Olmos, L.; Irwin, M.R. Inflammatory markers and sleep disturbance in major depression. *Psychosom. Med.* **2005**, *67*, 187–194. [CrossRef]
20. Fedewa, M.V.; Hathaway, E.D.; Ward-Ritaco, C.; Williams, T.D.; Dobbs, W.C. The effect of chronic exercise training on leptin: A systematic review and meta-analysis of randomized controlled trials. *Sports Med.* **2018**, *48*, 1437–1450. [CrossRef]
21. Moon, H.Y.; Kim, S.H.; Yang, Y.R.; Suh, P.-G. Macrophage migration inhibitory factor mediates the antidepressant actions of voluntary exercise. *Proc. Natl. Acad. Sci. USA* **2012**, *109*, 13094–13099. [CrossRef] [PubMed]
22. Gheysen, F.; Poppe, L.; DeSmet, A.; Swinnen, S.; Cardon, G.; De Bourdeaudhuij, I.; Chastin, S.; Fias, W. Physical activity to improve cognition in older adults: Can physical activity programs enriched with cognitive challenges enhance the effects? A systematic review and meta-analysis. *Int. J. Behav. Nutr. Phys. Act.* **2018**, *15*, 63. [CrossRef] [PubMed]
23. Ryan, M.P. The antidepressant effects of physical activity: Mediating self-esteem and self-efficacy mechanisms. *Psychol. Health* **2008**, *23*, 279–307. [CrossRef] [PubMed]
24. Brown, J.C.; Huedo-Medina, T.B.; Pescatello, L.S.; Ryan, S.M.; Pescatello, S.M.; Moker, E.; LaCroix, J.M.; Ferrer, R.A.; Johnson, B.T. The efficacy of exercise in reducing depressive symptoms among cancer survivors: A meta-analysis. *PLoS ONE* **2012**, *7*, e30955. [CrossRef]
25. Craft, L.L.; VanIterson, E.H.; Helenowski, I.B.; Rademaker, A.W.; Courneya, K.S. Exercise effects on depressive symptoms in cancer survivors: A systematic review and meta-analysis. *Cancer Epidemiol. Biomark. Prev.* **2012**, *21*, 3–19. [CrossRef]
26. Gonzalez, M.; Pascoe, M.C.; Yang, G.; de Manincor, M.; Grant, S.; Lacey, J.; Firth, J.; Sarris, J. Yoga for depression and anxiety symptoms in people with cancer: A systematic review and meta-analysis. *Psychooncology* **2021**, *30*, 1196–1208. [CrossRef]
27. Patsou, E.D.; Alexias, G.D.; Anagnostopoulos, F.G.; Karamouzis, M.V. Effects of physical activity on depressive symptoms during breast cancer survivorship: A meta-analysis of randomised control trials. *ESMO Open* **2017**, *2*, e000271. [CrossRef]
28. Vashistha, V.; Singh, B.; Kaur, S.; Prokop, L.J.; Kaushik, D. The effects of exercise on fatigue, quality of life, and psychological function for men with prostate cancer: Systematic review and meta-analyses. *Eur. Urol. Focus* **2016**, *2*, 284–295. [CrossRef]
29. Yi, L.J.; Tian, X.; Jin, Y.F.; Luo, M.J.; Jimenez-Herrera, M.F. Effects of yoga on health-related quality, physical health and psychological health in women with breast cancer receiving chemotherapy: A systematic review and meta-analysis. *Ann. Palliat. Med.* **2021**, *10*, 1961–1975. [CrossRef]
30. Page, M.J.; McKenzie, J.E.; Bossuyt, P.M.; Boutron, I.; Hoffmann, T.C.; Mulrow, C.D.; Shamseer, L.; Tetzlaff, J.M.; Akl, E.A.; Brennan, S.E.; et al. The PRISMA 2020 statement: An updated guideline for reporting systematic reviews. *Int. J. Surg.* **2021**, *88*, 105906. [CrossRef]

31. Smith, V.; Devane, D.; Begley, C.M.; Clarke, M. Methodology in conducting a systematic review of systematic reviews of healthcare interventions. *BMC Med. Res. Methodol.* **2011**, *11*, 15. [CrossRef] [PubMed]
32. Shea, B.J.; Grimshaw, J.M.; Wells, G.A.; Boers, M.; Andersson, N.; Hamel, C.; Porter, A.C.; Tugwell, P.; Moher, D.; Bouter, L.M. Development of AMSTAR: A measurement tool to assess the methodological quality of systematic reviews. *BMC Med. Res. Methodol.* **2007**, *7*, 10. [CrossRef] [PubMed]
33. Cohen, J. A power primer. *Psychol. Bull.* **1992**, *112*, 155–159. [CrossRef] [PubMed]
34. Higgins, J.P.; Thompson, S.G.; Deeks, J.J.; Altman, D.G. Measuring inconsistency in meta-analyses. *Bmj* **2003**, *327*, 557–560. [CrossRef]
35. Zabora, J.; BrintzenhofeSzoc, K.; Curbow, B.; Hooker, C.; Piantadosi, S. The prevalence of psychological distress by cancer site. *Psychooncology* **2001**, *10*, 19–28. [CrossRef]
36. Albert, P.R. Why is depression more prevalent in women? *J. Psychiatry Neurosci.* **2015**, *40*, 219–221. [CrossRef]
37. Vahdaninia, M.; Omidvari, S.; Montazeri, A. What do predict anxiety and depression in breast cancer patients? A follow-up study. *Soc. Psychiatry Psychiatr. Epidemiol.* **2010**, *45*, 355–361. [CrossRef]
38. Ganz, P.A. Psychological and social aspects of breast cancer. *Oncology* **2008**, *22*, 642–650, discussion 650.
39. Baucom, D.H.; Porter, L.S.; Kirby, J.S.; Gremore, T.M.; Keefe, F.J. Psychosocial issues confronting young women with breast cancer. *Breast Dis.* **2005**, *23*, 103–113. [CrossRef]
40. Helms, R.L.; O'Hea, E.L.; Corso, M. Body image issues in women with breast cancer. *Psychol. Health Med.* **2008**, *13*, 313–325. [CrossRef]
41. Snoj, Z.; Akelj, M.P.; Lièina, M.; Pregelj, P. Psychosocial correlates of progesterone receptors in breast cancer. *Depress. Anxiety* **2009**, *26*, 544–549. [CrossRef] [PubMed]
42. Kim, S.; Han, J.; Lee, M.Y.; Jang, M.K. The experience of cancer-related fatigue, exercise and exercise adherence among women breast cancer survivors: Insights from focus group interviews. *J. Clin. Nurs.* **2020**, *29*, 758–769. [CrossRef] [PubMed]
43. Huang, H.P. The effect of a 12-week home-based walking program on reducing fatigue in women with breast cancer undergoing chemotherapy: A randomised controlled study. *Int. J. Nurs. Stud.* **2019**, *99*, 103376. [CrossRef] [PubMed]
44. Schmitz, K.H.; Courneya, K.S.; Matthews, C.; Demark-Wahnefried, W.; Galvão, D.A.; Pinto, B.M. American College of Sports Medicine roundtable on exercise guidelines for cancer survivors. *Med. Sci. Sports Exerc.* **2010**, *42*, 1409–1426. [CrossRef]
45. Kok, R.M.; Reynolds, C.F., 3rd. Management of depression in older adults: A review. *JAMA* **2017**, *317*, 2114–2122. [CrossRef]
46. Loh, K.P.; Zittel, J.; Kadambi, S.; Pandya, C.; Xu, H.; Flannery, M.; Magnunson, A.; Bautista, J.; McHugh, C.; Mustian, K.; et al. Elucidating the associations between sleep disturbance and depression, fatigue, and pain in older adults with cancer. *J. Geriatr. Oncol.* **2018**, *9*, 464–468. [CrossRef]
47. Loh, K.P.; Kleckner, I.R.; Lin, P.J.; Mohile, S.G.; Canin, B.E.; Flannery, M.A.; Fung, C.; Dunne, R.F.; Bautista, J.; Culakova, E.; et al. Effects of a home-based exercise program on anxiety and mood disturbances in older adults with cancer receiving chemotherapy. *J. Am. Geriatr. Soc.* **2019**, *67*, 1005–1011. [CrossRef]
48. Gultyaeva, V.V.; Zinchenko, M.I.; Uryumtsev, D.Y.; Krivoschekov, S.G.; Aftanas, L.I. Exercise for depression treatment. Physiological mechanisms. *Zh. Nevrol. Psikhiatr. SS Korsakova* **2019**, *119*, 112–119. [CrossRef]
49. Ruddy, K.J.; Stan, D.L.; Bhagra, A.; Jurisson, M.; Cheville, A.L. Alternative exercise traditions in cancer rehabilitation. *Phys. Med. Rehabil.* **2017**, *28*, 181–192. [CrossRef]
50. Scott, J.M.; Zabor, E.C.; Schwitzer, E.; Koelwyn, G.J.; Adams, S.C.; Nilsen, T.S.; Moskowitz, C.S.; Matsoukas, K.; IYengar, N.M.; Dang, C.T.; et al. Efficacy of exercise therapy on cardiorespiratory fitness in patients with cancer: A systematic review and meta-analysis. *J. Clin. Oncol.* **2018**, *36*, 2297–2305. [CrossRef]
51. Gardner, J.R.; Livingston, P.M.; Fraser, S.F. Effects of exercise on treatment-related adverse effects for patients with prostate cancer receiving androgen-deprivation therapy: A systematic review. *J. Clin. Oncol.* **2014**, *32*, 335–346. [CrossRef] [PubMed]
52. Winters-Stone, K.M.; Dobek, J.C.; Bennett, J.A.; Dieckmann, N.F.; Maddalozzo, G.F.; Ryan, C.W.; Beer, T.M. Resistance training reduces disability in prostate cancer survivors on androgen deprivation therapy: Evidence from a randomised controlled trial. *Arch. Phys. Med. Rehabil.* **2015**, *96*, 7–14. [CrossRef]
53. Courneya, K.S.; Segal, R.J.; Mackey, J.R.; Gelmon, K.; Reid, R.D.; Friedenreich, C.M.; Ladha, A.B.; Proulx, C.; Vallance, J.K.H.; Lane, K.; et al. Effects of aerobic and resistance exercise in breast cancer patients receiving adjuvant chemotherapy: A multicenter randomised controlled trial. *J. Clin. Oncol.* **2007**, *25*, 4396–4404. [CrossRef] [PubMed]
54. Kleckner, I.R.; Kamen, C.; Gewandter, J.S.; Mohile, N.A.; Heckler, C.E.; Culakova, E.; Fung, C.; Janlesins, M.C.; Asare, M.; Lin, P.; et al. Effects of exercise during chemotherapy on chemotherapy-induced peripheral neuropathy: A multicenter, randomised controlled trial. *Support Care Cancer* **2018**, *26*, 1019–1028. [CrossRef] [PubMed]

biology MDPI

Article

In-Season Quantification and Relationship of External and Internal Intensity, Sleep Quality, and Psychological or Physical Stressors of Semi-Professional Soccer Players

Hadi Nobari [1,2,3,*], Roghayyeh Gholizadeh [3], Alexandre Duarte Martins [4,5,6], Georgian Badicu [7] and Rafael Oliveira [4,5,8]

1 Department of Physiology, Faculty of Sport Sciences, University of Extremadura, 10003 Cáceres, Spain
2 Sports Scientist, Sepahan Football Club, Isfahan 81887-78473, Iran
3 Department of Exercise Physiology, Faculty of Educational Sciences and Psychology, University of Mohaghegh Ardabili, Ardabil 56199-11367, Iran; r.gholizadeh@uma.ac.ir
4 Sports Science School of Rio Maior—Polytechnic Institute of Santarém, 2040-413 Rio Maior, Portugal; alexandremartins@esdrm.ipsantarem.pt (A.D.M.); rafaeloliveira@esdrm.ipsantarem.pt (R.O.)
5 Life Quality Research Centre, 2040-413 Rio Maior, Portugal
6 Comprehensive Health Research Centre (CHRC), Departamento de Desporto e Saúde, Escola de Saúde e Desenvolvimento Humano, Universidade de Évora, Largo dos Colegiais, 7000 Évora, Portugal
7 Department of Physical Education and Special Motricity, Faculty of Physical Education and Mountain Sports, Transilvania University of Braşov, 500068 Braşov, Romania; georgian.badicu@unitbv.ro
8 Research Centre in Sport Sciences, Health Sciences and Human Development, 5001-801 Vila Real, Portugal
* Correspondence: hadi.nobari1@gmail.com or hnobarin@alumnos.unex.es

Simple Summary: In this study, we analyse the relationship of the in-season variations of external, internal and well-being measures across different periods of a semi-professional soccer season (early-, mid- and end-season) and describe TM and TS for the entire period of the analysis. The main findings of our study revealed that increasing the training intensity affects the well-being of the players and consequently the training intensity management. Coaches and their staff should consider the results of this study, because despite the relationship between external and internal intensity, each has a unique effect on the perception of the player's training intensity management.

Abstract: The purpose of this study was two-fold: (a) to describe and analyse the relationship of the in-season variations of external and internal intensity metrics as well as well-being measures across different periods of a semi-professional soccer season (early-, mid- and end-season); and (b) to describe training monotony (TM) and training strain (TS) for 20 weeks in a semi-professional soccer season. Eighteen semi-professional players (age: 29 ± 4.1) from the Asian First League team participated in this study. The players were monitored for 20 consecutive weeks during in-season for external training intensity, internal training intensity and well-being parameters. The in-season was organized into three periods: early-season (weeks 1–7); mid-season (weeks 8–13); and end-season (weeks 14–20). Total distance (TD), high-speed running distance (HSRD), sprint distance, rate of perceived exertion (RPE), session-RPE (s-RPE), TM, TS, heart rate average and maximum, as well as sleep quality, stress and muscle soreness were collected. Results revealed that TD, HSRD and sprint distance (total values) were meaningfully greater during end-season than in the early-season. RPE showed a significantly highest value during the end-season (4.27 AU) than in early- (3.68 AU) and mid-season (3.65 AU), $p < 0.01$. TS showed significant differences between early-season with mid-season ($p = 0.011$) and end-season ($p < 0.01$), and the highest value occurred in week 17 during end-season (6656.51 AU), while the lowest value occurred in week 4 during early-season (797.17 AU). The average TD periods showed a moderate to large correlation with RPE, sleep and s-RPE at early-, mid- and end-season. Increasing the training intensity without considering the well-being of the players affects the performance of the team. Examining processes of the relationship between training intensity and other psychological indicators among players will probably be effective in training planning. Sports coaches and fitness professionals should be wary of changes in TM and TS that

Citation: Nobari, H.; Gholizadeh, R.; Martins, A.D.; Badicu, G.; Oliveira, R. In-Season Quantification and Relationship of External and Internal Intensity, Sleep Quality, and Psychological or Physical Stressors of Semi-Professional Soccer Players. *Biology* **2022**, *11*, 467. https://doi.org/10.3390/biology11030467

Academic Editor: Anne Delextrat

Received: 4 February 2022
Accepted: 16 March 2022
Published: 18 March 2022

affect players performance. Therefore, to better control the training, more consideration should be given by the coaches.

Keywords: load; heart rate; high-speed running; monotony; muscle soreness; sprint; sleep; strain; stress

1. Introduction

Nowadays, it is almost mandatory to monitor intensity and well-being to know the effects of exercise training programs on soccer players, and to individualize the training process in semi-professional soccer teams [1]. Through the season, there are several variations on training and well-being measures, and their monitoring is essential to apply the most successful strategy for recovery and competition [2].

In a recent study by Impellizzeri et al. [3], intensity monitoring definition was updated and defined in two dimensions: external (the physical demands imposed by the design and mode of exercise), and internal (the psychophysiological impact of external intensity). Usually, distances of different speed thresholds and accelerometry-based variables are the main measures reported as external intensity, while heart rate and rated perceived exertion (RPE) are the main measures reported as internal intensity [4].

Furthermore, well-being monitoring also represents a non-invasive valid and quick tool for collecting information associated with the status and readiness of the players to the training process and competition [5,6]. One example of an instrument that allows this monitoring process is the Hooper questionnaire [5], which include four categories: delayed onset muscle soreness (DOMS), fatigue, sleep, and stress.

Monitoring external, internal and well-being measures is part of daily strategies used to quantify the training session effects [6], but it can also allow the identification of intensity variations across the season [7]. In this sense, additional analyses can be performed to specific data obtained. For instance, two traditional indexes known as training monotony (TM) and training strain (TS) have been used to analyse such week variations. TM is used to analyse the intensity variability within the week, while TS is used to analyse the intensity variability multiplied by the accumulated intensity of the week [8]. The relationship between both indexes is supported by their formulas where TM is calculated by dividing the daily mean load by the standard deviation while TS is calculated through the product of weekly load by TM [8]. Usually, both indexes are based on the training duration multiplied by RPE (s-RPE) [8].

However, despite such a diversity of intensity measures, few studies have analysed the relationship between external, internal and well-being measures [9–11]. For instance, Haddad et al. [10] did not observe any relationship between the Hooper Index (HI) categories and RPE through submaximal exercises in junior soccer players, but Clemente et al. [9] showed negative correlations between s-RPE with DOMS, sleep, fatigue, and stress in weeks with two matches across training data from a full professional soccer season [10]. Moreover, Oliveira et al. [11] analysed associations in 10 in-season mesocycles (full-season) between all external, internal and wellbeing measures, and found negative correlations between stress and total distance [11]. In addition, the same authors found positive correlations between fatigue and s-RPE, between DOMS and s-RPE, and between HI total score and total distance [11].

The above-mentioned results were obtained in small sample sizes (n ranged between 17–35), which is common for soccer studies, and thus more research is needed to analyse the relationship between external, internal and well-being measures simultaneously, since there were multiple occasions where such analysis was not presented. Therefore, this study aims: (a) to describe and analyse the relationship of the in-season variations of external, internal and well-being measures across different periods of a semi-professional soccer season (early-, mid- and end-season); and (b) to describe TM and TS for the entire period of the analysis in a semi-professional soccer season.

2. Materials and Methods

2.1. Participants

In this study, eighteen male semi-professional soccer players from Iran's First League were examined and monitored (age, 29 ± 4.1 years; height, 179.6 ± 4.7 cm; body mass, 74.9 ± 3.9 kg). All players participated in the in-season (>80%) [9].

The exclusion criteria were adopted from a previous study, namely, players with injury or that did not participate in training for at least two consecutive weeks and non-field positions such as the goalkeepers due to differences in intensity in training and matches [12]. All participants were familiarised with the training protocols prior to investigation. The Ardabil University of Medical Sciences' ethical committee code IR.ARUMS.REC.1399.545 was authorized for this study, which followed the Declaration of Helsinki's guidelines.

2.2. Experimental Design

A descriptive longitudinal approach of 20 in-season consecutive weeks was used. Specifically, for the present study, all players participated in 47 training sessions and 20 matches. Only data from regular training sessions was considered for analysis which means that data from resistance training, competitions, rehabilitation and/or recuperation sessions was excluded. All session were planned by the coach and staff, and the researchers only controlled the initial and final 30 min of the sessions. The analysed period ranged from early season (30 October 2017) until the end of the season (18 March 2018). The present in-season was organized into three periods: early-season (weeks 1–7); mid-season (weeks 8–13); and end-season (weeks 14–20) (Figure 1). Players' weekly averages and accumulated values were used for analysis.

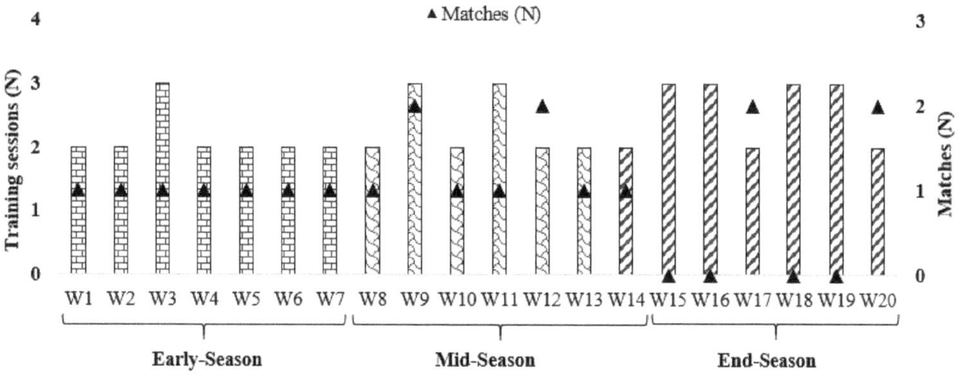

Figure 1. Weekly (W) distribution of training sessions and number of matches across the season.

Figure 1 shows the distributions of weeks per periods of the season, as well as the number of training sessions and matches.

2.3. External Intensity Monitoring

All training sessions were monitored using GPS (GPSPORTS systems Pty Ltd., Model: SPI HPU; Canberra, Australia). The GPS included 15 Hz position GPS and a tri-axial accelerometer, and this device has previously shown high validity and reliability [13]. For data collection, belts were placed on the players' shoulders and chests. At the end of the sessions, belts were collected and checked by the team's GPS manager. Then, devices entered the dock system to download the information to the Team AMS software. After this procedure and before next session, all belts were recharged. The SPI IQ Absolutes were adjusted for GPS default zone throughout the season.

The measures used for analysis were: TD; HSRD covered between 18 to 23 km/h^{-1}; and sprint distance covered above >23 km/h^{-1}. Data were considered in daily average values and the total of each period analysed, respectively.

2.4. Internal Intensity Monitoring

Players were monitored daily using a Borg's CR10 scale [14], adapted by Foster et al. [15]. This scale showed validity and reliability for quantifying session intensity [16].

Thirty minutes after each session, players individually provided their RPE value using a tablet to avoid non-valid scores. The RPE values provided were also multiplied by the training duration, to obtain the s-RPE [15,17]. Previously, all players were familiarized with the RPE scale.

Through s-RPE, TM (mean of training load during the seven days of the week divided by the SD of the training load of the seven days) [12,18,19] and TS (sum of the training load for all training sessions during a week multiplied by TM) [12,18,19] were calculated.

A flashing RED light was used to track HR. We placed each unit perpendicular to the bag, the logo on the unit was facing backwards and the RED light was on. HPUs are designed to automatically collect athlete's HR data in one session. In addition to the GPS receiver, the SPI Pro X unit consists of a tri-axial accelerometer for estimating the forces on the player, and an integrated HR monitor. The following variable was selected: HR average (HRavg). Then, weekly HRavg was calculated by the average value for the entire week for each period, respectively. The way this information was recorded was similar to previous studies [12,20,21]. Daily average data were used for RPE, s-RPE and HRavg.

2.5. Well-Being Monitoring

Approximately thirty minutes before sessions, players provided the HI scores [5] with the same procedures of the RPE. HI is a questionnaire that includes fatigue, stress, DOMS (scale of 1–7, in which 1 is very, very low and 7 is very, very high), and quality of sleep of the night that preceded the evaluation (scale of 1–7, in which 1 is very, very bad and 7 is very, very good). However, due to the purposes of the coach, fatigue was not considered in the present study. Daily average data was used for each category.

2.6. Statistical Analysis

Descriptive statistics were used to characterize the sample. Shapiro–Wilk was used to test normality of results. Results were presented as mean $\pm$ SD. The relationship between all variables at the different periods was verified using bivariate correlations [22] (Pearson's product–moment correlation coefficient (r)). The correlations' effect size (ES) were calculated using the following criteria: <0.1, (trivial); 0.1–0.3, (small); >0.3–0.5, (moderate); >0.5–0.7, (large); >0.7–0.9, (very large); and >0.9, (virtually perfect)

All variables obtained a normal distribution (Shapiro–Wilk, $p > 0.05$), a repeated measures ANOVA test was used with the Bonferroni post hoc test was used to compare variables for periods throughout the in-season. Statistical significance was set at $p \leq 0.05$. Hedge's g ES was also calculated to determine the magnitude of pairwise comparisons through the following formula: (mean 1–mean 2)/SD * pooled [23]. Then, the Hopkins threshold was applied: $g \leq 0.2$, (trivial); 0.2 to $\leq$0.6, (small); 0.6 to $\leq$1.2, (moderate); 1.2 to $\leq$2.0, (large); 2.0 to $\leq$4.0, (very large); and >4.0, (nearly perfect) [24,25]. All data were analysed using IBM SPSS Statistics (version 22, IBM Corporation (SPSS Inc., Chicago, IL, USA).

3. Results

Table 1 shows the differences between the early-, mid- and end-season for all measures.

To organize the results section, five sub-sections will address external, internal and well-being monitoring, correlations, monotony, and strain descriptions, respectively. To simplify the results description, only moderate to nearly perfect ES's will be described here.

Table 1. Descriptive statistics (mean ± standard deviation (SD)) of the external, and internal and well-being measures in the early-, mid- and end-season.

Measure	EarS (Mean ± SD)	MidS (Mean ± SD)	EndS (Mean ± SD)	*p*	Hedges' g (95% CI)
RPE (AU)	3.68 ± 0.62	3.65 ± 0.61	4.27 ± 0.60	EarS vs. MidS: 1.000 EarS vs. EndS: 0.009 MidS vs. EndS: 0.002	- −1.75 [−2.56, −0.99] large −1.87 [−2.70, −1.11] large
Avg TDur (Min)	58.91 ± 7.62	71.99 ± 3.72	77.44 ± 3.67	EarS vs. MidS: <0.01 EarS vs. EndS: <0.01 MidS vs. EndS: <0.01	−2.13 [−3.00, −1.34] very large −3.03 [−4.07, −2.11] very large −1.44 [−2.21, −0.73] large
Total TDur (Min)	989.61 ± 136.89	1180.22 ± 136.96	1846.77 ± 177.11	EarS vs. MidS: 0.001 EarS vs. EndS: <0.01 MidS vs. EndS: <0.01	−1.36 [−2.12, −0.65] large −5.29 [−6.84, −3.96] nearly perfect −4.11 [−5.39, −3.01] nearly perfect
s-RPE (AU)	235.89 ± 35.19	285.87 ± 26.96	333.27 ± 58.41	EarS vs. MidS: <0.01 EarS vs. EndS: <0.01 MidS vs. EndS: 0.028	−1.56 [−2.34, −0.83] large −1.97 [−2.82, −1.20] large −1.02 [−1.73, −0.34] moderate
Sleep	2.30 ± 0.75	2.13 ± 0.38	2.18 ± 0.44	EarS vs. MidS: 1.000 EarS vs. EndS: 1.000 MidS vs. EndS: 1.000	- - -
Stress	1.81 ± 0.49	1.50 ± 0.27	1.39 ± 0.29	EarS vs. MidS: 0.086 EarS vs. EndS: 0.032 MidS vs. EndS: 0.489	- 0.77 [0.09, 1.46] small -
DOMS	2.18 ± 0.71	2.12 ± 0.31	2.18 ± 0.36	EarS vs. MidS: 1.000 EarS vs. EndS: 1.000 MidS vs. EndS: 1.000	- - -
TM (AU)	3.88 ± 2.44	6.14 ± 3.86	3.93 ± 0.69	EarS vs. MidS: 0.120 EarS vs. EndS: 1.000 MidS vs. EndS: 0.087	- - -
TS (AU)	1988.93 ± 1210.84	4405.95 ± 2935.36	3948.96 ± 647.78	EarS vs. MidS: 0.011 EarS vs. EndS: <0.01 MidS vs. EndS: 1.000	−1.05 [−1.78, −0.37] moderate −1.97 [−2.82, −1.20] large -
Avg TD (Km)	5.62 ± 0.86	5.24 ± 0.86	5.14 ± 0.51	EarS vs. MidS: 0.165 EarS vs. EndS: 0.039 MidS vs. EndS: 1.000	- 0.66 [0.001, 1.35] moderate -
Total TD (Km)	108.30 ± 29.52	99.82 ± 0.86	133.86 ± 39.43	EarS vs. MidS: 0.493 EarS vs. EndS: 0.022 MidS vs. EndS: 0.003	- −0.71 [−1.40, −0,05] moderate −1.19 [−1.93, −0.49] moderate
Avg HSRD (Km)	0.72 ± 0.23	1.53 ± 0.29	3.07 ± 0.48	EarS vs. MidS: <0.01 EarS vs. EndS: <0.01 MidS vs. EndS: <0.01	−3.03 [−4.07, −2.10] very large −6.11 [−7.85, −4.62] nearly perfect −3.79 [−5.00, −2.75] very large
Total HSRD (Km)	14.06 ± 5.52	28.20 ± 9.47	77.95 ± 21.90	EarS vs. MidS: <0.01 EarS vs. EndS: <0.01 MidS vs. EndS: <0.01	−1.78 [−2.59, −1.03] large −3.91 [−5.14, −2.84] very large −2.88 [−3.89, −1.98] very large
Avg SD (Km)	0.61 ± 0.05	0.51 ± 0.03	0.56 ± 0.04	EarS vs. MidS: 0.079 EarS vs. EndS: 1.000 MidS vs. EndS: 0.638	- - -
Total SD (Km)	11.63 ± 4.14	9.55 ± 4.07	14.01 ± 4.00	EarS vs. MidS: 0.061 EarS vs. EndS: 0.202 MidS vs. EndS:0.002	- - −1.08 [−1.80, −0.39] moderate
HRavg (bpm)	137 ± 2	140 ± 9	135 ± 2	EarS vs. MidS: 1.000 EarS vs. EndS: 1.000 MidS vs. EndS: 1.000	- - -

Abbreviations: EarS, early-season; MidS, mid-season; EndS, end-season; AU, arbitrary units; RPE, rate of perceived exertion; Avg, average; TDur, Training duration; Min, minutes; s-RPE, session rate of perceived exertion; DOMS, delayed onset muscle soreness; TM, training monotony; TS, training strain; TD, total distance; Km, kilometres; HSRD, high-speed running distance; SD, speed distance; HRavg, heart rate average; Bpm, beats per minute.

3.1. External Intensity Monitoring

In relation to TD (average values), a significant difference was found between early- vs. end-season, while TD in total values showed two significant differences, early- vs. end-season ($p = 0.022$), and mid- vs. end-season ($p = 0.003$), both with moderate ES. The HSRD (in average and total values) showed significant differences between all periods of the in-season. The sprint distance shows a significant difference between mid- vs. end-season.

3.2. Internal Intensity Monitoring

The RPE showed the highest values in the end-season (4.27 AU), with significant differences between the early- and mid-season. Average and total training duration showed the lower values in the early-season (58.91 and 989.61 min, respectively). They both showed significant differences with mid- and end-season. Regarding to s-RPE, it showed higher values in end-season (333.27 AU) compared to early- and mid-season. TS showed significant differences between early-season with mid- and end-season.

3.3. Well-Being Monitoring

There were no meaningful differences for quality of sleep, stress, or DOMS.

3.4. Correlations of All Measures for Each Period

Tables 2–5 show the correlation coefficient of all measures in the study on the early-, mid-, end-season, and full-season, respectively.

3.5. Training Monotony and Training Strain Descriptions

Figure 2 shows an overall view of the weekly average for TM and TS calculated through the s-RPE across 20 weeks. Overall, Figure 2 shows that the highest TM occurred in week 12 during mid-season (9.60 AU), while the lowest value occurred in week 4 during early-season (1.92 AU). The highest TS occurred in week 17 during end-season (6656.51 AU), while the lowest value occurred in week 4 during early-season (797.17 AU).

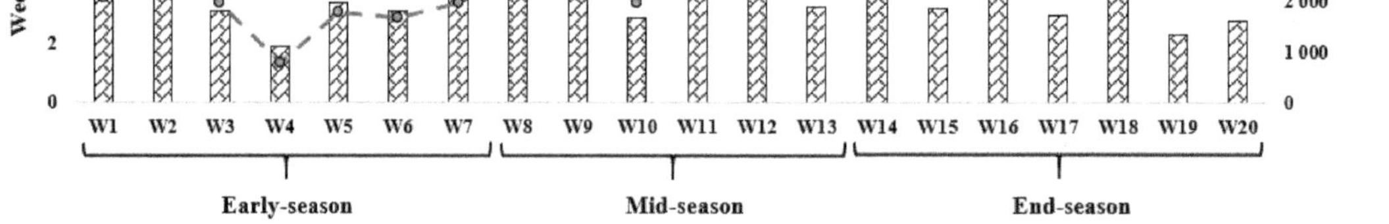

Figure 2. Training monotony (TM) and training strain (TS) variations calculated using the session rate of perceived exertion (s-RPE) across 20 weeks in different moments of a semi-professional soccer season.

Table 2. Correlation analysis between the measures in study on the early-season.

Measure	β0	β1	β2	β3	β4	β5	β6	β7	β8	β9	β10	β11	β12	β13	β14	β15
RPE (β0)	1.00															
Avg TDur (β1)	**0.687 #**	1.00														
Total TDur (β2)	**0.707 §**	**0.985 £**	1.00													
s-RPE (β3)	**0.783 §**	**0.528 #**	**0.542 #**	1.00												
Sleep (β4)	**0.519 #**	0.388	0.374	0.347	1.00											
Stress (β5)	−0.235	−0.120	−0.126	−0.371	0.270	1.00										
DOMS (β6)	0.338	0.263	0.234	0.364	0.465	0.051	1.00									
TM (β7)	0.047	0.109	0.129	−0.026	−0.088	0.175	0.117	1.00								
TS (β8)	0.136	0.200	0.215	0.132	0.007	0.157	0.244	**0.975 £**	1.00							
Avg TD (β9)	−0.047	−0.029	0.018	−0.080	−0.312	0.063	**−0.482 ***	−0.191	−0.267	1.00						
Total TD (β10)	−0.001	0.087	0.124	0.063	−0.248	0.103	−0.333	−0.065	−0.094	**0.930 £**	1.00					
Avg HSRD (β11)	−0.236	−0.314	−0.289	−0.295	**−0.485 ***	0.060	0.010	0.139	0.051	**0.613 #**	**0.586 #**	1.00				
Total HSRD (β12)	−0.179	−0.244	−0.215	−0.192	−0.430	0.112	−0.013	0.146	0.078	**0.701 §**	**0.733 §**	**0.970 £**	1.00			
Avg SD (β13)	0.027	0.020	−0.002	−0.052	−0.014	−0.231	0.036	−0.161	−0.164	0.425	0.339	**0.530 #**	**0.482 ***	1.00		
Total SD (β14)	0.061	0.067	0.066	0.027	−0.005	−0.097	0.032	−0.085	−0.079	**0.601 #**	**0.614 #**	**0.634 #**	**0.668 #**	**0.920 £**	1.00	
HRavg (β15)	−0.281	0.010	0.054	−0.150	−0.334	0.016	−0.331	−0.060	−0.098	0.461	0.441	0.154	0.209	0.076	0.154	1.00

Significant differences ($p \leq 0.05$) are highlighted in bold. Abbreviations: RPE, rate of perceived exertion; Avg, average; TDur, Training duration; s-RPE, session rate of perceived exertion; DOMS, delayed onset muscle soreness; TM, training monotony; TS, training strain; TD, total distance; HSRD, high-speed running distance; SD, speed distance; HRavg, heart rate average; *, moderate effect; #, large effect; §, very large effect; £, virtually perfect effect.

Table 3. Correlation analysis between the measures in study on the mid-season.

Measure	β0	β1	β2	β3	β4	β5	β6	β7	β8	β9	β10	β11	β12	β13	β14	β15
RPE (β0)	1.00															
Avg TDur (β1)	0.457	1.00														
Total TDur (β2)	**0.520 #**	0.439	1.00													
s-RPE (β3)	0.391	0.050	−0.005	1.00												
Sleep (β4)	**0.491 ***	0.404	**0.523 #**	−0.149	1.00											
Stress (β5)	0.241	−0.158	0.091	0.151	0.263	1.00										
DOMS (β6)	0.320	0.008	**0.499 ***	0.032	**0.518 #**	0.343	1.00									
TM (β7)	0.226	0.049	0.082	0.065	0.347	−0.229	0.022	1.00								
TS (β8)	0.285	0.039	0.093	0.234	0.300	−0.159	0.040	**0.969 £**	1.00							
Avg TD (β9)	0.320	0.444	−0.009	0.090	0.098	−0.284	−0.014	0.390	0.333	1.00						
Total TD (β10)	0.151	0.338	0.002	0.118	0.004	**−0.542 #**	−0.070	0.355	0.327	**0.852 §**	1.00					
Avg HSRD (β11)	0.263	0.104	0.080	0.122	0.096	−0.159	−0.019	−0.069	−0.128	0.403	0.378	1.00				
Total HSRD (β12)	0.184	0.270	0.082	0.255	−0.021	**−0.543 #**	−0.100	0.115	0.089	**0.646 #**	**0.839 §**	**0.725 §**	1.00			
Avg SD (β13)	0.280	0.326	−0.089	0.275	−0.148	−0.319	0.013	0.096	0.052	**0.533 #**	0.423	**0.472 ***	**0.537 #**	1.00		
Total SD (β14)	0.171	0.331	−0.038	0.309	−0.143	**−0.562 #**	−0.052	0.191	0.171	**0.671 #**	**0.825 §**	**0.507 #**	**0.878 §**	**0.790 §**	1.00	
HRavg (β15)	0.175	0.081	−0.146	0.272	0.151	−0.183	−0.194	**0.853 §**	**0.886§**	0.367	0.337	−0.020	0.132	0.189	0.239	1.00

Significant differences ($p \leq 0.05$) are highlighted in bold. Abbreviations: RPE, rate of perceived exertion; Avg, average; TDur, Training duration; s-RPE, session rate of perceived exertion; DOMS, delayed onset muscle soreness; TM, training monotony; TS, training strain; TD, total distance; HSRD, high-speed running distance; SD, speed distance; HRavg, heart rate average; *, moderate effect; #, large effect; §, very large effect; £, virtually perfect effect.

Table 4. Correlation analysis between the measures in study on the end-season.

Measure	β0	β1	β2	β3	β4	β5	β6	β7	β8	β9	β10	β11	β12	β13	β14	β15
RPE (β0)	1.00															
Avg TDur (β1)	0.005	1.00														
Total TDur (β2)	0.210	0.082	1.00													
s-RPE (β3)	**0.843 §**	−0.034	**0.483 ***	1.00												
Sleep (β4)	0.398	−0.264	0.168	0.298	1.00											
Stress (β5)	0.309	−0.137	0.328	**0.483 ***	0.418	1.00										
DOMS (β6)	0.342	−0.467	0.247	0.167	**0.656 #**	0.429	1.00									
TM (β7)	−0.312	0.103	0.221	−0.189	−0.402	−0.287	−0.285	1.00								
TS (β8)	0.085	0.204	**0.592 #**	0.256	−0.315	−0.152	−0.172	**0.711 §**	1.00							
Avg TD (β9)	0.404	−0.326	−0.299	0.236	0.153	−0.083	0.269	−0.361	−0.257	1.00						
Total TD (β10)	0.061	−0.361	−0.459	−0.054	−0.073	−0.244	0.029	−0.320	**−0.527 #**	**0.786 §**	1.00					
Avg HSRD (β11)	0.456	−0.346	−0.102	0.431	0.225	0.151	0.245	−0.456	−0.075	**0.706 §**	0.378	1.00				
Total HSRD (β12)	0.230	−0.363	−0.297	0.179	0.020	−0.057	0.109	**0.499 ***	−0.415	**0.841 §**	**0.867 §**	**0.751 §**	1.00			
Avg SD (β13)	0.233	−0.283	−0.080	0.144	0.284	0.097	0.223	0.038	0.187	0.073	−0.211	**0.495 ***	0.052	1.00		
Total SD (β14)	0.186	**−0.521 #**	−0.381	0.038	0.084	−0.064	0.252	−0.308	−0.339	**0.668 #**	**0.688 #**	**0.645 #**	**0.785 §**	**0.510 #**	1.00	
HRavg (β15)	0.249	0.121	−0.253	0.201	−0.066	0.169	−0.149	0.093	0.028	−0.017	−0.226	−0.104	−0.313	0.048	−0.275	1.00

Significant differences ($p \leq 0.05$) are highlighted in bold. Abbreviations: RPE, rate of perceived exertion; Avg, average; TDur, Training duration; s-RPE, session rate of perceived exertion; DOMS, delayed onset muscle soreness; TM, training monotony; TS, training strain; TD, total distance; HSRD, high-speed running distance; SD, speed distance; HRavg, heart rate average; *, moderate effect; #, large effect; §, very large effect.

Table 5. Correlation analysis between the measures in study on the full-season.

Measure	β0	β1	β2	β3	β4	β5	β6	β7	β8	β9	β10	β11	β12	β13	β14	β15
RPE (β0)	1.00															
Avg TDur (β1)	0.219	1.00														
Total TDur (β2)	0.355	0.454	1.00													
s-RPE (β3)	**0.772 §**	0.106	0.428	1.00												
Sleep (β4)	0.322	0.258	0.433	0.211	1.00											
Stress (β5)	−0.111	0.048	0.281	−0.008	0.426	1.00										
DOMS (β6)	0.161	0.031	0.375	0.250	0.373	0.090	1.00									
TM (β7)	0.088	−0.017	0.248	0.011	−0.373	−0.112	−0.003	1.00								
TS (β8)	0.202	0.079	0.350	0.197	−0.384	−0.157	0.073	**0.958 £**	1.00							
Avg TD (β9)	0.385	0.122	−0.127	0.086	−0.165	−0.357	−0.285	0.150	0.173	1.00						
Total TD (β10)	0.142	−0.036	−0.186	0.017	−0.326	−0.425	−0.316	0.151	0.143	**0.887 §**	1.00					
Avg HSRD (β11)	**0.496 ***	−0.312	−0.275	0.280	−0.175	−0.316	−0.048	−0.165	−0.111	**0.522 #**	0.466	1.00				
Total HSRD (β12)	0.275	−0.263	−0.291	0.181	−0.328	**−0.500 ***	−0.240	−0.069	−0.041	**0.671 #**	**0.814 §**	**0.813 §**	1.00			
Avg SD (β13)	0.390	0.059	−0.387	0.074	−0.126	**−0.528 #**	0.077	−0.116	−0.131	0.252	0.138	0.400	0.289	1.00		
Total SD (β14)	0.253	−0.022	−0.347	0.071	−0.324	**−0.640 #**	−0.062	−0.008	−0.024	**0.607 #**	**0.721 §**	**0.524 #**	**0.746 §**	**0.721 §**	1.00	
HRavg (β15)	0.154	0.149	−0.046	0.088	−0.394	−0.141	−0.189	**0.617 #**	**0.667 #**	0.405	0.279	0.088	0.084	0.030	0.098	1.00

Significant differences ($p \leq 0.05$) are highlighted in bold. Abbreviations: RPE, rate of perceived exertion; Avg, average; TDur, Training duration; s-RPE, session rate of perceived exertion; DOMS, delayed onset muscle soreness; TM, training monotony; TS, training strain; TD, total distance; HSRD, high-speed running distance; SD, speed distance; HRavg, heart rate average; *, moderate effect; #, large effect; §, very large effect; £, virtually perfect effect.

4. Discussion

The aim of the present study was to quantify and analyse the relationship of the external and internal training intensity metrics as well as the well-being measures in different periods of a semi-professional soccer season (early-, mid- and end-season); and to describe TM and TS for the entire period of the analysis in a semi-professional soccer season.

In the case of external training intensity metrics, it was observed that the total HSRD and sprint distance in comparison to the early and mid-season increased at the end-season. Previous studies had used various methods to examine these factors. According to the coaches, there could be variances in training and performance depending on the degree of play that led to differences in the values reported in different periods [26]. The number of training sessions and team competitions at the end-season was higher than the previous periods, which in turn can affect the high total HSRD and TD. The average HSRD showed a positive correlation with RPE changes. In many studies, changes in internal intensity exhibit high correlation with low speeds and low-intensity distance, but not with high speeds. Possible reasons include GPS error at high speeds, individual ability/requirement to reach high speeds, and the nonlinear relationship between speeds and internal intensity. One aspect that internal intensity does not take into account is moving at higher speeds (>14 $km \cdot h^{-1}$) and high accelerations (>2 $m \cdot s^{-2}$) [27,28]. According to previous results, RPE also showed the highest values in the end-season (4.27 AU), which shows a significant difference between the early and mid-season, and follows the same line of external intensity.

Analysis of internal intensity measured by psychological variables such as RPE is highly preferred because of its potential for integrating different types of stimuli and ease of use [8]. Various factors may affect RPE. As an example, situations such as scoring, scoring opportunities, ball control, tackle, good play on set, winning turnover, increasing ball possession or ability to block the attack, or even non-technical/tactical training can have an impact the perceived exertion of a player [29]. Therefore, given the team's matches and training seasons at the end-season compared to previous periods, a higher RPE value can be justified. Furthermore, s-RPE was considered an important global indicator of training intensity and intensity in team sports [29,30]. However, Haddad et al. suggested that s-RPE is not sufficient to identify health indicators such as subjective fatigue, DOMS, stress, and sleep level of young soccer players [10]. In this case, Hooper and Mackinnon [5] suggested a self-assessment-based psychometric questionnaire, which includes well-being related to sleep, stress, fatigue, and DOMS. It is considered one of the best questionnaires for estimating well-being and monitoring the perceived health of soccer players [6,19].

In the present study, no significant difference was observed in well-being measures between the periods of the season. Evaluating RPE was found very important instrument in correlating overtraining of athletes with physical demands on the body. However, changes in TM and TS were also not significant during the season. It is not clear why it fluctuates, exhibiting a W-shaped diagram, during the season. Several factors such as match position, match result, opponent quality, tactics system, and training program can affect these results. Contextual factors such as tactical formation and suspension of a match can affect the overall workload during a match, and further into the previous or next training session [21]. This result may have been a strategy of the coach to prepare the team for the next period to achieve better results. On the other hand, the rise in TS had led to a decline in TM in the coming weeks, which can be seen in the indicators of well-being (considering not being meaningful). The degree of sleep quality gradually decreased compared to the early-season, and the quality of sleep also improved with the increase in TS and decrease in TM in the last weeks of the end-season. Additionally, DOMS and stress were higher than the end-season. Past studies had also shown that increasing the training intensity and competitions impairs sleep quality [31,32]. On the other hand, the changes were consistent with previous literature, where the high diversity of TM and TS in the mid-season reflected that players prefer greater training intensity for motivation [33].

The average of TD showed a moderate to high correlation with RPE, sleep and s-RPE at early-, mid- and end-seasons. In a similar study, the daily intensity over the course of a week revealed a high and moderate correlation with peak power and change of direction at different periods of the seasons in elite youth soccer players [34]. Another study examined the daily training intensity and perceived wellness characteristics and showed that the amount of training intensity was linked to sleep perceived by elite football players. This condition was also reported in our study [6].

In the other part of results, there was a large association between TD and sprint distance and between TD and HSRD in the early-, mid- and end-season. It can be inferred that this high level of correlation indicates the effect of external training intensity indices on each other. Numerous studies have reported a negative correlation between training intensity and strength indices. However, the methods of measuring internal training intensity were different from the present study [35], which reinforces the need for more studies to confirm the present results. Additionally, a review article assessed the symptoms of perceived stress and it showed that both categories were sensitive to acute changes [36]. On the other hand, increasing HRavg is associated with increasing TM and TS, which indicates that increasing external intensity increases HRavg. On the other hand, the increase in TM leads to overtraining based on a previous study [37], which is one of the consequences of overtraining, and consequently increases the heart rate during training and competition.

In the early-, mid-, and end-season, there was a large and negative relationship between Avg TD and DOMS, as well as between sleep and Avg HSRD. Similar to the results of this study, previous research has linked perceived sleep, stress, fatigue, and DOMS to daily perceived intensity at the professional level [38].

Meanwhile, there are limitations to this study that need to be considered. First, we can point to the lack of pre-season information, which affects the overall results. Second, the number of athletes participating in this study was relatively small, which makes it difficult to generalize the results. Third, in future studies, changes in acute and chronic training can also be considered along with external and internal training intensity to obtain complete information. Fourth, stress from HI questionnaire was not considered in the present study which could present more details on data analysis. The final limitation of this study was the lack of internal and external load monitoring in resistance training and competitions sessions which should be considered in future studies.

5. Conclusions

Coaches and their staff should consider the results of this study. Despite the relationship between external and internal intensity, each metric had a unique effect on the perception of the player's training intensity management, and special attention should be paid to each player when monitoring a training session.

Sport coaches and fitness professionals should be wary of changes in TM and TS that affect players' good responses, because based on the results, increasing these metrics can have a negative effect on indicators such as DOMS and sleep.

This study shows the relationship between training intensity and other psychological indicators among players. Examining these processes will probably be effective in training planning. Therefore, in order to better control the training, more consideration should be given by the coaches, so that the team performance can be maximized, and better results can be obtained. The results serve as a useful tool for providing coaches and their staff information on determining perception factors.

Author Contributions: Conceptualization, H.N., R.G., A.D.M. and R.O.; methodology, H.N. and R.G.; data collection, H.N. and A.D.M.; analysis, H.N. and R.O.; writing—original draft preparation, A.D.M., G.B. and R.O.; writing—review and editing, R.G., R.O., G.B., H.N. and R.O. All authors have read and agreed to the published version of the manuscript.

Funding: This research was funded by the Portuguese Foundation for Science and Technology, I.P., Grant/Award Number UIDP/04748/2020.

Institutional Review Board Statement: The study was conducted in accordance with the Declaration of Helsinki, and approved by the Ardabil University of Medical Sciences authorized the ethical committee code IR.ARUMS.REC.1399.545 for this study.

Informed Consent Statement: Informed consent was obtained from all subjects involved in the study.

Data Availability Statement: The datasets generated and analysed during the current study are available from the corresponding author on reasonable request.

Acknowledgments: The authors thank Sepahan Football Club for your full cooperation in this study.

Conflicts of Interest: The authors declare no conflict of interest. The funders had no role in the design of the study; in the collection, analyses, or interpretation of data; in the writing of the manuscript, or in the decision to publish the results.

References

1. Gabbett, T.J.; Nassis, G.P.; Oetter, E.; Pretorius, J.; Johnston, N.; Medina, D.; Rodas, G.; Myslinski, T.; Howells, D.; Beard, A.; et al. The athlete monitoring cycle: A practical guide to interpreting and applying training monitoring data. *Br. J. Sports Med.* **2017**, *51*, 1451–1452. [CrossRef]
2. Brink, M.S.; Visscher, C.; Arends, S.; Zwerver, J.; Post, W.J.; Lemmink, K.A. Monitoring stress and recovery: New insights for the prevention of injuries and illnesses in elite youth soccer players. *Br. J. Sports Med.* **2010**, *44*, 809–815. [CrossRef]
3. Impellizzeri, F.M.; Marcora, S.M.; Coutts, A.J. Internal and External Training Load: 15 Years On. *Int. J. Sports Physiol. Perform.* **2019**, *14*, 270–273. [CrossRef]
4. Impellizzeri, F.; Rampinini, E.; Marcora, S. Physiological assessment of aerobic training in soccer. *J. Sports Sci.* **2005**, *23*, 583–592. [CrossRef]
5. Hooper, S.L.; MacKinnon, L.T. Monitoring Overtraining in Athletes. *Sports Med.* **1995**, *20*, 321–327. [CrossRef]
6. Moalla, W.; Fessi, M.S.; Farhat, F.; Nouira, S.; Wong, D.P.; Dupont, G. Relationship between daily training load and psychometric status of professional soccer players. *Res. Sports Med.* **2016**, *24*, 387–394. [CrossRef]
7. Manzi, V.; D'Ottavio, S.; Impellizzeri, F.M.; Chaouachi, A.; Chamari, K.; Castagna, C. Profile of Weekly Training Load in Elite Male Professional Basketball Players. *J. Strength Cond. Res.* **2010**, *24*, 1399–1406. [CrossRef]
8. Foster, C. Monitoring training in athletes with reference to overtraining syndrome. *Med. Sci. Sports Exerc.* **1998**, *30*, 1164–1168. [CrossRef]
9. Clemente, F.M.; Mendes, B.; Nikolaidis, P.; Calvete, F.; Carriço, S.; Owen, A.L. Internal training load and its longitudinal relationship with seasonal player wellness in elite professional soccer. *Physiol. Behav.* **2017**, *179*, 262–267. [CrossRef]
10. Haddad, M.; Chaouachi, A.; Wong, D.P.; Castagna, C.; Hambli, M.; Hue, O.; Chamari, K. Influence of fatigue, stress, muscle soreness and sleep on perceived exertion during submaximal effort. *Physiol. Behav.* **2013**, *119*, 185–189. [CrossRef]
11. Oliveira, R.; Brito, J.P.; Martins, A.; Mendes, B.; Marinho, D.A.; Ferraz, R.; Marques, M.C. In-season internal and external training load quantification of an elite European soccer team. *PLoS ONE* **2019**, *14*, e0209393. [CrossRef]
12. Nobari, H.; Praça, G.M.; Clemente, F.M.; Pérez-Gómez, J.; Vivas, J.C.; Ahmadi, M. Comparisons of new body load and metabolic power average workload indices between starters and non-starters: A full-season study in professional soccer players. *Proc. Inst. Mech. Eng. Part P J. Sports Eng. Technol.* **2021**, *235*, 105–113. [CrossRef]
13. Williams, J.; Tessaro, E. *Validity and Reliability of a 15 Hz GPS Device for Court-Based Sports Movements 2018*; Virginia Tech: Blacksburg, VA, USA, 2018.
14. Borg, G. Perceived exertion as an indicator of somatic stress. *Scand. J. Rehabil. Med.* **1970**, *2*, 92–98. [PubMed]
15. Foster, C.; Florhaug, J.A.; Franklin, J.; Gottschall, L.; Hrovatin, L.A.; Parker, S.; Doleshal, P.; Dodge, C. A new approach to monitoring exercise training. *J. Strength Cond. Res.* **2001**, *15*, 109–115. [PubMed]
16. Simpson, M.J.; Jenkins, D.G.; Scanlan, A.T.; Kelly, V. Relationships between External- and Internal-Workload Variables in an Elite Female Netball Team and Between Playing Positions. *Int. J. Sports Physiol. Perform.* **2020**, *15*, 841–846. [CrossRef]
17. Foster, C.; Hector, L.L.; Welsh, R.; Schrager, M.; Green, M.A.; Snyder, A.C. Effects of specific versus cross-training on running performance. *Eur. J. Appl. Physiol. Occup. Physiol.* **1995**, *70*, 367–372. [CrossRef]
18. Clemente, F.M.; Silva, R.; Castillo, D.; Los Arcos, A.; Mendes, B.; Afonso, J. Weekly Load Variations of Distance-Based Variables in Professional Soccer Players: A Full-Season Study. *Int. J. Environ. Res. Public Health* **2020**, *17*, 3300. [CrossRef]
19. Nobari, H.; Aquino, R.; Clemente, F.M.; Khalafi, M.; Adsuar, J.C.; Pérez-Gómez, J. Description of acute and chronic load, training monotony and strain over a season and its relationships with well-being status: A study in elite under-16 soccer players. *Physiol. Behav.* **2020**, *225*, 113117. [CrossRef]
20. Clemente, F.M.; Silva, R.; Chen, Y.-S.; Aquino, R.; Praça, G.M.; Paulis, J.C.; Nobari, H.; Mendes, B.; Rosemann, T.; Knechtle, B. Accelerometry-Workload Indices Concerning Different Levels of Participation during Congested Fixture Periods in Professional Soccer: A Pilot Study Conducted over a Full Season. *Int. J. Environ. Res. Public Health* **2021**, *18*, 1137. [CrossRef]
21. Meckel, Y.; Doron, O.; Eliakim, E.; Eliakim, A. Seasonal Variations in Physical Fitness and Performance Indices of Elite Soccer Players. *Sports* **2018**, *6*, 14. [CrossRef]

22. Hedges, L.V. Distribution theory for Glass's estimator of effect size and related estimators. *J. Educ. Stat.* **1981**, *6*, 107–128. [CrossRef]
23. Cohen, J.; Cohen, P.; West, S.G.; Aiken, L.S. *Applied Multiple Regression/Correlation Analysis for the Behavioral Sciences*; Routledge: London, UK, 2013.
24. Batterham, A.M.; Hopkins, W.G. Making Meaningful Inferences about Magnitudes. *Int. J. Sports Physiol. Perform.* **2006**, *1*, 50–57. [CrossRef] [PubMed]
25. Hopkins, W.G.; Marshall, S.W.; Batterham, A.; Hanin, J. Progressive Statistics for Studies in Sports Medicine and Exercise Science. *Med. Sci. Sports Exerc.* **2009**, *41*, 3–13. [CrossRef] [PubMed]
26. López, A.M.; Granero-Gil, P.; Ortega, J.P.; De Hoyo, M. The validity and reliability of a 5-hz GPS device for quantifying athletes' sprints and movement demands specific to team sports. *J. Hum. Sport Exerc.* **2017**, *12*, 156–166. [CrossRef]
27. Oliveira, R.; Brito, J.P.; Loureiro, N.; Padinha, V.; Ferreira, B.; Mendes, B. Does the distribution of the weekly training load account for the match results of elite professional soccer players? *Physiol. Behav.* **2020**, *225*, 113118. [CrossRef] [PubMed]
28. Sobolewski, E.J. The Relationships between Internal and External Load Measures for Division I College Football Practice. *Sports* **2020**, *8*, 165. [CrossRef]
29. Barrett, S.; McLaren, S.; Spears, I.; Ward, P.; Weston, M. The Influence of Playing Position and Contextual Factors on Soccer Players' Match Differential Ratings of Perceived Exertion: A Preliminary Investigation. *Sports* **2018**, *6*, 13. [CrossRef]
30. Borges, T.O.; Moreira, A.; Thiengo, C.R.; Medrado, R.G.S.D.; Titton, A.; Lima, M.R. Training intensity distribution of young elite soccer players. *Rev. Bras. Cineantropometria Desempenho Hum.* **2019**, *21*. [CrossRef]
31. Clemente, F.; Afonso, J.; Costa, J.; Oliveira, R.; Pino-Ortega, J.; Rico-González, M. Relationships between Sleep, Athletic and Match Performance, Training Load, and Injuries: A Systematic Review of Soccer Players. *Healthcare* **2021**, *9*, 808. [CrossRef]
32. Rabbani, A.; Baseri, M.K.; Reisi, J.; Clemente, F.M.; Kargarfard, M. Monitoring collegiate soccer players during a congested match schedule: Heart rate variability versus subjective wellness measures. *Physiol. Behav.* **2018**, *194*, 527–531. [CrossRef]
33. Nobari, H.; Alves, A.; Haghighi, H.; Clemente, F.; Carlos-Vivas, J.; Pérez-Gómez, J.; Ardigò, L. Association between Training Load and Well-Being Measures in Young Soccer Players during a Season. *Int. J. Environ. Res. Public Health* **2021**, *18*, 4451. [CrossRef] [PubMed]
34. Nobari, H.; Polito, L.F.T.; Clemente, F.M.; Pérez-Gómez, J.; Ahmadi, M.; Garcia-Gordillo, M.Á.; Silva, A.F.; Adsuar, J.C. Relationships Between Training Workload Parameters with Variations in Anaerobic Power and Change of Direction Status in Elite Youth Soccer Players. *Int. J. Environ. Res. Public Health* **2020**, *17*, 7934. [CrossRef] [PubMed]
35. Rössler, R.; Junge, A.; Chomiak, J.; Dvorak, J.; Faude, O. Soccer injuries in players aged 7 to 12 years: A descriptive epidemiological study over 2 seasons. *Am. J. Sports Med.* **2016**, *44*, 309–317. [CrossRef] [PubMed]
36. Saw, A.E.; Main, L.C.; Gastin, P.B. Monitoring the athlete training response: Subjective self-reported measures trump commonly used objective measures: A systematic review. *Br. J. Sports Med.* **2016**, *50*, 281–291. [CrossRef]
37. Tomar, R.; Allen, J.A. The Relationship of Training Parameters with Incidence of Injury, Sleep and Well Being of Young University Swimmers. *Int. J. Pharm. Res. Allied Sci.* **2019**, *8*, 47–60.
38. Nobari, H.; Kharatzadeh, M.; Khalili, S.M.; Pérez-Gómez, J.; Ardigò, L.P. Fluctuations of Training Load Variables in Elite Soccer Players U-14 throughout the Competition Season. *Healthcare* **2021**, *9*, 1418. [CrossRef]

biology

MDPI

Article

Prevalence of Positive Effects on Body Fat Percentage, Cardiovascular Parameters, and Cardiorespiratory Fitness after 10-Week High-Intensity Interval Training in Adolescents

Jarosław Domaradzki [1], Dawid Koźlenia [1,*] and Marek Popowczak [2]

[1] Unit of Biostructure, Faculty of Physical Education and Sport, Wroclaw University of Health and Sport Sciences, al. I.J. Paderewskiego 35, 51-612 Wroclaw, Poland; jaroslaw.domaradzki@awf.wroc.pl

[2] Unit of Team Sports Games, Faculty of Physical Education and Sport, Wroclaw University of Health and Sport Sciences, al. I.J. Paderewskiego 35, 51-612 Wroclaw, Poland; marek.popowczak@awf.wroc.pl

* Correspondence: dawid.kozlenia@awf.wroc.pl

Simple Summary: In this study, we aimed to analyze the prevalence of positive effects of high-intensity interval training (HIIT) on body composition, cardiovascular parameters, and cardiorespiratory fitness among adolescents. We investigated 52 boys and 89 girls from a secondary school, separated into an experimental group (EG) with HIIT intervention and a control group (CG). The measured parameters were body fat % (BFP), resting systolic blood pressure (SBP), diastolic blood pressure (DBP), and fitness index (FI). The results indicate that positive HIIT-induced changes in SBP, DBP, and FI were most common among boys, especially those with low body mass index. Our study also revealed relationships between changes in FI and BFP and BP parameters. The effectiveness of HIIT was confirmed concerning the prevalence of the positive changes in measured parameters. We suggest that HIIT should be implemented in PE lessons, although there is a need to look for a more efficient method for girls.

Abstract: Analysis of the interventions on cardiovascular disease risk factors focuses on quantitative changes, omitting assessment of positive effect frequency in individuals. The aim of this study was to assess the prevalence of positive effects of high-intensity interval training (HIIT) on body composition, cardiovascular parameters, and cardiorespiratory fitness among adolescents. A total of 52 boys and 89 girls from a secondary school were separated into an experimental group (EG) with HIIT and a control group (CG). Body fat % (BFP), resting systolic blood pressure (SBP), diastolic blood pressure (DBP), and fitness index (FI) changes were calculated. We assessed the influence and interaction of three factors: intervention (INT), sex (SEX), and body mass index (BMI_{status}) on the ratio of individuals with and without positive changes. We used log-linear models for interactions and multivariate correspondence analysis (MCA). The results indicate that HIIT affects the prevalence of positive changes in SBP, DBP, and FI. Interactions between factors suggest boys with low BMI get more benefit from the intervention than girls. The MCA indicates a relationship between FI and BFP and between BP parameters. The effectiveness of HIIT was confirmed concerning the prevalence of the positive changes in measured parameters. We suggest that HIIT should be implemented in PE lessons, although there is a need to look for a more efficient method for girls.

Keywords: adolescent; Tabata training; high-intensity interval training; body fat; blood pressure; cardiorespiratory fitness; physical education lessons; log-linear analysis; multivariate correspondence analysis

Citation: Domaradzki, J.; Koźlenia, D.; Popowczak, M. Prevalence of Positive Effects on Body Fat Percentage, Cardiovascular Parameters, and Cardiorespiratory Fitness after 10-Week High-Intensity Interval Training in Adolescents. *Biology* **2022**, *11*, 424. https://doi.org/10.3390/biology11030424

Academic Editors: Filipe Manuel Batista Clemente, Georgian Badicu and Eugenia Murawska-Ciałowicz

Received: 14 December 2021
Accepted: 9 March 2022
Published: 10 March 2022

1. Introduction

Obesity, high blood pressure (BP), and low cardiorespiratory fitness (CRF) in youth are associated the premature cardiovascular diseases and all-cause mortality in adulthood [1,2]. The literature has reported the prevalence of obesity in almost every fourth child and youth in the Western world; elevated BP (reflected in pre- and hypertension) in 11.2% of adolescents from developing countries; and low levels of CRF, which has declined over the past six decades [3–5].

There are strong associations between a high level of obesity, elevated BP, and lack of physical activity (PA) in children [6]. Epidemiological surveys demonstrated an increasing number of hypertension cases in youth linked to overweight/obesity in adolescents, particularly those who are not physically active [7]. A global growth in the prevalence of physical inactivity can be observed. About 80% of young people do not perform the minimum physical activity level recommended by the World Health Organization [8,9]. The same trend is observed in Poland [10]. One of the greatest global challenges is to improve the prevention and treatment of non-communicable diseases [11].

The effectiveness of the PA has been demonstrated in the regulation of body mass, resting blood pressure, also confirming its effectiveness in the prevention of obesity and hypertension [12,13]. The relevant setting to introduce PA seems to be physical education lessons [14]. Increasing the intensity of exercises in regular PE lessons at school seems to be the most accessible method [15]. Evidence suggests that implementing high-intensity exercises into physical education (PE) lessons positively correlates with improved blood pressure and body mass index [16–19]. High-intensity interval training (HIIT) seems to be the most appropriate method to increase exercise intensity in PE lessons. Its advantage is short intervention time with very intensive effort, which saves time but is enough to improve maximum oxygen uptake and affect cardiovascular parameters and body mass composition in adolescence [20–25]. As mentioned above, positive effects of HIIT intervention were observed in BP, body composition, and CRF [18,20–25]. However, some differences were observed in the relevant effects of sex and initial BMI [22–25]. This suggests differing effectiveness of HIIT intervention in various groups of adolescents. It remains unclear how the effectiveness of this type of intervention could differ due to sex and BMI, as well as the prevalence of positive effects on BP, body fat, and CRF.

Current studies on the effects of different interventions, including HIIT, on several aspects of biological condition and health of adolescents are primarily limited to the assessment of differences in measured variables (usually showing statistical significance of the difference in mean values) [26–28]. All studied groups included individuals who responded to experimental factors and individuals who did not respond. The question arises, what proportion of individuals experienced positive changes compared to those who experienced no such changes? The second question is: what factors moderate the mentioned relationship, and do these moderators interact with each other? The next question is: is there (and what type if yes) any correspondence between frequencies of individuals separated into different categories of factors (e.g., sex and body mass index)? To date, to our knowledge, there is a lack of an in-depth review of the effects of HIIT intervention on the prevalence of positive changes in adolescents, and there are, as yet, no answers to the questions posed above.

Therefore, the purpose of the present study was to assess the prevalence of positive changes in adolescents following high-intensity interval training (HIIT) implemented in physical education (PE) lessons in terms of body composition, cardiovascular parameters, and cardiorespiratory fitness according to sex and body mass index. Specifically, we aimed to (1) examine the potential interactions between factors that could affect positive changes in the mentioned parameters, (2) provide an overview of co-occurrences associated with the prevalence of positive changes in individuals following HIIT intervention, and (3) assess the structure and strength of similarities between categories.

2. Materials and Methods

2.1. Participants

The participants all participated in the same standard physical education program. The participants comprised six separate classes, of which three were randomly assigned to the experimental group (EG) and three to the control group (CG). Among 187 pupils from a secondary school, 141 subjects completed the study, comprising 52 boys (EG N = 31; CG N = 21; age, 16.24 (±0.34) years; body height, 176.74 (±6.07) cm; body mass, 65.42 (±12.51) kg) and 89 girls (EC N = 42; CG N = 47; age, 16.12 (±0.42) years; body height 164.38 (±6.54) cm; body mass 56.71 (±10.23) kg). Among the 46 excluded subjects, 10 were excluded due to medical contradiction, 17 were excluded from participating in additional sports training, and 19 were excluded during the intervention due to absence in physical education classes. There were no adverse effects observed. All participants were volunteers. A detailed description of the participants is presented in Table 1.

Table 1. Descriptive statistics of dependent variables (body fat percentage (BFP), resting systolic and diastolic blood pressure (SBP, DBP), and fitness index (FI)) in categories of factors (intervention (INT), sex (SEX), and body mass index intervals (BMI_{status}).

Factors				Outcomes			
		BMI_{status}		BFP	SBP	DBP	FI
INT	SEX	Category	Mean ± SD 95%CI	Mean ± SD 95%CI	Mean ± SD 95%CI	Mean ± SD 95%CI	Mean ± SD 95%CI
EG	M	L	18.36 ± 1.35 17.61–19.11	11.53 ± 2.93 9.90–13.15	121.40 ± 14.96 113.11–129.68	76.46 ± 5.93 73.17–79.75	44.81 ± 4.30 42.43–47.19
		M	21.00 ± 0.98 20.33–21.66	15.87 ± 4.60 12.77–18.96	123.63 ± 13.33 114.67–132.59	69.90 ± 6.93 65.25–74.56	45.14 ± 2.30 43.59–46.69
		H	28.26 ± 3.80 23.53–32.98	27.62 ± 6.67 19.32–35.91	128.40 ± 4.21 123.16–133.63	76.20 ± 9.49 64.40–87.99	42.65 ± 2.80 39.16–46.14
	F	L	18.62 ± 1.01 18.08–19.16	23.90 ± 2.88 22.37–25.44	116.87 ± 10.06 111.51–122.23	73.43 ± 8.81 68.73–78.13	42.28 ± 2.60 40.90–43.67
		M	21.47 ± 0.96 21.06–21.89	28.12 ± 5.32 25.82–30.43	116.69 ± 7.87 113.28–120.10	70.26 ± 6.46 67.46–73.05	44.32 ± 5.35 42.01–46.64
		H	23.97 ± 0.81 21.95–25.99	30.96 ± 3.00 23.51–38.42	117.01 ± 6.24 101.48–132.51	73.00 ± 6.08 57.88–88.11	42.64 ± 7.19 24.77–60.51
CG	M	L	18.43 ± 0.98 17.67–19.19	11.75 ± 3.17 9.31–14.19	116.00 ± 8.30 109.61–122.38	74.77 ± 5.51 70.53–79.01	43.21 ± 2.89 40.99–45.43
		M	21.30 ± 0.86 20.58–22.02	13.43 ± 3.14 10.80–16.06	122.25 ± 11.20 112.87–131.62	77.12 ± 5.74 72.32–81.92	45.36 ± 3.59 42.35–48.36
		H	25.55 ± 2.58 21.43–29.66	24.27 ± 10.00 8.34–40.20	128.75 ± 3.30 123.49–134.00	79.25 ± 10.71 62.19–96.30	41.93 ± 2.96 37.21–46.66
	F	L	18.41 ± 0.95 17.97–18.84	24.55 ± 3.86 22.79–26.31	113.28 ± 6.39 110.37–116.19	70.19 ± 5.52 67.67–72.70	44.08 ± 3.63 42.43–45.73
		M	21.57 ± 1.05 20.98–22.15	29.40 ± 3.31 27.56–31.24	116.46 ± 9.22 111.35–121.57	68.60 ± 7.64 64.36–72.83	44.44 ± 3.49 42.50–46.37
		H	26.21 ± 4.45 23.22–29.20	35.93 ± 5.83 32.01–39.85	117.72 ± 9.76 111.16–124.28	72.27 ± 9.33 65.99–78.54	45.79 ± 5.13 42.34–49.24

INT-intervention factor, categories: EG-experimental group, CG-control group; SEX-sex factor, categories: M-male, F-female; BMI_{status}-BMI factor, categories: L-low, M-medium, H-high; BFP-percentage of body fat; SBP-systolic blood pressure, DBP-diastolic blood pressure, FI-fitness index.

2.2. Procedures

The measurements were taken before and after the 10-week intervention on one day from 8:00 a.m. to 1:00 p.m. Participants were asked to excrete, avoid physical activity and excessive drinking of liquids, and keep their typical morning patterns directly before measurement.

2.3. Anthropometric Measurements

Body height was measured with an accuracy of 0.1 cm using anthropometers (GPM Anthropological Instruments, DKSH Ltd., Zurich, Switzerland). Bodyweight and body fat percentage (BF%) were measured using an InBody230 body composition analyzer (InBody Co. Ltd., Cerritos, CA, USA). This tool is characterized by very high reliability in men and women, as indicated by high intraclass correlation coefficients for BF% ($\geq$0.98), FM ($\geq$0.98), and FFM ($\geq$0.99) and low standard error of measurement [29]. BMI was calculated based on received body height and weight values. Obtained results were used to divide participants into three groups based on the following intervals: low category of BMI_{status} (BMI < 20), medium category of BMI_{status} (19.99 < BMI < 23.00), and high category of BMI_{status} (BMI > 22.99). Intervals were arbitrarily assumed regarding small numbers of individuals with very low or very high body mass index.

2.4. Fitness Index (FI) (Harvard Step Test)—Cardiorespiratory Fitness

The Harvard Step Test (HST) was used to evaluate aerobic capacity. The HST results of subjects allowed for calculation of the fitness index (FI) according to the following formula [30]: PEI = $(100 \times L)/(5.5 \times p)$, where L = duration of the test in seconds, $L < 300$ s, and p = heart rate within 1.5 min after the subject stopped the test. The reliability of the HST is acceptable at an intraclass correlations coefficient (ICC) of 0.63 [31].

The subjects had to step up and down on a stool with a height of 41.3 cm with a constant pace determined by a metronome. The process starts with the subjects stepping onto and off the step box at 30 cycles per minute. The test duration is a maximum of 300 s, or when the subject rejects the test due to fatigue. Resting heart rate and changes in pulse during exercise and recovery were measured (Polar H1 heart rate monitors, Polar Electro; Kempele, Finland). Heart rate monitors recorded pulse at 5 s intervals, which was transmitted to a smartwatch.

2.5. Resting Blood Pressure Measurements

Blood pressure was measured by an Omron BP710 automatic blood pressure monitor. The subjects had to sit quietly for 10 min. Next, the measurements were taken three times, separated into 10 min intervals. The analyzed results are the means of the three measurements.

2.6. Intervention

The intervention lasted 10 weeks. Participants followed the HIIT intervention during the one PE lesson (45 min) per week. First, a 10 min warmup with jogging and stretching exercises conducted. Next, the HIIT intervention was performed and lasted 14 min, divided into three sessions based on Tabata protocol (20 s work/10 s rest) separated by a 1 min break. In the first session, participants performed: pushups, high knees; in the second session: dynamic lunges, spider crawling; in the third session: plank-to-pushups and side squeezes. All exercises were played on a screen to ensure that workout and rest were implemented accurately. After the HIIT intervention, stretching and breathing exercises were performed to calm down. The control group participated in a standard physical education program.

The participants' heart rate was measured with a Polar H1 (Polar Electro, Kempele, Finland) and established the range of 75–80% HRmax (145–157 heartbeats/min) when performing HIIT. The Tanaka formula, HRmax = $208 - 0.7 \times$ "age" (age = 16 years in this study), was used to verify the intensity of the workout. The subjects achieved an HR of 156.2 ± 17.8 bpm (CI 95%: 123.0–184.0).

2.7. Statistical Analysis

Descriptive statistics for continuous measurements (BMI, BFP, SBP, DBP, and FI) are presented as means and SDs, with a 95% CI, and were calculated for boys and girls from experimental and control groups separated into BMI_{status} intervals.

The associations between the free factors and four outcomes were assessed in subsequent steps. The following categories of the elements were accepted: intervention (INT): experimental group (EG) with HIIT program vs. control group (CG) with standard PE lessons; sex (SEX): boys (M) vs. girls (F); BMI level (BMI_{status}): low category vs. medium category vs. high category. The outcomes were changes between post- and preintervention results, coded as positive (1) or lack of positive (0) changes). The changes were calculated as post- and preintervention differences. Positive change was defined as any change consistent with the direction defined as pro-health, including reducing body fat and blood pressure (lower postintervention result than preintervention measurement), as well as incrementing cardiorespiratory fitness (higher value of the postintervention development than preintervention measurement). A lack of positive change was defined as no difference or results opposite to abovementioned changes.

First, log-linear analysis was conducted to find the simplest model that fits the data concerning outcome. Log-linear analysis is a method for studying structural relationships between variables in a contingency table [32]. This method examines which variables interact and impact outcomes [33]. Considering the log-linear model, we assumed that the natural logarithm of the value of an expected quantity in the table of independence is a linear function of factors. In a two-way case, the unrestricted log-linear model has the following form [34]:

$$\log \pi_{ij} = \text{constant} + u_{1(i)} + u_{2(j)} + u_{12(ij)}$$

where π_{ij} denotes the probability for cell (i, j), and $\{u\}$ has to be constrained to identify the model.

An optimally designed log-linear model allows for the best quantitative prediction considering the smallest possible number of interactions. Pearson's $\chi 2$ and $\chi 2$ maximum likelihood statistics assess whether the expected quantities are significantly different from the observed quantities [35] in order to determine the order of interactions that must be included in the model. To consider which interactions of a given order should be included in the model, it is necessary to analyze partial and marginal dependence. Partial support informs the significance of the degree of interactions, assuming that all other effects of the same degree are included in the model. Marginal dependence reflects the influence of the exchange, provided that the model does not have any interactions in the same order. Marginal dependence can be verified using the χ^2 test of marginal interdependence [35]. Finally, we used Pearson's χ^2 to assess the model's fit to the data. To examine the nature of effects (main and interactions), marginal quantity tables were calculated to observe quantity [36].

Next, multiple correspondence analysis (MCA) was performed to examine co-occurrences between categories of factors and variables. The analysis base was a matrix of individual results in the Burt table. Correspondence analysis is an exploratory technique often used to take an in-depth look at the results obtained from chi-square or log-linear analysis. Correspondence analysis applies to a two-way (or more) crosstabulation that summarizes the distribution of perceived categories of obtained variables in different groups (factors). MCA aims to reduce multidimensional space to more diminutive synthetic dimensions (mainly two main dimensions), which represent original data in the best way [37]. The measurement of dimensions is inertia, which can be compared to variance in ANOVA. It is possible to indicate variables (categories) strongly correlated with each dimension using squared cosine (cos^2) to identify the nature of the synthetic dimension [38]. The result is a graph that plots data, visually showing the outcome of two or more data points.

The last step was to use cluster analysis (CIA) to graphically present the structure and similarities between categories linked together. In cluster analysis, we used Ward's method of linkage and Euclidean distances. The raw data were coordinates (row and column profiles) obtained from MCA. Grouping a set of objects in ClA can be conducted so that items in the same group are more similar than those in other groups [35]. The result is a dendrogram that visually presents similarities and dissimilarities between associations.

The significance level was set at $\alpha = 0.05$. Statistica V.13.0 statistical package (Tibco, 2020, Cracow, Poland) was used to analyze the study data.

3. Results

Descriptive statistics of the baseline values of measured parameters of the individuals in factor categories are presented in Table 1. Given that this work focused on qualitative rather than quantitative measured parameters, there was no calculation of statistics assessing differences between groups of adolescents.

The numbers of individuals who experienced positive or no positive changes (post-pre) in outcome after intervention are presented in Table 2. These frequencies were the starting point for building the Burt table used in correspondence analysis.

Table 2. Numbers and frequencies of individuals in categories of each factor (INT, SEX, and BMI_{status}) with positive (+) and no positive (−) changes after intervention in measured parameters (BFP, SBP, DBP, and FI).

FACTOR		DV							
		BFP		SBP		DBP		FI	
		−	+	−	+o	−	+	−	+
		N (%)	N (%)	N (%)	N (%)	N (%)	N (%)	N (%)	N (%)
INT	EG	29 (39.73)	44 (60.27)	12 (16.44)	61 (83.56)	26 (35.62)	47 (64.38)	19 (26.03)	54 (73.97)
	CG	20 (29.41)	48 (70.59)	42 (61.76)	26 (38.24)	20 (29.41)	48 (70.59)	32 (47.06)	36 (52.94)
SEX	M	18 (34.62)	31 (65.38)	18 (34.62)	34 (65.38)	21 (40.38)	31 (59.62)	16 (30.77)	36 (69.23)
	F	31 (34.83)	58 (65.17)	36 (40.45)	53 (59.55)	42 (47.19)	47 (52.81)	35 (39.33)	54 (60.67)
BMI_{status}	L	23 (37.70)	38 (62.30)	24 (39.34)	37 (60.66)	22 (36.07)	39 (63.93)	20 (32.79)	41 (67.21)
	M	20 (35.09)	37 (64.91)	19 (33.33)	38 (66.67)	32 (56.14)	25 (43.86)	21 (36.84)	36 (63.16)
	H	6 (26.09)	17 (73.91)	11 (47.83)	12 (52.17)	9 (39.13)	14 (60.87)	10 (43.48)	13 (56.52)

− lack of positive changes; + positive changes; INT-intervention factor, categories: EG-experimental group, CG-control group; SEX-sex factor, categories: M-male, F-female; BMI_{status}-BMI factor, categories: L-low, M-medium, H-high; BFP-percentage of body fat; SBP-systolic blood pressure; DBP-diastolic blood pressure; FI-fitness index.

Due to the small number of respondents in relation to the total number of categories of three factors (seven categories), too many empty classes were created, and a complete analysis of the interactions between all factors was impossible. Therefore, we decided to conduct analyses for two-factor designs (two factors included in the model simultaneously). The following models were tested: INT*SEX→DV and INT*BMI_{status} →DV. Considering that the dependent variables are changes involving HIIT intervention, the SEX*BMI→DV model was omitted in case individuals from the experimental group and control group would be mixed up in categories of SEX and BMI_{status} factors, which would not make sense in terms of the purposes of this paper.

At the onset of the log-linear analysis, we determined the specification of the models, specifically the order of interactions. The test results of interactions are presented in Table 3.

Regarding the information contained in Table 3, in both models (INT*SEX and HIIT* BMI_{status}), apart from the main components, second-order interactions (at most) should be included. Third-order and higher interactions were not statistically significant. Partial and marginal tests were used to assess which interactions should be included in the model (Table 4).

The essence of the method is the assessment of the relevance of the interactions, which confirms that the effect of one causal variable on an outcome depends on the state of a second causal variable. An optimally designed log-linear model allows for the best quantitative prediction considering the smallest possible number of interactions.

Table 3. Test results of interactions between factors (1–2) and dependent variables (BFP, SBP, DBP, and FI).

k-Factors	INT(1)*SEX(2)			INT(1)*BMI$_{status}$(2)		
	df	χ^2	p	df	χ^2	p
1	6	53.30	0.0000	7	66.28	0.0000
2	15	43.00	0.0002	20	48.21	0.0004
BFP	20	22.39	0.3197	30	17.47	0.9666
SBP	15	10.60	0.7807	25	16.32	0.9049
DBP	6	3.12	0.7952	11	6.84	0.8116
FI	1	0.94	0.3317	2	0.26	0.8777

INT-intervention factor, BMI$_{status}$-BMI factor, SEX-sex, BFP-percentage of body fat, SBP-systolic blood pressure, DBP-diastolic blood pressure, FI-fitness index.

Focusing only on interactions, the results of partial and marginal tests (Table 4) indicated the need to include two second-order interactions in the INT*SEX model: 41 and 61. Therefore, the strongest and most statistically significant influence of the HIIT intervention on SBP and FI was confirmed. In particular, there were more often positive changes in SBP and FI in the intervention group than in the control group. There was no interaction between INT and SEX. The impact of HIIT intervention on the prevalence of positive and lack of positive changes in SBP and FI was equal in boys and girls.

The same partial and marginal tests for the INT*BMI$_{status}$ model indicated the need to include three interactions apart from main components: 41, 61, and 52. Interpretation of the results is similar to that of INT and SEX factors, with a small difference. We observed an interaction between BMI$_{status}$ and DBP 52 The role of body-mass-to-height proportions was significant and induced positive changes in SBP. Detailed inspection of the marginal results indicated the highest rate of change in the low BMI$_{status}$ category (the ratio of positive-to-lack of changes was 1.5 to 1), whereas the lowest for the high BMI$_{status}$ category (the ratio of positive-to-lack of changes was 1 to 1).

The values of maximum likelihood statistics (L2), Pearson's chi-squared statistics (χ^2), and statistical non-significant p-values for both models (L2 = 46.64, p = 0.809, χ^2= 45.47, p = 0.841; L2 = 53.52, p = 0.996, χ^2 = 53.57, p = 0.996; respectively) confirmed models were well designed for the empirical data.

Implementation of the log-linear analysis allowed for a more detailed description of the interactions between the categories of factors and the amount of the positive and lack of positive changes after the intervention. Log-linear analysis was performed with a simple independence chi-squared test, which represents a more powerful tool to study dependence between qualitative data than an assessment based only on probability value [39].

The results of MCA, shown in Figures 1 and 2, present a 13-dimensional space of relationships reduced to two dimensions, the quality of presentation of which is acceptable. A projection of all variable associations in two-dimensional space describes over 40% of the total inertia (a measure of dispersion in categorical data that can be compared to variance for quantitative data) in the experimental group (first dimension showed 20,54% inertia, and the second dimension showed 18,95% inertia) and over 36% in the control group (19.20% and 17.12, respectively) (Figures 1 and 2). Table 5 shows marginal quantities concerning observed quantities in both models for effects of interactions.

Table 4. Results of the partial (χ^2 part) and marginal (χ^2 marg) tests between the factors (1*2) and dependent variables (BFP, SBP, DBP, and FI): main effects and interactions (only selected interactions are presented).

Effect	INT(1)*SEX(2)					INT(1)*BMI$_{status}$(2)				
	df	χ^2_{part}	p	χ^2_{marg}	p	df	χ^2_{part}	p	χ^2_{marg}	p
1	1	0.14	0.7038			1	0.13	0.7160		
2	1	7.97	0.0047			2	15.01	0.0006		
3 (BFP)	1	10.80	0.0010			1	9.87	0.0017		
4 (SBP)	1	6.33	0.0118			1	5.79	0.0161		
5 (DBP)	1	1.30	0.2538			1	1.19	0.2750		
6 (FI)	1	8.87	0.0029			1	8.11	0.0044		
12	1	0.97	0.3240	1.57	0.2103	2	2.19	0.3341	2.81	0.2456
13	1	1.73	0.1888	1.25	0.2628	1	1.54	0.2144	1.12	0.2904
14	1	22.86	0.0000	25.27	0.0000	1	19.72	0.0000	22.94	0.0000
15	1	2.01	0.1561	4.13	0.0422	1	2.59	0.1075	3.78	0.0518
16	1	4.57	0.0324	5.47	0.0194	1	4.15	0.0416	4.99	0.0255
23	1	0.01	0.9345	0.02	0.8848	2	0.24	0.8862	0.21	0.8994
24	1	0.00	0.9840	0.23	0.6306	2	0.61	0.7388	1.44	0.4863
25	1	0.19	0.6599	0.41	0.5210	2	4.83	0.0894	3.91	0.1417
26	1	0.28	0.5971	0.56	0.4545	2	0.83	0.6610	1.03	0.5975
34	1	0.13	0.7204	0.01	0.9235	1	0.12	0.7284	0.00	0.9695
35	1	0.01	0.9210	0.05	0.8298	1	0.00	0.9568	0.03	0.8564
36	1	1.29	0.2560	0.83	0.3634	1	1.31	0.2530	0.85	0.3554
45	1	0.76	0.3829	2.45	0.1174	1	0.97	0.3254	2.28	0.1309
46	1	0.00	0.9882	0.84	0.3588	1	0.01	0.9293	0.85	0.3579
56	1	0.15	0.6955	0.57	0.4514	1	0.14	0.7073	0.55	0.4603

INT-intervention factor, BMI$_{status}$-BMI factor, SEX-sex, BFP-percentage of body fat, SBP-systolic blood pressure, DBP-diastolic blood pressure, FI-fitness index.

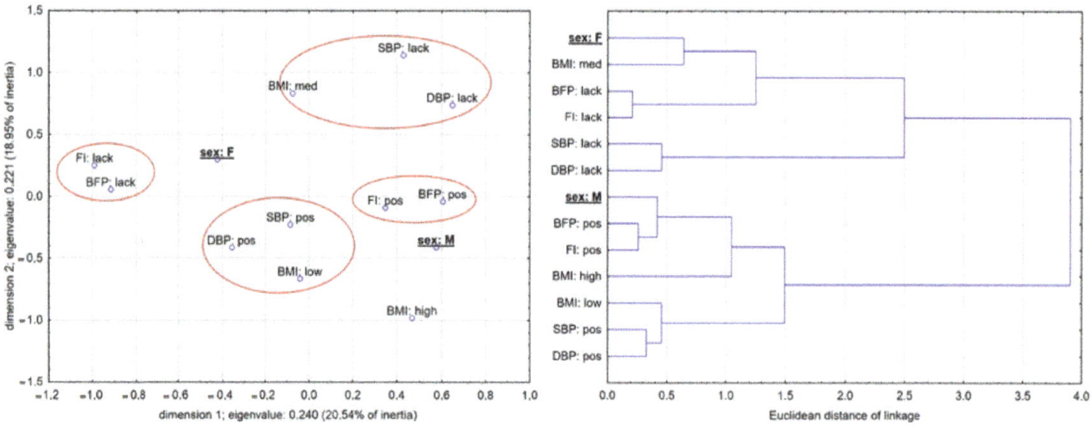

Figure 1. Results of intervention effects on the co-occurrence of changes in independent variables concerning sex and BMI$_{status}$. Taxonomical dendrogram illustrates similarities and distances (strength) of linkages.

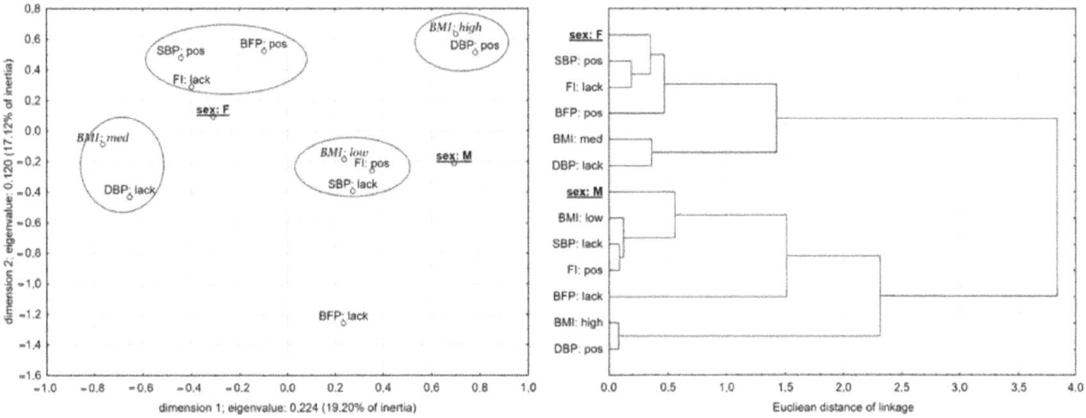

Figure 2. Co-occurrence of the changes in independent variables in the control group concerning sex and BMI_{status}. Taxonomical dendrogram illustrates similarities and distances (strength) of linkages.

Table 5. Table of marginal quantity concerning observed quantity in INT*SEX and INT-BMIstatus models for dependence: INT effect in SBP and INT effect in FI in the first model and INT effect in SBP, INT effect in FI, and BMIstatus effect in DBP in the second model.

	INT*SEX				INT*BMI_{status}						
	INT*SBP		INT*FI		INT*SBP		INT*FI		BMI_{status}*DBP		
	EG	CG	E	CG	EG	CG	EG	CG	L	M	H
0	20	50	27	40	24	54	31	44	30	40	17
1	69	34	62	44	73	38	66	48	47	33	22
all	89	84	89	84	97	92	97	92	77	73	39

INT-intervention factor, categories: EG-experimental group, CG-control group; SEX-sex factor, categories: B-boys, G-girls; BMI_{status}-BMI factor, categories: L-low, M-medium, H-high; BFP-percentage of body fat; SBP-systolic blood pressure; DBP-diastolic blood pressure; FI-fitness index.

A comparison of marginal quantity to observed quantity in the INT*SEX model for SBP and FI confirmed more frequent positive changes in the experimental group compared to the control group (in proportions of 3.5:1 for SBP and 2.5:1 for FI). In the case of the HIIT*BMI_{status} model, proportions of positive changes in SBP and FI in EG compared to CG were similar to those mentioned above. In the case of BMI_{status}, which interacts with DBP, for low-BMI_{status} and high-BMI_{status} categories of individuals, the ratio of positive changes to lack of positive changes was 1.5:1, whereas in medium-BMI_{status}, the opposite was true.

Figures 1 and 2 (EG and CG, respectively) present a clear separation between boys and girls (in both EG and CG) in the co-occurrence of changes in measured parameters. This indicates that SEX differentiation corresponds between categories of factors and variables independently of the intervention program. However, the picture of associations is quite different between EG and CG. In EG, the HIIT intervention response was higher in the male group. Girls more often exhibited a lack of changes in FI and BFP. Those categories of variables were very closely related. Boys from the low category of BMI_{status} more often exhibited positive changes in cardiovascular parameters. The structure and strength of similarities (based on distances between categories) are shown in dendrogram drawn using Ward's method and Euclidean linkage distances. The FI and BFP categories were strongly linked, with the closest connection among all variables. This confirms that improvement in cardiorespiratory fitness went side by side with improvement (decrease) in the percentage of body fat. Distances between categories in a male cluster were larger than those in female clusters.

In CG, there were no clear connections, although categories of changes in boys and girls were similarly separated (Figure 2). Associations in the boy group were stronger, with closer correspondences than those in the girl group. Girls more often noted positive changes in SBP and BFP, which were closely connected with to a lack of changes in FI. Girls from the medium category of BMI_{status} more often noted a lack of DBP changes. Boys from the high category of BMI_{status} more often exhibited positive changes in DBP. Boys in the low category of BMI_{status} exhibited positive changes in FI, which corresponded with a lack of changes in SBP. The taxonomical dendrogram shows a more tight but less coherent structure than that of EG. Distances between categories of the variables in boys were shorter than those in girls. The closest linkage was noted for positive changes in boys from the high category of BMI_{status} and in boys from the low category of BMI_{status} for the interrelationship between lack of changes in SBP and positive changes in FI.

4. Discussion

The main findings were that 10-week HIIT implemented in PE lessons improved resting blood pressure and cardiorespiratory fitness and slightly reduced body fat, which was reflected in the prevalence of positive changes in EG compared to CG. Sex moderated the impact of HIIT in such a way that positive changes were more often noted in boys than in girls. The factor of BMI_{status} interacted with HIIT differently in boys and girls. The most positive changes were observed in boys from the low category of BMI_{status}. In contrast, girls from the medium category of BMI_{status} exhibited the least positive changes. Secondly, a co-occurrence between some categories of variables was observed. Positive changes in FI were strongly related to positive changes in BFP, although mainly in boys, whereas the same co-occurrence related to FI and BFP but regarding a lack of positive changes was observed in girls.

Log-linear analysis allowed us to look deeper into the relationship between two factors potentially moderating INT effects. This is a unique approach compared to the use of simple independence chi-squared tests, which is mostly presented in the literature. Therefore, it is difficult to compare own results with those reported by others.

HIIT intervention has a broad influence on body composition and physical fitness. Our previous study [23] reported a significant decrease in mean values of body weight and BFP in response to HIIT implemented in physical education lessons. This effect was observed only in overweight subjects. Moreover, improvement in aerobic capacity was also observed. However, this effect was observed only among boys, which suggests sex as a factor differentiating HIIT effects. These results are convergent with the observations of other authors. In similar settings (PE lessons), Bogataj et al. [40] reported a positive impact of HIIT concerning body composition, with simultaneous improvement in physical fitness among obese girls. However, this effect was supported by additional nutrition intervention.

On the other hand, HIIT intervention effectively reduced waist and abdominal circumference directly associated with body fat [41]. Additionally, HIIT intervention is effective in cardiovascular improvement due to observation concern reduction of endothelial damage, which precedes atherosclerosis. A similar observation was noted by Tjonna et al. [42] independent of sex. Moreover, these authors noted improvement in blood pressure supported by improvement in metabolic parameters and body composition. In terms of cardiovascular status, a positive change in blood pressure was also observed in a study by Farah et al. [43].

Similarly to a previously cited study, a decrease in body weight was observed, with a simultaneous reduction in systolic and diastolic blood pressure among adolescents after HIIT. Sex is a factor that may affect HIIT effects. The literature often meets the approach with no respect to this factor, which may cause differentiative HIIT effects, as our observation confirms. This is supported by the results presented by Martinez-Vizcaino et al. [44]. Among boys, decreased body fat with free fat mass was noted. Reduce body fat was reported in girls, as well as a reduction in cardiometabolic risk through improvement in the blood lipid profile.

The observation mentioned above indicates simultaneous changes in body composition, cardiovascular parameters, and physical fitness in response to HIIT contributes evidence for interactions between these factors concerning sex and BMI_{status}. Moreover, accessed studies analyze the changes quantitatively. There is no analysis concerning the prevalence of HIIT effects in studied subjects. To our knowledge, our current research is the first with to employ this approach.

Contrary to log-linear analysis, conducted to identify interactions between factors affecting a single outcome, MCA was used to assess multidimensional associations between all categories of characteristics and outcomes. A clear pattern of correspondence between categories was discovered in EG. HIIT involved parallel changes in FI and BFP, but SEX moderated that co-occurrence. Boys were closely related to positive changes, but girls lacked positive changes. An interaction effect of INT and BMI_{status} was discovered as close a relationship between positive changes in SBP and DBP in boys from the low BMI_{status} category. Other interaction effects were associations of the lack of positive changes in resting BP parameters in girls, mainly from the medium of BMI_{status} category. There was no clear pattern in the control group participating in a standard PE program. Changes were multidirectional and independent of SEX and BMI_{status} factors.

The literature presents results of the exercise training influence on the cardiovascular risk factor and cardiorespiratory fitness [45–47]. Many results revealed changes in body fat, cardiovascular parameters, or cardiorespiratory fitness separately and didn't link associations in changes. Delgado-Floody et al. [28] first evaluated blood pressure changes and cardiorespiratory fitness improvements. No relationships were observed when testing the association between potential improvements in cardiorespiratory fitness and blood pressure improvements after the intervention. The same study showed decreases in percentage body fat ($\Delta-1.38\%$) in the experimental group, as well as SBP decreases ($\Delta-8.70$ mmHg). Our results are slightly different, showing parallel changes in FI and BFP (similar to observations by Delgado-Floody [28]) but no relationship between BFP and BP (nor SBP and DBP). In our study, the relationship between changes in SBP and DBP was quite different in boys and girls.

Interestingly, boys from the low category of BMI_{status} benefited more from the intervention than others. It is worth noting that a reduction in systolic BP is related to a 7% decrease in cardiovascular disease risk [48]. Similarly, positive changes in CRF linked to BFP can be associated with health and a decrement of 15% in risk of cardiovascular disease in healthy boys, as documented by Kodama et al. [49].

Sex differences in positive changes between males and females in BP and CRF were observed in previous studies. Burgomaster et al. [50] revealed no change in VO_2 max after HIIT, similar to earlier data [51] in men completing sprint training for eight weeks. In contrast, females exhibited changes in cardiorespiratory fitness [52]. Our results are contrary in the case of a frequency of girls compared to boys exhibiting no positive changes in either CRF or BFP and in both cardiovascular functions. Related results were obtained by Astorino et al. [53] concerning CRF and BP

The mechanism explaining positive changes in CRF, together with BFP in boys and opposed results in girls, as well as changes in resting blood pressure, is beyond the scope of this study. However, such a mechanism could be supposed to be associations with cardiac functions or O_2 pulse, different in both sexes. It is difficult to compare our own results with those of others because no studies, to our best knowledge, have examined differences in proportions of individuals with positive or a lack of positive changes after HIIT intervention. What's more, in our work, we presented co-occurrences between categories of factors and outcomes compared to multidimensional associations in quantitative data. No such analyses have been conducted until now.

Practical application refers to the recommendation to implement HIIT in PE lessons. We reported that regular 14 min intensive exercises (i.e., Tabata protocol) reduce health risk, particularly in boys (more prevalent changes), improving fitness and decreasing body fat and resting blood pressure. Measures related to the program were collected once a week.

A limitation of this study is the lack of prepubertal and peripubertal groups. Although studied groups of boys and girls were homogeneous in terms of age, adolescents who have finished their pubertal period could be at an advantage. Sexual maturation can significantly affect metabolic outcomes, so considering the effect of puberty on metabolism is vital for the validity of the results [19]. Controlling for sexual maturation (e.g., calculating maturity offset) would be a good solution. The second limitation was the lack of nutritional aspects, including investigating the effectiveness of interval protocols included in PE classes. Our study was limited to a once per week for a 10-week intervention during the school term. These conditions were forced by a need to simultaneously implement a standard PE program. Moreover, some limitations could be identified in a duration (14-min) intervention for one 45 min lesson. There was a need to conduct a warmup before HIIT and a cooldown after. On the other hand, due to possible side effects associated with the high-intensity effort with a higher volume of intervention, such an intervention could increase the risk of some side effects adolescents' aversion to participating in the study. There is a need for follow-up studies to assess the durability of changes. Cardiorespiratory fitness maximal oxygen uptake (VO_2 max) should be measured more objectively than can be achieved with the Harvard step-test, which has some limitations and does not assess CRF very precisely.

5. Conclusions

Our results bolster the importance of screening the effects of intervention in PE lessons in terms of percent of body fat, cardiovascular parameters, and cardiorespiratory fitness (which all are cardiovascular disease risk factors) not only as quantitative changes but also as the frequency of adolescents who benefit greatly from HIIT intervention comparing to peers participating in a standard PE program.

Positive changes are dependent on sex. Boys more often gained from intervention programs than girls, suggesting boys are more sensitive to HIIT influence. Interaction between BMI_{status} factor and intervention was observed within the male group. Boys with low BMI may benefit the most from HIIT intervention. Close associations were also revealed between positive changes in measured parameters: FI, BFP, and BP. This confirms the broad effect of HIIT as a valuable and effective tool to implement in PE lessons. However, there is a need to look for more efficient methods for girls, among whom fewer positive changes were noted.

The usefulness of multidimensional methods was confirmed. Log-linear models were efficient in investigating interactions between factors and outcomes. After applying MCA, the multidimensional space of categorical data may be efficiently reduced to less-dimensional space (mostly two-dimensional). Moreover, it is possible to study multidimensional relationships, called co-occurrence between categories of factors. Calculating distances using cluster analysis allowed us to assess the structure of linkages and similarities between groups of factors and dependent variable categories. The information obtained in this way might be a valuable clue to take practical action to decrease or increase the severity of the studied phenomena.

The conclusions mentioned above could be helpful for scientists studying the effects of interventions on cardiovascular disease risk factors. Interpreting the results from a public health point of view, inherent associations between the investigated variables can be a road map for authorities to plan their policies more efficiently.

Subsequent studies should focus on the prevalence of responders and non-responders in terms of body fat, blood pressure, and cardiorespiratory fitness after high-intensity interval training in adolescents.

Author Contributions: Conceptualization, J.D. and M.P.; methodology, J.D., M.P. and D.K.; software, J.D. and M.P.; validation, J.D; formal analysis, J.D. and M.P.; investigation, J.D. and M.P.; data curation, D.K., J.D. and M.P.; writing—original draft preparation and editing, J.D. and M.P.; writing—review, D.K. and M.P. All authors have read and agreed to the published version of the manuscript.

Funding: This research received no external funding.

Institutional Review Board Statement: The study was conducted following the ethical principles for medical research involving human subjects in the Declaration of Helsinki by the World Medical Association. Additionally, the study met the "ethical standards in sport and exercise science research" [54]. The Ethics Committee of Wroclaw University of Health and Sport Sciences approved the study (ECUPE No. 33/2018).

Informed Consent Statement: Study participants and their guardians were fully informed about the details of the study and gave written consent to confirm their willingness to participate in the study. Every subject could quit at any time without providing a reason.

Data Availability Statement: The data presented in this study are available upon request from the corresponding author.

Conflicts of Interest: The authors declare no conflict of interest.

References

1. Högström, G.; Nordström, A.; Nordström, P. High aerobic fitness in late adolescence is associated with a reduced risk of myocardial infarction later in life: A Nationwide Cohort Study in Men. *Eur. Heart J.* **2014**, *35*, 3133–3140. [CrossRef] [PubMed]
2. Ortega, F.B.; Ruiz, J.R.; Castillo, M.J.; Sjöström, M. Physical fitness in childhood and adolescence: A Powerful Marker of Health. *Int. J. Obes.* **2008**, *32*, 1–11. [CrossRef] [PubMed]
3. Ng, M.; Fleming, T.; Robinson, M.; Thomson, B.; Graetz, N.; Margono, C.; Mullany, E.C.; Biryukov, S.; Abbafati, C.; Abera, S.F.; et al. Global, regional, and national prevalence of overweight and obesity in children and adults during 1980–2013: A Systematic Analysis for the Global Burden of Disease Study 2013. *Lancet* **2014**, *384*, 766–781. [CrossRef]
4. Raghuveer, G.; Hartz, J.; Lubans, D.R.; Takken, T.; Wiltz, J.L.; Mietus-Snyder, M.; Perak, A.M.; Baker-Smith, C.; Pietris, N.; Edwards, N.M.; et al. Cardiorespiratory Fitness in Youth: An Important Marker of Health: A Scientific Statement from the American Heart Association. *Circulation* **2020**, *142*, e101–e118. [CrossRef] [PubMed]
5. Tomkinson, G.R.; Lang, J.J.; Tremblay, M.S. Temporal trends in the cardiorespiratory fitness of children and adolescents representing 19 high-income and upper middle-income countries between 1981 and 2014. *Br. J. Sports Med.* **2019**, *53*, 478–486. [CrossRef]
6. Falkner, B. Hypertension in children and adolescents: Epidemiology and Natural History. *Pediatric Nephrol.* **2010**, *25*, 1219–1224. [CrossRef]
7. Expert Panel on Integrated Guidelines for Cardiovascular Health and Risk Reduction in Children and Adolescents, & National Heart, Lung, and Blood Institute. Expert panel on integrated guidelines for cardiovascular health and risk reduction in children and adolescents: Summary Report. *Pediatrics* **2011**, *128*, S213–S256. [CrossRef]
8. Hallal, P.C.; Andersen, L.B.; Bull, F.C.; Guthold, R.; Haskell, W.; Ekelund, U.; Lancet Physical Activity Series Working Group. Global physical activity levels: Surveillance Progress, Pitfalls, and Prospects. *Lancet* **2012**, *380*, 247–257. [CrossRef]
9. World Health Organization. *Global Recommendations on Physical Activity for Health*; WHO: Geneva, Switzerland, 2010.
10. Charzewski, J.; Lewandowska, J.; Piechaczek, H.; Syta, A.; Łukaszewska, L. Kontrasty społeczne rozwoju somatycznego i aktywności fizycznej dzieci 13–15-letnich. *Studia I Monogr. AWF W Warszawie* **2003**, *97*, 117.
11. Halbert, J.A.; Silagy, C.A.; Finucane, P.; Withers, R.T.; Hamdorf, P.A.; Andrews, G.R. The effectiveness of exercise training in lowering blood pressure: A Meta-Analysis of Randomised Controlled Trials of 4 Weeks or Longer. *J. Hum. Hypertens.* **1997**, *11*, 641–649. [CrossRef]
12. Filho, N.S.; Reuter, C.P.; Silveira, J.F.; Borfe, L.; Renner, J.D.; Pohl, H.H. Low performance-related physical fitness levels are associated with clustered cardiometabolic risk score in schoolchildren: A Cross-Sectional Study. *Hum. Mov.* **2022**, *23*, 113–119. [CrossRef]
13. De Lima, T.R.; Silva, D.A.S. Association between sports practice and physical education classes and lifestyle among adolescents. *Hum. Mov.* **2019**, *20*, 48–55. [CrossRef]
14. Costigan, S.A.; Eather, N.; Plotnikoff, R.C.; Taaffe, D.R.; Lubans, D.R. High-intensity interval training for improving health-related fitness in adolescents: A Systematic Review and Meta-Analysis. *Br. J. Sports Med.* **2015**, *49*, 1253–1261. [CrossRef] [PubMed]
15. Heath, G.W.; Parra, D.C.; Sarmiento, O.L.; Andersen, L.B.; Owen, N.; Goenka, S.; Montes, F.; Brownson, R.C.; Lancet Physical Activity Series Working Group. Evidence-based intervention in physical activity: Lessons from Around the World. *Lancet* **2012**, *380*, 272–281. [CrossRef]
16. Carson, V.; Rinaldi, R.L.; Torrance, B.; Maximova, K.; Ball, G.D.; Majumdar, S.R.; Plotnikoff, R.C.; Veugelers, P.; Boulé, N.G.; Wozny, P.; et al. Vigorous physical activity and longitudinal associations with cardiometabolic risk factors in youth. *Int. J. Obes.* **2014**, *38*, 16–21. [CrossRef]
17. Hay, J.; Maximova, K.; Durksen, A.; Carson, V.; Rinaldi, R.L.; Torrance, B.; Ball, G.D.; Majumdar, S.R.; Plotnikoff, R.C.; Veugelers, P.; et al. Physical activity intensity and cardiometabolic risk in youth. *Arch. Pediatrics Adolesc. Med.* **2012**, *166*, 1022–1029. [CrossRef]
18. Foster, C.; Farland, C.V.; Guidotti, F.; Harbin, M.; Roberts, B.; Schuette, J.; Tuuri, A.; Doberstein, S.T.; Porcari, J.P. The Effects of High Intensity Interval Training vs Steady State Training on Aerobic and Anaerobic Capacity. *J. Sports Sci. Med.* **2015**, *14*, 747–755.

19. Camacho-Cardenosa, A.; Brazo-Sayavera, J.; Camacho-Cardenosa, M.; Marcos-Serrano, M.; Timón, R.; Olcina, G. Effects of High Intensity Interval Training on Fat Mass Parameters in Adolescents. Efectividad de un protocolo de entrenamiento interválico de alta intensidad en adolescentes sobre masa grasa corporal. *Rev. Esp. De Salud Publica* **2016**, *90*, e1–e9.
20. Ouerghi, N.; Fradj, M.; Bezrati, I.; Khammassi, M.; Feki, M.; Kaabachi, N.; Bouassida, A. Effects of high-intensity interval training on body composition, aerobic and anaerobic performance and plasma lipids in overweight/obese and normal-weight young men. *Biol. Sport* **2017**, *34*, 385–392. [CrossRef]
21. Heydari, M.; Freund, J.; Boutcher, S.H. The effect of high-intensity intermittent exercise on body composition of overweight young males. *J. Obes.* **2012**, *12*, 480467. [CrossRef]
22. Joyner, M.J.; Wallin, B.G.; Charkoudian, N. Sex differences and blood pressure regulation in humans. *Exp. Physiol.* **2016**, *101*, 349–355. [CrossRef] [PubMed]
23. Domaradzki, J.; Cichy, I.; Rokita, A.; Popowczak, M. Effects of Tabata Training During Physical Education Classes on Body Composition, Aerobic Capacity, and Anaerobic Performance of Under-, Normal- and Overweight Adolescents. *Int. J. Environ. Res. Public Health* **2020**, *17*, 876. [CrossRef] [PubMed]
24. Eddolls, W.; McNarry, M.A.; Stratton, G.; Winn, C.; Mackintosh, K.A. High-Intensity Interval Training Interventions in Children and Adolescents: A Systematic Review. *Sports Med.* **2017**, *47*, 2363–2374. [CrossRef] [PubMed]
25. Martin-Smith, R.; Cox, A.; Buchan, D.S.; Baker, J.S.; Grace, F.; Sculthorpe, N. High Intensity Interval Training (HIIT) Improves Cardiorespiratory Fitness (CRF) in Healthy, Overweight and Obese Adolescents: A Systematic Review and Meta-Analysis of Controlled Studies. *Int. J. Environ. Res. Public Health* **2020**, *17*, 2955. [CrossRef]
26. Buchan, D.S.; Ollis, S.; Young, J.D.; Cooper, S.M.; Shield, J.P.; Baker, J.S. High intensity interval running enhances measures of physical fitness but not metabolic measures of cardiovascular disease risk in healthy adolescents. *BMC Public Health* **2013**, *13*, 498. [CrossRef]
27. Cvetković, N.; Stojanović, E.; Stojiljković, N.; Nikolić, D.; Scanlan, A.T.; Milanović, Z. Exercise training in overweight and obese children: Recreational football and high-intensity interval training provide similar benefits to physical fitness. *Scand. J. Med. Sci. Sports* **2018**, *28*, 18–32. [CrossRef]
28. Delgado-Floody, P.; Latorre-Román, P.; Jerez-Mayorga, D.; Caamaño-Navarrete, F.; García-Pinillos, F. Feasibility of incorporating high-intensity interval training into physical education programs to improve body composition and cardiorespiratory capacity of overweight and obese children: A Systematic Review. *J. Exerc. Sci. Fit.* **2019**, *17*, 35–40. [CrossRef]
29. McLester, C.N.; Nickerson, B.S.; Kliszczewicz, B.M.; McLester, J.R. Reliability and Agreement of Various InBody Body Composition Analyzers as Compared to Dual-Energy X-Ray Absorptiometry in Healthy Men and Women. *J. Clin. Densitom.* **2020**, *23*, 443–450. [CrossRef]
30. Bajaj, A.; Appadoo, S.; Bector, C.; Chandra, S. Measuring physical fitness and cardiovascular efficiency using harvard step test approach under fuzzy environment. *ASAC* **2008**, *29*, 114.
31. Burnstein, B.D.; Steele, R.J.; Shrier, I. Reliability of fitness tests using methods and time periods common in sport and occupational management. *J. Athl. Train.* **2011**, *46*, 505–513. [CrossRef]
32. Dobosz, M. Computer aided statistical analysis of test results. *Akad. Oficyna Wydawnicza ELIT* **2004**, *11*, 364–375.
33. Daniel, A.; Freeman, J. *Applied Categorical Data Analysis*; Marcel Decker: New York, NY, USA, 1987.
34. Burke, P.J.; Knoke, D. *Log-Linear Models, Sage University Paper Series on Quantitative Applications in Social Sciences*; no. 07-020; Sage Publications: Beverly Hills, CA, USA; London, UK, 1986.
35. Stanisz, A. *Przystępny kurs statystyki z zastosowaniem Statistica PL na przykładach z medycyny, Analizy wielowymiarowe*; StatSoft: Kraków, Poland, 2007.
36. Van der Heijden, P.; De Falguerolles, A.; De Leeuw, J. A combined approach to contingency table analysis using correspondence analysis and log-linear analysis. *App. Stat.* **1989**, *38*, 249–292. [CrossRef]
37. Greenacre, M.J. *Correspondence Analysis in Practice*; Academic Press: London, UK, 1993.
38. Greenacre, M.J.; Blasius, J. (Eds.) *Multiple Correspondence Analysis and Related Methods*; Chapman and Hall/CRC: Boca-Raton, FL, USA, 2006.
39. Nyman, H.; Pensar, J.; Koski, T.; Corander, J. Context-specific independence in graphical log-linear models. *Comput. Stat.* **2016**, *31*, 1493–1512. [CrossRef]
40. Bogataj, Š.; Trajković, N.; Cadenas-Sanchez, C.; Sember, V. Effects of School-Based Exercise and Nutrition Intervention on Body Composition and Physical Fitness in Overweight Adolescent Girls. *Nutrients* **2021**, *13*, 238. [CrossRef]
41. Da Silva, M.R.; Waclawovsky, G.; Perin, L.; Camboim, I.; Eibel, B.; Lehnen, A.M. Effects of high-intensity interval training on endothelial function, lipid profile, body composition and physical fitness in normal-weight and overweight-obese adolescents: A Clinical Trial. *Physiol. Behav.* **2020**, *213*, 112728. [CrossRef]
42. Tjønna, A.E.; Stølen, T.O.; Bye, A.; Volden, M.; Slørdahl, S.A.; Odegård, R.; Skogvoll, E.; Wisløff, U. Aerobic interval training reduces cardiovascular risk factors more than a multitreatment approach in overweight adolescents. *Clin. Sci.* **2009**, *116*, 317–326. [CrossRef]
43. Farah, B.Q.; Ritti-Dias, R.M.; Balagopal, P.B.; Hill, J.O.; Prado, W.L. Does exercise intensity affect blood pressure and heart rate in obese adolescents? A 6-month multidisciplinary randomized intervention study. *Pediatric Obes.* **2014**, *9*, 111–120. [CrossRef]

44. Martínez-Vizcaíno, V.; Sánchez-López, M.; Notario-Pacheco, B.; Salcedo-Aguilar, F.; Solera-Martínez, M.; Franquelo-Morales, P.; López-Martínez, S.; García-Prieto, J.C.; Arias-Palencia, N.; Torrijos-Niño, C.; et al. Gender differences on effectiveness of a school-based physical activity intervention for reducing cardiometabolic risk: A Cluster Randomized Trial. *Int. J. Behav. Nutr. Phys. Act.* **2014**, *11*, 154. [CrossRef]
45. Pescatello, L.S.; MacDonald, H.V.; Lamberti, L.; Johnson, B.T. Exercise for Hypertension: A Prescription Update Integrating Existing Recommendations with Emerging Research. *Curr. Hypertens. Rep.* **2015**, *17*, 87. [CrossRef]
46. Olea, M.A.; Mancilla, R.; Martínez, S.; Díaz, E. Entrenamiento interválico de alta intensidad contribuye a la normalización de la hipertensión arterial. *Rev. Méd. Chile* **2017**, *145*, 1154–1159. [CrossRef]
47. Ramírez-Vélez, R.; Hernández-Quiñones, P.A.; Tordecilla-Sanders, A.; Álvarez, C.; Ramírez-Campillo, R.; Izquierdo, M.; Correa-Bautista, J.E.; Garcia-Hermoso, A.; Garcia , R.G. Effectiveness of HIIT compared to moderate continuous training in improving vascular parameters in inactive adults. *Lipids Health Dis.* **2019**, *18*, 981. [CrossRef]
48. Lewington, S.; Clarke, R.; Qizilbash, N.; Peto, R.; Collins, R.; Prospective Studies Collaboration. Age-specific relevance of usual blood pressure to vascular mortality: A Meta-Analysis of Individual Data for One Million Adults in 61 Prospective Studies. *Lancet* **2002**, *360*, 1903–1913. [CrossRef]
49. Kodama, S.; Saito, K.; Tanaka, S.; Maki, M.; Yachi, Y.; Asumi, M.; Sugawara, A.; Totsuka, K.; Shimano, H.; Ohashi, Y.; et al. Cardiorespiratory fitness as a quantitative predictor of all-cause mortality and cardiovascular events in healthy men and women: A Meta-Analysis. *JAMA* **2009**, *301*, 2024–2035. [CrossRef]
50. Burgomaster, K.A.; Howarth, K.R.; Phillips, S.M.; Rakobowchuk, M.; Macdonald, M.J.; McGee, S.L.; Gibala, M.J. Similar metabolic adaptations during exercise after low volume sprint interval and traditional endurance training in humans. *J. Physiol.* **2008**, *586*, 151–160. [CrossRef]
51. Thorstensson, A.; Sjödin, B.; Karlsson, J. Enzyme activities and muscle strength after "sprint training" in man. *Acta Physiol. Scand.* **1975**, *94*, 313–318. [CrossRef]
52. Talanian, J.L.; Galloway, S.D.; Heigenhauser, G.J.; Bonen, A.; Spriet, L.L. Two weeks of high-intensity aerobic interval training increases the capacity for fat oxidation during exercise in women. *J. Appl. Physiol.* **2007**, *102*, 1439–1447. [CrossRef]
53. Astorino, T.A.; Allen, R.P.; Roberson, D.W.; Jurancich, M. Effect of high-intensity interval training on cardiovascular function, VO2max, and muscular force. *J. Strength Cond. Res.* **2012**, *26*, 138–145. [CrossRef]
54. Harriss, D.J.; Atkinson, G. Ethical Standards in Sport and Exercise Science Research: 2016 Update. *Int. J. Sports Med.* **2015**, *36*, 1121–1124. [CrossRef]

biology

MDPI

Article

The Acute Effects of Normobaric Hypoxia on Strength, Muscular Endurance and Cognitive Function: Influence of Dose and Sex

Raci Karayigit [1,*], Mustafa Can Eser [1], Fatma Nese Sahin [1], Cengizhan Sari [2], Angela Sanchez-Gomez [3], Raul Dominguez [4] and Mitat Koz [5]

1 Faculty of Sport Sciences, Ankara University, Gölbaşı, Ankara 06830, Turkey; mceser@ankara.edu.tr (M.C.E.); nesesahin@ankara.edu.tr (F.N.S.)
2 Faculty of Sport Sciences, Muş Alparslan University, Muş 49001, Turkey; c.sari@alparslan.edu.tr
3 Department of Nursing Pharmacology and Physiotherapy, Faculty of Medicine and Nursing, University of Córdoba, 14000 Córdoba, Spain; asgomez@uco.es
4 Departamento de Motricidad Humana y Rendimiento Deportivo, Universidad de Sevilla, 41013 Sevilla, Spain; rdherrera@us.es
5 Physiotherapy and Rehabilitation Department, Faculty of Health Sciences, Eastern Mediterranean University, North Cyprus, Mersin, Famagusta 99628, Turkey; mitat.koz@emu.edu.tr
* Correspondence: rkarayigit@ankara.edu.tr; Tel.: +90-312-600-0100

Citation: Karayigit, R.; Eser, M.C.; Sahin, F.N.; Sari, C.; Sanchez-Gomez, A.; Dominguez, R.; Koz, M. The Acute Effects of Normobaric Hypoxia on Strength, Muscular Endurance and Cognitive Function: Influence of Dose and Sex. *Biology* **2022**, *11*, 309. https://doi.org/10.3390/biology11020309

Academic Editors: Filipe Manuel Clemente, Georgian Badicu and Eugenia Murawska-Ciałowicz

Received: 23 December 2021
Accepted: 14 February 2022
Published: 15 February 2022

Simple Summary: Hypoxic training is a novel method to increase resistance training adaptations. There is some evidence that resistance exercise performed in systemic hypoxia can lead to structural and functional adaptations of skeletal muscle. Studies have also demonstrated that normobaric hypoxia (i.e., normal pressure, low oxygen) increases intramuscular metabolic stress and type two (fast twitch) fiber recruitment, which leads to a greater morphological adaptations over time. However, to date, there is no research that has investigated the effects of different doses of acute normobaric hypoxia on strength and muscular endurance performance, nor has there been research investigating potential sex-based differences.

Abstract: The aim of this study was to examine the acute effects of different levels of hypoxia on maximal strength, muscular endurance, and cognitive function in males and females. In total, 13 males (mean ± SD: age, 23.6 ± 2.8 years; height, 176.6 ± 3.9 cm; body mass, 76.6 ± 2.1 kg) and 13 females (mean ± SD: age, 22.8 ± 1.4 years; height, 166.4 ± 1.9 cm; body mass, 61.6 ± 3.4 kg) volunteered for a randomized, double-blind, crossover study. Participants completed a one repetition strength and muscular endurance test (60% of one repetition maximum to failure) for squat and bench press following four conditions; (i) normoxia (900 m altitude; F_iO_2: 21%); (ii) low dose hypoxia (2000 m altitude; F_iO_2: 16%); (iii) moderate dose hypoxia (3000 m altitude; F_iO_2: 14%); and (iv) high dose hypoxia (4000 m altitude; F_iO_2: 12%). Heart rate, blood lactate, rating of perceived exertion, and cognitive function was also determined during each condition. The one repetition maximum squat ($p = 0.33$) and bench press ($p = 0.68$) did not differ between conditions or sexes. Furthermore, squat endurance did not differ between conditions ($p = 0.34$). There was a significant decrease in bench press endurance following moderate ($p = 0.02$; $p = 0.04$) and high ($p = 0.01$; $p = 0.01$) doses of hypoxia in both males and females compared to normoxia and low dose hypoxia, respectively. Cognitive function, ratings of perceived exertion, and lactate were also significantly different in high and moderate dose hypoxia conditions compared to normoxia ($p < 0.05$). Heart rate was not different between the conditions ($p = 0.30$). In conclusion, high and moderate doses of acute normobaric hypoxia decrease upper body muscular endurance and cognitive performance regardless of sex; however, lower body muscular endurance and maximal strength are not altered.

Keywords: hypoxic dose; resistance training; muscular endurance; sex difference; flanker

1. Introduction

Coaches and athletes are always looking for innovative training methods to gain an advantage. Since the 1968 Mexico Olympic games, hypoxic training has become popular especially among endurance athletes using Live High/Train Low (LHTL) or Live High/Train High (LHTH) strategies [1]. Beyond traditional LHTL or LHTH paradigms, a novel method of simulated altitude training entitled Live Low/Train High (LLTH) is becoming more popular for athletes [2]. Such an approach has been used to increase sea level exercise performance [3]. Recently, LLTH has been shown to augment physiological adaptations following resistance exercise, possibly via alterations in metabolic stress and greater intramuscular responses [4,5]. However, it is important to note that there is limited research examining responses to hypoxia following resistance exercise compared to aerobic or high intensity interval training.

Acute exposure to hypoxia may negatively alter resistance exercise performance. Mechanistically, these negative effects may be associated with diminished muscular and arterial oxygenation, electromyographic activity [6], neuromuscular activation [7], peak velocity [8], and increased expiratory parameters [8]. However, from a physiological perspective, the acute metabolic and neuromuscular stressors may augment hypertrophy over time [4,9–12]. Gains in muscle mass appear to be due to an enhancement in the recruitment of type II fibers, since type I fibers fatigue more quickly in a hypoxic environment [11]. Further, hypoxia-inducible factor-1 (HIF-1), a transcription factor, is upregulated, and is important to promote a slow to fast fiber-type transition in skeletal muscle [13]. However, the most probable mechanism by which hypoxia augments resistance training responses is due to elevated metabolic stress [4,9], leading to increases in motor unit recruitment [4]. The accumulation of metabolites [14,15] increases plasma growth hormone [16] and muscle cell swelling [10], which may be associated with elevated muscle protein synthesis.

In theory, the altered physiological and metabolic responses and acute performance changes may vary according to the dose of hypoxia. Presently, there is a lack of research investigating the optimal hypoxic dose required during strength training [17]. Campo et al. [18,19] reported that peak-mean power, RPE, and blood lactate during a resistance circuit training was significantly different in high hypoxia (F_iO_2 = 0.13; ~3.800) compared to moderate hypoxia (F_iO_2 = 0.16; ~2.100 m) and normoxia. In contrast, two studies [4,20] did not observe differences in force and muscular power during resistance exercise between high hypoxia (F_iO_2 = 0.13), moderate hypoxia (F_iO_2 = 0.16), and normoxia. Nonetheless, effort during resistance exercise could be greater in normobaric hypoxia than normoxia [21], which would influence performance. However, fixed number of repetitions was used in the test protocols of the aforementioned studies which hinder participants from reaching muscular failure. Further, resistance training to muscle failure could potentially increase hypertrophic adaptations by enhancing exercise-induced metabolic stress [22]. To date, only a few studies have investigated the effects of acute high and moderate hypoxia on leg extension [23], bench press [24], and biceps curl [25] exercise to failure. However, results were equivocal with studies showing either a negative impact of hypoxia [25] or no effect [23,24]. Further, these studies were only conducted on males [23–25] and untrained subjects [24].

Although relatively little is known about the acute responses to resistance exercise between sexes, females were reported to be more resistant to fatigue and faster to recover from fatiguing exercise than males in tasks utilizing low intensity loads and a slow repetition velocity in both concentric and eccentric phases [26]. Presently, there is no research examining sex-based responses to resistance exercise in hypoxia, which is warranted due to performance differences [27]. Females, compared to males, have greater fatigue resistance during training, and enhanced recovery, despite higher cardiovascular strain and RPE [28,29]. Females were also found to have lower lactate levels during hypoxic exercise (F_iO_2: 0.13), as well as higher glucose levels during recovery [30]. As such, the effect of hypoxia may differ in males and females.

Cerebral oxygen desaturation was significantly associated with impaired cognitive function (CF) [31]. As the severity of hypoxia increases, the level of deoxygenation increases. Moreover, two recent meta-analyses reported that attention, a parameter of complex cognitive processes, moderates acute muscular endurance [32] and strength [33] performance. However, currently there are no studies that have investigated CF before and after resistance exercise during hypoxia.

Therefore, the aim was to examine the acute effects of different doses of normobaric hypoxia (i.e., low [2000 m], moderate [3000 m], and high [4000 m]) on maximal strength, muscular endurance, and CF according to sex utilizing resistance exercise performed over multiple sets and to failure. We hypothesized that as the dose of hypoxia increased, maximal strength, muscular endurance, and CF will concomitantly decrease, while blood lactate and RPE will increase.

2. Materials and Methods

2.1. Participants

For this study, 13 males (mean ± SD: age, 23.6 ± 2.8 years; height, 176.6 ± 3.9 cm; body mass, 76.6 ± 2.1 kg) and 13 females (mean ± SD: age, 22.8 ± 1.4 years; height, 166.4 ± 1.9 cm; body mass, 61.6 ± 3.4 kg) who were healthy and were non-smokers volunteered to participate. The inclusion criteria included: (a) free from neuromuscular and musculoskeletal disorders, aged 18–30 years; and (b) able to perform a successful back squat and bench press exercise with load corresponding to 125% and 100% of their current body mass, respectively. Furthermore, participants had at least three years of resistance training experience, underwent training four times per week (which included squat and bench press in their routines), and were considered intermediately resistance trained [34]. All participants reported no previous exposure to an altitude of greater than 900 m within the last eight months, and were taking no substances that could affect the muscular and cognitive performance (i.e., creatine, anabolic steroids). Participants were requested to refrain from exercise, alcohol, and caffeine intake for 24 h before each session in order to maintain their customary sleep, to complete a 24 h dietary record before the first session, and to replicate this diet for 24 h before each subsequent session to standardize energy intake. Adherence to these instruction was verbally confirmed at the beginning of each session. This study was approved by Muş Alparslan University Scientific Research Ethics Committee (Approval no: 10776-6/31) and conducted in accordance with the Helsinki Declaration. Written informed consent was obtained from all participants prior to beginning the study.

2.2. Study Design

To investigate whether the degree of acute normobaric hypoxia affects strength (1RM), muscular endurance (repetitions to failure with 60% of 1RM), and cognitive function (Flanker Task; reaction time and response accuracy) in males and females, each participant performed the test protocol under four conditions with different O_2 availability: (1) normoxia (NORM; 900 m altitude; F_iO_2: 21%); (2) low hypoxia (LowH; 2000 m altitude; F_iO_2: 16%); (3) moderate hypoxia (ModH; 3000 m altitude; F_iO_2: 14%); and (4) high hypoxia (HighH; 4000 m altitude; F_iO_2: 12%). The study used a randomized, double-blind crossover research design. During each experimental testing session, participants wore a face mask that was connected to a hypoxic generator (Everest Summit II, Hypoxico, New York, NY, USA), which controlled the oxygen availability. Level of altitude was observed on the hypoxic generator's screen, which was hidden from participants to maintain blindness. To verify that hypoxic and normoxic conditions were provided, peripheral oxygen saturation (SpO_2) was measured by a pulse oximeter (Hypoxico Oxycon, New York, NY, USA) attached to the participants' finger. For this study, two familiarization sessions were conducted to ensure that the participants were able to squat and bench press with proper technique with the hypoxic generator mask. The testing sessions were well tolerated, and there were no reported side effects. In total, participants came to the laboratory on six occasions, which were separated by 72 h to allow for complete recovery. The first two sessions were familiarization sessions, which were similar to the experimental sessions, during which the participant's performed the back squat on a smith machine (Esjim, Eskişehir, Turkey) and the bench press exercise on a rack with safety bar. All sessions were supervised by a certified personal trainer. During the familiarization and warm-up, the personal trainer provided technical feedback, then participants completed a 1RM testing protocol and three sets with 60% of 1RM to failure for both back squat and bench press, respectively, while wearing the hypoxic generator's face mask. Bar grip position was recorded and replicated for subsequent sessions. These exercises were chosen to test lower and upper body's major muscle groups and their common inclusion in resistance training programs. Participants were also familiarized with and practiced the cognitive function (CF) test until they achieved consistent scores. Additionally, participants were also introduced to the 6–20 Borg scale [35] for measuring their ratings of perceived exertion (RPE). All sessions took place in the afternoon (14.00–17.00) in order to minimize the diurnal influence on muscle strength. On arrival at the laboratory, resting measures of heart rate (HR) (Polar Team 2, telemetric system, Kempele, Finland), capillary lactate concentration (LAC) (lactate scout, USA), and CF measurements were taken. After 5 min passive rest while wearing hypoxic generator's face mask for familiarization and 5 min warming-up on a treadmill, squat 1RM strength and 3 sets of 60% of 1RM muscular endurance tests were performed, respectively. Following 5 min passive rest, the same procedures were replicated for bench press. Immediately after the muscular test protocols, CF was measured again. HR, LA, RPE, and SpO_2 were recorded at different time points during the testing protocol (Figure 1). Total hypoxia exposure duration was ~40 min.

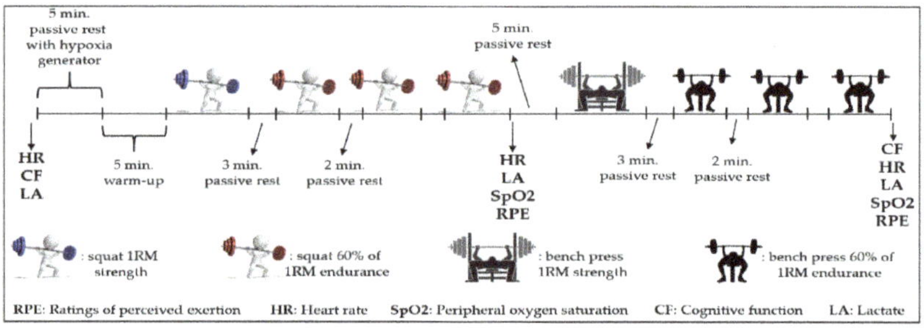

Figure 1. Schematic representation of the experimental sessions.

RM Strength and 60% of 1RM Muscular Endurance Test Protocol

1RM strength performance was identified in three to five steps according to Baechle and Earle protocol [34]. After 10 repetitions with 20 kg weights was completed, participants then rested passively for 1 min, which was followed by a further three to five repetitions with an added 10% and 20% more weight for the bench press and squat exercises, respectively. Following a 2 min passive rest period, a weight near to their predicted 1RM was used, and participants performed two to three repetitions. Following another 3 min of passive rest, the weight was further increased by another 10–20% for the squat and 5–10% for the bench press if the participant successfully lifted the weight. If the lift was unsuccessful, the weight was decreased by 5–10% for the squat and 2.5–5.0% for the bench press for another 1RM attempt. Further, 3 min was given after 1RM establishment and the weight was lowered to 60% of 1RM; thereafter, three sets of muscular endurance tests with 60% of 1RM with a two minutes passive rest period was carried out by the participants. Movement tempo during muscular endurance test was standardized to two seconds for both concentric and eccentric phases via a metronome. Total repetition number to failure was used as a muscular endurance performance. The test protocol was terminated according to three criteria: (1) voluntarily completion the repetition; (2) unable to perform repetitions synchronously with the metronome for three consecutive repetitions; and (3) unable to lift with proper technique and posture.

2.3. Cognitive Function

A modified version of Flanker task was used to measure cognitive performance and run on a Dell Computer via appropriate software (InqusitLab 5.0, Milliseconds) [36]. A yellow fixation star appeared in the center of the screen, followed by five black arrowheads in a line which appeared for 200 milliseconds on a white screen background with a response window of 1000 milliseconds. For this, two congruent (< < < < <) and two incongruent (< < > < < or > > < > >) tasks were provided in equal probability and participants were instructed to react as fast and accurately as possible to the direction of the target arrow (in the middle) by pressing corresponding response buttons with their left or right index fingers. The interstimulus time gap varied from 1000 to 1500 ms, and the trial order was random for each participant. Following 20 practice trials, participants performed 100 trials while wearing earplugs. The total duration of the cognitive performance test was approximately three minutes. Mean response accuracy (%) and reaction time (ms) were recorded to measure cognitive performance.

2.4. Statistical Analysis

All analyses were conducted with IBM SPSS statistics for Windows, version 22.0 (IBM Corp., Armonk, NY, USA). Normal distribution was confirmed by the Shapiro–Wilk test. A three-way repeated measures analysis of variance (ANOVA) was performed to investigate the main effects for (1) condition, (2) sex, and (3) time or sets. Furthermore, three-way ANOVA examined the interaction effect of condition × sex × time or set. A Mauchly's test was used to assess sphericity followed by the Greenhouse–Geisser adjustment, where appropriate. Relevant significant main or interaction effects were further examined in a Bonferroni post-hoc analysis. A 95% confidence interval (CI) was calculated, and partial eta squared (η^2) were reported with significant ANOVA main effects as a measure of effect size as trivial (<0.10), moderate (0.25–0.39) or large ($\geq$0.40) [37]. A two-way mixed model in consistency type of intraclass correlation coefficients (ICC) was calculated to determine the consistency of the four trials. Data is provided as mean ± SD and statistical significance was set at $p < 0.05$.

3. Results

3.1. Maximum Strength (1RM) and 60% of 1RM Muscular Endurance Performance

Hypoxia did not affect the 1RM squat ($p = 0.33$, $\eta^2 = 0.09$) or bench press strength ($p = 0.68$, $\eta^2 = 0.04$) (Figure 2). Although there was no main effect for condition in 60% of 1RM squat endurance ($p = 0.34$, $\eta^2 = 0.09$), it was detected that there was a statistical trend in the third set between HighH and NORM in males ($p = 0.059$) and females ($p = 0.085$), and between ModH and NORM in females ($p = 0.087$). Furthermore, repeated measures with ANOVA detected significant main effect for bench press 60% of 1RM endurance performance ($p = 0.01$, $\eta^2 = 0.28$). The Bonferroni post hoc test revealed that HighH ($p = 0.01$, 95% CI = −0.67–0.07; $p = 0.01$, 95% CI = −0.74–0.10) and ModH ($p = 0.04$, 95% CI = −0.75–0.02; $p = 0.02$, 95% CI = −0.81–0.06) had significantly lower repetition numbers during 60% of 1RM endurance test compared to LowH and NORM. Lastly, no significant condition × sex × set interaction was detected in the squat ($p = 0.69$, $\eta^2 = 0.05$) and bench press ($p = 0.96$, $\eta^2 = 0.01$) with 60% of 1RM endurance performance (Figures 3 and 4). ICC values during three sets of 60% of 1RM squat endurance performance were 0.96 and 0.97 for males and females, respectively. In the bench press with 60% of 1RM endurance, ICC were 0.96 and 0.97 for males and females, respectively.

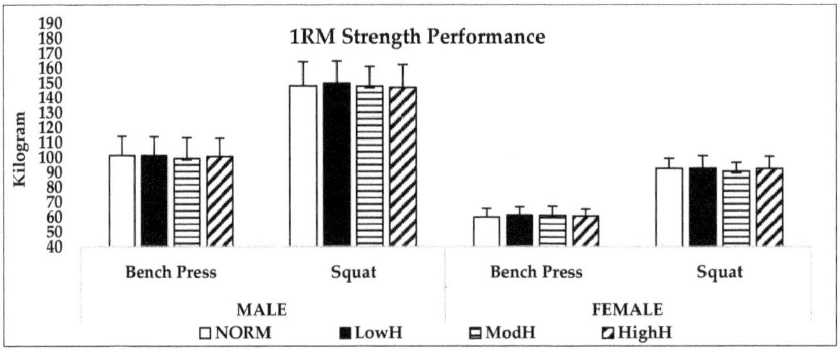

Figure 2. Mean (SD) bench press and squat strength (1-RM) performance. NORM: Normoxia (900 m); LowH: Low dose of hypoxia (2000 m); ModH: moderate dose of hypoxia (3000 m); HighH: high dose of hypoxia (4000 m).

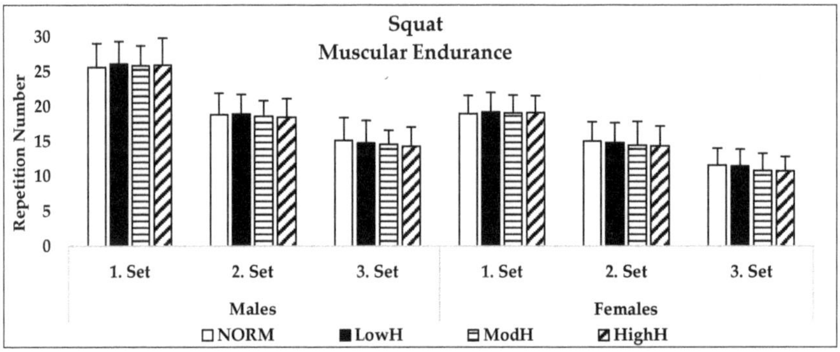

Figure 3. Mean (SD) squat 60% of 1RM endurance performance of males and females. NORM: Normoxia (900 m); LowH: Low dose of hypoxia (2000 m); ModH: moderate dose of hypoxia (3000 m); HighH: high dose of hypoxia (4000 m).

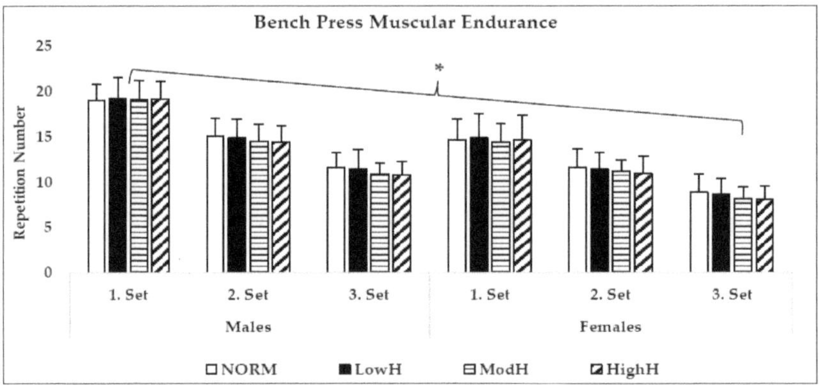

Figure 4. Mean (SD) bench press 60% of 1RM endurance performance of males and females. NORM: Normoxia (900 m); LowH: Low dose of hypoxia (2000 m); ModH: moderate dose of hypoxia (3000 m); HighH: high dose of hypoxia (4000 m). * ModH and HighH were significantly different than NORM and LowH.

3.2. Cognitive Function

Response accuracy for the congruent condition was not significantly different between conditions ($p = 0.56$, $\eta^2 = 0.05$). Further, no significant condition × sex × time interaction ($p = 0.07$, $\eta^2 = 0.18$) was found. However, hypoxia affected reaction time for the congruent condition ($p = 0.03$, $\eta^2 = 0.21$). Post-hoc analysis revealed that NORM was significantly faster than ModH ($p = 0.04$, 95% CI = −15.63–0.35) and HighH ($p = 0.01$, 95% CI = −0.67–0.07).

Response accuracy for the incongruent condition was not significantly different between conditions ($p = 0.09$, $\eta^2 = 0.17$). However, hypoxia affected reaction time for the incongruent condition ($p = 0.01$, $\eta^2 = 0.43$). The Bonferroni analysis demonstrated that NORM was significantly lower than ModH ($p = 0.01$, 95% CI =−22.77–3.86) and HighH ($p = 0.01$, 95% CI = −36.14–10.77). Furthermore, LH was also significantly different than HighH ($p = 0.01$, 95% CI = −29.26–4.85) (Table 1).

Table 1. Cognitive function parameters.

	Males				Females			
	Pre Test		Post Test		Pre Test		Post Test	
	M	SD	M	SD	M	SD	M	SD
	Response Accuracy [%]-Congruent Task							
NORM	96.41	1.4	95.83	1.6	96.58	1.7	96.66	1.3
LowH	96.16	2.3	95.83	2.0	96.41	2.1	97.08	1.3
ModH	96.83	2.3	96.50	1.7	97.16	2.0	96.25	1.9
HighH	96.00	1.8	96.33	1.6	97.33	1.2	95.16	1.6
	Response Accuracy [%]-Incongruent Task							
NORM	93.25	2.8	93.33	2.0	94.58	2.4	95.50	1.7
LowH	92.91	1.9	93.16	2.2	95.08	1.3	94.83	1.4
ModH	93.00	3.1	92.33	2.6	94.50	1.7	94.91	1.2
HighH	93.41	1.8	91.16	2.0	94.50	1.5	94.83	1.8
	Reaction Time [ms]-Congruent Task							
NORM	493.45	42.2	480.77	46.2	533.40	44.3	517.70	47.7
LowH	496.80	38.9	485.53	46.9	541.04	39.1	518.55	30.3
ModH	497.90	42.8	494.21 *	42.6	530.83	42.4	534.35 *	34.8
HighH	491.95	40.0	502.68 *	35.9	531.91	32.7	542.35 *	38.2
	Reaction Time [ms]-Incongruent Task							
NORM	531.81	31.5	516.94	25.3	608.10	49.9	596.34	55.1
LowH	546.70	31.4	528.46 #	34.2	615.10	44.5	588.52 #	45.0
ModH	541.22	27.3	535.80 *	36.6	603.97	51.4	625.48 *	59.1
HighH	542.53	29.7	567.42 *	25.0	607.01	40.3	630.07 *	34.2

NORM: Normoxia (900 m); LowH: Low dose of hypoxia (2000 m); ModH: Moderate dose of hypoxia (3000 m); HighH: High dose of hypoxia (4000 m). *: significantly different than NORM; #: significantly different than HighH.

3.3. Heart Rate, RPE, Lactate and Oxygen Saturation

There was no difference between conditions ($p = 0.30$, $\eta^2 = 0.10$) in heart rate. However, RPE values were significantly different between conditions ($p = 0.01$, $\eta^2 = 0.33$). Post-hoc analysis showed that ModH ($p = 0.02$, 95% CI = 0.10–0.98) and HighH ($p = 0.01$, 95% CI = 0.31–1.51) were significantly higher than NORM. HighH was also significantly higher than LowH ($p = 0.03$, 95% CI = 0.05–1.44). There was a significant main effect on lactate between conditions ($p = 0.01$, $\eta^2 = 0.57$). Post-hoc analysis showed that, after the bench press exercise, HighH was significantly higher than NORM ($p = 0.01$, 95% CI = 0.17–0.45) and LowH ($p = 0.01$, 95% CI = 0.17–0.38). ModH was also significantly higher than NORM ($p = 0.01$, 95% CI = 0.12–0.33) and LowH ($p = 0.01$, 95% CI = 0.11–0.27). Lastly, as expected, oxygen saturation values were significantly different between conditions ($p = 0.01$, $\eta^2 = 0.97$): as hypoxic dose increases, in turn, oxygen saturation decreases (Table 2).

Table 2. Lactate, heart rate, ratings of perceived exertion, and oxygen saturation values.

		NORM		LowH		ModH		HighH	
		Lactate (mmol/L)							
		M	SD	M	SD	M	SD	M	SD
Males	Pretest	1.2	0.1	1.23	0.2	1.14	0.1	1.15	0.1
	Postsquat	6.56	1.1	6.45	1.0	6.54	1.1	6.61	1.0
	Postbench	7.11	1.2	7.23 #	1.0	7.98 *	1.2	7.94 *	0.9
Females	Pretest	1.26	0.1	1.2	0.1	1.11	0.1	1.13	0.1
	Postsquat	5.66	0.7	5.7	0.4	5.83	0.4	5.91	0.7
	Postbench	6.01	0.7	6.13 #	0.8	6.58 *	0.5	6.91 *	0.8
		Heart Rate (beat/min)							
Males	Pretest	64.75	4.9	63.50	5.31	64.33	3.6	64.58	4.8
	Postsquat	173.25	8.6	171.33	5.39	172.33	4.7	171.33	7.4
	Postbench	176.08	8.4	175.08	10.4	175.33	8.1	176.08	7.6
Females	Pretest	64.75	3.5	60.50	3.3	62.66	4.0	64.16	3.2
	Postsquat	159.83	8.7	157.75	8.4	158.41	9.0	157.91	6.4
	Postbench	162.33	7.8	163.50	8.8	160.83	9.7	160.58	9.2
		Ratings of Perceived Exertion (6–20)							
Males	Postsquat	13.41	1.2	13.66 #	1.0	13.58 *	1.1	14.00 *	1.2
	Postbench	15.91	1.9	16.33 #	1.6	17.00 *	2.3	17.66 *	2.1
Females	Postsquat	13.50	1.7	13.83 #	1.5	14.08 *	1.6	13.91 *	1.6
	Postbench	17.41	1.9	17.08 #	2.1	17.75 *	1.9	18.33 *	1.7
		Oxygen Saturation (%)							
Males	Postsquat	94.83	1.9	90.75	3.2	85.16	3.7	80.91	2.3
	Postbench	94.33	1.8	89.58	2.9	83.91	3.8	79.83	2.5
Females	Postsquat	95.08	1.5	90.16	2.7	86.41	2.6	81.83	2.4
	Postbench	94.16	2.2	90.08	2.6	85.75	3.9	80.66	2.2

NORM: Normoxia (900 m); LowH: Low dose of hypoxia (2000 m); ModH: Moderate dose of hypoxia (3000 m); HighH: High dose of hypoxia (4000 m). Pretest: before test protocol; Postsquat: Immediately after three sets of squat endurance; Postbench: Immediately after three sets of bench press endurance *: significantly different than NORM; #: significantly different than HH.

4. Discussion

To our knowledge, no study to date has examined the relationship between the effect of hypoxia dose and sex on acute resistance training performance. As such, the purpose of this study was to examine the acute effects of different doses of normobaric hypoxia on strength, muscular endurance, and cognitive function according to sex. The main findings demonstrate that moderating effects of hypoxia on strength, muscular endurance, and cognitive function did not differ between sex. Further, high (4000 m) and moderate (3000 m) doses of hypoxia significantly altered bench press endurance performance, cognitive function, RPE, and blood lactate levels. However, 1RM strength (both squat and bench press) and squat endurance performance was not influenced by hypoxia.

4.1. Strength (1RM) Performance

Our data indicate that 1RM squat and bench press strength performance was not different between conditions. This finding is consistent with Smith et al. [23], who concluded that moderate (F_iO_2 = 15.4%) and severe (F_iO_2 = 12.9%) hypoxia did not affect 1RM leg extension strength. Another study by Feriche et al. [38] reported that 1RM bench press strength did not show a difference between normobaric hypoxia (F_iO_2 = 15.7%) and normoxia. It seems that hypoxia has no influence on 1RM strength. Moreover, Girard et al. [25] alleged that, at higher training loads, the effects of hypoxia on muscular performance tends to be smaller. On the other hand, the method of strength measurement (1RM) in the current and previous [23,38] studies may be disputable due to not being sensitive enough to detect subtle effects of hypoxia on 1RM strength. It may be asserted that potential negative effects of hypoxia on 1RM strength can be lost in large inter-day 1RM strength variations [39]. In the present study, the ICC values during 1RM strength measurements showed high consistency (0.97–0.99) between conditions. Further studies are needed to examine the influence of hypoxia on strength performance measured via more sensitive devices, such as an isokinetic dynamometer or gauge.

4.2. *Muscular Endurance (60% of 1RM) Performance*

In the current study, high (F_iO_2 = 12%) and moderate (F_iO_2 = 14%) doses of hypoxia had a negative impact on 60% of 1RM bench press repetition to failure performance. However, low (F_iO_2 = 16%) dose hypoxia did not change neither squat nor bench press muscular endurance performance compared to normoxia. Hence, resistance exercise with high or moderate doses of hypoxia might develop more fatigue due to the increase in metabolic stress than the fatigue that develops in low hypoxia or normoxia (Table 2). This fact might be explained by exacerbated perturbations of cellular homeostasis in active skeletal muscles [19]. In this respect, high (F_iO_2 = 13%), but not low (F_iO_2 = 16%), doses of hypoxia increased lactate and diminished blood HCO_3-, causing a reduction in blood pH [19]. This indicates that a high dose of hypoxia relies more on non-oxidative ATP phosphorylation [40] and increases intracellular acidosis, which are known to contribute to muscular fatigue [41].

Decrement in bench press performance with hypoxia in a dose-dependent manner was also shown by Campo et al. [18], who reported that peak and mean force were significantly lower in high dose hypoxia (F_iO_2 = 13%) when compared with low dose hypoxia (F_iO_2 = 16%) and normoxia. However, in the same study [18], a dose effect was not observed in the half squat exercise. Further, dose of hypoxia was previously shown as a moderating factor on muscular performance during high intensity resistance circuit training [19]. In support, a meta-analysis concluded that repetition to failure performance occurs earlier during resistance exercise with high but not low hypoxia [42]. Conversely, high (F_iO_2 = 13%) or moderate (F_iO_2 = 16%) hypoxic stimulus did not alter back squat and deadlift force or power variables [20]. Discrepancies in the type of test protocols (circuit training vs. one/two exercise; at an intensity between 60 vs. 85% 1RM) may vary the influence of various doses of hypoxia on resistance exercise performance.

The repetition to failure test protocol required participants to use maximum effort, thus, level of effort could be greater than protocols which use a fixed number of repetitions [4,18–20,38]. There are only a few studies that have used repetition to failure at a given percentage of 1RM to measure muscular endurance performance in hypoxia. For example, three sets of 70% 1RM leg extension to failure performance did not change between normoxia, moderate (F_iO_2 = 15.4%), and high (F_iO_2 = 12.9%) hypoxia [23]. Secondly, three sets of 75% 1RM to bench press failure performance was not different between normoxia and high hypoxia (F_iO_2 = 13%) [24]. In contrast, in their very well-designed study, Girard et al. [25] demonstrated that hypoxia (F_iO_2 = 12.9%) negatively impacts resistance exercise to failure performance at light (30% of 1RM) loads. The reason for the contrasting results can be elucidated by hypoxia exposure duration. The ergolytic effect of hypoxia was proposed to be dependent on exposure duration [43]. In addition, exposure duration, as a moderating factor, was also found to be significant by a meta-analysis [44] that quantifies the effect of acute hypoxia on exercise performance. They concluded that the negative effect of hypoxia increases in parallel with the prolongation of the exercise protocol [44]. In this regard, we can speculate that studies which found no ergolytic effect on repetition to failure performance may not have had enough hypoxia exposure duration during their exercise protocols (~12–15 min) [23,24]. Therefore, it appears that relatively short durations (~20 min) in hypoxia during squat endurance likely influenced the results in our test protocol. Further, bench press endurance test was performed after squat exercise and was negatively affected by hypoxia, during which exposure time was nearly 40 min.

This finding may not only be due to the morphological and neuromuscular differences between the lower (squat) and upper (bench press) extremities [18]; it may be due to the prolonged exposure to total hypoxia due to the bench press test being performed after the squat test. Contrary to our argument, Walden et al. [45] reported that various exposure durations (20 min vs. 30 min) to hypoxia (F_iO_2 = 13%) had no influence on the bench press and shoulder press endurance. More research is required in this topic to make confirm our speculation.

Females have a greater resistance to hypoxia [46] and recover faster from fatiguing exercise, in contrast to their male counterparts [26]. Latella et al. [47] found that the mechanism in which corticospinal excitability is modulated appears to be sex-specific during resistance training. Lastly, apparent differences between sexes in terms of hormonal status and lean body mass might alter the results for the resistance exercise performance during hypoxia. For this reason, it is difficult to generalize the recommendations from hypoxia studies, all of which were conducted on men [4,18–20,23–25,38,45]. To the best of our knowledge, this is the first study investigating the effects of hypoxia during resistance exercise in females. In contrast to the aforementioned hypothesis, our results showed that men and women respond similarly to the different doses of hypoxia. In support, previously from our laboratory, sprint interval training performance was not different between sexes both at moderate (2500 m) and high (3500 m) hypoxia [48]. However, in another study, despite similar oxygen desaturation levels, men exhibited higher sympathetic responses to very high hypoxia (F_iO_2 = 9.6%) compared to women [49]. Direct comparisons in this regard are not possible, as there is no study directly comparing sexes in a hypoxic resistance exercise setting. However, RPE values in our study were approximately 17–18 at the end of test protocol. We can speculate that near maximal RPE values, in turn, high intensity nature of our test protocol may create a "ceiling effect" that makes any appreciable differences between sexes extremely hard to distinguish, especially during hypoxia.

4.3. Cognitive Performance

In our study, cognitive performance was found to be lower in high and moderate hypoxia than low hypoxia and normoxia. These findings are in line with some [50,51], but not all, studies [52,53]. A recent meta-analysis [54] concluded that various characteristics (e.g., cognitive task type, exercise type/intensity, and hypoxia level) moderated the effects of hypoxia and exercise on cognitive function. Similar to our results, acute exposure to severe hypoxia decreased cognitive function compared to moderate and low hypoxic conditions in a dose-dependent manner [55]. Additionally, cognitive function has various domains, such as information processing, executive function, and memory [50–55]. The Flanker task which was used in the current study measures executive function that enable an individual to focus attention [36]. Since attentional focus is already reported to moderate acute muscle performance [32,33], decrements in bench press endurance performance during high and moderate hypoxia may be associated with equivalent impairment in cognitive function.

Women have greater resistance to hypoxia and higher SpO_2 levels than men [30]. In addition, women have greater basal cerebral blood flow than men during hypoxia, possibly associated with estrogen [56]. However, in the current study, we did not observe any sex differences in cognitive performance. This result is consistent with Lefferts et al. [50] suggesting that cognitive performance of the Flanker task was lower in hypoxia (F_iO_2 = 12.5%) during 55% HRmax aerobic exercise in both men and women. In contrast, short term (~20 min) severe hypoxia (F_iO_2 = 12%) improved the cognitive performance of the Go/NoGo task during exercise, with 45% peak power output in women [57]. Discordance between the previous studies [52,53,57] and ours are related to exercise type (aerobic vs. resistance) and measurement time of cognitive performance (during vs. after) [54].

4.4. Heart Rate, RPE, Lactate

Our results demonstrated that high and moderate hypoxia significantly increased RPE and capillary lactate after bench press exercise, but not after squat exercise. Further, heart rate values were not different between conditions. This mechanism can be explained by the following: just as the repetition number of bench press decreases as exposure duration to hypoxia increases, blood lactate responses may also increase due to prolonged exposure to hypoxia.

Our RPE and lactate outcomes are in support of others [4,18,24,45]. In support, RPE and blood lactate were higher in high hypoxia (F_iO_2 = 13%) than in moderate hypoxia (F_iO_2 = 16%) and normoxia [19]. However, Scott et al. [20] demonstrated no differences in RPE during 5 × 5 repetitions at 80% 1RM during high (F_iO_2 = 13%) hypoxia. These conflicting results can be explained by the type of test protocol (circuit vs. traditional vs. repetition to failure). Nevertheless, RPE is a useful tool to determine the intensity of hypoxic resistance exercise, as demonstrated by the current study.

4.5. Strengths and Limitations

In the current study, we used a double-blind design. The scientific gold standard design of a double-blind, randomized, crossover trials are rarely carried out in hypoxic resistance exercise studies [1,9,25]. Moreover, we used repetitions to failure, which ensures maximal or near maximal effort and is a potent strategy to induce muscle hypertrophy, possibly due to an increase in metabolic stress [22]. Lastly, for the first time in the literature, we directly compared men and women participants' acute responses to various doses of hypoxia. However, our study is not without limitations. One of the limitations is that squat and bench exercises are not performed in a random order. Secondly, we did not assess women participants' menstrual cycle, which may impact physical and cognitive performance [57]. However, the four experimental conditions were completed in nine days, which would have minimized the impact of the menstrual cycle.

5. Conclusions

Overall, the current study found that acute hypoxia does not alter resistance exercise responses between sexes differently. More specifically, high (4000 m) and moderate (3000 m) hypoxia significantly decreased bench press endurance and cognitive performance, and significantly increased RPE and blood lactate, as compared to low hypoxia (2000 m) and normoxia. Athletes and coaches should carefully design resistance exercise trainings under high and moderate hypoxia ($\geq$30 min in duration) due to the impaired effects on upper body muscular endurance and perceived exertion. Nevertheless, metabolic stress induced by hypoxic resistance training can improve long term hypertrophic adaptations [40].

Author Contributions: Conceptualization, R.K., R.D. and A.S.-G.; methodology, R.K., M.C.E., C.S. and F.N.S.; formal analysis, F.N.S. and M.K.; investigation, R.K., M.C.E. and C.S.; data curation, F.N.S. and M.K.; writing—original draft preparation, R.K. and M.C.E.; writing—review and editing, R.D., A.S.-G. and M.K.; visualization, R.K., M.C.E., C.S. and R.D.; supervision, M.K., R.D., F.N.S. and A.S.-G. All authors have read and agreed to the published version of the manuscript.

Funding: This research received no external funding.

Institutional Review Board Statement: The study was conducted according to the guidelines of the Declaration of Helsinki and approved by Muş Alparslan University Scientific Research Ethics Committee (Approval no: 10776-6/31).

Informed Consent Statement: Informed consent was obtained from all subjects involved in the study.

Data Availability Statement: The data presented in this study are available on request from the corresponding author. The data are not publicly available due to restrictions of privacy.

Conflicts of Interest: The authors declare no conflict of interest.

References

1. Millet, G.P.; Roels, B.; Schmitt, L.; Woorons, X.; Richalet, J.P. Combining hypoxic methods for peak performance. *Sports Med.* **2010**, *40*, 1–25. [CrossRef] [PubMed]
2. McLean, B.D.; Gore, C.J.; Kemp, J. Application of 'live low-train high' for enhancing normoxic exercise performance in team sport athletes. *Sports Med.* **2014**, *44*, 1275–1287. [CrossRef] [PubMed]
3. Girard, O.; Brocherie, F.; Goods, P.S.R.; Millet, G.P. An updated panorama of "living low-train high" altitude/hypoxic methods. *Front. Sports Act. Living* **2020**, *2*, 26. [CrossRef] [PubMed]
4. Scott, B.R.; Slattery, K.M.; Sculley, D.V.; Lockhart, C.; Dascombe, B.J. Acute physiological responses to moderate-load resistance exercise in hypoxia. *J. Strength Cond. Res.* **2017**, *31*, 1973–1981. [CrossRef] [PubMed]
5. Li, J.; Li, Y.; Atakan, M.M.; Kuang, J.; Hu, Y.; Bishop, D.J.; Yan, X. The molecular adaptive responses of skeletal muscle to high-intensity exercise/training and hypoxia. *Antioxidants* **2020**, *9*, 656. [CrossRef]
6. Smith, K.J.; Billaut, F. Influence of cerebral and muscle oxygenation on repeated-sprint ability. *Eur. J. Appl. Physiol.* **2010**, *109*, 989–999. [CrossRef]
7. Amann, M.; Pegelow, D.F.; Jacques, A.J.; Dempsey, J.A. Inspiratory muscle work in acute hypoxia influences locomotor muscle fatigue and exercise performance of healthy humans. *Am. J. Physiol. Regul. Integr. Comp. Physiol.* **2007**, *293*, R2036–R2045. [CrossRef]
8. Jagim, A.R.; Dominy, T.A.; Camic, C.L.; Wright, G.; Doberstein, S.; Jones, M.T.; Oliver, J.M. Acute effects of the elevation training mask on strength performance in recreational weight lifters. *J. Strength Cond. Res.* **2018**, *32*, 482–489. [CrossRef]
9. Feriche, B.; García-Ramos, A.; Morales-Artacho, A.J.; Padial, P. Resistance training using different hypoxic training strategies: A basis for hypertrophy and muscle power development. *Sports Med. Open* **2017**, *3*, 12. [CrossRef]
10. Loenneke, J.; Fahs, C.; Rossow, L.; Abe, T.; Bemben, M. The anabolic benefits of venous blood flow restriction training may be induced by muscle cell swelling. *Med. Hypotheses* **2012**, *78*, 151–154. [CrossRef]
11. Manimmanakorn, A.; Hamlin, M.J.; Ross, J.J.; Taylor, R.; Manimmanakorn, N. Effects of low-load resistance training combined with blood flow restriction or hypoxia on muscle function and performance in netball athletes. *J. Sci. Med. Sport* **2013**, *16*, 337–342. [CrossRef] [PubMed]
12. Nishimura, A.; Sugita, M.; Kato, K.; Fukuda, A.; Sudo, A.; Uchida, A. Hypoxia increases muscle hypertrophy induced by resistance training. *Int. J. Sports Physiol. Perform.* **2010**, *5*, 497–508. [CrossRef] [PubMed]
13. Chaillou, T. Skeletal muscle fiber type in hypoxia: Adaptation to high-altitude exposure and under conditions of pathological hypoxia. *Front. Physiol.* **2018**, *9*, 1450. [CrossRef] [PubMed]
14. Loenneke, J.; Wilson, G.; Wilson, J. A mechanistic approach to blood flow occlusion. *Int. J. Sports Med.* **2010**, *31*, 1–4. [CrossRef] [PubMed]
15. Schoenfeld, B.J. The mechanisms of muscle hypertrophy and their application to resistance training. *J. Strength Cond. Res.* **2010**, *24*, 2857–2872. [CrossRef] [PubMed]
16. Takarada, Y.; Nakamura, Y.; Aruga, S.; Onda, T.; Miyazaki, S.; Ishii, N. Rapid increase in plasma growth hormone after low-intensity resistance exercise with vascular occlusion. *J. Appl. Physiol.* **2000**, *88*, 61–65. [CrossRef]
17. Namboonlue, C.; Hamlin, M.J.; Sirasaporn, P.; Manimmanakorn, N.; Wonnabussapawich, P.; Thuwakum, W.; Sumethanurakkhakun, W.; Manimmanakorn, A. Optimal degree of hypoxia combined with low-load resistance training for muscle strength and thickness in athletes. *J. Phys. Educ. Sport* **2020**, *20*, 828–838.
18. Ramos-Campo, D.J.; Rubio-Arias, J.Á.; Freitas, T.T.; Camacho, A.; Jiménez-Diaz, J.F.; Alcaraz, P.E. Acute physiological and performance responses to high-intensity resistance circuit training in hypoxic and normoxic conditions. *J. Strength Cond. Res.* **2017**, *31*, 1040–1047. [CrossRef]
19. Ramos-Campo, D.J.; Rubio-Arias, J.A.; Dufour, S.; Chung, L.; Ávila-Gandía, V.; Alcaraz, P.E. Biochemical responses and physical performance during high-intensity resistance circuit training in hypoxia and normoxia. *Eur. J. Appl. Physiol.* **2017**, *117*, 809. [CrossRef]
20. Scott, B.R.; Slattery, K.M.; Sculley, D.V.; Hodson, J.A.; Dascombe, B.J. Physical performance during high-intensity resistance exercise in normoxic and hypoxic conditions. *J. Strength Cond. Res.* **2015**, *29*, 807–815. [CrossRef]
21. Desplanches, D.; Hoppeler, H.; Linossier, M.T.; Denis, C.; Claassen, H.; Dormois, D.; Lacour, J.R.; Geyssant, A. Effects of training in normoxia and normobaric hypoxia on human muscle ultrastructure. *Pflug. Arch.* **1993**, *425*, 263–267. [CrossRef]
22. Vieira, J.G.; Sardeli, A.V.; Dias, M.R.; Filho, J.E.; Campos, Y.; Sant'Ana, L.; Leitao, L.; Reis, V.; Wilk, M.; Novaes, J.; et al. Effects of resistance training to muscle failure on acute fatigue: A systematic review and meta-analysis. *Sports Med.* **2021**. Online ahead of print. [CrossRef] [PubMed]
23. Smith, C.M.; Salmon, O.F.; Jenkins, J.R. Effect of moderate and severe hypoxic exposure coupled with fatigue on psychomotor vigilance testing, muscle tissue exygenation, and muscular performance. *Curr. Res. Phyisol.* **2021**, *4*, 223–251.
24. Martinez-Guardado, I.; Sanchez-Urena, B.; Olcina, G.; Camacho-Cardenosa, A.; Camacho-Cardenosa, M.; Timon, R. Bench press performance during an intermittent hypoxic resistance training to muscle failure. *J. Sports Med. Phys. Fit.* **2019**, *59*, 1138–1143. [CrossRef]
25. Girard, O.; Mariotti-Nesurini, L.; Malatesta, D. Acute performance and physiological responses to upper-limb multi-set exercise to failure: Effects of external resistance and systemic hypoxia: Local and systemic hypoxia combined. *Eur. J. Sport Sci.* **2021**, 1–28, Online ahead of print.

26. Hunter, S.K. Sex differences in fatigability of dynamic contractions. *Exp. Physiol.* **2016**, *101*, 250–255. [CrossRef]
27. Hively, K.; El-Alayli, A. "You throw like a girl:" The effect of stereotype threat on women's athletic performance and gender stereotypes. *Psychol. Sport Exerc.* **2014**, *15*, 48–55. [CrossRef]
28. Laurent, C.; Green, J.; Bishop, P.; Sjökvist, J.; Schumacker, R.; Richardson, M.; Curtner-Smith, M. Effect of gender on fatigue and recovery following maximal intensity repeated sprint performance. *J. Sports Med. Phys. Fit.* **2010**, *50*, 243–253.
29. Laurent, C.M.; Vervaecke, L.S.; Kutz, M.R.; Green, J.M. Sex-specific responses to self-paced, high-intensity interval training with variable recovery periods. *J. Strength Cond. Res.* **2014**, *28*, 920–927. [CrossRef] [PubMed]
30. Sandoval, D.A.; Matt, K.S. Effects of the oral contraceptive pill cycle on physiological responses to hypoxic exercise. *High Alt. Med. Biol.* **2003**, *4*, 61–72. [CrossRef] [PubMed]
31. Slater, J.P.; Guarino, T.; Stack, J.; Vinod, K.; Bustami, R.T.; Brown, J.M., III; Rodriguez, A.L.; Magovern, C.J.; Zaubler, T.; Freundlich, K. Cerebral oxygen desaturation predicts cognitive decline and longer hospital stay after cardiac surgery. *Ann. Thorac. Surg.* **2009**, *87*, 36–45. [CrossRef] [PubMed]
32. Grgic, J.; Mikulic, P. Effects of attentional focus on muscular endurance: A meta-analysis. *Int. J. Environ. Res. Public Health* **2021**, *19*, 89. [CrossRef]
33. Grgic, J.; Mikulic, I.; Mikulic, P. Acute and long-term effects of attentional focus strategies on muscular strength: A meta-analysis. *Sports* **2021**, *9*, 153. [CrossRef] [PubMed]
34. Baechle, T.R.; Earle, R.W. *Essentials of Strength Training and Conditioning*; Human Kinetics: Champaign, IL, USA, 2008.
35. Borg, G. Perceived exertion: A note on history and models. *Med. Sci. Sports Exerc.* **1973**, *5*, 90–93. [CrossRef]
36. Eriksen, B.A.; Eriksen, C.W. Effects of noise letters upon the identification of a target letter in a nonsearch task. *Percept. Psychophys.* **1974**, *16*, 143–149. [CrossRef]
37. Cohen, J. A power primer. *Psychol. Bull.* **1992**, *112*, 155–159. [CrossRef]
38. Feriche, B.; Garcia-Ramos, A.; Calderon-Soto, C.; Drobnic, F.; Bonitch-Gongora, J.G.; Galilea, P.A.; Riera, J.; Padial, P. Effect of acute exposure to moderate altitude on muscle power: Hypobaric hypoxia vs. normobaric hypoxia. *PLoS ONE* **2014**, *9*, e114072. [CrossRef]
39. Bazzuchi, I.; Felici, F.; Macaluso, A.; De Vito, G. Differences between young and older women in maximal force, force fluctuations, and surface EMG during isometric knee extension and elbow flexion. *Muscle Nerve* **2004**, *30*, 626–635. [CrossRef]
40. Scott, B.R.; Slattery, K.M.; Dascombe, B.J. Intermittent hypoxic resistance training: Is metabolic stress the key moderator? *Med. Hypotheses* **2015**, *84*, 145–149. [CrossRef]
41. Bowtell, J.L.; Cooke, K.; Turner, R.; Mileva, K.N.; Sumners, D.P. Acute physiological and performance responses to repeated sprints in varying degrees of hypoxia. *J. Sci. Med. Sport* **2014**, *17*, 399–403. [CrossRef]
42. Cerqueira, M.S.; Lira, M.; Barboza, J.A.M.; Burr, J.F.; Wanderley E Lima, T.B.; Maciel, D.G.; De Brito Vieira, W.H. Repetition failure occurs earlier during low-load resistance exercise with high but not low blood flow restriction pressures. a systematic review and meta-analysis. *J. Strength Cond. Res.* **2021**. Online ahead of print. [CrossRef] [PubMed]
43. Wyatt, F. Physiological responses to altitude: A brief review. *J. Exerc. Physiol. Online* **2014**, *17*, 90–96.
44. Deb, S.K.; Brown, D.R.; Gough, L.A.; Mclellan, C.P.; Swinton, P.A.; Sparks, S.A.; Mcnaughton, L.R. Quantifying the effects of acute hypoxic exposure on exercise performance and capacity: A systematic review and meta-regression. *Eur. J. Sport Sci.* **2018**, *18*, 243–256. [CrossRef] [PubMed]
45. Walden, T.P.; Girard, O.; Jonson, A.M.; Dempsey, A.R.; Marston, K.J.; Scott, B.R. Sessional work-rate does not affect the magnitude to which simulated hypoxia can augment acute physiological responses during resistance exercise. *Eur. J. Appl. Physiol.* **2020**, *120*, 2159–2169. [CrossRef] [PubMed]
46. Mage, D.T.; Donner, M. Female resistance to hypoxia: Does it explain the sex difference in mortality rates? *J. Women's Health* **2006**, *15*, 786–794. [CrossRef]
47. Koozehchian, M.; Karayigit, R.; Mabrey, G.; Newton, A. The effect of different doses of acute normobaric hypoxia on sprint interval training and cognitive performance. *FASEB J.* **2021**, *35* (Suppl. 1). conference paper. [CrossRef]
48. Haseler, L.J.; Hogan, M.C.; Richardson, R.S. Skeletal muscle phosphocreatine recovery in exercise-trained humans is dependent on O_2 availability. *J. Appl. Physiol.* **1999**, *86*, 2013–2018. [CrossRef]
49. Botek, M.; Krejci, J.; McKune, A. Sex differences in autonomic cardiac control and oxygen saturation response to short-term normobaric hypoxia and following recovery: Effect of aerobic fitness. *Front. Endocrinol.* **2018**, *23*, 697. [CrossRef]
50. Lefferts, W.K.; Babcock, M.C.; Tiss, M.J.; Ives, S.J.; White, C.N.; Brutsaert, T.D.; Heffernan, K.S. Effect of hypoxia on cerebrovascular and cognitive function during moderate intensity exercise. *Physiol. Behav.* **2016**, *165*, 108–118. [CrossRef]
51. Sun, S.; Loprinzi, P.D.; Guan, H.; Zou, L.; Kong, Z.; Hu, Y.; Shi, Q.; Nie, J. The effects of high-intensity interval exercise and hypoxia on cognition in sedentray young adults. *Medicine* **2019**, *55*, 43.
52. Ando, S.; Hatamoto, Y.; Sudo, M.; Kiyonaga, A.; Tanaka, H.; Higaki, Y. The effects of exercise under hypoxia on cognitive function. *PLoS ONE* **2013**, *8*, e63630. [CrossRef] [PubMed]
53. Komiyama, T.; Sudo, M.; Higaki, Y.; Kiyonaga, A.; Tanaka, H.; Ando, S. Does moderate hypoxia alter working memory and executive function during prolonged exercise? *Physiol. Behav.* **2015**, *139*, 290–296. [CrossRef]
54. Jung, M.; Zou, L.; Yu, J.J.; Ryu, S.; Kong, Z.; Yang, L.; Kang, M.; Lin, J.; Li, H.; Smith, L.; et al. Does exercise have protective effect on cognitive function unders hypoxia? A systematic review with meta-analysis. *J. Sport Health Sci.* **2020**, *9*, 562–577. [CrossRef] [PubMed]

55. Ochi, G.; Kanazawa, Y.; Hyodo, K.; Suwabe, K.; Shimizu, T.; Fukuie, T.; Byun, K.; Soya, H. Hypoxia-induced lowered executive function depends on arterial oxygen desaturation. *J. Physiol. Sci.* **2018**, *68*, 847–853. [CrossRef] [PubMed]
56. Peltonen, G.L.; Harrell, J.W.; Aleckson, B.P.; LaPlante, K.M.; Crain, M.K.; Schrage, W.G. Cerebral blood flow regulation in women across mentrual phase: Differential contribution of cyclooxygenase to basal, hypoxic, and hypercapnic vascular tone. *Am. J. Physiol. Regul. Integr. Comp. Physiol.* **2016**, *311*, R222–R231. [CrossRef]
57. Lei, O.K.; Kong, Z.; Loprinzi, P.D.; Shi, Q.; Sun, S.; Zou, L.; Hu, Y.; Nie, J. Severe hypoxia does not offset the benefits of exercise on cognitive function in sedentary young women. *Int. J. Environ. Res. Public Health* **2019**, *16*, 1003. [CrossRef]

biology

MDPI

Article

Effect of Concurrent Resistance Training on Lower Body Strength, Leg Kick Swimming, and Sport-Specific Performance in Competitive Swimmers

Sofiene Amara [1,2,*], Tiago Manuel Barbosa [3,4], Oussama Gaied Chortane [1,2], Raouf Hammami [1,5], Ahmed Attia [1,2], Sabri Gaied Chortane [1,6] and Roland van den Tillaar [7,*]

1 Higher Institute of Sport and Physical Education of Ksar-Said, University of La Manouba, Tunis 2010, Tunisia; oussama.gaeid@gmail.com (O.G.C.); raouf.cnmss@gmail.com (R.H.); attiaahmed3@yahoo.fr (A.A.); sabrigaied1@gmail.com (S.G.C.)
2 Research Unit (UR17JS01) Sports Performance, Health & Society, Higher Institute of Sport and Physical Education of Ksar Saîd, Universite de la Manouba, Tunis 2010, Tunisia
3 Research Center in Sport, Health and Human Development, 5000-801 Vila Real, Portugal; barbosa@ipb.pt
4 Department of Sports Sciences, Instituto Politécnico de Bragança, Campus Sta., 5301-856 Bragança, Portugal
5 Research Laboratory: Education, Motor Skills, Sports and Health (LR19JS01), Higher Institute of Sport and Physical Education of Sfax, University of Sfax, Sfax 3029, Tunisia
6 Laboratory of Cardio-Circulatory, Respiratory, Metabolic and Hormonal Adaptations to Muscular Exercise, Faculty of Medicine Ibn El Jazzar, Sousse 4002, Tunisia
7 Department of Sport Sciences and Physical Education, Nord University, 7600 Levanger, Norway
* Correspondence: coachsofieneamara@gmail.com (S.A.); roland.v.tillaar@nord.no (R.v.d.T.)

Simple Summary: Resistance training in and out of the water aims to improve swimming performance. Previous studies have shown that dry land resistance training has positive effects on improving strength and therefore this could optimize swimming performance. The present study investigated the effect of 9 weeks of combined resistance training (aquatic and dry land resistance) on maximum lower body strength, leg kick, and swimming performance in competitive swimmers. The results demonstrated that 9 weeks of combined resistance training could improve the maximum lower body strength and leg kick swimming performance. These improvements can be the essential factors that subsequently positively affected swimming start and turn performance.

Abstract: The present study investigated the effect of 9 weeks of combined resistance training (aquatic and dry land resistance) on maximum lower body strength, leg kick, and swimming performance in competitive swimmers. Twenty-two male national competitive swimmers were randomly assigned into two groups: experimental group (EG: age = 16.2 ± 0.3 years) or control group (CG: age = 16.3 ± 0.3 years). The EG performed a combined resistance training while the CG group completed their usual training. One repetition maximum (1RM) back squat, 30 m leg kick, and swimming performance (100 m front crawl, start and turn) were evaluated in pre and post test. The findings showed a significant increase in 1RM back squat (d = 1.90; 14.94 ± 1.32%) after 9 weeks of combined resistance training. In addition, ours results revealed a significant improvement in 30 m leg kick swimming (d = 2.11; 5.84 ± 0.16%) and in all swimming, start and turn performances (d = 1.83 to 2.77; 2.69 ± 0.18% to 15.14 ± 1.06%) in EG. All dependent variables remained unchanged in the CG. To sum up, 9 weeks of combined resistance training can improve the maximum lower body strength and leg kick swimming performance. These improvements can be the essential factors that subsequently positively affected swimming, start and turn performances. Combined resistance training is an effective training that can be incorporated by coaches and swimmers into their programs to improve strength, leg kick swimming, and, subsequently, swimming performance in competitive swimmers.

Keywords: dry land training; one repetition maximum; back squat; water parachute; aquatic training; swimming performance

Citation: Amara, S.; Barbosa, T.M.; Chortane, O.G.; Hammami, R.; Attia, A.; Chortane, S.G.; van den Tillaar, R. Effect of Concurrent Resistance Training on Lower Body Strength, Leg Kick Swimming, and Sport-Specific Performance in Competitive Swimmers. *Biology* **2022**, *11*, 299. https://doi.org/10.3390/biology11020299

Academic Editor: Johannes Vogel

Received: 3 January 2022
Accepted: 10 February 2022
Published: 12 February 2022

1. Introduction

Over the years, the 100 m race times have been improved probably due to better aquatic and dry land training programs [1,2]. Concurrent resistance training on dry land and in water showed to be an effective approach to improve swimming performance [1,3,4]. Amara et al. [1] noted that nine weeks of concurrent resistance training could improve maximal upper body strength (12.11 ± 1.79%) and sprint performance in front crawl (4.2 ± 0.2% to 7.1 ± 0.2%) in male competitive swimmers (age = 16.5 ± 0.30 years). Moreover, Lopes et al. [3] reported a significant improvement in sprint swimming performance (4.0% to 4.3%) after eight weeks of dry land strength combined with swimming training in university swimmers of national level (age = 20.55 ± 1.76 years). However, most studies focus upon increasing strength in the upper body, while improving the performance of the lower limbs is also an important factor in determining swimming performance [1,5]. Morouço et al. [5] showed that a relative contribution of leg kick was 29.7% for male swimmers and 33.4% for females. In the same context, Bartolomeu et al. [6] revealed that swim velocity with leg kick was 59% compared with full front crawl velocity in competitive swimmers (age = 14.20 ± 1.71 years). Furthermore, the start and turn are among the most important factors in determining swimming performance these days in competition settings. Morais et al. [7] showed that the start and the turn phases combined accounted for 31% to 32% of the final race time in the four swimming strokes. Notwithstanding, lower body strength and power are found to be two very important underlying factors determining the performance of start and turn in competitive swimmers [8,9]. Thus, it is necessary to know more about the effect of combined resistance training on maximum lower body strength, leg kick swimming, and swimming performance.

Lower limbs strength routines (e.g., countermovement jumps (CMJ), squat, and leg extension) are heavily used in several dry land swimming training programs [3,10]. For instance, Lopes et al. [3] showed that eight weeks of dry land training, including full squat and CMJ exercises, could improve CMJ performance (6.77%) in national competitive swimmers. Moreover, Sammoud et al. [10] revealed that 8 weeks of dry land training based on plyometric jump training (CMJ) could optimize the performance of CMJ (24.5%) and the performance of 25 m front crawl swimming (6.33%) in prepubertal female swimmers (age = 10.0 ± 0.6 years). These authors [3,10] showed that lower body dry land training optimized strength and swimming performance, but the information about the transfer of strength gain in water remains insufficient. Hence, it is necessary to understand the effect of dry land training on leg kick swimming performance.

Resistance training in water with aquatic equipment (e.g., water parachute) is also an effective approach to improve the propulsive force [1,11]. Gourgoulis et al. [11] noted that 11 weeks of training with water parachute improved the performance in the 50, 100, and 200 m front crawl (3.2% to 7.3%) in competitive swimmers. Amara et al. [1] reported that the 50 m front crawl performance was improved after 9 weeks of resistance training in water with water parachute (4.22 ± 0.18%) in young competitive swimmers. Amara et al. [1] also showed that an aquatic resistance training program can improve the performance in the 50 m front crawl arm-pull (7.1 ± 0.23%). The question arises here, could resistance training with this aquatic equipment improve leg kick performance? If so, how efficient is the transfer of dry land strength improvement into in-water leg kick performance?

Training in water with additional resistance by the water parachute is shown to be an effective training to improve swimming performance [11]. However, information about the effect of combined resistance training (additional aquatic resistance in and dry land resistance) on leg kick and swimming performance remains insufficient. Therefore, the objective of this study was to investigate the effect of lower body strength and additional aquatic resistance training (using water parachute) on strength, leg kick, and swimming performance in competitive swimmers. We hypothesized that improving lower body strength could improve leg kick performance and subsequently optimize swimming performance.

2. Materials and Methods

2.1. Experimental Approach to the Problem

The present study was applied to examine the effect of combined resistance training (aquatic and dry land resistance training) during a nine-week intervention period on maximum lower body strength (1RM back squat) and swimming performance (100 m front crawl, 30 m leg kick swimming, start and turn performance) compared to standard training. All swimmers participated in a week of familiarization for resistance training in dry land and in the water. All dependent variables were measured pre and post test. All dry land training and testing were performed in the weight room, and all swimming training and testing were performed in a 50 m indoor pool with 27.1 and 25.9 °C water and air temperatures (respectively) and 64% relative humidity. All swimmers were asked to avoid all other intensive activities and avoid all stimulating nutrients during the entire experimental period.

2.2. Subjects

Twenty-two male national competitive swimmers were randomly assigned into two groups: experimental group (EG: n = 11, age = 16.5 ± 0.3 years; height: 174 ± 9.80 cm; body mass = 72.7 ± 5.30 kg) or control group (CG: n = 11, age = 16.1 ± 0.3 years; height: 175 ± 9.70 cm body mass 73.6 ± 5.25 kg). An a priori power analysis (G * Power 3.1.9.3, Heinrich-Heine-Universität Düsseldorf, Düsseldorf, Germany) yielded a sample size of at least 9 swimmers per group to detect large effects (d = 1.29), assuming a power of 0.8 and alpha of 0.05. All subjects had three years of experience in resistance training and 5 years of experience in resistance aquatic training. The best time of the fastest swimmer in 100 m front crawl was 57.08 s. All swimmers were informed about all research-related risks and potential benefits. All subjects and parents read and signed written informed consent. This study was approved by an institutional review board of Higher Institute of Sport and Physical Education of Ksar Said, University of Manouba, Tunisia (Research Unit of Sports Performance, Health & Society, UR17JS01) and conducted according to the last declaration of Helsinki.

2.3. Procedures

This study was integrated into the preparatory phase of the winter competition (September to December). All swimmers were conditioned to training and have started the training season at the same time (3 months before the experimental protocol). The study design consisted of a pretest followed-up by 9 weeks of intervention and a one-week pos test (Figure 1).

2.3.1. Aquatic Resistance Training

The swimming training program was composed of six sessions per week with an hourly volume between 90 and 120 min, and swimming distance between 4000 and 6000 m per session. All-out sets of specific resistance training using only a leg kick (4 sessions in a week) were included in both training programs (EG and CG) immediately after the warm-up (800 m of aerobic training (55% to 80% of maximum heart rate). On Monday and Thursday, swimmers completed 3 sets × 6 reps × 15 m with 60 s and 5 min of rest between repetitions and sets, respectively. On Tuesday and Friday, swimmers completed 2 sets × 4 reps × 25 m with 90 s and 5 min of rest between repetitions and sets, respectively [1,11]. To eliminate the effect of arm stroke, specific resistance training was conducted with the aid of a kickboard, which was held by both hands. The experimental group completed the specific aquatic training with additional resistance using a small water parachute (288 cm^2). The water parachute was attached to the swimmer's back by a 2 m-long rigid rope with a belt that was fasted around the swimmer's waist [1] (Figure 2).

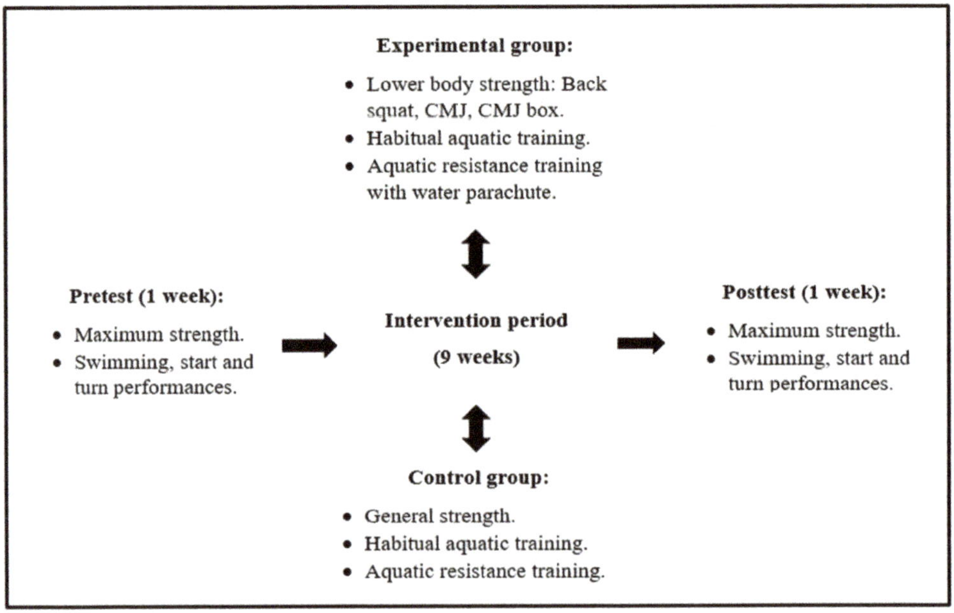

Figure 1. Overview of the study design. CMJ: countermovement jump; CMJ box: countermovement jump box.

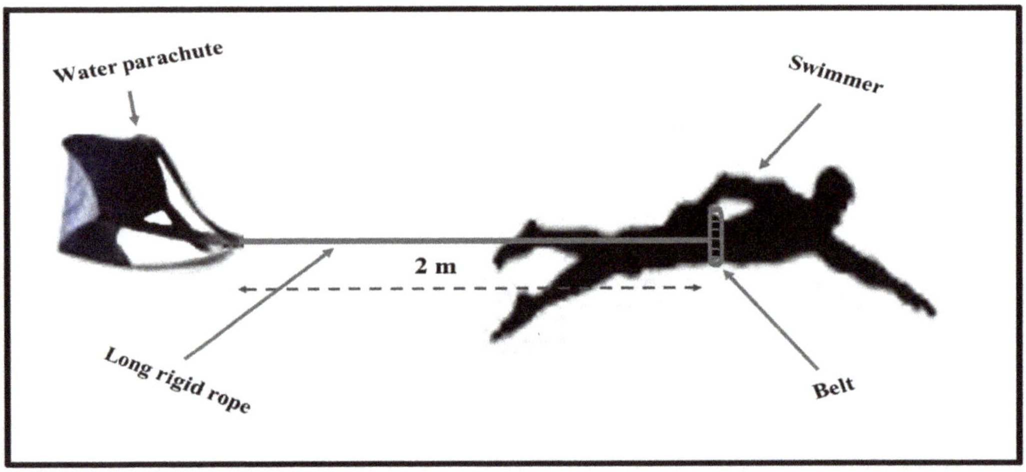

Figure 2. Set-up of water parachute during aquatic resistance training.

2.3.2. Lower Body Strength Training

Two non-consecutive sessions of lower body strength per week (separated by 48 h) were set up. Two strength and conditioning coaches designed and conducted the dry land training. Each session started with a 20 min standard warm-up featuring aerobic exercises (ergocycle and treadmill) and two sets of 8 to 10 reps of back squat with lower load (20% to 30% 1RM back squat). Subsequently, subjects performed three lower body strength exercises, which were back squat, countermovement jump (CMJ), and CMJ box with moderate contraction velocity and complete motion angle. The back squat exercise was performed with an intensity between 60% and 80% of 1RM. The sets varied between 2 and 3 and repetitions between 6 and 10. A custom power rack and a standard Olympic weightlifting bar calibrated and certified (20 kg) were used to perform the back squat and to better control the technique of performing the exercise. Moreover, CMJ and CMJ box exercises consisted of 6 to 8 sets with 6 to 10 repetitions. The recovery between sets and exercises was fixed at 2 min (Table 1) [1,12,13]. The CG was invited to follow their usual training based on general body strength (2 sessions per week: 60 to 75 min per session). General body strength consisted of a general warm-up (cardiorespiratory adaptation and muscle stimulation), general strength exercises (pump-action, medicine ball throw, abdominal exercises), and muscle stretching.

Table 1. Detailed resistance training conducted by the experimental group.

Week (Sessions)	Exercises	Sets × Repetition × Intensity; Recovery between Sets and Exercises (3 min)
1 (1/2)	Back squat CMJ CMJ box	2 × 10 × 60% 1RM 2 × 6 2 × 6
2 (3/4)	Back squat CMJ CMJ box	2 × 10 × 65% 1RM 2 × 8 2 × 8
3 (5/6)	Back squat CMJ CMJ box	2 × 10 × 70% 1RM 2 × 10 2 × 10
4 (7/8)	Back squat CMJ CMJ box	2 × 8 × 75% 1RM 3 × 8 3 × 8
5 (9/10)	Back squat CMJ CMJ box	3 × 6 × 80% 1RM 3 × 10 3 × 10
6 (11/12)	Back squat CMJ CMJ box	2 × 8 × 75% 1RM 3 × 6 3 × 6
7 (13)	Back squat CMJ CMJ box	3 × 6 × 80% 1RM 3 × 10 3 × 10
8 (14)	Back squat CMJ CMJ box	3 × 8 × 75% 1RM 3 × 8 3 × 8
9 (15)	Back squat CMJ CMJ box	3 × 8 × 70% 1RM 3 × 6 3 × 6

1RM: one repetition maximum; CMJ: countermovement jump; CMJ box: countermovement jump box.

2.3.3. Maximum Lower Body Strength Test

On the first day of testing (at 10.00 am), subjects visited the weight room to determine the 1RM back squat test. The strength test was performed using a custom power rack and a standard Olympic weight lifting bar calibrated and certified (20 kg). Two strength and conditioning coaches controlled the execution of the exercises and conducted all measurements of the maximum back squat strength test.

All subjects completed a warm-up on an ergocycle for 3 min, followed-up by 5 min of overall static stretching. Thereafter, subjects performed 1 set of 8 reps and 1 set of 3 reps at 50% and 70% of their estimated 1RM back squat, respectively. The load was gradually increased (10% to 20%), 2 to 3 repetitions and 2 to 4 min of rest were performed. Thereafter, a small increase in the load (5%) and 2 to 4 min of rest were carried out to reach the 1RM squat. The test was finished when the subjects failed to complete the squat, and the last successful attempt presents the 1RM back squat [14,15]. The intraclass correlation coefficient (ICC) for the pretest and posttest reliability was 0.93.

2.3.4. Swimming Performance Measurements

Swimming performance measurements were established on the second day of tests also at 10.00 am. The 100 m front crawl performance and the 30 m leg flutter kick swimming were measured by two qualified timekeepers per stopwatch (SEIKO S120-4030, Tokyo, Japan) and noted in seconds. The starting signal was given by the timers, and diving start was performed in the 100 m front crawl test, whereas a push-off start was used in the 30 m leg kick swimming [1]. The pretest and posttest reliability for 100 m front crawl was ICC = 0.91, and for the 30 m leg kick swimming, ICC = 0.93.

To determine the start and turn performances, three cameras (SNC VB 603, Sony, Tokyo, Japan; f = 50 Hz, full HD, 1080 p) were set at well-defined points laterally to the swimming pool about 5 m above the water surface and about 10 m away from the swimming lane to film the 100 m front crawl test (Figure 3) [1]. A video analysis system (Kinovea, version 0.8.15, Joan Charmant & Contrib., Kinovea.org, accessed on 3 January 2022) was used to evaluate: (i) start performances: Block time (the time lag between the starting signal and the instant the swimmer's feet left the block) and the 15 m time (the time lag between the starting signal and the swimmer's head reaching the 15 m mark). (ii) turn performance: the total turn (the time lag between reaching the 45 m mark and the 15 m mark of the following split) [7,16]. The pretest and posttest reliability of start and turn performances were $0.89 \leq ICC \leq 0.95$.

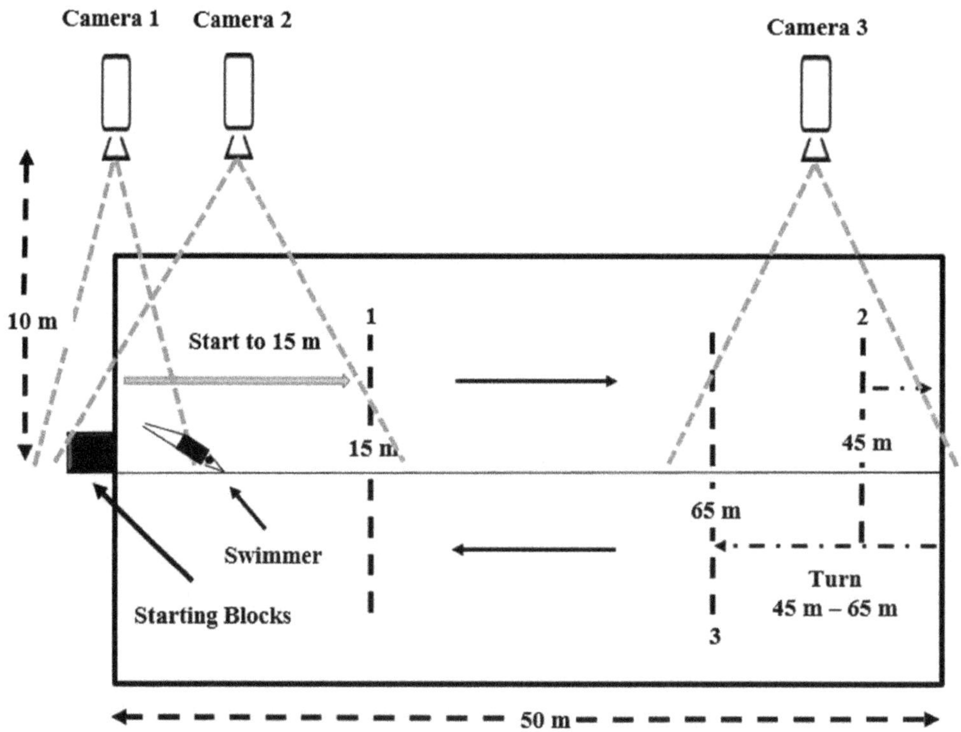

Figure 3. Set-up to analyze the start and turn performances.

2.4. Statistical Analyses

SPSS 26.0 (SPSS Inc., Chicago, IL, USA) was used for statistical analysis. All data are presented as mean and SD, mean difference, partial difference in percentage, and 95% confidence interval. The baseline between-group differences were computed through independent sample *t*-tests. Normality and sphericity of the data were tested and confirmed using Shapiro–Wilk test and Mauchly test, respectively. Pre and post test reliability was assessed using the intraclass correlation coefficient ($ICC_{2,1}$) [17]. The standard error of measurement (SEM) was determined by SD of the pretest × $\sqrt{(1\text{-ICC})}$ [18]. Minimum detectable change (MDC) of all dependent variables was derived using SEM × $\sqrt{2}$ × 1.96, and MDC % was defined as ((MDC/mean of pretest) × 100) [18]. Two-way ANOVA was used to determine the change of strength and swimming performance between pre and post test. The effect size (ES) was assessed by converting partial Eta-squared to Cohen's d [19]. ES was classified as trivial (d < 0.25), small (0.25 ≤ d < 0.50), moderate (0.50 ≤ d < 1) and large (d ≥ 1) [20]. The level of significance was established at $p \leq 0.05$.

3. Results

No significant difference in the baseline values between both groups in anthropometrics, swimming performance, and maximum strength test were shown ($p > 0.05$). Significant main effects of time, group, and time × group interaction were shown in all dependent variables ($p < 0.05$).

A significant improvement in 1RM back squat (14.94 ± 1.32%, $p < 0.001$, ES = 1.90 (large)) in EG was found while the 1RM back squat weight remained unchanged in CG ($p > 0.05$, Table 2).

Table 2. Changes in the maximum lower body strength, swimming, start and turn performances between the pre and post test in the experimental (EG) and control groups.

Variables	Groups	Pre Test	Post Test	*p*-Value	Effect (95% CI)	Δ (%)	MDC (%)	Effect Size
1RM back squat (kg)	EG	63.46 ± 4.99	72.91 ± 5.45	<0.001	5.27 (1.88 to 8.66)	14.94 ± 1.32	3.89	1.90 (large)
	CG	62.18 ± 5.72	63.64 ± 6.04	0.568		2.32 ± 0.73	(6.19)	0.46 (small)
100 m front crawl (s)	EG	59.56 ± 1.50	56.93 ± 1.51	0.001	−1.10 (−2.03 to −0.17)	4.41 ± 1.39	1.25	1.83 (large)
	CG	59.49 ± 1.55	59.21 ± 1.55	0.678		0.47 ± 0.02	(2.10)	0.19 (trivial)
30 m Leg kick swimming (s)	EG	20.97 ± 0.61	19.75 ± 0.61	<0.001	−0.65 (−1.02 to −0.28)	5.84 ± 0.16	0.44	2.11 (large)
	CG	21.12 ± 0.60	20.90 ± 0.60	0.398		1.04 ± 0.05	(2.09)	0.39 (small)
Time 15 m (s)	EG	6.65 ± 0.07	6.47 ± 0.07	<0.001	−0.07 (−0.11 to −0.03)	2.69 ± 0.18	0.06 (0.90)	2.77 (large)
	CG	6.65 ±0.07	6.61 ± 0.06	0.159		0.60 ± 0.22		0.66 (moderate)
Block time (s)	EG	0.78 ± 0.05	0.66 ± 0.05	<0.001	−0.05 (−0.08 to −0.02)	15.14 ± 1.06	0.06 (7.69)	2.61 (large)
	CG	0.78 ± 0.05	0.75 ± 0.05	0.216		3.39 ±0.65		0.57 (moderate)
Total turn (s)	EG	11.48 ± 0.19	11.15 ± 0.18	0.001	−0.14 (−0.25 to −0.03	2.88 ± 0.08	0.11	1.85 (large)
	CG	11.48 ± 0.17	11.42 ± 0.17	0.434		0.51 ± 0.04	(0.96)	0.36 (small)

MDC: minimal detectable change; Δ: delta change (pre test to post test).

In addition, significant improvement was found in 100 m front crawl (4.41 ± 1.39%, $p = 0.001$, ES = 1.83 (large)) and in 30 m leg kick swimming (5.84 ± 0.16%, $p < 0.001$, ES = 2.11 (large)) after 9 weeks of intervention period in the experimental group, but not in the control group (Table 2).

Also a significant improvement in all start performances (15 m time: 2.69 ± 0.18%, $p < 0.001$, ES = 2.77 (large); block time: 15.14 ± 1.06, $p < 0.001$, ES = 2.61 (large)) and in turn performance (total turn: 2.88 ± 0.08, $p = 0.001$, ES = 1.85 (large)) in the experimental group, while no significant changes in any of these parameters were observed in the control group ($p > 0.05$, Table 2).

4. Discussion

The aim of the present study was to investigate the effect of concurrent resistance training in water (parachute) and dry land on maximum lower body strength, leg kick, and swimming performances in competitive swimmers. The main results showed significant increases in strength and all swimming variables after nine weeks of combined resistance training, while the control groups did not have positive improvements in any of the selected variables.

Combined resistance training for nine weeks increased the 1RM back squat by 14.94 ± 1.32%. This improvement is in accordance with previous studies conducting dry land interventions between 8 and 10 weeks and intensity between 60% and 80% 1RM [2,21,22]. Garrido et al. [21] also showed that eight weeks of resistance dry land training that included leg extension (intensities between 50% and 75% of the 1RM), CMJ and CMJ box exercises with 2 to 3 sets and 5 to 8 repetitions improved the performance of the 6RM leg extension (55.6%) in young competitive swimmers (12.08 ± 0.76 years). Likewise, Kubo et al. [22] showed 11.3 ± 8.6% improvement in the 1RM half squat performance after 10 weeks of lower limb training (full squat and half squat exercises) in healthy male subjects (age = 20.9 years). In addition, Amara et al. [2] reported that nine weeks of different strength training load including leg extension exercise (high load: 5 to 6 sets and 3 to 5 repetitions; moderate load: 4 to 5 sets and 3 to 5 repetitions; low load: 3 to 4 sets and 3 to 5 repetitions) with intensity between 85% and 95% 1RM leg extension improved the 1RM leg extension (9.87% to 19.82%) in male competitive swimmers. Hence, 8–10 weeks of training at 60–80% of 1RM yields improvements in maximum and power strength of lower limb [2,21,22]. As such, the dry land program design for this research is as effective as others reported in the literature.

According to the previous literature, this study is the first investigation that studied the effect of combined resistance training on leg kick swimming performance; for this reason, it is challenging to benchmark these results against previous findings. Only, Konstantaki et al. [23] noted that training during six weeks based on leg kick training (3 sessions per week) could improve the performance of the 200 m leg kick swimming (6 ± 2%) in male competitive swimmers. Furthermore, the concurrent aquatic and lower limb resistance training improved the 30 m leg kick swimming performance (5.84 ± 0.16%), which is probably caused by the improvement in maximum lower body strength as suggested by the increased 1-RM squat results. More specifically, the transfer of force gain from the lower limb to the leg kick swimming was clearly evident in the EG (1.22 s) better than in the CG (0.22 s) with a difference of ≈1 s, and this explains the effective role of resistance lower limb training in improving the performance of the leg kick swimming in the water.

The 100 m front crawl times increased by approximately 2.63 s (4.41 ± 1.39%) after the concurrent resistance training. In fact, the increase in the maximum strength and the leg kick swimming performance in the EG (9.45 kg, 1.22 s, respectively) compared in the CG (1.46 kg, 0.22 s, respectively) could explain that the swimmers of EG make suitable use of the transfer of gain of force to improve the 100 m swimming performance (2.63 s) compared in CG (0.28 s) with a difference of ≈ 2.35 s. This present investigation is a complement to the previous study developed by Amara et al. [1]. Whereas Amara et al. [1] revealed an improvement of 25 m front crawl arm-pull (1.05 s) due to the transfer of force gain after the increase in maximum upper limb performance (5.45 kg), and subsequently, the sprint swimming velocity was optimized (0.16 ms^{-1}). On the other hand, periodization and training planning plays an important role in further improving front crawl performance. Whereas, 8 to 9 weeks of concurrent resistance training during the preparatory phase could be sufficient to cope with the new imposed aquatic and in dry land training load and subsequently the improvement sprint swimming performance (4.22% to 6.82%) [1,3].

A start and turn performance improvement (2.96% to 15.14%) was shown after 9 weeks of concurrent resistance training, in which the contribution of the legs is predominant [5]. An increase in the force of pushing by the legs at the starting block level (start) and at the wall level (turn) due to the improvement in the maximum force of the lower limb can explain these present results [24,25]. More specifically, the neuromuscular adaptations represented by the learning and coordination of the back squat exercise during the 9 weeks of training, the competitive level of the swimmers, which could favor the specificity of the adaptation to the training, and the transfer of gain of strength may be indirect evidence that justifies the results obtained at the start and turn level [26]. In addition, this study does not present any specific technical start and turn training; therefore, there is no change in the start and turn technique in the two groups. This could confirm the important contribution of maximum lower limb strength in the optimization of start and turn performance.

To sum up, 9 weeks of concurrent training included sets of aquatic resistance and dry land training with 1 to 2 sessions per week, 2 to 3 sets, 6 to 10 repetitions, and intensity between 60% and 80% of 1RM could improve the maximum lower limb strength and ultimately the optimization of swimming performance.

This study has some methodological limitations that warrant discussion. Perhaps the combined resistance training is a limitation because we do not know exactly what the partial contribution of each type of training (dry land vs. in water) is to strength improvement. The small sample size could also be a limiting factor in this study. In addition, the present study only includes male swimmers. Future studies must include female counterparts. Added to that, it is required to reproduce the combined resistance training reported here in other swim strokes (breaststroke, butterfly, and backstroke) and competitive levels. The effect of concurrent resistance training on certain physiological variables (cardiorespiratory adaptation to physical exertion) may also be a future field of study.

5. Conclusions

This study showed that 9 weeks of concurrent resistance training can improve the maximum lower limb strength. In addition, our findings showed that suitable transfer of strength gain to leg kick swimming, start and turn performance after the concurrent resistance training ultimately improved the 100 m front crawl performance. Concurrent resistance training included dry land exercises (back squat, CMJ, and CMJ box) with 1 to 2 sessions per week, 2 to 3 sets, 6 to 10 repetitions, and intensity between 60% and 80% of 1RM and aquatic resistance training with water parachute is an effective strategy that can be incorporated by coaches and swimmers into their training programs to improve sprint swimming performance in competitive swimmers.

Author Contributions: Conceptualization, S.A. and R.v.d.T.; formal analysis, S.A. and R.H.; methodology, S.A., A.A., and S.G.C.; writing—original draft, S.A. and R.H. and writing—review and editing, O.G.C., T.M.B., R.v.d.T., and R.H. All authors have read and agreed to the published version of the manuscript.

Funding: This research received no external funding.

Institutional Review Board Statement: The study was conducted according to the guidelines of the Declaration of Helsinki and approved by the ethical advisory committee of the Research Unit (UR17JS01) Sports Performance, Health and Society, Higher Institute of Sport and Physical Education of Ksar Saîd, Universite de la Manouba, Tunis, 2010, Tunisia.

Informed Consent Statement: Informed consent was obtained from all subjects and guardians involved in the study.

Data Availability Statement: The data presented in this study are available on reasonable request from the corresponding author.

Acknowledgments: The authors thank all the subjects who participated in this study.

Conflicts of Interest: The authors declare no conflict of interest.

References

1. Amara, S.; Barbosa, T.M.; Negra, Y.; Hammami, R.; Khalifa, R.; Chortane, S.G. The Effect of Concurrent Resistance Training on Upper Body Strength, Sprint Swimming Performance and Kinematics in Competitive Adolescent Swimmers. A Randomized Controlled Trial. *Int. J. Environ. Res. Public Health* **2021**, *18*, 10261. [CrossRef] [PubMed]
2. Amara, S.; Crowley, E.; Sammoud, S.; Negra, Y.; Hammami, R.; Chortane, O.G.; Khalifa, R.; Chortane, S.G.; van den Tillaar, R. What Is the Optimal Strength Training Load to Improve Swimming Performance? A Randomized Trial of Male Competitive Swimmers. *Int. J. Environ. Res. Public Health* **2021**, *18*, 11770. [CrossRef] [PubMed]
3. Lopes, T.J.; Neiva, H.P.; Gonçalves, C.A.; Nunes, C.; Marinho, D.A. The effects of dry-land strength training on competitive sprinter swimmers. *J. Exerc. Sci. Fit* **2021**, *19*, 32–39. [CrossRef]
4. Amara, S.; Chortane, O.G.; Negra, Y.; Hammami, R.; Khalifa, R.; Chortane, S.G.; van den Tillaar, R. Relationship between Swimming Performance, Biomechanical Variables and the Calculated Predicted 1-RM Push-up in Competitive Swimmers. *Int. J. Environ. Res. Public Health* **2021**, *18*, 11395. [CrossRef] [PubMed]
5. Morouço, P.G.; Marinho, D.A.; Izquierdo, M.; Neiva, H.; Marques, M.C. Relative Contribution of Arms and Legs in 30 s Fully Tethered Front Crawl Swimming. *BioMed Res. Int.* **2015**, *2015*, 563206. [CrossRef]
6. Bartolomeu, R.F.; Costa, M.J.; Barbosa, T.M. Contribution of limbs' actions to the four competitive swimming strokes: A nonlinear approach. *J. Sports Sci.* **2018**, *36*, 1836–1845. [CrossRef]
7. Morais, J.E.; Marinho, D.A.; Arellano, R.; Barbosa, T.M. Start and turn performances of elite sprinters at the 2016 European Championships in swimming. *Sports Biomech.* **2019**, *18*, 100–114. [CrossRef]
8. West, D.J.; Owen, N.J.; Cunningham, D.J.; Cook, C.J.; Kilduff, L.P. Strength and power predictors of swimming starts in international sprint swimmers. *J. Strength Cond. Res.* **2011**, *25*, 950–955. [CrossRef]
9. Keiner, M.; Wirth, K.; Fuhrmann, S.; Kunz, M.; Hartmann, H.; Haff, G.G. The influence of upper- and lower-body maximum strength on swim block start, turn, and overall swim performance in sprint swimming. *J. Strength Cond. Res.* **2021**, *35*, 2839–2845. [CrossRef]
10. Sammoud, S.; Negra, Y.; Bouguezzi, R.; Hachana, Y.; Granacher, U.; Chaabene, H. The effects of plyometric jump training on jump and sport-specific performances in prepubertal female swimmers. *J. Exerc. Sci. Fit* **2021**, *19*, 25–31. [CrossRef]
11. Gourgoulis, V.; Valkoumas, I.; Boli, A.; Aggeloussis, N.; Antoniou, P. Effect of an 11-Week In-Water Training Program with Increased Resistance on the Swimming Performance and the Basic Kinematic Characteristics of the Front Crawl Stroke. *J. Strength Cond. Res.* **2019**, *33*, 95–103. [CrossRef]
12. Crowley, E.; Harrison, A.J.; Lyons, M. The Impact of Resistance Training on Swimming Performance: A Systematic Review. *Sports Med.* **2017**, *47*, 2285–2307. [CrossRef] [PubMed]
13. Haycraft, J.; Robertson, S. The effects of concurrent aerobic training and maximal strength, power and swim-specific dry-land training methods on swim performance: A review. *J. Aust. Strength Cond. Res.* **2015**, *23*, 91–99.
14. Pereira, S.M.; Ruschel, C.; Hubert, M.; Machado, L.; Roesler, H.; Fernandes, R.; Vilas-Boas, J.P. Kinematic, kinetic and EMG analysis of four front crawl flip turn techniques. *J. Sports Sci.* **2015**, *33*, 2006-15. [CrossRef] [PubMed]
15. Matuszak, M.E.; Fry, A.C.; Weiss, L.W.; Ireland, T.R.; McKnight, M.M. Effect of rest interval length on repeated 1 repetition maximum back squats. *J. Strength Cond. Res.* **2003**, *17*, 634–637.
16. Puig-Diví, A.; Escalona-Marfil, C.; Padullés-Riu, J.M.; Busquets, A.; Padullés-Chando, X.; Marcos-Ruiz, D. Validity and reliability of the Kinovea program in obtaining angles and distances using coordinates in 4 perspectives. *PLoS ONE* **2019**, *14*, e0216448. [CrossRef]
17. Weir, J.P. Quantifying test-retest reliability using the intraclass correlation coefficient and the SEM. *J. Strength Condit. Res.* **2015**, *19*, 231–240.
18. Sainani, K.L. Reliability statistics. *PM&R* **2017**, *9*, 622–628.
19. Cohen, J. *Statistical Power Analysis for the Social Sciences*; Routledge: New York, NY, USA, 1988. [CrossRef]
20. Rhea, M.R. Determining the magnitude of treatment effects in strength training research through the use of the effect size. *J. Strength Condit. Res.* **2004**, *18*, 918–920.
21. Garrido, N.; Marinho, D.A.; Reis, V.M.; van den Tillaar, R.; Costa, A.M.; Silva, A.J.; Marques, M.C. Does combined dry land strength and aerobic training inhibit performance of young competitive swimmers? *J. Sports Sci. Med.* **2010**, *9*, 300–310.
22. Kubo, K.; Ikebukuro, T.; Yata, H. Effects of squat training with different depths on lower limb muscle volumes. *Eur. J. Appl. Physiol.* **2019**, *119*, 1933–1942. [CrossRef] [PubMed]
23. Konstantaki, M.; Winter, E.M. The effectiveness of a leg-kicking training program on performance and physiological measures of competitive swimmers. *Int. J. Sports Sci. Coach.* **2007**, *2*, 37–48. [CrossRef]
24. Bishop, D.C.; Smith, R.J.; Smith, M.F.; Rigby, H.E. Effect of plyometric training on swimming block start performance in adolescents. *J. Strength Cond. Res.* **2009**, *23*, 2137–2143. [CrossRef] [PubMed]
25. Potdevin, F.J.; Alberty, M.E.; Chevutschi, A.; Pelayo, P.; Sidney, M.C. Effects of a 6-week plyometric training program on performances in pubescent swimmers. *J. Strength Cond. Res.* **2011**, *25*, 80–86. [CrossRef]
26. Folland, J.P.; Williams, A.G. The adaptations to strength training: Morphological and neurological contributions to increased strength. *Sports Med.* **2007**, *37*, 145–168. [CrossRef] [PubMed]

biology

MDPI

Article

Is the Secret in the Gut? SuperJump Activity Improves Bone Remodeling and Glucose Homeostasis by GLP-1 and GIP Peptides in Eumenorrheic Women

Sonya Vasto [1], Alessandra Amato [2], Patrizia Proia [2,*] and Sara Baldassano [1,*]

[1] Department of Biological, Chemical and Pharmaceutical Sciences and Technologies (STEBICEF), University of Palermo, Viale delle Scienze, 90128 Palermo, Italy; sonya.vasto@unipa.it
[2] Sport and Exercise Sciences Research Unit, Department of Psychological, Pedagogical and Educational Sciences, University of Palermo, 90128 Palermo, Italy; alessandra.amato02@unipa.it
* Correspondence: patrizia.proia@unipa.it (P.P.); sara.baldassano@unipa.it (S.B.)

Simple Summary: We previously showed that SuperJump activity, an innovative workout training performed on an elastic minitrampoline, exerts osteogenic action in women. The present study analyzed whether the gut peptides (GLP-1, GIP, GLP-2, PYY, ghrelin) are involved in the mechanism of action. This is because there is a link between gut peptides and bone. In fact, ingestion of a meal induces secretion of the gut peptides that act by decreasing bone resorption and blood glucose level. After 20 weeks of SuperJump activity GLP-1 and GIP levels were significantly increased while fasting insulin, glucose, insulin resistance, were significantly reduced. The study suggests that GLP-1, and GIP are involved in the mechanism of action that improves bone health and blood glucose level following 20 weeks of SuperJump activity in women.

Citation: Vasto, S.; Amato, A.; Proia, P.; Baldassano, S. Is the Secret in the Gut? SuperJump Activity Improves Bone Remodeling and Glucose Homeostasis by GLP-1 and GIP Peptides in Eumenorrheic Women. *Biology* **2022**, *11*, 296. https://doi.org/10.3390/biology11020296

Academic Editors: Georgian Badicu, Filipe Manuel Clemente and Eugenia Murawska-Ciałowicz

Received: 29 November 2021
Accepted: 9 February 2022
Published: 11 February 2022

Abstract: We showed that twenty weeks of SuperJump activity, an innovative workout training performed on an elastic minitrampoline, reduced bone resorption and increased bone formation in eumenorrheic women acting on the key points of the regulation of bone metabolism. The present study analyzed whether the gastrointestinal hormones are involved in the mechanism of action and if it has an impact on glucose homeostasis. The control group was composed of twelve women, similar to the exercise group that performed SuperJump activity for twenty weeks. The analysis was performed on blood samples and investigated GLP-1, GIP, GLP-2, PYY, ghrelin, glucose, insulin, insulin resistance, β-cell function, and insulin sensitivity. The results showed that the activity contributes to raising the GLP-1and GIP levels, and not on GLP-2, PYY, and ghrelin, which did not change. Moreover, SuperJump activity significantly reduced fasting insulin, glucose, insulin resistance, and increased insulin sensitivity but did not affect beta cell function. These data suggest that GLP-1, and GIP are involved in the mechanism of action that improves bone and glucose homeostasis following 20 weeks of SuperJump activity in eumenorrheic women.

Keywords: biological mechanisms; physical health; sports and exercise physiology; glucagon-like peptide-1; glucose-dependent insulinotropic polypeptide

1. Introduction

The gastrointestinal tract is the body's largest endocrine organ secreting hormones which in turn regulate whole-body homeostasis. Therefore, many dysmetabolic conditions such as insulin resistance or higher risk of fractures are accompanied by altered secretion of gut peptides [1,2]. Gastrointestinal secretion of gut peptides is stimulated by nutrients when these reach the intestinal L cells, but levels of the gut hormones seem to be influenced by an exercise bout [3–6], suggesting that physical exercise could modulate gut peptides release.

There is a high link between gut peptides and bone. In fact, ingestion of a meal induces secretion of gut peptides that act by decreasing bone resorption [7]. The responsible gut peptides appear to include glucagon-like peptide-1 (GLP-1), glucose-dependent insulinotropic polypeptide (GIP), glucagon-like peptide-2 (GLP-2), peptide YY (PYY), and ghrelin [8]. GLP-1 and GIP are also known as incretin hormones due to their role in regulating glucose homeostasis by acting on insulin release. However, other gut peptides such as GLP-2 and PYY influence glucose metabolism [9–13].

Physical exercise is indispensable to improve bone health [14,15] and glucose metabolism [16,17]. It can even replace glucose-lowering medication [3]. However, thus far, how exercise improves glucose homeostasis in humans is not fully understood and the influence of exercise on beta cell adaptations remains to be clarified.

SuperJump is an innovative activity performed on an elastic minitrampoline that can be used to be fit, maintain well-being, and counteract a sedentary lifestyle due to home confinement such as during COVID-19 [18]. We have previously shown that 20 weeks of SuperJump training reduced bone resorption and increased bone formation in eumenorrheic women acting on the key points of the regulation of bone metabolism [19]. In this manuscript, it was hypothesized that gastrointestinal hormones are involved in the metabolic pathway underlying bone remodeling following SuperJump exercise in eumenorrheic women. Furthermore, since gastrointestinal hormones impact on glucose metabolism, it was secondarily hypothesized that SuperJump may have effects on glucose homeostasis and beta cell function. Therefore, the aim of the study was to investigate whether the gastrointestinal hormones, and specifically GLP-1, GIP, GLP-2, PYY, and ghrelin, are involved in the mechanism of action that influences bone remodeling following 20 weeks of SuperJump activity and whether these changes would also impact on glucose homeostasis.

2. Materials and Methods

2.1. Subjects and Experimental Design

This study is part of a larger project (TRAMP2021). As previously described in [19], from an initial number of forty-two women, due to lack of inclusion criteria or withdrawal, twenty-four eumenorrheic women were randomized into two groups, the exercise group and the non-exercise (control) group for a total of twelve women in each group. Inclusion and exclusion criteria are summarized in Table 1. Briefly, during the first visit, the participants underwent anthropometric measurement and completed a habitual dietary intake assessment [20]. Before starting the activity, a blood sample (BASE) was collected; the second sample of blood was collected at the end of the 20 weeks (W20) (Figure 1).

Table 1. Inclusion and exclusion criteria of the study.

Inclusion Criteria	Exclusion Criteria
Women living in Italy	Bone fracture within the previous year
Age: 18–40 years	Self-reported long (>35 days) or short (<24 days) or irregular menstrual cycles
Currently injury free	Use of medication or suffering from any condition known to affect bone metabolism
Body mass index between 18.5 and 28 kg/m^2	Pregnancy, breastfeeding
Menstrual cycle interval between 24 and 35 days	Current smokers
	Use of any type of hormonal contraception within the past six months
	Calcium or vitamin D supplementation in the preceding six months
	Participation in moderate and high impact-activity for ≥3 h·week before enrolling in the study

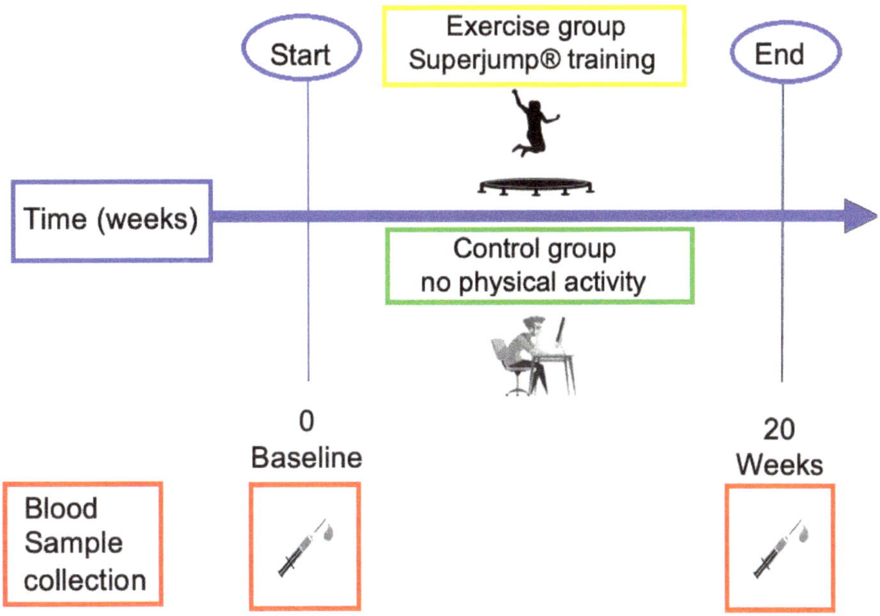

Figure 1. Overview of the experimental design. Blood samples were collected at baseline, at time 0, and after twenty weeks in the two groups of study (control group and exercise group). In the exercise group, SuperJump activity was performed for a total of 20-weeks, three times a week, 60 min each session. The control group did not perform physical activity.

2.1.1. Workout Characteristic

The exercise group performed SuperJump training (CoalSport, Rome, Italy). The intensity was 65–75% HR max and the frequency was three times a week for a total of 20 weeks. The session time was 60 min. The SuperJump training session was performed by the whole exercise group on the same days, at the same time, by the exercise group together. The training sessions were carried out on the mini trampoline and led by experienced instructors. Each session was divided into five min warm-up, a central phase with full body jumping exercises, and five min cool-down phase. The central phase was a circuit of 10 exercises, 50 s each, with 10 s of active recovery each time. The circuit was repeated five times per training session. The training session was entirely performed on the mini trampoline including the recovery phase during which the subjects continued to jump on the trampoline at the minimum intensity that allowed them to perform the jump (just lift both feet off the trampoline together). The ten resistance exercises were: (1) isometric lateral raises; (2) curl, 3) oblique; (4) adductors/ abductor; (5) triceps; (6) front lifts; (7) split jump alternating drill; (8) Pull to the chin; (9) jumping jack single arm; (10) standing Russian twist. All exercises were performed with dumbbells, with the weight allowing the subject to carry out the exercise for 50 s. SuperJump is a moderate-to-vigorous activity (sRPE = 3.1 ± 1.2); during the training session, the subjects spend 47.1 ± 34.4% of the time on moderate intensity (64 ± 76.9 % of HRmax) and 34.6 ± 39.6% of the session time on vigorous intensity (77 ± 95.9 %of HRmax) [18]. The performance and the effects of SuperJump were recently studied [18,19] and the characteristics of this activity, which include resistance exercise and impact activities such as during the active recovery phase, classify it among activities with osteogenic potential. The rationale of the SuperJump protocol was to undertake both resistance exercises and impact activities to exert osteogenic effects. Indeed, it has been demonstrated that resistance exercise has a better osteogenic potential than just aerobic exercise [8]. The active recovery phase was important to prolong "the impact stimulus" of the activity over time. The impact of a physical exercise is the combination of force magnitude and the speed at which the force is applied [21]. Activities with the most osteogenic potential have ground reaction forces (GRF) greater than 3.5 times BW (per leg), with peak force occurring in less than 0.1 s [22]. Comparing three main activities such as walking, running, jumping, the last has the greatest benefits to bone mineralization [23,24]. It also seems important that not only the characteristic of the movement but also the number of repetitions, in fact 50 jumps in a session [25], does not seem to have an osteogenic effect compared to 100 jumps [26,27]. The control group did not perform physical activity during the time of the study. Physical activity was intended as structured activity and excluded daily life activities (e.g., physically heavy work) and journeys on foot or by bike to go to work.

2.1.2. Anthropometry

Body composition, specifically lean mass and fat mass, was measured by electrical bioimpedance measurements (InBody 320 Body Composition Analyzer). Body weight and barefoot standing height were measured by using an electronic scale and a wall-mounted stadiometer, respectively (Gima 27335 and Gima 27088, Italy). Body mass index (BMI) was reported as weight (kilograms) per standing height (meters squared).

2.2. Blood Sample Collection

Blood samples were collected by a specialist the morning after overnight fasting. For plasma samples, we used a tube containing EDTA while for serum, we allowed it to clot in serum tubes at room temperature for 30 min before being centrifuged under the same conditions and were centrifuged at 3000 rpm for 10 min.

2.3. Assays

To measure gut peptides, plasma samples were collected in pre-chilled EDTA-containing tubes with apoprotein (0.6 TIU/mL blood) and dipeptidyl peptidase IV inhibitor (10 μL/mL blood). Plasma was obtained by centrifugation at 5 °C for 10 min at 3000 rpm for measurements of GLP-1, GIP, ghrelin, PYY, and GLP-2. All samples were immediately stored at −80 °C until analyzed. All samples were analyzed in duplicate. As previously reported [20], human plasma peptide samples were analyzed using the following enzyme immunoassay kit: EZGLPHS-35K for active GLP-1, EZHGIP-54K for total GIP, EZGRT-89K for total ghrelin, EZHPPYYT66K for total PYY, and EZGLP-237K for GLP-2, all from Millipore. The inter- and intra-assay coefficients of variation for total ghrelin were 6.62% and 1.32%; 6.62% and 5.15% for GLP-2; 11.5% and 4.5% for active GLP-1; 7.41% and 2.27% for total PYY; and 3.37% and 6.45% for total GIP. All samples were measured in one assay to avoid inter-assay variation. Glucose, insulin, total cholesterol, HDL-cholesterol, and triglycerides were measured by standard commercial assays supplied by Roche Diagnostics performed on the Roche COBAS c501. The HOMA2 computer model was used to estimate insulin resistance (HOMA2-IR), β-cell function (HOMA2-%B), and insulin sensitivity (HOMA2-%S) from fasting insulin and glucose concentrations calculated by the HOMA2 calculator for specific insulin version 2.2.3, available from http://www.dtu.ox.ac.uk/homacalculator, accessed on 28 November 2021. The method is an updated HOMA model and has been used extensively to measure insulin resistance β-cell function and insulin sensitivity [28,29].

2.4. Ethics

The study conducted in accordance with the Declaration of Helsinki was approved by ethics committee 1 of the University of Palermo, Policlinico Giaccone Hospital, approval number 2–2020–27. Before the start of the study, all subjects involved provided written informed consent. In addition, the clinical study was registered on Clinicaltrials.gov under number NCT04942691.

2.5. Statistics

Based on the results of previous studies on exercise and gut peptides [3,30], the study was powered to detect a change in GLP-1 of 30% (SD 20%) considering a Type I error (α) = 0.05 (two-sided), and Type II error (β) = 0.20 (power of 80%). An a priori power calculation determined that ten subjects were required to achieve 80% power at $p < 0.05$ by using G Power software. The comparison between the groups was performed by one-way ANOVA followed by Tukey's posttest. A $p < 0.05$ was considered to be statistically significant by using GraphPad Prism software.

3. Results

The cohort under investigation did not show significant differences in body mass index or composition among the groups (control vs. exercise) or within the groups (time zero vs. 20 weeks). Instead, a significant difference was observed in triglyceride levels in the exercise group at W20 compared with BASE, while no differences were reported in total cholesterol, HDL-cholesterol, and LDL-cholesterol between the groups (control vs. exercise) or within the groups (time zero vs. 20 weeks) (Table 2). Moreover, we previously showed that there was significant change in the markers of bone remodeling after 20 weeks of training in the exercise group. The marker of bone formation, osteocalcin, increased from 16.2 ± 5 (BASE) to 22.2 ± 6 μg/L (W20). The marker of bone resorption, CTX, decreased from 0.44 ± 0.1 (BASE) to 0.29 ± 0.1 μg/L (W20). PTH decreased from 44 ± 15 (BASE) to 34 ± 11 ng/L (W20). Calcitonin, vitamin D, and phosphate concentrations did not change while there was a significant increase in calcium and potassium concentrations [19].

Table 2. Characteristics of the subjects measured baseline and after 20 weeks (20 W) in the two groups of women.

Subjects Charact	Control Group			Exercise Group		
	BASE	20 W		BASE	20 W	
	Mean ± SD	Mean ± SD	*p*-Value	Mean ± SD	Mean ± SD	*p*-Value
BMI (kg/m^2)	22.5 ± 2.7	23.7 ± 2.9	$p > 0.05$	22.8 ± 2.4	22.8 ± 2.8	$p > 0.05$
LM %	74.4 ± 5.8	76.6 ± 5.2	$p > 0.05$	73.2 ± 5.9	73.7 ± 7.2	$p > 0.05$
FM %	25.6 ± 5.8	23.9 ± 6.2	$p > 0.05$	26.8 ± 6	26.3 ± 7.2	$p > 0.05$
TRIG (mg/dL)	91 ± 14	89 ± 26	$p > 0.05$	80 ± 21	55 ±18	$p = 0.02$
Total Chol (mg/dL)	182 ± 18	188 ± 27	$p > 0.05$	182 ± 23	184 ± 24	$p > 0.05$
HDL-Chol (mg/dL)	77 ± 12	75 ± 11	$p > 0.05$	77 ± 15	80 ± 14	$p > 0.05$
LDL-Chol	93 ± 24	98 ± 18	$p > 0.05$	100 ± 14	89 ± 16	$p > 0.05$

Abbreviation: Charact, Characteristics; BMI, Body Mass Index; LM, Lean Mass; FM, Fat Mass; TRIG, Triglycerides; Chol, Cholesterol.

3.1. Incretins

In the control group, there were no significant changes in plasma GLP-1 and GIP levels at W20 compared with BASE within the group (Figure 2). In the exercise group, GLP-1 was significantly increased at W20 compared with BASE (Figure 2A). GLP-1 concentrations at W20 increased by 58% from BASE (5.7 ± 1.7 vs. 3.6 ± 0.7 pmol/L). The levels detected were within the normal range. In addition, the GIP level was significantly increased. Specifically, in the exercise group, GIP increased by 102% at W20 compared to BASE (64.3 ± 21 vs. 31.9 ± 12 pg/mL) (Figure 2B) and the concentrations detected were within the physiological range. There was a significant change in the endogenous levels of GLP-1 and GIP in the exercise group at W20 compared to the control group (Figure 2A,B).

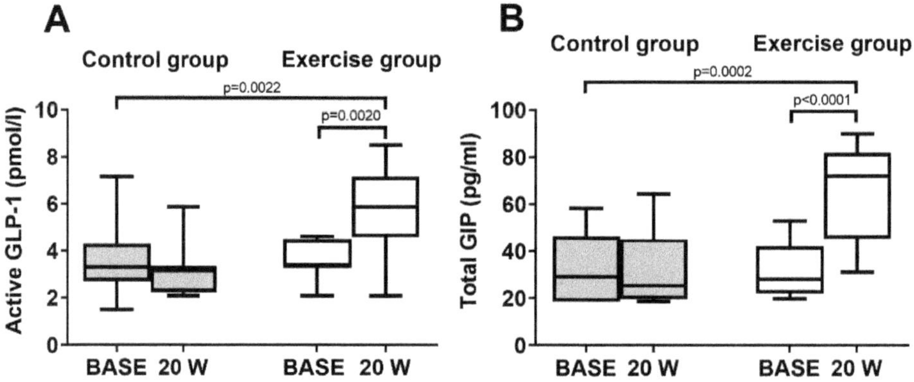

Figure 2. Endogenous incretin levels measured baseline (BASE) and after 20 weeks (W20) in the control group and exercise group. (**A**) Box and whisker plot of GLP-1. (**B**) Box and whisker plot of GIP.

3.2. Other Gut Hormones

In the control group, there were no changes in GLP-2 (2.1 ± 0.25 vs. 2 ± 0.26 ng/mL) PYY (58 ± 11 vs. 56 ± 12 pg/mL) and ghrelin levels (1081 ± 491 vs. 1036 ± 376 pg/mL) at W20 compared with BASE. Moreover, in the exercise group, SuperJump training did not affect plasma GLP-2 (2.1 ± 0.20 vs. 1.9 ± 0.16 ng/mL), PYY (55 ± 11 vs. 52 ± 6 pg/mL), and ghrelin concentrations (947 ± 351 vs. 1014 ± 458 pg/mL) at W20 compared to the baseline. Additionally, the comparison between the two groups (control vs. exercises) showed no significant changes in GLP-2, PYY, and ghrelin (Figure 3A–C).

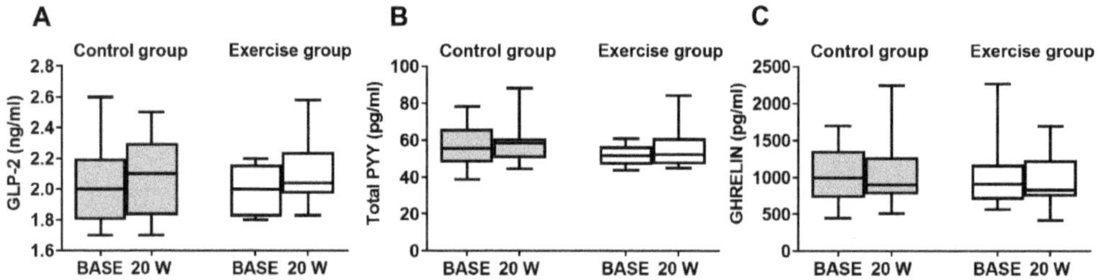

Figure 3. Endogenous gut peptide levels measured at baseline (BASE) and after 20 weeks (W20) in the control group and exercise group. (**A**) Box and whisker plot of GLP-2. (**B**) Box and whisker plot of PYY. (**C**) Box and whisker plot of ghrelin.

3.3. Markers of Glucose Homeostasis

In the control group, there was no difference in fasting glucose (88 ± 3.1 vs. 90 ± 4.2 mg/dL), insulin (8.1 ± 2.8 vs. 8.6± 3.0 mUI/L), or insulin resistance (1.0 ± 0.4 vs. 1.2 ± 0.3) at W20 compared with BASE. In the exercise group, SuperJump training significantly reduced fasting glucose (80 ± 7.4 vs. 88 ± 4.0 mg/dL), insulin (4.7 ± 1.9 vs. 7.9 ± 2.6 mUI/L), and insulin resistance (0.6 ± 0.2 vs. 1.0± 0.3) at W20 compared with BASE. The comparison between the groups (control vs. exercise) showed a significant reduction in fasting insulin, glucose, and insulin resistance in the exercise group at W20 compared to the control group (Figure 4A–C). There was no difference in β-cell function in the control group or the exercise group at W20 compared with BASE (Figure 4D). There was no difference in insulin sensitivity in the control group while in the exercise group, there was a significant increase in insulin sensitivity at W20 compared with BASE. The comparison between the groups (control vs. exercises) showed a significant increase in insulin sensitivity at W20 in the exercise group compared to the control group (Figure 4E).

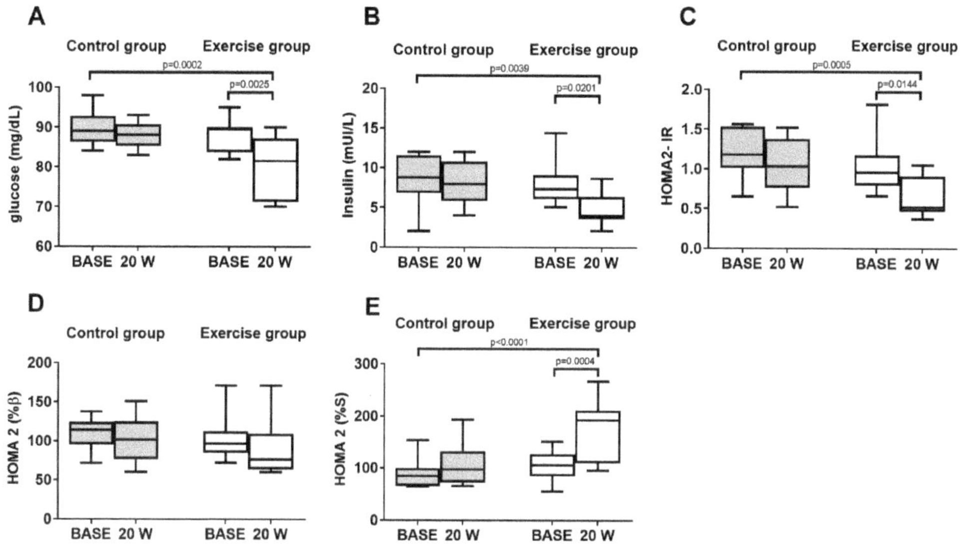

Figure 4. Markers of glucose homeostasis measured at baseline (BASE) and after 20 weeks (W20) in the control group and exercise group. (**A**) Box and whisker plot of fasting glucose. (**B**) Box and whisker plot of fasting insulin. (**C**) Box and whisker plot of insulin resistance. (**D**) Box and whisker plot of β-cell function. (**E**) Box and whisker plot of insulin sensitivity.

4. Discussion

In previous studies, the endogenous levels of GLP-1 and GIP following physical activity have been measured at the end of the single training session [2] and have not investigated the potential link between gut peptides, bone remodeling, and physical activity.

This study shows that the gut peptides GLP-1 and GIP are involved in the mechanism of action that influences bone remodeling and ameliorates glucose homeostasis following 20 weeks of SuperJump training in eumenorrheic women.

We previously showed that SuperJump activity exerts osteogenic action in eumenorrheic women. In fact, after 20 weeks of SuperJump training, the levels of the marker of bone resorption CTX were significantly reduced while the levels of the marker of bone formation osteocalcin were increased. We found that PTH, calcium, and potassium were involved in the mechanism of action [19]. The present study showed that the SuperJump exercise program for 20 weeks significantly increased endogenous GLP-1 and GIP levels, suggesting that these two incretins are part of the mechanism of action by which this type of high impact activity influences bone remodeling in eumenorrheic women. This was confirmed by the lack of changes in the endogenous level of GLP-1 or GIP in the control group of sedentary women. We observed an increase in the endogenous levels of GLP-1 and GIP that was positive for bone remodeling because it is within the normal physiological range. In fact, in women, treatment with the long-acting agonist of the GLP-1R, liraglutide, increased P1NP (bone formation marker) and bone mineral content and reduced the bone loss, indicating that GLP-1 acted by increasing bone formation [31]. In ovariectomized rats, the treatment with liraglutide increased bone mineral density and improved trabecular thickness, number, and volume [32]. Moreover, the activation of GLP-1R decreased P1NP secretion and increased cell viability in osteoblasts [33]. Additionally, GIP exerts an anti-resorptive action and anabolic effect [34]. GIP stimulated the expression of P1NP and of

ALP activity [35] and reduced the level of CTX, the marker of bone resorption [36]. The GIP receptor is expressed in osteoblast and osteoclast derived cell lines. Therefore, the loss of function for the GIP receptor gene, in women carrying the gene polymorphism E354Q, was correlated with decreased bone mineral density and increased risk of fractures [37].

Regarding GLP-2, previous studies have shown that GLP-2 administered subcutaneously in postmenopausal women reduced CTX, markers of bone resorption, and had a minimal effect on bone formation [38]. In our study, GLP-2 levels did not differ between the exercise and control groups, suggesting that it is not involved in the mechanism of action that impacts on bone remodeling in exercising women. We cannot exclude that 20 weeks of SuperJump training were not sufficient to induce differences in the endogenous levels of the peptide. However, on the basis of previous studies, supraphysiological doses of exogenous GLP-2 are necessary to reduce bone resorption [1]. Thus, changes within the physiological range may not be sufficient to see an effect and this may account for the lack of differences. However, thus far, it is still unknown whether GLP-2 affects bone metabolism directly or indirectly by involving other intestinal factors. In fact, the GLP-2 receptor has not been identified in human osteoclasts or in any other bone-related cell types [34].

Evidence from human studies indicates that PYY modulates bone homeostasis [1]. The PYY increases were associated with low bone mineral density in women with weight alteration [39,40] and absence of menstrual periods [41]. In our study, we did not find any difference in PYY concentration, ruling out an involvement of PYY. This is probably because our study population was constituted of eumenorrheic women with no weight alteration. We did not also find any differences in circulating ghrelin levels in the groups of study. Ghrelin is first a regulator of energy metabolism but seems to influence bone [42]. However, the basal concentration of ghrelin is inversely associated with body mass index. In fact, reduced ghrelin levels were found in obese people [43].

Thus far, we know that physical activity ameliorates glucose homeostasis, but it is still unclear how it acts to do so. Here, we suggest that GLP-1 and GIP could be part of the physiological mechanism of action that improves glucose homeostasis following a high impact physical activity. In fact, the higher endogenous GLP-1 and GIP level in the exercise group following 20 weeks of SuperJump activity improved glucose metabolism. In the exercise group, reduced fasting glucose, insulin and insulin resistance, and increased insulin sensitivity was observed. This was confirmed by the lack of changes in fasting glucose, insulin, or insulin sensitivity in the control group of the study. This agrees with the reduced gut peptide responses reported in sedentary obese people that develop insulin resistance [44]. The observation of elevated GLP-1 and reduced insulin could be counterintuitive in consideration of the ability of GLP-1 to stimulate insulin release. Thus, it is necessary to point out that the incretin effect is defined as the increase in insulin response after an oral ingestion of glucose. In fact, GLP-1 induces insulin secretion via the GLP-1R in a glucose-regulated manner [45]. The blood samples in the groups of study were obtained after an overnight fast. Thus, it is possible to hypothesize that fasting levels of insulin and glucose were lower thanks to a regulatory mechanism of the peptide on beta-cells that could be sensitized to secrete the minimum amount of insulin required to have an accurate glycemic control. In fact, insulin sensitivity was increased. Further studies are required.

A key strength of the present study was to analyze the effects of chronic exercise (20 weeks of SuperJump exercise) on endogenous peptides with respect to previous studies that have focused on the effects of acute exercise on the secretion of gastrointestinal hormones [46]. In fact, to our knowledge, the endogenous levels of GLP-1 and GIP following physical activity have only been measured acutely, at the end of the training session [2,46–51], and not after several weeks of training program like in our study. Moreover, the mechanism of action and therefore the potential link between gut peptides, bone remodeling, glucose metabolism, and physical activity has not been investigated. For GLP-1, the studies measured total and not the active form of GLP-1 l such as in our study [51]. However, GLP-1, similar to us, showed an increase in basal GLP-1 levels [46,51,52]. These investigations were conducted after acute exercise not only in normal weight, but also in obese trained women and suggest that endogenous levels of GLP-1 are very sensitive to physical activity. For GIP, the studies have been conducted in obese/diabetic cohort of patients or following a glucose tolerance test [53–55]. These studies were unconcluded and showed a decrease, increase, or no change in the GIP concentrations. We are conscious that we did not compare the SuperJump group with another group that performed other forms of exercise such as a different high impact or strength training. Thus, we do not know the effects of other forms of chronic exercise on basal gut peptide release and further studies are necessary. In fact, the exercise protocol characteristics such as the age, fitness level, BMI, and the exercise protocols such as duration and intensity could differently impact on gut peptide release. which is a limitation of the study. We also do not know whether the observed effects were mediated by GLP-1 or GIP. We may suppose that the observed effects are mediated by a synergistic action of the two peptides, but further studies are necessary to clarify the point.

5. Conclusions

In conclusion, the study points out the ability of physical activity by increasing endogenous GLP-1 and GIP levels to ameliorate bone and glucose metabolism, suggesting that the peptides are involved in the physiological mechanism of action that improves bone and glucose homeostasis following 20 weeks of SuperJump activity in eumenorrheic women.

Author Contributions: Conception and design of the work, A.A., S.V., S.B. and P.P.; Analysis and interpretation of the data, S.V., S.B. and P.P.; Drafted the work, S.B.; Substantively revised work, S.V., S.B., A.A. and P.P. All authors have read and agreed to the published version of the manuscript.

Funding: This work was supported by a grant from Ministero dell'Istruzione, dell'Università e della Ricerca," Italy (FFR-D15-162221) and from Elijha sunshine s.r.l. The APC was covered by Elijha sunshine s.r.l.

Institutional Review Board Statement: The study was approved by the University of Palermo ethics committee Palermo 1, Policlinico Giaccone Hospital (2-2020-27) and is registered at https://clinicaltrials.gov/, accessed on 28 November 2021, (NCT04942691).

Informed Consent Statement: Informed consent was obtained from all subjects involved in the study.

Data Availability Statement: The datasets during and/or analyzed during the current study are available from the corresponding author on reasonable request.

Acknowledgments: The authors would like to thank all the participants enrolled in the study for their significant contribution.

Conflicts of Interest: The authors declare no conflict of interest.

References

1. Schiellerup, S.P.; Skov-Jeppesen, K.; Windeløv, J.A.; Svane, M.; Holst, J.; Hartmann, B.; Rosenkilde, M.M. Gut Hormones and Their Effect on Bone Metabolism. Potential Drug Therapies in Future Osteoporosis Treatment. *Front. Endocrinol.* **2019**, *10*, 75. [CrossRef] [PubMed]
2. Zouhal, H.; Sellami, M.; Saeidi, A.; Slimani, M.; Abbassi-Daloii, A.; Khodamoradi, A.; El Hage, R.; Hackney, A.C.; Ben Abderrahman, A. Effect of physical exercise and training on gastrointestinal hormones in populations with different weight statuses. *Nutr. Rev.* **2019**, *77*, 455–477. [CrossRef] [PubMed]
3. Martins, C.; Morgan, L.M.; Bloom, S.R.; Robertson, M.D. Effects of exercise on gut peptides, energy intake and appetite. *J. Endocrinol.* **2007**, *193*, 251–258. [CrossRef] [PubMed]
4. Burns, S.F.; Broom, D.R.; Miyashita, M.; Mundy, C.; Stensel, D.J. A single session of treadmill running has no effect on plasma total ghrelin concentrations. *J. Sports Sci.* **2007**, *25*, 635–642. [CrossRef] [PubMed]
5. Broom, D.R.; Batterham, R.L.; King, J.A.; Stensel, D.J. Influence of resistance and aerobic exercise on hunger, circulating levels of acylated ghrelin, and peptide YY in healthy males. *Am. J. Physiol. Regul. Integr. Comp. Physiol.* **2009**, *296*, R29–R35. [CrossRef] [PubMed]
6. King, J.A.; Wasse, L.K.; Broom, D.R.; Stensel, D.J. Influence of brisk walking on appetite, energy intake, and plasma acylated ghrelin. *Med. Sci. Sports Exerc.* **2010**, *42*, 485–492. [CrossRef]
7. Bjarnason, N.H.; Henriksen, E.E.G.; Alexandersen, P.; Christgau, S.; Henriksen, D.B.; Christiansen, C. Mechanism of Circadian Variation in Bone Resorption. *Bone* **2002**, *30*, 307–313. [CrossRef]

8. Proia, P.; Amato, A.; Drid, P.; Korovljev, D.; Vasto, S.; Baldassano, S. The Impact of Diet and Physical Activity on Bone Health in Children and Adolescents. *Front Endocrinol.* **2021**, *12*, 704647. [CrossRef]
9. Baldassano, S.; Amato, A. GLP-2: What do we know? What are we going to discover? *Regul. Pept.* **2014**, *194–195*, 6–10. [CrossRef]
10. Baldassano, S.; Amato, A.; Caldara, G.F.; Mulè, F. Glucagon-like peptide-2 treatment improves glucose dysmetabolism in mice fed a high-fat diet. *Endocrine* **2016**, *54*, 648–656. [CrossRef]
11. Baldassano, S.; Amato, A.; Terzo, S.; Caldara, G.; Lentini, L.; Mulè, F. Glucagon-like peptide-2 analog and inflammatory state in obese mice. *Endocrine* **2020**, *68*, 695–698. [CrossRef]
12. Baldassano, S.; Gasbjerg, L.; Kizilkaya, H.; Rosenkilde, M.; Holst, J.; Hartmann, B. Increased Body Weight and Fat Mass After Subchronic GIP Receptor Antagonist, but Not GLP-2 Receptor Antagonist, Administration in Rats. *Front. Endocrinol.* **2019**, *10*, 492. [CrossRef] [PubMed]
13. Baldassano, S.; Accardi, G.; Vasto, S. Beta-glucans and cancer: The influence of inflammation and gut peptide. *Eur. J. Med. Chem.* **2017**, *142*, 486–492. [CrossRef] [PubMed]
14. Amato, A.; Baldassano, S.; Cortis, C.; Cooper, J.; Proia, P. Physical activity, nutrition, and bone health. *Hum mov.* **2018**, *19*, 1–10. [CrossRef]
15. Amato, A.; Proia, P.; Caldara, G.F.; Alongi, A.; Ferrantelli, V.; Baldassano, S. Analysis of Body Perception, Preworkout Meal Habits and Bone Resorption in Child Gymnasts. *Int. J. Environ. Res. Public Health* **2021**, *18*, 2184. [CrossRef]
16. Fäldt, J.; Wernstedt, I.; Fitzgerald, S.M.; Wallenius, K.; Bergström, G.; Jansson, J.O. Reduced exercise endurance in interleukin-6-deficient mice. *Endocrinology* **2004**, *145*, 2680–2686. [CrossRef] [PubMed]
17. Færch, K.; Torekov, S.S.; Vistisen, D.; Johansen, N.B.; Witte, D.R.; Jonsson, A.; Pedersen, O.; Hansen, T.; Lauritzen, T.; Sandbæk, A.; et al. GLP-1 Response to Oral Glucose Is Reduced in Prediabetes, Screen-Detected Type 2 Diabetes, and Obesity and Influenced by Sex: The ADDITION-PRO Study. *Diabetes* **2015**, *64*, 2513–2525. [CrossRef] [PubMed]
18. Iannaccone, A.; Fusco, A.; Jaime, S.J.; Baldassano, S.; Cooper, J.; Proia, P.; Cortis, C. Stay home, stay active with superjump®: A home-based activity to prevent sedentary lifestyle during COVID-19 outbreak. *Sustainability* **2020**, *12*, 135. [CrossRef]
19. Vasto, S.; Amato, A.; Proia, P.; Caldarella, R.; Cortis, C.; Baldassano, S. Dare to Jump: The Effect of New High Impact Activity SuperJump on Bone Remodeling. A New Tool to Be Fit During COVID-19 Home Confinement. *Biol. Sport* **2022**, *39*, 1011–1019. [CrossRef]
20. Baldassano, S.; Accardi, G.; Aiello, A.; Buscemi, S.; Di Miceli, G.; Galimberti, D.; Candore, G.; Ruisi, P.; Caruso, C.; Vasto, S. Fibres as functional foods and the effects on gut hormones: The example of β-glucans in a single arm pilot study. *J. Funct. Foods* **2018**, *47*, 264–269. [CrossRef]
21. Gunter, K.; Almstedt, H.; Janz, K. Physical Activity in Childhood May Be the Key to Optimizing Lifespan Skeletal Health. *Exerc. Sport Sci. Rev.* **2012**, *40*, 13–21. [CrossRef] [PubMed]
22. Hind, K.; Burrows, M. Weight-bearing exercise and bone mineral accrual in children and adolescents: A review of controlled trials. *Bone* **2007**, *40*, 14–27. [CrossRef] [PubMed]
23. MacKelvie, K.J.; Khan, K.M.; Petit, M.A.; Janssen, P.A.; McKay, H.A. A school-based exercise intervention elicits substantial bone health benefits: A 2-year randomized controlled trial in girls. *Pediatrics* **2003**, *112*, e447. [CrossRef] [PubMed]
24. MacKelvie, K.J.; Petit, M.A.; Khan, K.M.; Beck, T.J.; McKay, H.A. Bone mass and structure are enhanced following a 2-year randomized controlled trial of exercise in prepubertal boys. *Bone* **2004**, *34*, 755–764. [CrossRef] [PubMed]
25. Wiebe, P.N.; Blimkie, C.J.; Farpour-Lambert, N.; Briody, J.; Marsh, D.; Kemp, A.; Cowell, C.; Howman-Giles, R. Effects of single-leg drop-landing exercise from different heights on skeletal adaptations in prepubertal girls: A randomized controlled study. *Pediatr. Exerc. Sci.* **2008**, *20*, 211–228. [CrossRef]
26. Fuchs, R.K.; Bauer, J.J.; Snow, C.M. Jumping improves hip and lumbar spine bone mass in prepubescent children: A randomized controlled trial. *J. Bone Miner. Res.* **2001**, *16*, 148–156. [CrossRef] [PubMed]
27. Gunter, K.; Baxter-Jones, A.D.; Mirwald, R.L.; Almstedt, H.; Fuller, A.; Durski, S.; Snow, C. Jump starting skeletal health: A 4-year longitudinal study assessing the effects of jumping on skeletal development in pre and circum pubertal children. *Bone* **2008**, *42*, 710–718. [CrossRef]
28. Wallace, T.M.; Levy, J.C.; Matthews, D.R. Use and abuse of HOMA modeling. *Diabetes Care* **2004**, *27*, 1487–1495. [CrossRef]
29. Ghasemi, A.; Tohidi, M.; Derakhshan, A.; Hasheminia, M.; Azizi, F.; Hadaegh, F. Cut-off points of homeostasis model assessment of insulin resistance, beta-cell function, and fasting serum insulin to identify future type 2 diabetes: Tehran Lipid and Glucose Study. *Acta Diabetol.* **2015**, *52*, 905–915. [CrossRef]
30. Druce, M.R.; Wren, A.M.; Park, A.J.; Milton, J.E.; Patterson, M.; Frost, G.; Ghatei, M.A.; Small, C.; Bloom, S.R. Ghrelin increases food intake in obese as well as lean subjects. *Int. J. Obes.* **2005**, *29*, 1130–1136. [CrossRef]
31. Iepsen, E.W.; Lundgren, J.R.; Hartmann, B.; Pedersen, O.; Hansen, T.; Jørgensen, N.R.; Jensen, J.E.; Holst, J.J.; Madsbad, S.; Torekov, S.S. GLP-1 Receptor Agonist Treatment Increases Bone Formation and Prevents Bone Loss in Weight-Reduced Obese Women. *J. Clin. Endocrinol. Metab.* **2015**, *100*, 2909–2917. [CrossRef] [PubMed]
32. Lu, N.; Sun, H.; Yu, J.; Wang, X.; Liu, D.; Zhao, L.; Sun, L.; Zhao, H.; Tao, B.; Liu, J. Glucagon-like peptide-1 receptor agonist Liraglutide has anabolic bone effects in ovariectomized rats without diabetes. *PLoS ONE* **2015**, *10*, e0132744. [CrossRef] [PubMed]
33. Yamada, C.; Yamada, Y.; Tsukiyama, K.; Yamada, K.; Udagawa, N.; Takahashi, N.; Tanaka, K.; Drucker, D.J.; Seino, Y.; Inagaki, N. The murine glucagon-like peptide-1 receptor is essential for control of bone resorption. *Endocrinology* **2008**, *149*, 574–579. [CrossRef]

34. Pacheco-Pantoja, E.; Ranganath, L.; Gallagher, J.; Wilson, P.; Fraser, W. Receptors and effects of gut hormones in three osteoblastic cell lines. *BMC Physiol.* **2011**, *11*, 12. [CrossRef] [PubMed]
35. Bollag, R.J.; Zhong, Q.; Phillips, P.; Min, L.; Zhong, L.; Cameron, R.; Mulloy, A.L.; Rasmussen, H.; Qin, F.; Ding, K.H.; et al. Osteoblast-derived cells express functional glucose-dependent insulinotropic peptide receptors. *Endocrinology* **2000**, *141*, 1228–1235. [CrossRef] [PubMed]
36. Nissen, A.; Christensen, M.; Knop, F.K.; Vilsbøll, T.; Holst, J.J.; Hartmann, B. Glucose-dependent insulinotropic polypeptide inhibits bone resorption in humans. *J. Clin. Endocrinol. Meta.b* **2014**, *99*, E2325–E2329. [CrossRef]
37. Torekov, S.S.; Harsløf, T.; Rejnmark, L.; Eiken, P.; Jensen, J.B.; Herman, A.P.; Hansen, T.; Pedersen, O.; Holst, J.J.; Langdahl, B.L. A Functional Amino Acid Substitution in the Glucose-Dependent Insulinotropic Polypeptide Receptor (GIPR) Gene is Associated With Lower Bone Mineral Density and Increased Fracture Risk. *J. Clin. Endocrinol. Metab.* **2014**, *99*, E729–E733. [CrossRef]
38. Askov-Hansen, C.; Jeppesen, P.B.; Lund, P.; Hartmann, B.; Holst, J.J.; Henriksen, D.B. Effect of glucagon-like peptide-2 exposure on bone resorption: Effectiveness of high concentration versus prolonged exposure. *Regul. Pept.* **2013**, *181*, 4–8. [CrossRef]
39. Utz, A.L.; Lawson, E.A.; Misra, M.; Mickley, D.; Gleysteen, S.; Herzog, D.B.; Klibanski, A.; Miller, K.K. Peptide YY (PYY) levels and bone mineral density (BMD) in women with anorexia nervosa. *Bone* **2008**, *43*, 135–139. [CrossRef]
40. Russell, M.; Stark, J.; Nayak, S.; Miller, K.K.; Herzog, D.B.; Klibanski, A.; Misra, M. Peptide YY in adolescent athletes with amenorrhea, eumenorrheic athletes and non-athletic controls. *Bone* **2009**, *45*, 104–109. [CrossRef]
41. Scheid, J.L.; Toombs, R.J.; Ducher, G.; Gibbs, J.C.; Williams, N.I.; de Souza, M.J. Estrogen and peptide YY are associated with bone mineral density in premenopausal exercising women. *Bone* **2011**, *49*, 194–201. [CrossRef] [PubMed]
42. Ibrahim Abdalla, M.M. Ghrelin—Physiological Functions and Regulation. *Eur. Endocrinol.* **2015**, *11*, 90–95. [CrossRef] [PubMed]
43. McLaughlin, T.; Abbasi, F.; Lamendola, C.; Frayo, R.S.; Cummings, D.E. Plasma ghrelin concentrations are decreased in insulin-resistant obese adults relative to equally obese insulin-sensitive controls. *J. Clin. Endocrinol. Metab.* **2004**, *89*, 1630–1635. [CrossRef] [PubMed]
44. Koliaki, C.; Liatis, S.; Dalamaga, M.; Kokkinos, A. The Implication of Gut Hormones in the Regulation of Energy Homeostasis and Their Role in the Pathophysiology of Obesity. *Curr. Obes. Rep.* **2020**, *9*, 255–271. [CrossRef] [PubMed]
45. Meloni, A.R.; DeYoung, M.B.; Lowe, C.; Parkes, D.G. GLP-1 receptor activated insulin secretion from pancreatic β-cells: Mechanism and glucose dependence. *Diabetes Obes. Metab.* **2013**, *15*, 15–27. [CrossRef]
46. Schubert, M.M.; Sabapathy, S.; Leveritt, M.; Desbrow, B. Acute exercise and hormones related to appetite regulation: A meta-analysis. *Sports Med.* **2014**, *44*, 387–403. [CrossRef] [PubMed]
47. Hallworth, J.R.; Copeland, J.L.; Doan, J.; Hazell, T.J. The Effect of Exercise Intensity on Total PYY and GLP-1 in Healthy Females: A Pilot Study. *J. Nutr. Metab.* **2017**, *2017*, 4823102. [CrossRef]
48. Ueda, S.Y.; Miyamoto, T.; Nakahara, H.; Shishido, T.; Usui, T.; Katsura, Y.; Yoshikawa, T.; Fujimoto, S. Effects of exercise training on gut hormone levels after a single bout of exercise in middle-aged Japanese women. *Springerplus* **2013**, *2*, 83. [CrossRef]
49. Howe, S.M.; Hand, T.M.; Larson-Meyer, D.E.; Austin, K.J.; Alexander, B.M.; Manore, M.M. No Effect of Exercise Intensity on Appetite in Highly-Trained Endurance Women. *Nutrients* **2016**, *8*, 223. [CrossRef]
50. Larson-Meyer, D.E.; Palm, S.; Bansal, A.; Austin, K.J.; Hart, A.M.; Alexander, B.M. Influence of running and walking on hormonal regulators of appetite in women. *J. Obes.* **2012**, *2012*, 730409. [CrossRef]
51. Unick, J.L.; Otto, A.D.; Goodpaster, B.H.; Helsel, D.L.; Pellegrini, C.A.; Jakicic, J.M. Acute effect of walking on energy intake in overweight/obese women. *Appetite* **2010**, *55*, 413–419. [CrossRef]
52. Lund, M.T.; Taudorf, L.; Hartmann, B.; Helge, J.W.; Holst, J.J.; Dela, F. Meal induced gut hormone secretion is altered in aerobically trained compared to sedentary young healthy males. *Eur. J. Appl. Physiol.* **2013**, *113*, 2737–2747. [CrossRef] [PubMed]
53. Martins, C.; Kulseng, B.; Rehfeld, J.F.; King, N.A.; Blundell, J.E. Effect of chronic exercise on appetite control in overweight and obese individuals. *Med. Sci. Sports Exerc.* **2013**, *45*, 805–812. [CrossRef]
54. Kelly, K.R.; Brooks, L.M.; Solomon, T.P.; Kashyap, S.R.; O'Leary, V.B.; Kirwan, J.P. The glucose-dependent insulinotropic polypeptide and glucose-stimulated insulin response to exercise training and diet in obesity. *Am. J. Physiol. Endocrinol. Metab.* **2009**, *296*, E1269–E1274. [CrossRef] [PubMed]
55. Larsen, P.S.; Donges, C.E.; Guelfi, K.J.; Smith, G.C.; Adams, D.R.; Duffield, R. Effects of Aerobic, Strength or Combined Exercise on Perceived Appetite and Appetite-Related Hormones in Inactive Middle-Aged Men. *Int. J. Sport Nutr. Exerc. Metab.* **2017**, *27*, 389–398. [CrossRef] [PubMed]

biology

MDPI

Article

Vibrating Exercise Equipment in Middle-Age and Older Women with Chronic Low Back Pain and Effects on Bioelectrical Activity, Range of Motion and Pain Intensity: A Randomized, Single-Blinded Sham Intervention Study

Grzegorz Zurek [1], Martyna Kasper-Jędrzejewska [2,*], Iwona Dobrowolska [2], Agata Mroczek [2], Gerda Delaunay [1], Kuba Ptaszkowski [3] and Tomasz Halski [2]

[1] Department of Biostructure, University School of Physical Education, I.J. Paderewskiego 35, 51-612 Wroclaw, Poland; grzegorz.zurek@awf.wroc.pl (G.Z.); gerda.delaunay@awf.wroc.pl (G.D.)
[2] Institute of Health Sciences, University of Opole, Katowicka 68, 45-060 Opole, Poland; iwona.dobrowolska@uni.opole.pl (I.D.); agata.mroczek@uni.opole.pl (A.M.); tomasz.halski@uni.opole.pl (T.H.)
[3] Department of Clinical Biomechanics and Physiotherapy in Motor System Disorders, Faculty of Health Science, Wroclaw Medical University, Grunwaldzka 2, 50-355 Wroclaw, Poland; kuba.ptaszkowski@umed.wroc.pl
* Correspondence: martyna.kasperjedrzejewska@uni.opole.pl

Citation: Zurek, G.; Kasper-Jędrzejewska, M.; Dobrowolska, I.; Mroczek, A.; Delaunay, G.; Ptaszkowski, K.; Halski, T. Vibrating Exercise Equipment in Middle-Age and Older Women with Chronic Low Back Pain and Effects on Bioelectrical Activity, Range of Motion and Pain Intensity: A Randomized, Single-Blinded Sham Intervention Study. *Biology* **2022**, *11*, 268. https://doi.org/10.3390/biology11020268

Academic Editors: Filipe Manuel Clemente, Georgian Badicu and Eugenia Murawska-Ciałowicz

Received: 2 December 2021
Accepted: 7 February 2022
Published: 8 February 2022

Simple Summary: Physical activity is often recommended as part of the management of chronic low back pain, which is one of the most common musculoskeletal disorders. Vibrating exercise equipment is used despite little scientific evidence to support its effectiveness in the prevention and treatment of musculoskeletal problems. The aim of this study was to evaluate the efficiency of using vibrating exercise equipment in women with chronic low back pain. Here, 92 women aged 49–80 years were assigned to one of two groups: the experimental and the control group. The intervention consisted of aerobic exercises with specific handheld equipment. Both groups performed physical activity twice weekly for 10 weeks. The erector spinae muscles' bioelectrical activity, the lumbar range of motion and pain intensity were measured in all participants at baseline and after 10 weeks. Compared with baseline measures, there was a significant decrease in the bioelectrical activity of the erector spinae muscles during flexion movement, rest at maximum flexion, extension movement and rest in a prone position; an increase in the lumbar range of motion and a decrease in pain intensity following a program of physical activity with vibrating exercise equipment. No significant changes were found in intergroup comparisons; however, physical activity with vibrating exercise equipment could be a prospective strategy for increasing lumbar range of motion and decreasing pain and erector spinae muscle activity in people with chronic low back pain.

Abstract: Background: Physical activity is often recommended as part of the management of chronic low back pain, which is one of the most common musculoskeletal disorders. Vibrating exercise equipment is used despite little scientific evidence to support its effectiveness in the prevention and treatment of musculoskeletal problems. The aim of this study was to evaluate the efficiency of using vibrating exercise equipment in women with chronic low back pain. Here, 92 women aged 49–80 years were assigned to one of two groups: the experimental and the control group. The intervention consisted of aerobic exercises with specific handheld equipment. Both groups performed physical activity twice weekly for 10 weeks. The erector spinae muscles' bioelectrical activity, the lumbar range of motion and pain intensity were measured in all participants at baseline and after 10 weeks. Compared with baseline measures, there was a significant decrease in the bioelectrical activity of the erector spinae muscles during flexion movement, rest at maximum flexion, extension movement and rest in a prone position; an increase in the lumbar range of motion and a decrease in pain intensity following a program of physical activity with vibrating exercise equipment. No significant changes were found in intergroup comparisons; however, physical activity with vibrating exercise equipment could be a prospective strategy for increasing lumbar range of motion and decreasing pain and erector spinae muscle activity in people with chronic low back pain. Chronic low back pain (CLBP) is one of the most common musculoskeletal disorders. Physical activity (PA) is often recommended as part of the management of CLBP, but to date, no one particular exercise has been shown to be superior. Vibrating exercise equipment (VEE) is widely available and used despite little scientific evidence to support its effectiveness in the prevention and treatment of musculoskeletal problems. The aim of this study was to evaluate the efficiency of using VEE compared with sham-VEE in women with CLBP. **Methods:** A randomized (1:1 randomization scheme) single-blinded sham-controlled intervention study was conducted. Through simple randomization, 92 women aged 49–80 years were assigned to one of two groups: VEE (the experimental group) and sham-VEE (the control group). The VEE and sham-VEE intervention consisted of aerobic exercises with specific handheld equipment. Both groups performed physical activity twice weekly for 10 weeks. The erector spinae muscles' bioelectrical activity (using an eight-channel electromyograph MyoSystem 1400L), lumbar range of motion (Schober's test) and pain intensity (visual analog scale) were measured in all participants at baseline and after 10 weeks. **Results**: There was a significant decrease in the bioelectrical activity of the erector spinae muscles during flexion movement (left: Me = 18.2 before; Me = 14.1 after; $p = 0.045$; right: Me = 15.4 before; Me = 12.6 after; $p = 0.010$), rest at maximum flexion (left: Me = 18.1 before; Me = 12.5 after; $p = 0.038$), extension movement (right: Me = 21.8 before; Me = 20.2 after; $p = 0.031$) and rest in a prone position (right: Me = 3.5 before; Me = 3.2 after; 0.049); an increase in lumbar range of motion (Me = 17.0 before; Me = 18.0 after; $p = 0.0017$) and a decrease in pain intensity (Me = 4.0 before; Me = 1.0 after; $p = 0.001$) following a program of PA in the VEE group. **Conclusions:** No significant changes were found in intergroup comparisons. The beneficial changes regarding decreased subjective pain sensation in the VEE and sham-VEE groups may be due to participation in systematic physical activity. However, PA with vibrating exercise equipment could be a prospective strategy for increasing lumbar range of motion and for decreasing pain and erector spinae muscle activity in people with CLBP.

Keywords: vibrating exercise equipment; women; chronic low back pain; surface electromyography

---

1. Introduction

Chronic low back pain (CLBP) is one of the most common health complaints in adults [1]. CLBP affects independence, the mental state and physical activity (PA) [2,3]. Conservative treatment of CLBP often involves PA in the form of specific training sessions [4,5] or exercises [6]. To date, there are no clear guidelines to suggest the superiority of any one exercise method [7]. A 2017 review by Geneen et al. [8] found that there is poor scientific evidence to support the use of PA in reducing pain in the general population, and the authors suggested the need for continuing research into new solutions for conservative treatment of chronic pain [8]. Other studies have found that regular PA in the older adults alleviates anxiety and depression, and improves cognitive function while also enhancing mobility, balance and upper limb function [9]. Furthermore, PA also prevents the formation of fibrosis in muscle tissue and has a role to play in promoting the anti-inflammatory properties of macrophages, reducing tissue sensitivity and metabolism. Use of PA also has a positive impact on myofascial tissue and supports the concept of "physical activity as medicine" [10]. Mora and Valencia [11] believe that for PA to have optimum health benefits in older adults, it must consist of aerobic, strength, flexibility and proprioceptive exercises. Moreover, the PA must be tailored to take past injuries, hip and knee function, chronic disease, prescribed medications and nutrition into account [11]. The use of vibration stimuli in medicine and rehabilitation is becoming increasingly popular [12]. Vibration provides a strong proprioceptive stimulus [13] which can improve body balance and gait re-education in older adults [14,15], and can reduce pain by stimulating the afferent alpha and beta motoneurons and inhibiting nociceptive pain fibers [16]. Vibration stimuli have been shown to lead to improvements in muscle strength [17] and muscle mass [18,19], a reduction in obesity [20], and an improvement of gait in neurological patients [21,22]. Moreover, vibration can be applied directly (local vibration) to the muscle [17,23] or indirectly using vibrating platforms (e.g., whole-body vibration, WBV). The use of WBV in conjunction with exercise has been studied in randomized clinical trials [24–26], but thus far, no comprehensive review evaluating specific interventions and protocols, especially in the case of older adults has appeared in the Cochrane Library. A review published in 2019 by Leite et al. [27] suggested there is not enough evidence to support the use of WBV in clinical practice among people with various disabilities [27]. Furthermore, no standard clinical guidelines are available for WBV physical activity in relation to curing CLBP [28]. There also may be a role for exercise using vibrating exercise equipment (VEE) as an alternative to exercises on a WBV platform [29–31]. VEE applies local vibration from the hand to the rest of the body. The frequency produced by VEE was measured at the soft grip on the top of the rings during movement. It ranged from 0 to about 460 Hertz and showed a mean amplitude at about 60 Hertz (information provided by the manufacturer). The frequency is not dependent on any electronic motor or device but instead is created by natural and dynamic movements created by the swinging of the rings in different directions; therefore, it does not achieve a constant frequency. The movement always starts at 0 Hertz (the resting position) and reaches its highest frequency at the midpoint of the movement until changing direction to the resting point again [29,30]. However, little is known about the efficacy of combining this specific exercise equipment with vibration to enhance low back pain improvements. We identified two studies on the use of VEE in people with cancer. Crevenna et al. [30] assessed the quality of life and the strength of the upper limb muscles in women with breast cancer before and after a 12-month intervention in the form of physical activity with VEE. It was found that exercising with the VEE was safe and improved the quality of life and upper limb muscle strength. In a pilot intervention study, Cenik et al. [29] used the same device in women with breast cancer and evaluated participants' quality of life, upper limb muscle strength, body composition and 6-min walk test. After 3 months of physical activity with the vibration-generating device, there was an improvement in all the measured parameters. These projects are not directly comparable with our study; however, they provide evidence that the equipment used in our project is safe and well tolerated. Both of the aforementioned studies showed encouraging results and their authors recommended further research on the effectiveness of VEE.

Recent World Health Organization data have confirmed the rapid aging of the population around the world [32]. In Poland, the long-term senior policy for 2014–2020 encouraged the creation of physical activity programs as part of the so-called concept of healthy aging. These programs aim to be diverse and innovative, and function to encourage older people to participate in various forms of physical activity. So far, to our knowledge, the bioelectrical activity of the erector spinae muscles, lumbar range of motion (ROM) and reported low back pain in middle-aged and older women has not been studied in conjunction with exercises using vibrating equipment. Further research is needed into the benefits of a combined exercise intervention program involving muscular strength, flexibility and aerobic fitness for CLBP patients, as the literature has supported the use of each of these fitness areas individually, but more research should be conducted combining all three [33]. Despite encouraging evidence on the usefulness of exercise involving vibration in people with CLBP, the practical application of such research findings in this area remains limited. Programs and exercises utilizing vibration equipment are typically not well reported or not reported at all.

Wilke et al. [34] have carried out a systematic review in order to find evidence of myofascial continuity between the trunk and upper limb muscles. The analysis revealed the presence of three myofascial connectivities between the trunk and the upper limbs: the dorsal arm chain, the ventral arm chain and the lateral arm chain. Two myofascial continuities start on the dorsal arm chain: both latissimus dorsi muscles and the infraspinatus and teres minor muscles fuse with the triceps brachii. At the elbow, the triceps brachii muscle merges with the small anconeus muscle, which then connects to the extensor carpi ulnaris of the lower arm [34]. It seems that chronic low back pain and methods of their treatment, due to fascial connections with other regions of the body [35], should be considered from the aspect of the body as a whole, in agreement with the biotensegrity biomechanical theory [36,37]. The evidence supports this theory, pointing toward functional and clinically relevant myofascial continuity [38,39]. Several authors [28,40–42] have shown that adding WBV to specific exercise could increase muscle activation of the lumbar–abdominal muscles in young patients with CLBP. According to VEE, during the swinging movements of the arms, the four steel balls inside the tube not only create vibration but also momentum due to the acceleration of the VEE and the inertia of the steel balls. In the resting position, the VEE weighs about 0.5 kg, which can change to about 5 kg through a dynamic swinging movement. We conclude that the momentum and vibration created by the swinging movement could affect the lumbar–abdominal muscles, which are important factors for CLBP. Therefore, the main goal of this study was to compare the ES muscles' bioelectrical activity, lumbar ROM and low back pain in middle-aged and older women participating in a 10-week physical activity program with VEE and sham-VEE. In addition, the correlations among these parameters was evaluated. The main hypothesis was that the intervention of PA with VEE would decrease the resting and functional bioelectrical activity of the erector spinae muscles, increase the lumbar ROM and reduce low back pain. The present study has clinical relevance. If there were intergroup differences, then the simple handheld swinging-ring system could be used by physiotherapists in individuals with CLBP. If it were possible to prove the beneficial influence of the local vibration generated by the handheld swinging-ring system in people with CLBP, the device could be used in the work of a physiotherapist. This would be both economically important (reducing labor costs) and easy to use by the patient, as he or she would be able to do the exercises themselves.

2. Materials and Methods

2.1. Study Methodology

A randomized intervention study assessed the resting and functional bioelectrical activity of the erector spinae muscles before and after a 10-week PA program using VEE (experimental group). A control group was used, which performed the same 10-week PA program, but with sham non-vibrating equipment (sham-VEE). The project was carried out between September 2016 and June 2017, and was based on the Consolidated Standards of Reporting Trials (CONSORT) guidelines. The study was approved by the Bioethical Committee at Opole Medical School in Poland (24 October 2016, No. 44/2016) and registered on the Australian New Zealand Clinical Trials Registry platform (1 December 2016, number ACTRN12616001661460). The study was carried out in the Functional Research Laboratory of the Physiotherapy Department and used the gym of the Opole Medical School, Poland.

2.2. Study Participants

The target group included women between 50 and 80 years of age recruited from the University of the Third Age of Health and Beauty of Seniors of the Opole Medical School, Poland. Written informed consent was obtained from all participants included. The inclusion criteria were: aged over 49, a history of bilateral chronic pain in the lumbar spine lasting >6 months (they had completed the standard care procedure for acute LBP) and provision of signed voluntary consent to participate in the project. Exclusion criteria were: age <50 years old, acute pain in the lumbar spine requiring pharmacological treatment lasting <6 months, lack of pain in the lumbar spine, unstable hypertension, an inability to perform standing exercises, hip endoprosthesis and a lack of voluntary consent to participate in the project. Based on the criteria, 92 women were included in the project. The estimated sample size was calculated on the basis of the pilot study. The means and standard deviations of lumbar ROM before and after the intervention were used in the analysis of estimating the sample size. On the basis of the parameters, an estimated sample size equal to 39 participants in each group was obtained. In addition, the risk of losing patients in the follow-up assessment (10%) was assumed. The final sample size equaled a minimum of 43 participants in each group. The estimation of the sample size was performed using Statistica 13 (TIBCO Software Inc.). Participants were randomly assigned to the experimental group (VEE group, n = 43) or the control group (sham-VEE group, n = 49). Randomization was carried out using computer-generated random numbers (simple randomization). The participants were randomly assigned to the groups in a 1:1 ratio. Both groups were subjected to a systematic PA intervention (a 60-min exercise session twice a week for 10 weeks) with the use of equipment. The equipment used in the experimental group was a handheld device which generated vibrations and momentum when it was set in motion. The same equipment was used in the control group, but it did not generate vibration and momentum during movement (sham-VEE). Both pieces of equipment were identical in weight and external structure. The participants were not aware which intervention they were receiving (single-blinded).

2.3. Methods

The bioelectrical activity of the right and left erector spinae muscles in the lumbar spine of each participant was measured using an eight-channel electromyograph (MyoSystem 1400 L, Noraxon, Scottsdale, AZ, USA), MyoResearch software (Noraxon) and compatible, disposable, self-adhesive Ag/AgCl electrodes. The participants assumed a comfortable, casual position lying face down. The study was conducted in accordance with SENIAM guidelines [43].

The electromyographic (surface electromyography, sEMG) signals were subjected to standard post hoc processing with rectification (purification) and smoothing using an RMS calculation algorithm. The sEMG recording frequency was set to the range of 10 to 450 Hz with a high-pass filter cutoff of 10 Hz and a low-pass filter cutoff of 500 Hz. The level of common-mode rejection was a minimum of 100 dB, and the input impedance for the sEMG channels was higher than 100 mΩ. The system had high sensitivity in terms of recording sEMG signals (1 µV). The erector spinae (ES) muscles' bioelectrical activity was measured during "rest" (45 s of ES activity at rest before any functional measurements), and at "functional tone" (measurement in the standing position lasting 45 s, during which the patient performed lumbar flexion, with maintenance of rest at maximal flexion and extension). Normalization to maximal or submaximal contractions has not been considered a solution for low back pain patients, and thus, non-normalized EMG amplitudes were preferred in our study [44].

The original Schober's test was used to measure the lumbar ROM. The assessment was performed at the level of L5, with 2 points marked 5 cm below and 10 cm above L5. The participants were then asked to perform a forward bend, flexing the torso and touching their toes if possible while keeping the knees straight. The distance between the 2 points was measured in this position [45].

Participants were asked to rate the intensity of pain on a visual analog scale (VAS). Scores were measured by asking each participant to mark the intensity of pain on a 10 cm line, annotated with "no pain" at one end and "as bad as it could possibly be" at the other end [46].

2.4. Intervention

The PA with VEE and sham-VEE is detailed in Table A2 (Appendix A). These were overseen by physiotherapists trained in using vibrating equipment. The VEE is a handheld spiral tube with 4 steel balls within it. Movement in the sagittal, frontal or horizontal plane (e.g., a swinging motion) sets the metal balls in motion. The movement of the balls creates a vibration of 60 Hz, which is transmitted through the handles of the tool. The weight of the static equipment is 0.5 kg; however, it can reach up to 5 kg during movement due to centrifugal forces.

The handheld vibration exercise equipment used in the study consisted of 4 steel balls (26 g each, diameter = 24 mm) located inside a spiraled tube made of soft (65%) and hard (35%) PVC (internal groove protruding, helix pitch 6.2 mm) and a grip with cushioning elements. The device name "Smovey" comes from the three words: "swing", "move" and "smiley" (Appendix A, Figure A1).

Each participant was instructed how to hold the equipment properly and how to perform the correct swinging movement in different planes whilst maintaining proper body posture. The PA comprised simple movements of moderate intensity. The level of the intensity (perceived exertion) was based on the subjective assessment of each participant expressed during the exercise using the Borg scale. Perceived exertion was measured by the Borg RPE scale, which contains both verbal anchors and a numerical scale. The numbers range from 6 to 20, while the verbal anchors start at 6, which is labeled as "least effort", 7 is "very, very light", 9 is "very light", 11 is "fairly light", 13 is "somewhat hard", 15 is "hard", 17 is "very hard", 19 is "very, very hard" and 20 is "maximum effort". After each set of exercises, the physiotherapist asked the participants to rate her level of effort in performing the exercises on the Borg scale ("How hard you feel your body has worked?") [47]. Participants performed the PA regime for ~60 min twice per week for a total duration of 10 weeks. The exercises were performed in the same order for both groups. Each participant had their own exercise mat and set of exercise equipment for each session. Each exercise was demonstrated by a physiotherapist, and the participants were asked to copy each movement. Participants were reminded at each session of the correct starting position and tonicity, the importance of keeping slightly bent knee joints, the correct way to hold the equipment (keeping the wrists stiff) and the correct range of arm movement. The classes were always held at the same time. Participants were dressed in loose sports clothing and footwear that did not constrict movement. The demonstrator set the pace of the exercises.

2.5. Statistical Analysis

Statistica 13 software was used to perform the statistical analysis (StatSoft, Inc., Tulsa, OK, USA). Due to the lack of a normal distribution in the obtained results, the medians, quartiles (Q1, Q3) and range of variability were calculated for each measurable variable. The frequency of occurrence (percent) was calculated for qualitative variables. All the tested quantitative variables were checked by means of the Shapiro–Wilk test to determine the type of distribution. Intragroup comparisons (before vs. after the intervention) were performed using the Wilcoxon test. The differences between the results obtained in the experimental group and the control group were determined using the non-parametric Mann–Whitney U-test. For all comparisons, a level of $\alpha = 0.05$ was assumed. Correlations between quantitative variables were analyzed using the Spearman correlation coefficient.

3. Results

In total, 120 participants were enrolled in the study. Based on the inclusion and exclusion criteria, 92 women were included in the project. They were randomly assigned to one of the two groups: the experimental group (n = 43) or the control group (n = 49). All 92 subjects completed the 10-week exercise program (Figure 1) following the intention to treat principle.

3.1. Characteristics of the Participants

Table 1 shows the characteristics of the experimental group and the control group. The age in the experimental group ranged from 50 to 76 (mean = 66.0 years old) and that in the control group ranged from 56 to 80 (mean = 66.0 years old). There were no statistical differences in the characteristics of the two groups.

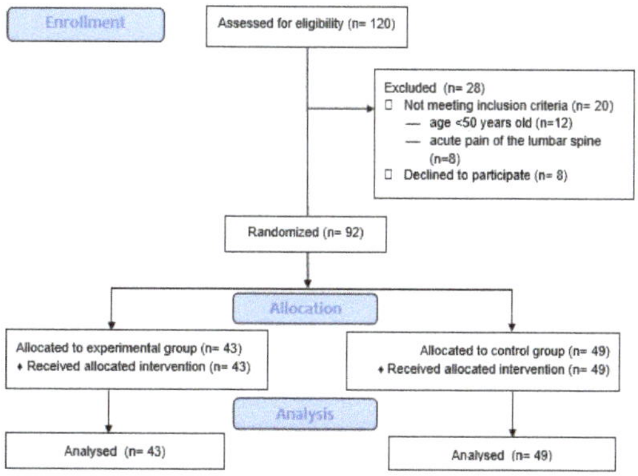

Figure 1. Flow chart of study enrollment, allocation and analysis.

Table 1. Characteristics of the participants.

	VEE (Experimental) Group n = 43					Sham-VEE (Control) Group n = 49					*p*-Value
	Me	Min	Max	Q1	Q3	Me	Min	Max	Q1	Q3	
Age (years)	66.0	50.0	76.0	63.0	69.0	66.0	56.0	80.0	64.0	69.0	0.68 *
Height (cm)	160.0	146.0	176.0	156.0	165.0	160.0	149.0	171.0	155.0	164.0	0.89 *
Weight (kg)	75.4	47.1	107.5	64.2	85.0	70.6	49.3	177.7	63.0	77.0	0.17 *
BMI (kg/m^2)	28.4	19.1	42.3	25.2	32.5	27.7	19.3	42.7	25.8	29.2	0.24 *

* Mann–Whitney U-test; Me—median; Mi—minimum; Ma—maximum; Q1—lower quartile; Q3—upper quartile.

3.2. Erector Spinae sEMG Results

A comparison of the results of the sEMG measurements is presented in Table 2. Compared with the baseline measures, there was a statistically significant decrease in the flexion sEMG of the left and right ES muscles in the VEE group following the PA intervention. Moreover, there was a decrease in the sEMG activity of rest at maximal flexion in the left ES muscle and in the sEMG activity of extension of the right ES observed in the VEE group. The sEMG of the right ES muscle in the resting position also decreased significantly following the intervention. There were no statistically significant changes in sEMG following the PA intervention observed in the sham-VEE group. Moreover, there were no statistically significant differences in the change in sEMG between the VEE and sham-VEE groups.

3.3. Lumbar ROM Results

The measurements of lumbar ROM are presented in Table 3. Compared with the baseline measures, there was a statistically significant improvement in the ROM following the PA intervention in the VEE group. There were no statistically significant changes in ROM following the PA intervention observed in the sham-VEE group. There were no statistically significant differences observed between the VEE and sham-VEE groups.

3.4. Pain Scores (VAS)

Both the VEE and sham-VEE groups demonstrated a decrease in pain intensity following the intervention, which is shown in Table 4. There was a statistically significant decrease in VAS in the VEE group, from 4 to 1 following the exercise intervention. Similarly, the level of pain expressed on the VAS scale in the sham-VEE group decreased from 5 to 1 after the end of the study, which was also statistically significant. There was no statistical difference in VAS score between the groups either before or after the intervention.

Table 2. Intragroup and intergroup comparisons of the changes in erector spinae sEMG measured before and after the PA intervention (μV).

	VEE (Experimental) Group										Sham-VEE (Control) Group										*p*-Value *	*p*-Value **	*p*-Value ***	*p*-Value ****
	Before					After					Before					After								
	Me	Min	Max	Q1	Q3	Me	Min	Max	Q1	Q3	Me	Min	Max	Q1	Q3	Me	Min	Max	Q1	Q3				
Flexion sEMG—left ES	18.2	7.6	57.3	12	19.9	14.1	4.5	49.9	8.8	18.5	14.7	8.4	33.4	11.7	17.6	13.7	7.8	28.1	10.9	17.5	0.045	0.34	0.07	0.70
Flexion sEMG—right ES	15.4	6.6	39.9	12	19.8	12.6	5.6	30	9.1	16.7	13.9	6.5	45.2	10.2	17.3	12.3	4.6	30.4	10	16	0.010	0.22	0.30	0.87
Rest (in maximum flexion)—left ES	18.1	3.1	40.6	12	20.2	12.5	2.8	53.3	8	17.5	14.7	3.5	27.9	11.4	18.9	12.2	3.5	30.8	9.2	16.8	0.038	0.10	0.28	0.73
Rest (in maximum flexion)—right ES	15.5	2.7	32.9	10.3	20.6	11.1	2.3	29.4	7.2	16.9	13	3.2	46.3	9.9	17.8	10.9	3.1	30	7.7	17	0.06	0.14	0.31	0.95
Extension—left ES	22.9	10.2	47.4	17.2	31.6	20.3	4.5	54.1	13.4	28.5	18.2	3.8	48.3	14.4	25.6	18.5	8.3	58.9	14.7	25.1	0.06	0.57	0.13	0.89
Extension—right ES	21.8	8.5	38.9	15.3	29.1	20.2	5.9	45.4	14.2	24.1	17.7	8.1	48.6	13.9	23.6	15	7.8	43.9	12.3	23.3	0.031	0.64	0.07	0.29
Rest—left erector spinae	5.7	2.1	19.4	4	7.1	5.3	2.3	18.1	4.4	7.1	5.7	2.3	20.1	4.6	8.3	5.3	2.4	16.1	4.3	7	0.78	0.16	0.34	0.88
Rest—right erector spinae	3.5	1.9	16.2	2.8	4.9	3.2	1.5	7.3	2.5	4.6	3.4	1.6	18.5	2.7	4.8	3.2	1.9	11	2.5	4.4	0.049	0.37	0.52	0.88

* Experimental group: before vs. after (Wilcoxon test); ** control group: before vs. after (Wilcoxon test); *** comparison of "before" results: experimental group vs. control group (Mann–Whitney U-test); **** comparison of "after" results: experimental group vs. control group (Mann–Whitney U-test); Me—median; Mi—minimum; Ma—maximum; Q1—quartile lower; Q3—quartile upper; left ES—left erector spinae, right ES—right erector spinae.

Table 3. Intragroup and intergroup comparisons of the changes in lumbar ROM measured by Schober's test before and after the PA intervention (cm).

	VEE (Experimental) Group										Sham-VEE (Control) Group										*p*-Value *	*p*-Value **	*p*-Value ***	*p*-Value ****
	Before					After					Before					After								
	Me	Min	Max	Q1	Q3	Me	Min	Max	Q1	Q3	Me	Min	Max	Q1	Q3	Me	Min	Max	Q1	Q3				
Distance between points in a standing position (cm)	13.0	10.0	16.0	12.0	14.0	14.0	12.0	18.0	10.0	15.0	13.0	10.0	15.0	12.0	14.0	13.0	10.0	15.0	12.0	14.0	0.0015	0.11	0.79	0.68
Distance between points in a flexing position (cm)	17.0	13.0	21.0	15.5	19.0	18.0	15.0	23.0	17.0	19.5	17.0	13.0	22.0	16.0	19.0	18.0	14.0	22.0	16.0	19.0	0.0017	0.09	0.88	0.29

* Experimental group: before vs. after (Wilcoxon test); ** control group: before vs. after (Wilcoxon test); *** comparison of "before" results: experimental group vs. control group (Mann–Whitney U-test); **** comparison of "after" results: experimental group vs. control group (Mann–Whitney U-test); Me—median; Mi—minimum; Ma—maximum; Q1—lower quartile; Q3—upper quartile.

Table 4. Intragroup and intergroup comparisons of the changes in VAS scores before and after the PA intervention.

	VEE (Experimental) Group										Sham-VEE (Control) Group										*p*-Value *	*p*-Value **	*p*-Value ***	*p*-Value ****
	Before					After					Before					After								
	Me	Min	Max	Q1	Q3	Me	Min	Max	Q1	Q3	Me	Min	Max	Q1	Q3	Me	Min	Max	Q1	Q3				
VAS	4.0	1.0	10.0	2.0	5.0	1.0	0.0	8.0	0.0	4.0	5.0	1.0	10.0	3.0	8.0	1.0	0.0	9.0	0.0	5.0	0.001	<0.001	0.07	0.90

* Experimental group: before vs. after (Wilcoxon test); ** control group: before vs. after (Wilcoxon test); *** comparison of "before" results: experimental group vs. control group (Mann–Whitney U-test); ****comparison of "after" results: experimental group vs. control group (Mann–Whitney U-test); Me—median; Mi—minimum; Ma—maximum; Q1—lower quartile; Q3—upper quartile.

4. Discussion

The aim of the study was to compare the effect of a 10-week exercise program, using either VEE or sham-VEE, on the bioelectrical activity of the erector spinae muscles in the lumbar spine, lumbar range of motion and reported pain in middle-aged and older women with chronic low back pain. The study also investigated whether bioelectric activity correlated with the ROM of the lumbar spine and reported subjective pain levels. It was hypothesized that after the end of the 10-week PA intervention with VEE, the bioelectrical activity of the erector spinae muscles would be significantly lower compared with their activity before the intervention, particularly in the experimental group and in comparisons between the groups. It was also hypothesized that a reduction in the measured bioelectrical activity of the erector spinae muscles would correlate with increased lumbar ROM and reduced pain intensity. The results showed a significant reduction in the bioelectrical activity of the erector spinae muscles following the PA intervention in the experimental group; however, there was no statistical difference when comparing the experimental group with the control group. Moreover, there was no significant correlations between the measured bioelectrical activity of the erector spinae muscles and ROM. To our knowledge, the study was the first randomized sham trial that has been conducted to evaluate the effects of exercise with VEE on these parameters. The study aimed to answer the question of whether PA with VEE is more effective than PA with sham-VEE in middle-aged and older women suffering from CLBP.

4.1. Erector Spinae sEMG

It appears that the reduction in the bioelectrical activity of the ES muscles in our study after the 10-week intervention in the experimental group is a positive phenomenon. This would suggest that the use of VEE has a beneficial stimulatory effect on the nervous system responsible for ES muscle innervation, as well as possibly improving the blood supply to the muscle due to changes in its tone. However, measurements of bioelectrical activity of the ES muscles in patients with CLBP are still inconsistent [48,49]. Lima et al. [44] found that those suffering from CLBP had increased muscle activity, which was possibly caused by excessive stimulation of the nervous system. Participants with CLBP showed an increase in back muscle activity compared with asymptomatic participants, regardless of the type of functional task [44]; moreover, increased trunk muscle activity has been shown to be a key feature in the presence of pain [50]. Therefore, the reduction in resting and functional sEMG as demonstrated in our study may have resulted in a decrease in ES muscle hyperactivity, which is one of the factors contributing to pain onset and persistence in study participants. This study hypothesized that the use of VEE during PA would modify the level of bioelectric activity of the ES through myofascial connections between the upper and lower back [51]. The ES muscles are located between the lamina superficialis and the lamina profunda of the thoracolumbar fascia, which functions to carry mechanical loads [52] along with the back extensors and gluteal muscles [53]. The thoracolumbar fascia also contributes to movement coordination, stability and proprioception, and aids in promoting sliding and reducing friction during movement. Any trauma or pain may alter the sliding mechanism within the fascial plane [54]. Therefore, if fascial stiffness is increased, the nociceptors could be sensitized, causing the underlying muscles to be stiffer [55]. The studies by Nowotny et al., (2018) and Daneau et al., (2019) suggested that the reduction in the endurance and strength of the paraspinal muscles, which results in development of pain in this region, may be related to an increased number of Type II muscle fibers in patients with LBP pain compared with people who did not report pain in this region [56,57].

A review of literature showed that this type of vibration (particularly related to the VEE) and its use in exercise has not been extensively studied; therefore, it was difficult to compare our study with any other literature. A supplementary stimulus was also generated by the noise of the four steel balls rolling in the spiral tube, which generated an auditory feedback corresponding to the intensity of the vibration stimuli. Goossens et al. [13] conducted an experiment in which local vibration from 20 to 60 Hz applied to the triceps surae and back muscles was chosen as the "stimulation condition" to control the simultaneous activation of vibrotactile skin receptors. This study used magnetic resonance imaging (MRI) to evaluate brain activity when the local vibrating stimulus was applied to the erector spinae muscles in people with and without non-specific low back pain (NSLBP). The results indicated that patients with NSLBP were more cautious of movement and needed more time to complete tasks compared with the control group. In addition, MRI of patients with NSLBP showed activation of the right S2 cortex and the right primary auditory cortex (Heschl's gyri), areas that are important for proprioception, during stimulation of the erector spinae muscles. Although there were no significant differences in the processing of proprioceptive information between participants with NSLBP and healthy participants, correlations of brain behavior showed that in order to maintain optimal proprioception to respond to postural changes, patients with NSLBP may experience increased activation of the regions responsible for sensory processing. The use of vibration stimuli also triggered increased activation of brain areas involved in threat detection and fear processing in some participants, which was associated with poorer proprioceptive posture control [13]. The results of our study indicated no statistically significant differences between the VEE and sham-VEE groups, which may suggest that the vibration generated by the equipment used did not affect the erector spinae muscles in the aspects of neurophysiology studied.

4.2. Lumbar ROM

The reduction in ROM that occurs with age plays an important role in physical function [58]. Some authors suggested that people with CLBP are highly fearful and not only guard themselves during flexion and extension movements, but they also fear that pain will significantly limit their ROM, for example, during flexion–extension movements [59]. In this study, there was a statistically significant improvement in lumbar ROM after the PA intervention in the experimental group. The increase in ROM associated with the reduction in pain intensity within the painful lumbar spine may be associated with a restoration of soft tissue architecture in this region of the body and a potential reduction in the number of Type II muscle fibers [56,57]. According to Langevin et al., (2011), a potential consequence of chronic low back pain is fibrosis and adhesions that may inhibit independent motion of the adjacent connective tissue layers, which can restrict movement [60]. Since mobility range is closely related to soft tissue flexibility and because of the fact that mobility range decreases not only due to pain but also age [61], any increase in flexibility will be of particular value to middle-aged and older adults, as those who maintain better flexibility are more likely to be independent when performing functional everyday activities [61].

4.3. Pain Scores (VAS)

This study showed a significant reduction in subjective pain sensations within the lumbar spine in both the experimental and the control group. If we assume that pain increases ES muscles' bioelectrical activity, which, in turn, further increases pain, then reducing this subjective perception of pain may be the key element in breaking the vicious cycle of CLBP [62]. There are various theories regarding the impact of pain on the bioelectrical activity of muscles [63]. The results of research conducted by Arendt-Nielsen et al. [64] showed that there was no correlation between VAS reduction and a decrease in the erector spinae muscles' bioelectrical activity. Our results showed some discrepancy between objective measurements of the bioelectric activity of the ES and subjective measurement of pain sensation. The change in the bioelectrical activity of the ES in the VEE group was simultaneously supported by a significant reduction in pain; however, a reduction in pain without a significant change in the bioelectric activity of the ES muscles was also observed in the sham-VEE group. From the point of view of the patient's quality of life, such a significant reduction in pain in both groups is obviously a beneficial phenomenon. The results of the study also did not reveal whether PA with VEE or with sham-VEE was superior. This may partly stem from the fact that the participants were properly educated on how to perform each exercise, particularly the importance of maintaining correct posture during the PA. Maintaining correct posture, emphasizing the lumbar lordosis, during PA makes it possible to keep the various intricate structures of the back and spine healthy [65]. PA with the equipment proposed in our intervention study focused on the activation of the deep torso muscles, targeting the restoration of control and coordination of these muscles with the view to progressing to more complex and functional tasks that integrate the activation of the deep torso muscles as a whole.

4.4. Clinical Implications

The study aimed to answer the question of whether PA with VEE is more effective than PA with sham-VEE in middle-aged and older women suffering from CLBP. A surprising finding was that there was no difference between the VEE and sham-VEE physical activity groups (intergroup comparison). Only the VEE group had significant decreases in the ES muscles' bioelectrical activity and increases in lumbar ROM. Both groups had significant decreases in VAS. These findings do not support our hypothesis that the intervention of PA with VEE, more than PA with sham-VEE, would decrease the resting and functional bioelectrical activity of the erector spinae muscles, increase the lumbar ROM and reduce low back pain because it provides a vibration effect. Thus, at this phase of the study, we are unable to confirm that in middle-aged and older women with CLBP, the local vibration generated by a simple handheld swinging-ring system during a physical activity program is more effective compared with exercises in which it does not occur. For planning physical activity in middle-aged and older women with chronic low back pain, exercise with VEE can be included but, based on our results, the beneficial changes observed in this study may result from systematic physical activity, not from the equipment used (no statistical differences in intergroup comparisons). In the case that an individual possesses exercise equipment such as VEE, this can only enrich aerobic training and expand the spectrum of exercises that can be performed, especially when aerobic exercise is routinely recommended to improve physical function in aging individuals [66,67].

4.5. Study Limitations

There was no long-term follow-up of the participants to see if any benefit was sustained long-term. The physiotherapists involved in this project were not blinded as to whether the devices used were vibrating or non-vibrating. The VEE used in the project could have been described better in technical terms, focusing on how the vibrations are generated—the project was based only on the information available from the manufacturer. Future work could, however, seek to explore approaches to the normalization of sEMG data, accounting for factors such as activities of daily living and regular physical activity, training participants to control the deep muscles of the trunk to stabilize the spine during perturbations, different training methods and other training equipment or vibration applied to the lower extremities. Another weakness was that the vibration never reached a constant frequency, which might be required (or devices with a higher frequency), which could be tested in future studies. Moreover, the weakness was that there were no minimum pain score criteria for LBP (range from 1 to 10), and including middle-aged and older participants with low chronic back pain scores (1–3) may have adversely impacted the overall pain-related outcome measures. Future studies should include participants with greater chronic LBP than scores of 1 or 2.

5. Conclusions

Application of local vibration from the hand to the rest of the body did not result in significant changes in the bioelectric activity of the ES, lumbar ROM or pain intensity in intergroup comparisons. The beneficial changes observed in this study regarding decreased subjective pain sensation in the VEE and sham-VEE groups may be due to participation in systematic physical activity rather than the vibration-generating equipment used. Physical activity with VEE increased lumbar ROM and decreased pain and the erector spinae muscles' activity in middle-aged and older women with chronic low back pain. In future, well-designed studies with a large sample size should be conducted to assess the possible effects of the manual swinging-ring system. This will allow further exploration and validation of the benefits of PA with VEE for this age group.

Author Contributions: G.Z. participated in the design of the study, contributed to data interpretation and to the writing of the manuscript, and critically revised the manuscript. M.K.-J. participated in the design of the study, conducted the study, contributed to data analysis and interpretation, and to the writing of the manuscript. I.D. participated in the design of the study, conducted the study and analyzed the data. A.M. supervised the collection of data and contributed to data analysis and interpretation. G.D. participated in the design and to the writing of the manuscript. K.P. supervised the collection of data. T.H. critically revised the manuscript. All authors meet the criteria for authorship stated in the Uniform Requirements for Manuscripts Submitted to Biomedical Journals. All authors have read and agreed to the published version of the manuscript.

Funding: This research received no external funding.

Institutional Review Board Statement: All procedures performed in this study involving human participants were in accordance with the ethical standards of the institutional and/or national research committee and with the 1964 Helsinki Declaration and its later amendments or comparable ethical standards. This experimental protocol was approved by the Bioethical Committee at Opole Medical School (24/10/2016, No. 44/2016). This study was registered in the Australian New Zealand Clinical Trials Registry platform (01/12/2016, No. ACTRN 12616001661460).

Informed Consent Statement: Written informed consent was obtained from all participants included.

Data Availability Statement: The datasets used and/or analyzed during this study are available from the corresponding author on reasonable request.

Acknowledgments: We would like to thank all the participants who invested themselves as best as possible, and thank the Opole Medical School for providing the rooms in which the project was conducted free of charge.

Conflicts of Interest: The authors declare no conflict of interest.

Appendix A

Figure A1. Smovey rings.

Table A1. Correlation between the bioelectric activity (μV) of the erector spinae muscles and the ROM of the lumbar spine in the experimental and control groups.

		Experimental Group				Control Group			
		Lumbar ROM Before		Lumbar ROM After		Lumbar ROM Before		Lumbar ROM After	
		r_s	*p*-Value	r_s	*p*-Value	r_s	*p*-Value	r_s	*p*-Value
Before	Baseline	0.239019	0.102	0.123192	0.404	0.198600	0.2017	0.109459	0.4847
	Functional	0.106851	0.470	0.189222	0.198	−0.089447	0.5684	0.088912	0.1571
After	Baseline	0.086178	0.560	−0.057862	0.696	0.190182	0.2219	0.003511	0.9822
	Functional	0.005431	0.971	−0.008733	0.953	0.011753	0.9404	0.050751	0.7464

r_s = Spearman's correlation coefficient

Table A2. Overview of the exercise classes.

Course of the Class	Name and Description of Exercise	Time (min)
Welcome and introduction	The group is welcomed and the equipment is distributed. The participants are asked about their wellbeing and pain level, Participants are reminded of the principles of using the equipment, and the importance of maintaining correct and normal breathing during exercising.	5
Warm-up 5 min	Starting position (SP) = Participants stand with their arms by their sides. Shoulder circles are performed forward and backward.	1
	SP as above, with shoulder circles forward and backward, combined with a light stepping motion of the feet	1
	SP as above, marching the feet, swinging each alternate arm forward and backward.	1
	SP as above, marching the feet, with parallel arm movement forward and backward.	1
	SP with the arms by the sides holding the Smovey in each hand. The hands are moved inward and outward	1
Main part 32 min	SP as above, marching (four steps forward, four steps backward) with alternate arm movements and with lifting the knees up on every fourth step.	2
	SP as above, stepping out to the side with alternating abduction of each arm to the side.	2
	SP as above, raising each alternate leg upward, with alternating arm movement to the side.	2
	SP as above, with a step with raised knees, with movement of the arms up and down on the outside.	2
	SP as above, stepping with alternating arm movement front to back.	2
	SP as above, stepping with alternate arm movement forward and backward.	2
	SP as above, stepping forward and backward with weight transfer, with arms moving forward and backward, crossing at the front at the height of the chest.	2
	SP as above, diagonal movement of the legs forward and backward, with parallel shoulder movement sideways.	2
	SP as above, the Smovey joined together, held behind with both hands at the height of the hips, slight arm movement sideways.	2
	SP = standing in straddle position, arms at the height of the shoulders, crossing and abduction of the arms	2
	SP = lying on the side, arm movement above the head and then toward legs. Uppermost leg is lifted upward. Performed both right and left.	2

Table A2. *Cont.*

Course of the Class	Name and Description of Exercise	Time (min)
Main part 32 min	SP as above, arms kept straight along the torso, the Smovey joined together, forward and backward movement of the arm and the leg in opposite directions. Performed right and left.	2
	SP as above, the knee of the upper leg is bent and moved forward, massage of the buttock with the Smovey. Performed right and left.	2
	SP = lying on the back, the Smovey is held behind the head with both hands, moving arms forward once to the right and once to the left of the torso.	2
	SP as above, left leg is bent at the knee, the heel is leaned against the ground, right leg lifted straight up and down with parallel arm movement toward the right leg. Repeated right and left.	2
	SP as above, legs are bent at an angle of 90°, the Smovey is held with both hands at the height of the chest. Moving legs sideways while moving arms in the opposite direction.	2
Cool down 5 min	SP standing with the Smovey held in each hand, bending forward with arms directed forward.	1
	SP = standing in straddle position, the Smovey is held with both hands, lifting the hands up and bending forward, moving the Smovey toward the floor.	1
	SP as above but bending to the sides.	1
	SP as above, the Smovey is held on the shoulders, with rocking sideways movements.	1
	SP as above, raising arms up while breathing in and lowering arms while breathing out.	1
Organizational matters	The end of the class, asking the participants about their perceived intensity of exercise using the Borg scale, saying goodbye.	3

References

1. Vancampfort, D.; Stubbs, B.; Koyanagi, A. Physical chronic conditions, multimorbidity and sedentary behavior amongst middle-aged and older adults in six low- and middle-income countries. *Int. J. Behav. Nutr. Phys. Act.* **2017**, *14*, 147. [CrossRef] [PubMed]
2. Garcia, W.J.; Johnson, A.; Keldermans, D.; Tang, B. Exercise and Low Back Pain in the Older Adult: Current Recommendations. *J. Allied Health* **2019**, *48*, 302–307. [PubMed]
3. Jesus-Moraleida, F.R.; Silva, J.P.; Pereira, D.S.; Dias, J.M.D.; Dias, R.C.; Ferreira, M.L.; Hayden, J.A.; Pereira, L.S.M. Exercise therapy for older adults with low-back pain. *Cochrane Database Syst. Rev.* **2016**. [CrossRef]
4. de Oliveira, N.T.B.; Ricci, N.A.; Dos Santos Franco, Y.R.; Salvador, E.M.E.S.; Almeida, I.C.B.; Cabral, C.M.N. Effectiveness of the Pilates method versus aerobic exercises in the treatment of older adults with chronic low back pain: A randomized controlled trial protocol. *BMC Musculoskelet. Disord.* **2019**, *20*, 250. [CrossRef]
5. Colgrove, Y.M.; Gravino-Dunn, N.S.; Dinyer, S.C.; Sis, E.A.; Heier, A.C.; Sharma, N.K. Physical and Physiological Effects of Yoga for an Underserved Population with Chronic Low Back Pain. *Int. J. Yoga* **2019**, *12*, 252–264. [CrossRef]
6. Akodu, A.K.; Akindutire, O.M. The effect of stabilization exercise on pain-related disability, sleep disturbance, and psychological status of patients with non-specific chronic low back pain. *Korean J. Pain* **2018**, *31*, 199–205. [CrossRef]
7. Psycharakis, S.G.; Coleman, S.G.S.; Linton, L.; Kaliarntas, K.; Valentin, S. Muscle Activity During Aquatic and Land Exercises in People With and Without Low Back Pain. *Phys Ther.* **2019**, *99*, 297–310. [CrossRef]
8. Geneen, L.J.; Moore, R.A.; Clarke, C.; Martin, D.; Colvin, L.A.; Smith, B.H. Physical activity and exercise for chronic pain in adults: An overview of Cochrane Reviews. *Cochrane Database Syst. Rev.* **2017**. [CrossRef]
9. Levin, O.; Netz, Y.; Ziv, G. The beneficial effects of different types of exercise interventions on motor and cognitive functions in older age: A systematic review. *Eur Rev. Aging Phys. Act.* **2017**, *14*, 20. [CrossRef]
10. Hides, J.; Hodges, P.; Lambrecht, G. State-of-the-Art Exercise Concepts for Lumbopelvic and Spinal Muscles—Transferability to Microgravity. *Front. Physiol.* **2019**, *10*, 837. [CrossRef]
11. Mora, J.C.; Valencia, W.M. Exercise and Older Adults. *Clin. Geriatr. Med.* **2018**, *34*, 145–162. [CrossRef] [PubMed]
12. Musumeci, G. The Use of Vibration as Physical Exercise and Therapy. *J. Funct. Morphol. Kinesiol.* **2017**, *2*, 17. [CrossRef]

13. Goossens, N.; Janssens, L.; Caeyenberghs, K.; Albouy, G.; Brumagne, S. Differences in brain processing of proprioception related to postural control in patients with recurrent non-specific low back pain and healthy controls. *Neuroimage Clin.* **2019**, *23*, 101881. [CrossRef]
14. Awan, R.; Khan, N.; Malik, S. The Effect of WBV on Balance, Mobility and Strength in Aging Adults: A Systematic Review. *Biol. Syst. Open Access* **2017**, *6*, 179. [CrossRef]
15. Sitjà-Rabert, M.; Martínez-Zapata, M.J.; Fort Vanmeerhaeghe, A.; Rey Abella, F.; Romero-Rodríguez, D.; Bonfill, X. Effects of a whole body vibration (WBV) exercise intervention for institutionalized older people: A randomized, multicentre, parallel, clinical trial. *J. Am. Med. Dir. Assoc.* **2015**, *16*, 125–131. [CrossRef]
16. Saggini, R.; Carmignano, S.M.; Palermo, T.; Bellomo, R.G. Mechanical Vibration in Rehabilitation: State of the Art. *J. Nov. Physiother.* **2016**, *6*. [CrossRef]
17. Wu, S.; Ning, H.-T.; Xiao, S.-M.; Hu, M.-Y.; Wu, X.-Y.; Deng, H.-W.; Feng, H. Effects of vibration therapy on muscle mass, muscle strength and physical function in older adults with sarcopenia: A systematic review and meta-analysis. *Eur. Rev. Aging Phys. Act.* **2020**, *17*, 14. [CrossRef]
18. Chang, S.-F.; Lin, P.-C.; Yang, R.-S.; Yang, R.-J. The preliminary effect of whole-body vibration intervention on improving the skeletal muscle mass index, physical fitness, and quality of life among older people with sarcopenia. *BMC Geriatr.* **2018**, *18*, 17. [CrossRef]
19. Lai, C.-C.; Tu, Y.-K.; Wang, T.-G.; Huang, Y.-T.; Chien, K.-L. Effects of resistance training, endurance training and whole-body vibration on lean body mass, muscle strength and physical performance in older people: A systematic review and network meta-analysis. *Age Ageing* **2018**, *47*, 367–373. [CrossRef]
20. Zago, M.; Capodaglio, P.; Ferrario, C.; Tarabini, M.; Galli, M. Whole-body vibration training in obese subjects: A systematic review. *PLoS ONE* **2018**, *13*, e0202866. [CrossRef]
21. Choi, W.; Han, D.; Kim, J.; Lee, S. Whole-Body Vibration Combined with Treadmill Training Improves Walking Performance in Post-Stroke Patients: A Randomized Controlled Trial. *Med. Sci. Monit.* **2017**, *23*, 4918–4925. [CrossRef] [PubMed]
22. Celletti, C.; Suppa, A.; Bianchini, E.; Lakin, S.; Toscano, M.; La Torre, G.; Di Piero, V.; Camerota, F. Promoting post-stroke recovery through focal or whole body vibration: Criticisms and prospects from a narrative review. *Neurol. Sci.* **2020**, *41*, 11–24. [CrossRef]
23. Souron, R.; Besson, T.; Millet, G.Y.; Lapole, T. Acute and chronic neuromuscular adaptations to local vibration training. *Eur. J. Appl. Physiol.* **2017**, *117*, 1939–1964. [CrossRef]
24. Taani, M.H.; Siglinsky, E.; Libber, J.; Krueger, D.; Binkley, N.; Kovach, C.R.; Buehring, B. Semi-Recumbent Vibration Exercise in Older Adults: A Pilot Study of Methodology, Feasibility, and Safety. *Gerontol. Geriatr. Med.* **2019**, *5*, 2333721419881552. [CrossRef]
25. Delafontaine, A.; Vialleron, T.; Fischer, M.; Laffaye, G.; Chèze, L.; Artico, R.; Genêt, F.; Fourcade, P.C.; Yiou, E. Acute Effects of Whole-Body Vibration on the Postural Organization of Gait Initiation in Young Adults and Elderly: A Randomized Sham Intervention Study. *Front. Neurol.* **2019**, *10*, 1023. [CrossRef] [PubMed]
26. Yang, F.; King, G.A.; Dillon, L.; Su, X. Controlled whole-body vibration training reduces risk of falls among community-dwelling older adults. *J. Biomech.* **2015**, *48*, 3206–3212. [CrossRef] [PubMed]
27. Leite, H.R.; Camargos, A.C.R.; Mendonça, V.A.; Lacerda, A.C.R.; Soares, B.A.; Oliveira, V.C. Current evidence does not support whole body vibration in clinical practice in children and adolescents with disabilities: A systematic review of randomized controlled trial. *Braz. J. Phys. Ther.* **2019**, *23*, 196–211. [CrossRef]
28. Dong, Y.; Wang, H.; Zhu, Y.; Chen, B.; Zheng, Y.; Liu, X.; Qiao, J.; Wang, X. Effects of whole body vibration exercise on lumbar-abdominal muscles activation for patients with chronic low back pain. *BMC Sports Sci. Med. Rehabil.* **2020**, *12*, 78. [CrossRef] [PubMed]
29. Cenik, F.; Keilani, M.; Galid, A.; Crevenna, R. First exercise group for Turkish breast cancer patients in Vienna—A pilot project to include Turkish migrants. *Disabil. Rehabil.* **2020**, *42*, 20–25. [CrossRef]
30. Crevenna, R.; Cenik, F.; Galle, A.; Komanadj, T.S.; Keilani, M. Feasibility, acceptance and long-term exercise behaviour in cancer patients: An exercise intervention by using a swinging-ring system. *Wien. Klin. Wochenschr.* **2015**, *127*, 751–755. [CrossRef]
31. Germann, D.; El Bouse, A.; Shnier, J.; Abdelkader, N.; Kazemi, M. Effects of local vibration therapy on various performance parameters: A narrative literature review. *J. Can. Chiropr. Assoc.* **2018**, *62*, 170–181. [PubMed]
32. GHO | By Category | Life Expectancy and Healthy Life Expectancy—Data by Country. Available online: https://apps.who.int/gho/data/view.main.SDG2016LEXv?xml:lang=en (accessed on 10 June 2020).
33. Gordon, R.; Bloxham, S. A Systematic Review of the Effects of Exercise and Physical Activity on Non-Specific Chronic Low Back Pain. *Healthcare* **2016**, *4*, 22. [CrossRef] [PubMed]
34. Wilke, J.; Krause, F. Myofascial chains of the upper limb: A systematic review of anatomical studies. *Clin. Anat.* **2019**, *32*, 934–940. [CrossRef] [PubMed]
35. Ajimsha, M.S.; Shenoy, P.D.; Gampawar, N. Role of fascial connectivity in musculoskeletal dysfunctions: A narrative review. *J. Bodyw. Mov. Ther.* **2020**, *24*, 423–431. [CrossRef] [PubMed]
36. Dischiavi, S.L.; Wright, A.A.; Hegedus, E.J.; Bleakley, C.M. Biotensegrity and myofascial chains: A global approach to an integrated kinetic chain. *Med. Hypotheses* **2018**, *110*, 90–96. [CrossRef]
37. Scarr, G. Biotensegrity: What is the big deal? *J. Bodyw. Mov. Ther.* **2020**, *24*, 134–137. [CrossRef]
38. Stecco, C.; Pirri, C.; Fede, C.; Yucesoy, C.A.; De Caro, R.; Stecco, A. Fascial or Muscle Stretching? A Narrative Review. *Appl. Sci.* **2021**, *11*, 307. [CrossRef]

39. Fan, C.; Fede, C.; Gaudreault, N.; Porzionato, A.; Macchi, V.; DE Caro, R.; Stecco, C. Anatomical and functional relationships between external abdominal oblique muscle and posterior layer of thoracolumbar fascia. *Clin. Anat.* **2018**, *31*, 1092–1098. [CrossRef]
40. Chen, B.; Dong, Y.; Guo, J.; Zheng, Y.; Zhang, J.; Wang, X. Effects of Whole-Body Vibration on Lumbar-Abdominal Muscles Activation in Healthy Young Adults: A Pilot Study. *Med. Sci. Monit.* **2019**, *25*, 1945–1951. [CrossRef]
41. Zheng, Y.-L.; Wang, X.-F.; Chen, B.-L.; Gu, W.; Wang, X.; Xu, B.; Zhang, J.; Wu, Y.; Chen, C.-C.; Liu, X.-C.; et al. Effect of 12-Week Whole-Body Vibration Exercise on Lumbopelvic Proprioception and Pain Control in Young Adults with Nonspecific Low Back Pain. *Med. Sci. Monit.* **2019**, *25*, 443–452. [CrossRef]
42. Chow, D.H.K.; Lee, T.Y.; Pope, M.H. Effects of whole body vibration on spinal proprioception in healthy individuals. *Work* **2018**, *61*, 403–411. [CrossRef] [PubMed]
43. Hermens, H.J.; Freriks, B.; Merletti, R.; Stegeman, D.F.; Blok, J.H.; Rau, G.; Klug, C.; Hägg, G.; Blok, W.J. *European Recommendations for Surface Electromyography: Results of the SENIAM Project*; Roessingh Research and Development b.v.: Enschede, The Netherlands, 1999.
44. Lima, M.; Ferreira, A.S.; Reis, F.J.J.; Paes, V.; Meziat-Filho, N. Chronic low back pain and back muscle activity during functional tasks. *Gait Posture* **2018**, *61*, 250–256. [CrossRef] [PubMed]
45. Rezvani, A.; Ergin, O.; Karacan, I.; Oncu, M. Validity and Reliability of the Metric Measurements in the Assessment of Lumbar Spine Motion in Patients With Ankylosing Spondylitis. *Spine* **2012**, *37*, E1189–E1196. [CrossRef] [PubMed]
46. Price, D.D.; Staud, R.; Robinson, M.E. How should we use the visual analogue scale (VAS) in rehabilitation outcomes? II: Visual analogue scales as ratio scales: An alternative to the view of Kersten et al. *J. Rehabil. Med.* **2012**, *44*, 800–804. [CrossRef]
47. Yu, F.; Bil, K. Correlating Heart Rate and Perceived Exertion during Aerobic Exercise in Alzheimer's Disease. *Nurs. Health Sci.* **2010**, *12*, 375–380. [CrossRef]
48. Becker, S.; Bergamo, F.; Schnake, K.J.; Schreyer, S.; Rembitzki, I.V.; Disselhorst-Klug, C. The relationship between functionality and erector spinae activity in patients with specific low back pain during dynamic and static movements. *Gait Posture* **2018**, *66*, 208–213. [CrossRef]
49. Russo, M.; Deckers, K.; Eldabe, S.; Kiesel, K.; Gilligan, C.; Vieceli, J.; Crosby, P. Muscle Control and Non-specific Chronic Low Back Pain. *Neuromodulation* **2018**, *21*, 1–9. [CrossRef]
50. Hemming, R.; Sheeran, L.; van Deursen, R.; Sparkes, V. Investigating differences in trunk muscle activity in non-specific chronic low back pain subgroups and no-low back pain controls during functional tasks: A case-control study. *BMC Musculoskelet. Disord.* **2019**, *20*, 459. [CrossRef]
51. Willard, F.H.; Vleeming, A.; Schuenke, M.D.; Danneels, L.; Schleip, R. The thoracolumbar fascia: Anatomy, function and clinical considerations. *J. Anat.* **2012**, *221*, 507–536. [CrossRef]
52. Nelson-Wong, E.; Glinka, M.; Noguchi, M.; Langevin, H.; Badger, G.J.; Callaghan, J.P. Acute Surgical Injury Alters the Tensile Properties of Thoracolumbar Fascia in a Porcine Model. *J. Biomech. Eng.* **2018**, *140*, 1010121–1010127. [CrossRef]
53. Sharma, S.K.; Saiyad, S.; Bid, D.N. Role of Latissimus Dorsi and Lower Trapezius in Chronic Mechanical Low Back Pain due to Thoraco-lumbar Dysfunction. *Indian J. Physiother. Occup. Ther.* **2013**, *7*, 219–224. [CrossRef]
54. Fan, C.; Guidolin, D.; Ragazzo, S.; Fede, C.; Pirri, C.; Gaudreault, N.; Porzionato, A.; Macchi, V.; De Caro, R.; Stecco, C. Effects of Cesarean Section and Vaginal Delivery on Abdominal Muscles and Fasciae. *Medicina* **2020**, *56*, 260. [CrossRef] [PubMed]
55. Fede, C.; Albertin, G.; Petrelli, L.; Sfriso, M.M.; Biz, C.; De Caro, R.; Stecco, C. Hormone Receptor Expression in Human Fascial Tissue. *Eur. J. Histochem.* **2016**, *60*, 2710. [CrossRef] [PubMed]
56. Nowotny, A.H.; Calderon, M.G.; de Souza, P.A.; Aguiar, A.F.; Léonard, G.; Alves, B.M.O.; Amorim, C.F.; da Silva, R.A. Lumbar stabilisation exercises versus back endurance-resistance exercise training in athletes with chronic low back pain: Protocol of a randomised controlled trial. *BMJ Open Sport Exerc. Med.* **2018**, *4*, e000452. [CrossRef] [PubMed]
57. Daneau, C.; Cantin, V.; Descarreaux, M. Effect of Massage on Clinical and Physiological Variables During Muscle Fatigue Task in Participants With Chronic Low Back Pain: A Crossover Study. *J. Manip. Physiol. Ther.* **2019**, *42*, 55–65. [CrossRef]
58. Jung, H.; Yamasaki, M. Association of lower extremity range of motion and muscle strength with physical performance of community-dwelling older women. *J. Physiol Anthr.* **2016**, *35*, 1–9. [CrossRef]
59. da Silva, R.A.; Vieira, E.R.; Léonard, G.; Beaulieu, L.-D.; Ngomo, S.; Nowotny, A.H.; Amorim, C.F. Age- and low back pain-related differences in trunk muscle activation during one-legged stance balance task. *Gait Posture* **2019**, *69*, 25–30. [CrossRef]
60. Langevin, H.M.; Fox, J.R.; Koptiuch, C.; Badger, G.J.; Greenan- Naumann, A.C.; Bouffard, N.A.; Konofagou, E.E.; Lee, W.-N.; Triano, J.J.; Henry, S.M. Reduced thoracolumbar fascia shear strain in human chronic low back pain. *BMC Musculoskelet. Disord.* **2011**, *12*, 203. [CrossRef]
61. Nuzzo, J.L. The Case for Retiring Flexibility as a Major Component of Physical Fitness. *Sports Med.* **2020**, *50*, 853–870. [CrossRef]
62. Vulfsons, S.; Chervonenko, S.; Haddad, M.; Weisman, M.H.; Lavi, N.; Dar, G. Decreased amplitude of surface electromyo- graphic recordings of muscle activation along the posterior myofascial kinematic chain in subjects with chronic nonspecific low back pain compared to healthy subjects. *J. Back Musculoskelet. Rehabil.* **2018**, *31*, 785–793. [CrossRef]
63. Shamsi, M.; Sarrafzadeh, J.; Jamshidi, A.; Arjmand, N.; Ghezelbash, F. Comparison of spinal stability following motor control and general exercises in nonspecific chronic low back pain patients. *Clin. Biomech.* **2017**, *48*, 42–48. [CrossRef] [PubMed]
64. Arendt-Nielsen, L.; Graven-Nielsen, T.; Svarrer, H.; Svensson, P. The influence of low back pain on muscle activity and coordination during gait: A clinical and experimental study. *Pain* **1996**, *64*, 231–240. [CrossRef]

65. Suh, J.H.; Kim, H.; Jung, G.P.; Ko, J.Y.; Ryu, J.S. The effect of lumbar stabilization and walking exercises on chronic low back pain. *Medicine* **2019**, *98*, e16173. [CrossRef] [PubMed]
66. Sbardelotto, M.L.; Costa, R.R.; Malysz, K.A.; Pedroso, G.S.; Pereira, B.C.; Sorato, H.R.; Silveira, P.C.L.; Nesi, R.T.; Grande, A.J.; Pinho, R.A. Improvement in muscular strength and aerobic capacities in elderly people occurs independently of physical training type or exercise model. *Clinics* **2019**, *74*, e833. [CrossRef] [PubMed]
67. Lee, P.G.; Jackson, E.A.; Richardson, C.R. Exercise Prescriptions in Older Adults. *Am. Fam. Physician* **2017**, *95*, 425–432. [PubMed]

biology

MDPI

Article

Effects of 8 Weeks of High-Intensity Interval Training and Spirulina Supplementation on Immunoglobin Levels, Cardio-Respiratory Fitness, and Body Composition of Overweight and Obese Women

Hadi Nobari [1,2,3,*], Elham Eyni Gandomani [3], Jalil Reisi [3,*], Reyhaneh Vahabidelshad [4], Katsuhiko Suzuki [5], Stella Lucia Volpe [6] and Jorge Pérez-Gómez [7]

1 Department of Exercise Physiology, Faculty of Educational Sciences and Psychology, University of Mohaghegh Ardabili, Ardabil 56199-11367, Iran
2 Department of Physiology, School of Sport Sciences, University of Extremadura, 10003 Cáceres, Spain
3 Department of Exercise Physiology, Faculty of Sport Sciences, University of Isfahan, Isfahan 81746-7344, Iran; elham_eynii@yahoo.com
4 Faculty of Motor Sciences, Université Libre de Bruxelles, 1070 Brussels, Belgium; reyhanevahabi@yahoo.com
5 Faculty of Sport Sciences, Waseda University, Saitama 359-1192, Japan; katsu.suzu@waseda.jp
6 Department of Human Nutrition, Foods and Exercise, Virginia Polytechnic Institute and State University, Blacksburg, VA 24061, USA; stellalv@vt.edu
7 HEME Research Group, Faculty of Sport Sciences, University of Extremadura, 10003 Cáceres, Spain; jorgepg100@gmail.com
* Correspondence: hadi.nobari1@gmail.com (H.N.); jalil_reisi@yahoo.com (J.R.)

Citation: Nobari, H.; Gandomani, E.E.; Reisi, J.; Vahabidelshad, R.; Suzuki, K.; Volpe, S.L.; Pérez-Gómez, J. Effects of 8 Weeks of High-Intensity Interval Training and Spirulina Supplementation on Immunoglobin Levels, Cardio-Respiratory Fitness, and Body Composition of Overweight and Obese Women. *Biology* **2022**, *11*, 196. https://doi.org/10.3390/biology11020196

Received: 28 December 2021
Accepted: 24 January 2022
Published: 26 January 2022

Simple Summary: Overweight and obese, like other forms of malnutrition, have been shown to affect immune function through changing immunoglobin or cardio-respiratory fitness levels and cell-mediated immune responses. Although calorie restriction and exercise are the most common therapies for obesity or overweight, it is unclear what kind of supplementation these people should take or how much exercise they should perform. Hence, in this study, we examined the effect of 8 weeks of high-intensity interval training (HIIT) with spirulina supplementation on the humoral immunity, cardio-respiratory fitness, and body composition of overweight and obese women. The results demonstrated that spirulina supplementation with HIIT not only decreased fat free mass but also boosted immunoglobin-A, which plays an important role in the immune system.

Abstract: Our study examined the effect of 8 weeks of high-intensity interval training (HIIT) and spirulina supplementation on the humoral immunity, cardio-respiratory fitness, and body composition of overweight and obese women. Thirty sedentary women (height: 161.7 ± 2.8 cm, body mass: 75.8 ± 8.4 kg, body mass index [BMI]: 28.8 ± 2.5 kg/m^2, age: 25.1 ± 6.7 years) were divided into three groups: placebo with HIIT group, spirulina group (SG), or combined group (CG). Exercise groups performed HIIT for 8 weeks, with three sessions per week and four to seven repetitions in each session of 30 s running and 30 s walking; the intensity was established at 90% of the maximum heart rate. Supplementation groups received 6 g of spirulina powder per day. Fasting blood samples were collected before and after 8 weeks to determine the concentrations of immunoglobulins (IgA and IgG). There was a significant group-by-time interaction for fat free mass (FFM; $p = 0.001$, f = 8.52, $\eta p^2 = 0.39$) and IgA ($p = 0.036$, f = 3.86, $\eta p^2 = 0.22$). The post hoc analysis revealed that CG reduced FFM significantly ($p = 0.012$, $g = -0.55$) after training. CG and SG showed significantly greater IgA concentrations after 8 weeks ($p = 0.02$, $g = 0.70$ and $p = 0.001$, $g = 0.34$, respectively). We conclude that spirulina supplementation with HIIT affects the body composition (lower FFM) but also boosts IgA, which plays an important role in the immune system.

Keywords: antioxidant; body fat; IgA; immunomodulation; nutritional supplement; obesity; physical activity

1. Introduction

Overweight and obesity refer to the abnormal or excessive accumulation of fat that may lead to an increased risk of chronic disease. The World Health Organization (WHO) defines a body mass index (BMI) $\geq$ 25 kg/m^2 as overweight and a BMI $\geq$ 30 kg/m^2 as obesity [1,2]. Research has shown that, with each unit of increase in BMI, the risk of cardiovascular disease increases by 8%. However, for each unit of a metabolic equivalent task increase in physical activity, the risk of cardiovascular disease decreases. Excess body fat is a condition associated with an impaired immune system and greater susceptibility to developing an infectious disease [3–5].

Spirulina maxima is used as a nutritional supplement because of its phytochemical content (phenolic compounds, carotenoids, and tocopherols) and essential nutrients (proteins, *n*-3 and *n*-6 fatty acids) [6,7]. It has been suggested that spirulina might help to increase lean body mass because of its high protein content, particularly of the branched-chain amino acids, leucine, valine, and isoleucine. As a result of this, athletes have used spirulina to improve the body composition and physical performance [8,9]. Among the species of spirulina that are safe for consumption are *Spirulina maxima*, arthrospira fusiformis, and platensis, and the latter is the most commonly used and studied in the scientific literature [10]. As previously stated, spirulina has a high protein content (50% to 70% of its dry weight) [11], all of the essential amino acids, most of the vitamins and minerals, and it confers numerous health benefits, such as antioxidant, immunomodulatory, anti-inflammatory, and antiviral activities [7,12]. Many athletes have consumed spirulina for these health benefits, and it was suggested that the Chinese and Cuban Olympic teams consumed spirulina daily for many years to improve their athletic performance [13]. Hernández-Lepe et al. gave 4.5 g of spirulina per day to their participants for six weeks [14].

High-intensity interval training (HIIT), which is alternating between periods of high intensity and recovery, has become a popular training method due to its time efficiency. HIIT is effective for improving fasting blood glucose concentrations and reducing blood pressure in overweight or obese populations [15,16]. Recent studies have clearly shown that intense intermittent exercise is better for reducing fat than endurance exercise [17–19]. In addition, HIIT has been shown to reduce blood pressure in individuals who are overweight or obese [20]. HIIT has been shown to have benefits in young and older individuals on body weight, the regulation of physiological parameters such as blood pressure, improvement in aerobic capacity as measured by maximum oxygen consumption (VO_{2max}), and reductions in glucose and triglyceride concentrations and fat free mass, with an increased lower limb muscle power [21]. HIIT is also effective for improving fasting glucose concentrations and reducing blood pressure in individuals who are overweight or obese [22]. In a study of 20 healthy untrained overweight/obese males, the following 12 weeks of HIIT reported that the BMI and fat mass percentage were significantly decreased [23]. HIIT also increase cardiopulmonary fitness. The best indicator for assessing cardiorespiratory fitness is the measurement of VO_{2max} [24]. Gillen et al. [25] showed that short-term low-volume HIIT is a time-efficient strategy to improve the body composition and muscle oxidative capacity in women who are overweight or obese. In addition, Andreato et al. [26] reported that HIIT can be used as a secondary method for the treatment of obesity in adults.

The immune system contains complex mechanisms that are of particular importance in the body's defense against pathogenic microorganisms, bacteria, parasites, and viruses. The immune system is divided into two arms: innate immunity (natural or non-special) and adaptive immunity (acquired or special), where the acquired immunity is divided into humoral and cellular parts. Cellular immunity includes T cells (e.g., CD_8, CD_4, and CD_3) and B cells (e.g., CD_{22}, CD_{20}, and CD_{19}). Humoral immunity includes immunoglobulins (e.g., IgM, IgA, IgD, IgE, and IgG) [27]. IgA is the major immunoglobulin in mucous secretions, such as saliva and tears, and it is thought to provide a front line of defense against pathogens and antigens present on mucosal surfaces, such as the airways. IgA is able to inhibit the binding of viruses and bacteria to the mucosal epithelium and viral replication [28]. Controversial results have been observed regarding the impact of HIIT on the immune system. The majority of the studies have shown that HIIT suppressed the immune system, whereas others have reported that this training did not affect the immune system [27,29,30]. In a few cases, it has been reported that HIIT improved immune function [31,32]. IgG is vital for the proper functioning of the body's defense system, staying healthy, and fighting pathogens [33]. Owen et al. [34] showed that high-intensity soccer training might cause a significant decrease in s-IgA values post-exercise compared to low-intensity training. Lee et al. [35] indicated that there was a reduced trend of IgA in male adults after 12 weeks of judo training alone, or combined with resistance training or with interval training. The initial laboratory examination of humoral immunity consists of measuring the levels of various immunoglobulin (IgG and IgA) in serum [36]. The mean values for IgG were from 720 to 1038 mg/100 mL in the females that we can compare with our study [37]. In our study, overweight and obese women were recruited because there are few studies about HIIT and spirulina supplementation on the humoral immune system, cardio-respiratory fitness, and body composition in this population. Therefore, we hypothesized that 8 weeks of HIIT and spirulina supplementation could affect the humoral immune system, VO_{2max}, and body composition in overweight and obese women.

2. Materials and Methods

2.1. Participants

Thirty women (mean $\pm$ standard deviation); 25.1 $\pm$ 6.7 years of age; height: 161.7 $\pm$ 2.8 cm, body mass: 75.8 $\pm$ 8.4 kg; with a BMI between 25 to 35 kg/m^2, were divided into three groups: placebo with HIIT (PH, n = 10), spirulina group (SG, n = 10), and combined group (CG, n = 10). PH performed HIIT for 8 weeks, 3 sessions per week, and received a placebo per day. SG received 6 g of spirulina powder per day and did not participate in any regular training. CG performed the same HIIT and received 6 g of spirulina powder per day [38]. Exclusion and enrollment criteria for all groups were: (1) participants who were physically active; (2) participants who were on a weight loss diet; (3) participants with medical conditions for physical activity; (4) participants with a BMI lower than 25 or above 35 kg/m^2; (5) they should not have any specific illness or diet; (6) those who missed more than three practice sessions were excluded from the study (Figure 1). All participants signed and accepted the informed consent according to the recommendations of the Helsinki Declaration for Human Research. Our study is a part of the master's thesis of the University of Isfahan registered with the code IR.UI.REC.1397.145.

Within 24 h following the treatment, subjects were advised not to ingest alcohol, caffeine, theine, hot liquids, or smoke. In addition, subjects were advised not to ingest medications, performance-enhancing capsules, or other supplements during the study [39]. We instructed the participants to maintain their usual dietary intake and not to lower their energy intake, and also asked them to maintain their usual physical activity during the study [40]. Based on the recommendation, participants maintained their usual dietary intake and physical activity levels.

2.2. Sample Size

Using the statistical method investigated and G-Power software (University of Dusseldorf, Dusseldorf, Germany), we calculated the design's power and sample size. This included the following: a priori and F tests are used to calculate the achieved power; ANOVA: repeated measurements, within-interaction analysis; err prob for $\alpha = 0.05$; minimum effect size = 0.35; number of groups = 3; number of measures = 2; and err prob for $1\text{-}\beta = 0.80$. With real or actual power, there is an 82.2% chance of successfully rejecting the null hypothesis of no difference in variables in the study with 24 participants.

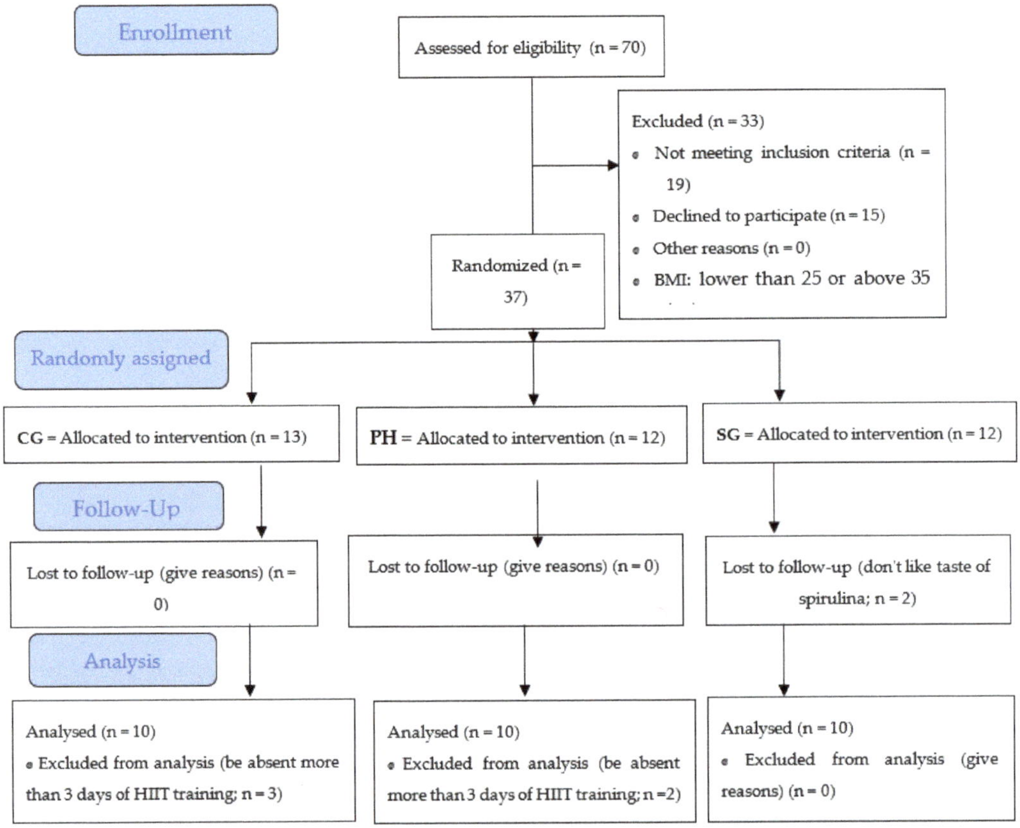

Figure 1. Flow diagram of how to enter, experimental course, and analysis of participants. Abbreviation; PH = placebo and high-intensity interval training (HIIT); CG = combined group (spirulina and HIIT); SG = spirulina group; BMI = body mass index.

2.3. Experimental Approach to the Problem

This research was a quasi-experimental, double-blinded design with baseline and post-intervention measurements. Before starting the research, age, height, and BMI were assessed (Table 1), and BF% was measured by the thickness of the subcutaneous fat layer using skinfold analysis. In the next stage, participants performed a shuttle run test to assess aerobic fitness according to guidelines [41] of VO_{2max} measurement. Participants received oral and written information about supplement. Blood samples were collected at two different time points: before and after the supplementation and training periods. The samples were transferred to the laboratory immediately after each collection and centrifuged there.

Table 1. Characteristics of the participants.

Group	Age (Year)	Height (cm)	Body Mass (kg)	BMI (kg/m^2)	90% HR_{max}
PH	26 ± 8	162 ± 4	73 ± 5	27.6 ± 1.9	~171 bpm
CG	24 ± 6	163 ± 3	76 ± 11	28.8 ± 4.3	~172 bpm
SG	24 ± 6	160 ± 2	77 ± 9	29.9 ± 1.2	~172 bpm

Data are presented in mean ± standard deviation. Abbreviation: PH = placebo and high-intensity interval training (HIIT); CG = combined group (spirulina and HIIT); SG = spirulina group; BMI = body mass index; 90% HR_{max} = 90% of maximum heart rate; bpm = beats per minute.

2.4. Measurement of Fat Free Mass

Fat free mass (FFM) was calculated with the following formula for women, with weight (W) in kilograms and height (H) in centimeters: FFM = (0.29569 × W) + (0.41813 × H) − 43.2933 [42,43]. To measure body fat percentage, seven subcutaneous fat thicknesses were used by the Jackson and Pollock method [44–46]. Seven points were included: triceps, chest, subscapular, suprailiac, axilla, abdominal, and thigh [47]. Data were collected by Lafayette Instrument Company (Lafayette, IN, USA) with an accuracy of 0.1 mm. All measurements were performed by one person on the right side of the body. The person who took the skinfold measurements had taken several skinfold measurements over many years [48–50]. The technical measurement error was considered according to previous studies [51,52].

2.5. Measurement of Cardio-Respiratory Fitness

VO_{2max} was measured by 20-m shuttle-run test. The maximum field test consisted of reciprocating runs between two lines 20 m apart at a speed adjusted to a pre-recorded audible alarm [53]. The initial speed set to start the test was 8.5 km/h^{-1}, which was increased to 0.5 km/h^{-1} after each minute. Participants were instructed to continue the test to the last step as much as possible. The test ended when the person could not keep up with the running speed, or when the person was unable to reach the 20-m area within each lane three times in a row in accordance with the audible warning. The velocity obtained during the last step that was fully performed was considered as the maximum test velocity, and was calculated as VO_{2max} by placing it in the following formula [41]: VO_{2max} (mL·kg^{-1}·min^{-1}) = 6 (x) − 24.4 X. X is the maximum aerobic speed, which is determined by the running speed at the highest level [54].

2.6. Measurement of Blood Samples

To measure IgA and IgG 24 h before and 24 h after the study period, 10 mL blood samples were taken from the left vein of the participants between 9:00 and 11:00 in the morning, blood samples were collected into pipes containing solution of acidic anticoagulant EDTA K2, and, after plasma centrifugation, their plasma was separated, where the resulting plasma was kept at −20 °C. IgA and IgG concentrations were measured by using Hitachi device, laboratory turbidometry, and Pars test kits. The turbidometry method is based on a complex formation resulting from the reaction between immunoglobulins and its specific antiserum. The amount of turbidity generated is directly related to the amount of immunoglobulins. The minimum volume required to measure IgG and IgA by a turbidometric device is 50 µL.

2.7. Exercise Protocol

Participants participated in HIIT exercises three times a week for 8 weeks, with an intensity of 90% of maximum heart rate. Exercises started from 24 min in the first session (5 to 10 min of warm-up, 30 s of exercise (running), and 30 s of active rest (walking), with 4 repetitions and 5 to 10 min of cool down), and 27 min in the last session (5 to 10 min of warm-up, 30 s of exercise, and 30 s of active rest, with 7 repetitions and 5 to 10 min of cool down) [55]. The participants of the two training groups performed the training protocol at a distance of 20 m according to Glaister et al. [56]. In the round-robin test practice protocol, participants first ran at maximum speed from the starting point (cone 1) to cone 2 in lane

A. After returning in the opposite direction on route B, they ran 20 m towards cone 3 with maximum speed, and finally, after returning, they ran again at maximum speed on route C (cone 1) to complete the distance of 40 m. Participants continued to perform this at maximum speed until the 30-s period of the training protocol ended, and, after a 30-s break, they repeated the training protocol. Exercise progressed by increasing the number of 30-s repetitions from four times in the first and second weeks to five times in the third and fourth weeks, to six times, in the fifth and sixth weeks, and to seven times in the seventh and 8-week of practice, like previous study training protocol [56]. The intensity of training in all stages of the protocol was 90% of the maximum heart rate. The heart rate was measured by a control instructor using a Polar pacemaker made in Finland, and the maximum predictable heart rate was estimated with the formula of Tanaka (208 − 0.7 × age (years)) [57]. All participants participated in the exercises until they were completed. In this study, the participants in the supplement group did not have any regular exercise.

2.8. Supplementation

In the present study, spirulina algae powder was prepared from Isfahan Green Agate Company (Isfahan, Iran). The participants in the CG and SG received 6 g per day of water-soluble spirulina powder half an hour before a meal, and the participants in the PH received a green coloring food dissolved in water; this was close to other relevant human studies, as they received 8 g of spirulina per day [58].

2.9. Statistical Methods

Data analysis of the present study has been carried out at both descriptive and inferential concentrations. The distribution between the data was also examined by Shapiro–Wilk test. Equality of variance in different groups was also assessed by Levin test. To determine possible group differences pre-training calculated with a one-way analysis of variance (ANOVA), a 2 × 3 ANOVA with repeated measures (time [pre- vs. post-training] × group [CG vs. SG vs. PH]) was used to determine differences between groups, and then we used the suitable Tukey post hoc test when a significant group-by-time interaction was discovered. Hedge's g effect size with 95% confidence interval was also calculated to determine the magnitude of pairwise comparisons for pre- and post-test, and was defined as trivial (<0.2), small (≥0.02), moderate (≥0.05), and large (≥0.08) [59]. If the results of the one-way ANOVA and effect sizes were similar for each group (i.e., FFM), then the percentage changes were computed and assessed. Significance of statistical analysis was used at the level of $p < 0.05$. All statistical calculations were performed using SPSS (Version 25.0; IBM SPSS Inc., Chicago, IL, USA).

3. Results

Table 2 shows the mean and standard deviation of the changes in VO_{2max}, IgA, and IgG. At the baseline, there were no differences observed between groups in the above variables ($p > 0.05$). There was no significant main effect of time for IgA ($p = 0.073$, f = 3.48, $\eta p^2 = 0.11$); however, there was a meaningful group-by-time interaction ($p = 0.036$, f = 3.86, $\eta p^2 = 0.22$). The post hoc analysis found that IgA (CG, $p = 0.02$, $g = 0.70$ and SG, $p = 0.001$, $g = 0.34$) was significantly greater post-test versus pre-test.

Table 3 shows the mean and standard deviation in the anthropometric and body composition. At the baseline, there were no differences observed between groups in all variables, except the waist-to-hip ratio (f = 4.39, $p < 0.02$). Based on the analysis, there was no significant main effect of time for FFM ($p = 0.36$, f = 0.86, $\eta p^2 = 0.03$), whereas there was a meaningful group-by-time interaction ($p = 0.001$, f = 8.52, $\eta p^2 = 0.39$). The post hoc analysis indicated that FFM (kg) (CG, $p = 0.012$, $g = -0.54$) was significantly reduced. In addition, in the SG group, this variable increased but was not significant ($p > 0.05$). However, there were significant main effects of time for the rest of the variables, but there were no significant group-by-time interactions for changes in these variables.

Table 2. Changes in VO_{2max} and immunoglobulins.

Variables	Groups	Pre-Training	Post-Training	CI 95% for Difference		Hedge's g	% Changes
				Lower	Upper		
VO_{2max} ($mL \cdot kg^{-1} \cdot min^{-1}$)	PH	21.8 ± 3.4	23.4 ± 1.8	−4.21	0.93	0.57	9.21
	CG	23.2 ± 2.1	25.6 ± 3.1	−4.85	0.05	0.88	11.03
	SG	21.3 ± 1.9	21.9 ± 3.0	−2.91	1.73	0.23	2.58
IgA (mg/lit)	PH	161.7 ± 86.2	149.2 ± 50.8	−53.97	78.97	−0.16	−2.31
	CG	173.1 ± 33.7	212.7 ± 67.8 *	−89.90	10.70	0.70	21.23
	SG	171.1 ± 47.9	187.5 ± 44.6 *	−59.86	27.06	0.33	11.82
IgG (mg/lit)	PH	1253.5 ± 413.6	1227.1 ± 316.1	−319.46	372.26	−0.06	−3.21
	CG	1309.3 ± 285.3	1461.5 ± 240.1	−399.91	95.51	0.55	13.28
	SG	1301.0 ± 317.8	1460.4 ± 272.2	−437.38	118.58	0.51	14.17

Abbreviation: VO_{2max} = maximum oxygen consumption; IgA = immunoglobulin A; IgG = immunoglobulin G; HIIT = high-intensity interval training; CI = confidence interval; PH = placebo and HIIT; CG = combined group; SG = spirulina group. * to reflect significance pre- vs. post-test at the level of $p < 0.05$.

Table 3. Changes in anthropometric and body composition.

Variables	Groups	Pre-Training	Post-Training	CI 95% for Difference		Hedge's g	% Changes
				Lower	Upper		
BMI (kg/m^2)	PH	27.6 ± 1.9	26.3 ± 1.9	−0.48	3.08	−0.65	−4.64
	CG	28.8 ± 4.3	27.7 ± 4.6	−3.08	5.26	−0.23	−3.92
	SG	29.9 ± 4.0	29.6 ± 4.1	−3.49	4.09	−0.07	−1.06
WHR (cm)	PH	0.80 ± 0.05	0.79 ± 0.06	−0.04	0.06	−0.17	−1.15
	CG	0.73 ± 0.03	0.72 ± 0.03	−0.01	0.03	−0.31	−1.79
	SG	0.88 ± 0.18	0.80 ± 0.09	−0.05	0.21	−0.53	−6.74
Body mass (kg)	PH	73.3 ± 5.3	70.2 ± 4.0	−1.33	7.45	−0.62	−4.03
	CG	76.7 ± 11.1	74.1 ± 11.1	−7.83	13.01	−0.22	−3.40
	SG	77.3 ± 8.9	76.1 ± 10.0	−7.64	10.10	−0.12	−1.73
FFM (kg)	PH	19.5 ± 2.1	19.1 ± 2.4	−1.76	2.46	−0.14	−1.89
	CG	18.8 ± 0.8	18.3 ± 0.96 *	−0.32	1.36	−0.54	−2.76
	SG	19.7 ± 2.3	20.1 ± 2.1	−2.47	1.67	0.17	2.22
BF (%)	PH	31.7 ± 3.0	30.4 ± 3.0	−1.39	4.09	−0.43	−4.20
	CG	32.0 ± 2.4	31.0 ± 3.0	−1.48	3.62	−0.37	−3.43
	SG	33.9 ± 3.9	31.3 ± 2.9	−0.64	5.82	−0.71	−7.29

Abbreviation: BMI = body mass index; WHR = waist–hip ratio; FFM = free fat mass; BF% = body fat percentage; HIIT = high-intensity interval training; CI = confidence interval; PH = placebo and HIIT; CG = combined group spirulina and HIIT; SG = spirulina group. * to reflect significance pre- vs. post-test at the level of $p < 0.05$.

4. Discussion

Our purpose was to examine the effect of 8-week HIIT and spirulina supplementation on the humoral immune function, cardio-respiratory fitness, and body composition of overweight and obese women. We found that 8-week HIIT with 6 g of spirulina supplementation per day significantly improved IgA (CG: 212.7 ± 67.8 and SG: 187.5 ± 44.6) compared to the baseline (CG: 173.1 ± 33.7 and SG: 171.1 ± 47.9) and FFM (18.3 ± 1.0) compared to the baseline (18.8 ± 0.8) in CG. Although IgG did not change significantly, the percentage change in the spirulina groups was illustrated as being between 13 to 15%, with a medium-to-high effect size. The study of Jahani et al. [60] that examined the effect of HIIT and probiotic supplementation on immune cells, C-reactive protein, and IgA showed that intense intermittent exercise increases IgA, which is consistent with the results of the present study. Spirulina has a powerful stimulatory effect on the immune system through increasing the phagocytic activity of macro-phages, inducing the accumulation of natural killers cells in tissues, stimulating antibody and cytokine production, and activating and mobilizing T or B cells [61]. Previous studies demonstrated that spirulina diminishes the negative effect of different agents of Ig concentrations [62,63] and leukocyte numbers [64].

The researchers concluded that exercise-induced changes in serum Ig concentrations may be due to the participation of extravascular proteins, increased lymphocytes after exercise, a combination of changes in the plasma volume and extravascular flow, and changes in the subject's circadian cycle [65]. Factors involved in immunity include sex, age, race, smoking, strenuous or moderate physical activity, alcohol consumption, obesity, pregnancy, hormonal factors, and common microflora in each individual's digestive tract [66,67]. Therefore, one of the factors involved in the difference between the results of other studies and the findings of the present study can be enumerated in environmental and genetic factors that were beyond the control of the researchers [68]. Saeedy et al. [31] showed that HIIT, along with zinc supplementation, significantly improved IgA. Qieqeshlaq et al. [68] showed that HIIT and probiotic increased IgA significantly. Spirulina has a hypolipidemic activity and decreases the concentrations of liver profiles [69].

Mohebi et al. [70] showed that 8-week high-intensity resistance training decreased IgG concentration significantly in untrained men. The contradiction with the present study may be due to the difference between detailed training and participants. Santoso et al. [71] showed that there was a change between the IgG level of pre- and post-test of breathing arts sports treatment, and that this change increased significantly after the respiratory exercise. Although the IgG concentrations did not alter considerably, the spirulina groups showed a percentage change of 13 to 15% with a medium to large effect size. The phycocyanin in spirulina increases biological activity against infectious diseases by maintaining the function of the mucosal immune system and reducing allergic inflammation by suppressing specific antibodies, and injecting it produces IgA antibodies. Spirulina polysaccharide also activates innate immune cells and increases antibodies [72]. The polysaccharides and phycocyanins in spirulina help to both increase the number of antigens through physical activity and increase the immune system. Therefore, intense intermittent exercise with spirulina supplementation has a greater effect on strengthening the immune system [73,74].

Spirulina supplementation decreased the FFM of overweight and obese women significantly. Hunter et al. showed that, after 4 weeks of resistance training, the amount of FFM decreased significantly [75]. The results of their research is consistent with our results, which is probably due to the similar training duration [76].

Our present study had some limitations. We did not control the participants' diets. Furthermore, we could have had a longer duration for our study. In addition, by observing large percentage changes in the variable of IgG, we did not see significant changes, which was probably due to individual effects and large changes in some participants. We strongly recommend that individual differences, resting energy expenditure, physical activity levels, and dietary intake be considered in future studies. Additionally, more studies with different ages, as well as with women and men, may help to delineate the effects of HIIT and spirulina supplementation on the immune function, body composition, and exercise performance. Finally, another limitation of the study could be the field training protocol that has been performed. This could be considered in future studies by increasing the control of the training intensity with heart rate and considering the session calibration of the device used.

5. Conclusions

In this study, the effect of an 8-week period of HIIT combined with spirulina supplementation on the humoral immune system and body composition of overweight and obese women was investigated. The data in the present study demonstrated the effectiveness of spirulina supplementation and HIIT concurrently in making significant changes in IgA concentrations and FFM. Taking spirulina with HIIT for overweight and obese women may be helpful not only for losing FFM but also for boosting IgA, which plays an important role in the immune system.

Author Contributions: Conceptualization, E.E.G., H.N. and J.R.; methodology, E.E.G., H.N., R.V. and J.R.; software, H.N., S.L.V. and R.V.; interpretation of data, H.N. and R.V., critical revision, H.N., J.P.-G. and K.S.; revised manuscript, R.V., S.L.V. and H.N., writing—original draft preparation, H.N., R.V.

and E.E.G.; writing—review and editing, H.N., K.S. and S.L.V. All authors have read and agreed to the published version of the manuscript.

Funding: There is no external funding.

Institutional Review Board Statement: The study has been approved by the Ethics Committee of the University of Isfahan IR.UI.REC.1397.145.

Informed Consent Statement: Informed consent was obtained from all participants involved in the study.

Data Availability Statement: The datasets used and/or analyzed during the current study are available from the corresponding author on reasonable request.

Conflicts of Interest: The authors declare no conflict of interest.

References

1. WHO. *Obesity: Preventing and Managing the Global Epidemic. Report of a WHO Consultation*; World Health Organization Technical Report Series nr 894; WHO: Geneva, Switzerland, 2000; pp. 1–253.
2. WHO Expert Consultation. Appropriate body-mass index for Asian populations and its implications for policy and intervention strategies. *Lancet* **2004**, *363*, 157–163. [CrossRef]
3. Huttunen, R.; Syrjänen, J. Obesity and the risk and outcome of infection. *Int. J. Obes. Lond.* **2013**, *37*, 333–340. [CrossRef] [PubMed]
4. Kanneganti, T.-D.; Dixit, V.D. Immunological complications of obesity. *Nat. Immunol.* **2012**, *13*, 707–712. [CrossRef] [PubMed]
5. Kaspersen, K.A.; Pedersen, O.B.; Petersen, M.S.; Hjalgrim, H.; Rostgaard, K.; Møller, B.K.; Juul-Sørensen, C.; Kotzé, S.; Dinh, K.M.; Erikstrup, L.T.; et al. Obesity and risk of infection: Results from the Danish Blood Donor Study. *Epidemiology* **2015**, *26*, 580–589. [CrossRef] [PubMed]
6. Miranda, M.; Cintra, R.; Barros, S.; Mancini-Filho, J. Antioxidant activity of the microalga Spirulina maxima. *Braz. J. Med. Biol. Res.* **1998**, *31*, 1075–1079. [CrossRef] [PubMed]
7. Deng, R.; Chow, T.-J. Hypolipidemic, Antioxidant, and Antiinflammatory Activities of Microalgae Spirulina. *Cardiovasc. Ther.* **2010**, *28*, e33–e45. [CrossRef] [PubMed]
8. Milasius, K.; Malickaite, R.; Dadeliene, R. Effect of spirulina food supplement on blood morphological parameters, biochemical composition and on the immune function of sportsmen. *Biol. Sport* **2009**, *26*, 157–172. [CrossRef]
9. Hernández-Lepe, M.A.; Olivas-Aguirre, F.J.; Gómez-Miranda, L.M.; Hernández-Torres, R.P.; Manríquez-Torres, J.J.; Ramos-Jiménez, A. Systematic physical exercise and Spirulina maxima supplementation improve body composition, cardiorespiratory fitness, and blood lipid profile: Correlations of a randomized double-blind controlled trial. *Antioxidants* **2019**, *8*, 507. [CrossRef]
10. Thengodkar, R.R.M.; Sivakami, S. Degradation of chlorpyrifos by an alkaline phosphatase from the cyanobacterium Spirulina platensis. *Biodegradation* **2010**, *21*, 637–644. [CrossRef]
11. Ramírez-Rodrigues, M.M.; Estrada-Beristain, C.; Metri-Ojeda, J.; Pérez-Alva, A.; Baigts-Allende, D.K. Spirulina platensis Protein as Sustainable Ingredient for Nutritional Food Products Development. *Sustainability* **2021**, *13*, 6849. [CrossRef]
12. Wu, Q.; Liu, L.; Miron, A.; Klímová, B.; Wan, D.; Kuča, K. The antioxidant, immunomodulatory, and anti-inflammatory activities of Spirulina: An overview. *Arch. Toxicol.* **2016**, *90*, 1817–1840. [CrossRef]
13. Zovko Koncic, M.; Tomczyk, M. New insights into dietary supplements used in sport: Active substances, pharmacological and side effects. *Curr. Drug Targets* **2013**, *14*, 1079–1092. [CrossRef]
14. Hernández-Lepe, M.A.; López-Díaz, J.A.; Juárez-Oropeza, M.A.; Hernández-Torres, R.P.; Wall-Medrano, A.; Ramos-Jiménez, A. Effect of Arthrospira (Spirulina) maxima supplementation and a systematic physical exercise program on the body composition and cardiorespiratory fitness of overweight or obese subjects: A double-blind, randomized, and crossover controlled trial. *Mar. Drugs* **2018**, *16*, 364. [CrossRef]
15. Clark, T.; Morey, R.; Jones, M.D.; Marcos, L.; Ristov, M.; Ram, A.; Hakansson, S.; Franklin, A.; McCarthy, C.; De Carli, L.; et al. High-intensity interval training for reducing blood pressure: A randomized trial vs. moderate-intensity continuous training in males with overweight or obesity. *Hypertens. Res* **2020**, *43*, 396–403. [CrossRef]
16. Francois, M.E.; Little, J.P. Effectiveness and safety of high-intensity interval training in patients with type 2 diabetes. *Diabetes Spectr.* **2015**, *28*, 39–44. [CrossRef]
17. Gibala, M.J.; Little, J.P.; MacDonald, M.J.; Hawley, J.A. Physiological adaptations to low-volume, high-intensity interval training in health and disease. *J. Physiol.* **2012**, *590*, 1077–1084. [CrossRef]
18. D'Amuri, A.; Sanz, J.M.; Capatti, E.; Di Vece, F.; Vaccari, F.; Lazzer, S.; Zuliani, G.; Nora, E.D.; Passaro, A. Effectiveness of high-intensity interval training for weight loss in adults with obesity: A randomised controlled non-inferiority trial. *BMJ Open Sport Exerc. Med.* **2021**, *7*, e001021. [CrossRef]
19. Nobari, H.; Ahmadi, M.; Sá, M.; Pérez-Gómez, J.; Clemente, F.M.; Adsuar, J.C.; Minasian, V.; Afonso, J. The effect of two types of combined training on bio-motor ability adaptations in sedentary females. *J. Sports Med. Phys. Fit.* **2021**, *61*, 1317–1325. [CrossRef]

20. Jurio-Iriarte, B.; Maldonado-Martín, S. Effects of different exercise training programs on cardiorespiratory fitness in overweight/obese adults with hypertension: A pilot study. *Health Promot. Pract.* **2019**, *20*, 390–400. [CrossRef]
21. Sculthorpe, N.F.; Herbert, P.; Grace, F. One session of high-intensity interval training (HIIT) every 5 days, improves muscle power but not static balance in lifelong sedentary ageing men: A randomized controlled trial. *Medicine* **2017**, *96*, e6040. [CrossRef]
22. Batacan, R.B.; Duncan, M.J.; Dalbo, V.J.; Tucker, P.S.; Fenning, A.S. Effects of high-intensity interval training on cardiometabolic health: A systematic review and meta-analysis of intervention studies. *Br. J. Sports Med.* **2017**, *51*, 494–503. [CrossRef]
23. Khammassi, M.; Ouerghi, N.; Hadj-Taieb, S.; Feki, M.; Thivel, D.; Bouassida, A. Impact of a 12-week high-intensity interval training without caloric restriction on body composition and lipid profile in sedentary healthy overweight/obese youth. *J. Exerc. Rehabil.* **2018**, *14*, 118–125. [CrossRef]
24. Ribeiro, R.L.; de Oliveira Silva, J.I.; Dantas, M.G.B.; Menezes, E.S.; Arruda, A.; Schwingel, P. High-intensity interval training applied in Brazilian Jiu-jitsu is more effective to improve athletic performance and body composition. *J. Combat Sports Martial Arts* **2015**, *6*, 1–5. [CrossRef]
25. Gillen, J.B.; Percival, M.E.; Ludzki, A.; Tarnopolsky, M.A.; Gibala, M.J. Interval training in the fed or fasted state improves body composition and muscle oxidative capacity in overweight women. *Obesity* **2013**, *21*, 2249–2255. [CrossRef]
26. Andreato, L.; Esteves, J.V.; Coimbra, D.R.; Moraes, A.J.P.; de Carvalho, T. The influence of high-intensity interval training on anthropometric variables of adults with over-weight or obesity: A systematic review and network meta-analysis. *Obes. Rev.* **2019**, *20*, 142–155. [CrossRef]
27. Shirvani, H.; Ghahreman Tabrizi, K.; Sobhani, V. Effects of high intensity intermittent exercise on serum Immunoglobulin's and Complement system response in youth soccer players. *J. Birjand Univ. Med. Sci.* **2013**, *20*, 233–243.
28. Rahimi, R.; Ghaderi, M.; Mirzaei, B.; Ghaeni, S.; Faraji, H.; Vatani, D.S.; Rahmani-Nia, F. Effects of very short rest periods on immunoglobulin A and cortisol responses to resistance exercise in men. *J. Hum. Sport Exerc.* **2010**, *5*, 146–157. [CrossRef]
29. Lira, F.S.; Yamashita, A.S.; Uchida, M.C.; Zanchi, N.E.; Gualano, B.; Martins, E.; Caperuto, E.C.; Seelaender, M. Low and moderate, rather than high intensity strength exercise induces benefit regarding plasma lipid profile. *Diabetol. Metab. Syndr.* **2010**, *2*, 31. [CrossRef]
30. Azarbayjani, M.A.; Nikbakht, H. The effect of continuous and intermittent training on resting level and acute response of salivary IgA and total protein in male basketball players. *J. Shahrekord Univ. Med. Sci.* **2010**, *12*, 1–11.
31. Saeedy, M.; Bijeh, N.; Moazzami, M. The effect of six weeks of high-intensity interval training with Zinc supplementation on some humoral immunity markers in female futsal players. *Ann. Appl. Sport Sci.* **2018**, *6*, 11–19. [CrossRef]
32. Nobari, H.; Saedmocheshi, S.; Chung, L.H.; Suzuki, K.; Maynar-Mariño, M.; Pérez-Gómez, J. An Overview on How Exercise with Green Tea Consumption Can Prevent the Production of Reactive Oxygen Species and Improve Sports Performance. *Int. J. Env. Res. Public Health* **2021**, *19*, 218. [CrossRef] [PubMed]
33. Zwetsloot, K.A.; John, C.S.; Lawrence, M.M.; Battista, R.A.; Shanely, R.A. High-intensity interval training induces a modest systemic inflammatory response in active, young men. *J. Inflamm. Res.* **2014**, *7*, 9–17. [CrossRef] [PubMed]
34. Owen, A.L.; Wong, D.P.; Dunlop, G.; Groussard, C.; Kebsi, W.; Dellal, A.; Morgans, R.; Zouhal, H. High-intensity training and salivary immunoglobulin a responses in professional top-level soccer players: Effect of training intensity. *J. Strength Cond. Res.* **2016**, *30*, 2460–2469. [CrossRef] [PubMed]
35. Lee, N.; Kim, J.; Hyung, G.A.; Park, J.H.; Kim, S.J.; Kim, H.B.; Jung, H.S. Training effects on immune function in judoists. *Asian J. Sports Med.* **2015**, *6*, e24050. [CrossRef]
36. Oyeyinka, G.; Salimonu, L.S.; Williams, A.I.; Johnson, A.O.; Ladipo, O.A.; Osunkoya, B.O. Range of normal serum immunoglobulin (IgG, IgA and IgM) values in Nigerians. *Afr. J. Med. Med Sci.* **1984**, *13*, 169–176.
37. Bayram, R.O.; Özdemir, H.; Emsen, A.; Türk Dağı, H.; Artaç, H. Reference ranges for serum immunoglobulin (IgG, IgA, and IgM) and IgG subclass levels in healthy children. *Turk. J. Med. Sci.* **2019**, *49*, 497–505. [CrossRef]
38. Kalafati, M.; Jamurtas, T.; Nikolaidis, M.G.; Paschalis, V.; Theodorou, A.A.; Sakellariou, G.K.; Koutedakis, Y.; Kouretas, D. Ergogenic and antioxidant effects of spirulina supplementation in humans. *Med. Sci. Sports Exerc.* **2010**, *42*, 142–151. [CrossRef]
39. Faelli, E.; Bisio, A.; Codella, R.; Ferrando, V.; Perasso, L.; Panascì, M.; Saverino, D.; Ruggeri, P. Acute and Chronic Catabolic Responses to CrossFit® and Resistance Training in Young Males. *Int. J. Environ. Res. Public Health* **2020**, *17*, 7172. [CrossRef]
40. Codella, R.; Chirico, A.; Lucidi, F.; Ferrulli, A.; La Torre, A.; Luzi, L. The immune-modulatory effects of exercise should be favorably harnessed against COVID-19. *J. Endocrinol. Investig.* **2021**, *44*, 1119–1122. [CrossRef]
41. Gibson, A.S.C.; Broomhead, S.; Lambert, M.; Hawley, J. Prediction of maximal oxygen uptake from a 20-m shuttle run as measured directly in runners and squash players. *J. Sports Sci.* **1998**, *16*, 331–335. [CrossRef]
42. Pearce, C. Is Using a Patient's Estimated Lean Body Mass a Suitable Method of Predicting Their Eventual Therapeutic Units of Botulinum Toxin for the Treatment of Cervical Dystonia? 2016. Available online: https://rde.dspace-express.com/bitstream/handle/11287/610852/Is+using+a+patient%20s+lean+body+mass+a+suitable+method+of+predicting+their+eventual+therapeutic+dose+of+botulinum+toxin+for+the+treatment+of+cervical+dystonia.pdf?sequence=2 (accessed on 16 December 2021).
43. Hume, R. Prediction of lean body mass from height and weight. *J. Clin. Pathol.* **1966**, *19*, 389–391. [CrossRef]
44. Nobari, H.; Silva, A.F.; Clemente, F.M.; Siahkouhian, M.; García-Gordillo, M.Á.; Adsuar, J.C.; Pérez-Gómez, J. Analysis of fitness status variations of Under-16 soccer players over a season and their relationships with maturational status and training load. *Front. Physiol.* **2021**, *11*, 1840. [CrossRef]

45. Nobari, H.; Cholewa, J.M.; Pérez-Gómez, J.; Castillo-Rodríguez, A. Effects of 14-weeks betaine supplementation on pro-inflammatory cytokines and hematology status in professional youth soccer players during a competition season: A double blind, randomized, placebo-controlled trial. *J. Int. Soc. Sports Nutr.* **2021**, *18*, 42. [CrossRef]
46. Jackson, A.S.; Pollock, M.L.; Ward, A. Generalized equations for predicting body density of women. *Med. Sci. Sports Exerc.* **1980**, *12*, 175–181. [CrossRef]
47. Wagner, D.R. Ultrasound as a tool to assess body fat. *J. Obes.* **2013**, *2013*, 280713. [CrossRef]
48. Arazi, H.; Mirzaei, B.; Nobari, H. Anthropometric profile, body composition and somatotyping of national Iranian cross-country runners. *Turk. J. Sport Exerc.* **2015**, *17*, 35–41. [CrossRef]
49. Rahmat, A.J.; Arsalan, D.; Bahman, M.; Hadi, N. Anthropometrical profile and bio-motor abilities of young elite wrestlers. *Phys. Educ. Stud.* **2016**, *20*, 63–69. [CrossRef]
50. Wannamethee, S.G.; Shaper, A.G.; Whincup, P.H. Body fat distribution, body composition, and respiratory function in elderly men. *Am. J. Clin. Nutr.* **2005**, *82*, 996–1003. [CrossRef]
51. Nobari, H.; Alves, A.R.; Clemente, F.M.; Pérez-Gómez, J.; Clark, C.C.T.; Granacher, U.; Zouhal, H. Associations between variations in accumulated workload and physiological variables in young male soccer players over the course of a season. *Front. Physiol.* **2021**, *12*, 233. [CrossRef]
52. Nobari, H.; Aquino, R.; Clemente, F.M.; Khalafi, M.; Adsuar, J.C.; Pérez-Gómez, J. Description of acute and chronic load, training monotony and strain over a season and its relationships with well-being status: A study in elite under-16 soccer players. *Physiol. Behav.* **2020**, *225*, 113117. [CrossRef]
53. Jafari, M.; Pouryamehr, E.; Fathi, M. The effect of eight weeks high intensity interval training (HIIT) on E-selection and P-selection in young obese females. *Int. J. Sport Stud. Health* **2018**, *1*, e64336. [CrossRef]
54. Kilding, A.E.; Aziz, A.R.; Teh, K. Measuring and predicting maximal aerobic power in international-level intermittent sport athletes. *J. Sports Med. Phys. Fit.* **2006**, *46*, 366.
55. Bahram, M.E.; Mogharnasi, M. The Effect of Twelve weeks High Intensity Training Interval (HIIT) on Leptin Levels and Obesity Dependent Factors among Female Students Suffering Overweight. *J. Sport Biosci.* **2015**, *6*, 451–465.
56. Glaister, M.; Hauck, H.; Abraham, C.S.; Merry, K.L.; Beaver, D.; Woods, B.; McInnes, G. Familiarization, reliability, and comparability of a 40-m maximal shuttle run test. *J. Sports Sci. Med.* **2009**, *8*, 77.
57. Camarda, S.R.d.A.; Tebexreni, A.S.; Páfaro, C.N.; Sasai, F.B.; Tambeiro, V.L.; Juliano, Y.; de Barros Neto, T.L. Comparison of maximal heart rate using the prediction equations proposed by Karvonen and Tanaka. *Arq. Bras. Cardiol.* **2008**, *91*, 311–314. [CrossRef]
58. Nobari, H.; Fashi, M.; Eskandari, A.; Pérez-Gómez, J.; Suzuki, K. Potential Improvement in Rehabilitation Quality of 2019 Novel Coronavirus by Isometric Training System; Is There "Muscle-Lung Cross-Talk"? *Int. J. Environ. Res. Public Health* **2021**, *18*, 6304. [CrossRef]
59. Hopkins, B.; Cole, B.L.; Mason, T.L. A critique of the usefulness of inferential statistics in applied behavior analysis. *Behav. Anal.* **1998**, *21*, 125–137. [CrossRef]
60. Jahani, Q.G.R.; Abkar, A.R.; Heidari, H. The Effect of High-Intensity Intermittent Training (HIIT) and Consumption of Probiotic Supplement on Immune Cells, C-Reactive Protein, and IgA in Young Football Player. *Qom Univ. Med. Sci. J.* **2016**, *10*, 36–46.
61. Khafaga, A.F.; El-Sayed, Y.S. Spirulina ameliorates methotrexate hepatotoxicity via antioxidant, immune stimulation, and proinflammatory cytokines and apoptotic proteins modulation. *Life Sci.* **2018**, *196*, 9–17. [CrossRef]
62. Hayashi, O.; Hirahashi, T.; Katoh, T.; Miyajima, H.; Hirano, T.; Okuwaki, Y. Class specific influence of dietary Spirulina platensis on antibody production in mice. *J. Nutr. Sci. Vitaminol.* **1998**, *44*, 841–851. [CrossRef] [PubMed]
63. Tominaga, A.; Konishi, Y.; Taguchi, T.; Fukuoka, S.; Kawaguchi, T.; Noda, T.; Shimizu, K. Autonomous cure of damaged human intestinal epithelial cells by TLR2 and TLR4-dependent production of IL-22 in response to Spirulina polysaccharides. *Int. Immunopharmacol.* **2013**, *17*, 1009–1019. [CrossRef] [PubMed]
64. Uzar, E.; Sahin, O.; Koyuncuoglu, H.R.; Uz, E.; Bas, O.; Kilbas, S.; Yilmaz, H.R.; Yurekli, V.A.; Kucuker, H.; Songur, A. The activity of adenosine deaminase and the level of nitric oxide in spinal cord of methotrexate administered rats: Protective effect of caffeic acid phenethyl ester. *Toxicology* **2006**, *218*, 125–133. [CrossRef] [PubMed]
65. Leicht, C.A.; Goosey-Tolfrey, V.L.; Bishop, N.C. Exercise intensity and its impact on relationships between salivary immunoglobulin A, saliva flow rate and plasma cortisol concentration. *Eur. J. Appl. Physiol.* **2018**, *118*, 1179–1187. [CrossRef] [PubMed]
66. De Souza, D.C.; Matos, V.; Dos Santos, V.O.A.; Medeiros, I.F.; Marinho, C.S.R.; Nascimento, P.R.P.; Dorneles, G.P.; Peres, A.; Müller, C.H.; Krause, M.; et al. Effects of High-Intensity Interval and Moderate-Intensity Continuous Exercise on Inflammatory, Leptin, IgA, and Lipid Peroxidation Responses in Obese Males. *Front. Physiol.* **2018**, *9*, 567. [CrossRef] [PubMed]
67. Nobari, H.; Nejad, H.; Kargarfard, M.; Mohseni, S.; Suzuki, K.; Adsuar, J.C.; Pérez-Gómez, J. The Effect of Acute Intense Exercise on Activity of Antioxidant Enzymes in Smokers and Non-Smokers. *Biomolecules* **2021**, *11*, 171. [CrossRef]
68. Nobari, H.; Kargarfard, M.; Minasian, V.; Cholewa, J.M.; Pérez-Gómez, J. The effects of 14-week betaine supplementation on endocrine markers, body composition and anthropometrics in professional youth soccer players: A double blind, randomized, placebo-controlled trial. *J. Int. Soc. Sports Nutr.* **2021**, *18*, 20. [CrossRef]

69. Szulinska, M.; Gibas-Dorna, M.; Miller-Kasprzak, E.; Suliburska, J.; Miczke, A.; Walczak-Gałezewska, M.; Stelmach-Mardas, M.; Walkowiak, J.; Bogdanski, P. Spirulina maxima improves insulin sensitivity, lipid profile, and total antioxidant status in obese patients with well-treated hypertension: A randomized double-blind placebo-controlled study. *Eur. Rev. Med. Pharm. Sci.* **2017**, *21*, 2473–2481.
70. Mohebbi, H.; Azizi, M.; Moradiani, H. Effect of 8 weeks low and high intensity resistance training on leukocyte count, Igg, cortisol and lactate concentration in untrained men. *World Appl. Sci. J.* **2012**, *16*, 949–954.
71. Santoso, D.A.; Asnar, E. Changes of Interleukin-6 (IL-6) and Immunoglobulin G (igG) in Respiratory Excercise. *Indian J. Forensic Med. Toxicol.* **2020**, *14*, 1501–1506.
72. Chu, W.-L.; Van Quynh, L.; Radhakrishnan, A.K. Effect of Spirulina (Arthrospira) supplementation on the immune response to tetanus toxoid vaccination in a mouse model. *J. Diet. Suppl.* **2013**, *10*, 229–240. [CrossRef]
73. Seyidoglu, N.; Inan, S.; Aydin, C. A Prominent Superfood: Spirulina platensis. In *Superfood Functional Food. The Development of Superfoods Their Roles as Medicine*; Shiomi, N., Ed.; InTechOpen: London, UK, 2017; pp. 1–27.
74. Ötleş, S.; Pire, R. Fatty acid composition of Chlorella and Spirulina microalgae species. *J. AOAC Int.* **2001**, *84*, 1708–1714. [CrossRef] [PubMed]
75. Hunter, G.R.; Fisher, G.; Neumeier, W.H.; Carter, S.J.; Plaisance, E.P. Exercise training and energy expenditure following weight loss. *Med. Sci. Sports Exerc.* **2015**, *47*, 1950. [CrossRef] [PubMed]
76. Vella, C.A.; Taylor, K.; Drummer, D. High-intensity interval and moderate-intensity continuous training elicit similar enjoyment and adherence levels in overweight and obese adults. *Eur. J. Sport Sci.* **2017**, *17*, 1203–1211. [CrossRef] [PubMed]

biology

MDPI

Article

Long-Term Combined Effects of Citrulline and Nitrate-Rich Beetroot Extract Supplementation on Recovery Status in Trained Male Triathletes: A Randomized, Double-Blind, Placebo-Controlled Trial

José Burgos [1,2], Aitor Viribay [3], Julio Calleja-González [4], Diego Fernández-Lázaro [5,6], Jurgi Olasagasti-Ibargoien [7,8], Jesús Seco-Calvo [9,†] and Juan Mielgo-Ayuso [10,*]

1 Department of Nursing and Physiotherapy, University of León, 24071 León, Spain; joseburgos88@hotmail.com
2 Burgos Nutrition, Physiology, Nutrition and Sport, 26007 Logroño, Spain
3 Glut4Science, Physiology, Nutrition and Sport, 01004 Vitoria-Gasteiz, Spain; aitor@glut4science.com
4 Department of Physical Education and Sport, Faculty of Education and Sport, University of the Basque Country, 01007 Vitoria-Gasteiz, Spain; julio.calleja.gonzalez@gmail.com
5 Department of Cellular Biology, Histology and Pharmacology, Faculty of Health Sciences, University of Valladolid, 42003 Soria, Spain; diego.fernandez.lazaro@uva.es
6 Neurobiology Research Group, Faculty of Medicine, University of Valladolid, 47005 Valladolid, Spain
7 Faculty of Education and Sports, University of Deusto, 20012 Donostia-San Sebastian, Spain; jurgi.olasagasti@deusto.es
8 Health, Physical Activity and Sports Science Laboratory (HealthPASS), Departament of Physical Activity and Sport, Faculty of Education and Sport, University of Deusto, 48007 Bilbao, Spain
9 Physiotherapy Department, Institute of Biomedicine (IBIOMED), University of Leon, Campus de Vegazana, 24071 Leon, Spain; dr.seco.jesus@gmail.com
10 Department of Health Sciences, Faculty of Health Sciences, University of Burgos, 09001 Burgos, Spain
* Correspondence: jfmielgo@ubu.es
† Researcher of Basque Country University.

Citation: Burgos, J.; Viribay, A.; Calleja-González, J.; Fernández-Lázaro, D.; Olasagasti-Ibargoien, J.; Seco-Calvo, J.; Mielgo-Ayuso, J. Long-Term Combined Effects of Citrulline and Nitrate-Rich Beetroot Extract Supplementation on Recovery Status in Trained Male Triathletes: A Randomized, Double-Blind, Placebo-Controlled Trial. *Biology* **2022**, *11*, 75. https://doi.org/10.3390/biology11010075

Academic Editors: Filipe Manuel Clemente, Georgian Badicu, Eugenia Murawska-Ciałowicz and Carlo Reggiani

Received: 30 November 2021
Accepted: 31 December 2021
Published: 4 January 2022

Simple Summary: Recovery is one of the main elements in achieving adequate athletic performance. Various supplements have been used for this purpose. Citrulline (CIT) and Nitrate-Rich Beetroot Extract (BR) are so-called nitric oxide precursor supplements that have shown an ergogenic effect on sports performance when used on a short-term, individual basis. These supplements appear to have other pathways that may promote athletic performance. The purpose of this study was to assess the effect of a co-supplementation for 9 weeks of 3 g/day of CIT plus 2.1 g/day of BR (300 mg/day of nitrates) on recovery by exercise-induced muscle damage markers (EIMD), anabolic/catabolic hormones and distance covered in the Cooper test (CP). Thirty-two male triathletes were randomized into 4 groups of 8 in this double-blind, placebo-controlled trial: placebo group, CIT group, BR group and CIT-BR group. Blood samples and CP were collected at baseline and after 9 weeks. The main conclusions were the combination of 3 g/day of CIT plus 2.1 g/day of BR (300 mg/day of NO_3^-) supplementation for 9 weeks did not present any benefit for EIMD. However, CIT-BR improved recovery status by preventing an increase in cortisol and showing an increase in Testosterone/Cortisol ratio and distance covered in the CP.

Abstract: Citrulline (CIT) and nitrate-rich beetroot extract (BR) are widely studied ergogenic aids. Nevertheless, both supplements have been studied in short-term trials and separately. To the best of the authors' knowledge, the effects of combining CIT and BR supplementation on recovery status observed by distance covered in the Cooper test, exercise-induced muscle damage (EIMD) and anabolic/catabolic hormone status have not been investigated to date. Therefore, the main purpose of this research was to assess the effect of the long-term (9 weeks) mixture of 3 g/day of CIT plus 2.1 g/day of BR (300 mg/day of nitrates (NO_3^-)) supplementation on recovery by distance covered in the Cooper test, EIMD markers (urea, creatinine, AST, ALT, GGT, LDH and CK) and anabolic/catabolic hormones (testosterone, cortisol and testosterone/cortisol ratio (T/C)) in male trained triathletes. Thirty-two triathletes were randomized into four different groups of eight triathletes in this double-blind, placebo-controlled trial: placebo group (PLG), CIT group (CITG; 3 g/day of CIT), BR group (BRG; 2.1 g/day of BR (300 mg/day of NO_3^-)) and CIT-BR group (CIT-BRG; 3 g/day of CIT plus 2.1 g/day of BR (300 mg/day of NO_3^-)). Distance covered in the Cooper test and blood samples were collected from all participants at baseline (T1) and after 9 weeks of supplementation (T2). There were no significant differences in the interaction between group and time in EIMD markers (urea, creatinine, AST, ALT, GGT, LDH and CK) ($p > 0.05$). However, significant differences were observed in the group-by-time interaction in distance covered in the Cooper test ($p = 0.002$; $\eta^2p = 0.418$), cortisol ($p = 0.044$; $\eta^2p = 0.247$) and T/C ($p = 0.005$; $\eta^2p = 0.359$). Concretely, significant differences were observed in distance covered in the Cooper test percentage of change ($p = 0.002$; $\eta^2p = 0.418$) between CIT-BRG and PLG and CITG, in cortisol percentage change ($p = 0.049$; $\eta^2p = 0.257$) and in T/C percentage change ($p = 0.018$; $\eta^2p = 0.297$) between CIT-BRG and PLG. In conclusion, the combination of 3 g/day of CIT plus 2.1 g/day of BR (300 mg/day of NO_3^-) supplementation for 9 weeks did not present any benefit for EIMD. However, CIT + BR improved recovery status by preventing an increase in cortisol and showing an increase in distance covered in the Cooper test and T/C.

Keywords: triathlon; performance; ergogenic aids; muscle fatigue; recovery; hormones

1. Introduction

Prolonged and strenuous exercise produces organic stress [1] that could decrease athletic performance [2–4]. As a consequence of this status, there are several alterations in biochemical parameters of exercise-induced muscle damage (EIMD) [5] as well as anabolic/catabolic hormone alterations which could hinder endogenous exercise adaptations [6]. Therefore, in addition to an adequate training program, it could be essential to include different strategies to delay or reduce muscle fatigue and improve adaptation to training [7]. In this sense, supplementation with nitrate-rich beetroot extract (BR) and citrulline (CIT) has been proposed to achieve these goals, partly because they are precursors of nitric oxide (NO) [8–11].

The NO produces vasodilation by increasing the blood level in muscles and improving their efficiency in muscle contraction and relaxation processes [12]. Moreover, NO regulates force generation and satellite cell activation [13]. In the long term, NO can regulate muscle function and even affect skeletal muscle recovery due to its antioxidant effect and the constant increase in muscle blood flow which, together with an adequate supply of essential amino acids, would allow better muscle fueling [14] and could prevent EIMD [15,16]. Moreover, decreased blood flow to the testis could reduce testosterone synthesis [17]. It has also been shown in animal models that NO enhancement resulted in a significant reduction of ACTH-mediated cortisol production [18]. Consequently, although this mechanism is speculative, increasing NO could improve blood flow in the testis and promote testosterone synthesis by vasodilator effect [14,19] and could be successful in maintaining an anabolic state, decreasing muscular damage and metabolic stress [2,9].

On the one hand, BR supplementation is widely used by athletes as a precursor of NO [20]. When the athletes digest BR, its nitrates (NO_3^-) are transformed into nitrites (NO_2^-) which are partially reduced to NO by the action of stomach acids and subsequently absorbed in the intestine and passed into the bloodstream [21]. Moreover, BR is rich in other compounds such as phenolic acids, flavonoids, carotenoids and betalains, which have antioxidant effects [22]. Therefore, although the mechanisms for potential improvements in muscle recovery following EIMD after NO_3^- supplementation are not clear, it would be expected that long-term BR supplementation could attenuate EIMD after prolonged, strenuous exercise [22,23] based on the effects of NO and additional compounds. Moreover, long-term BR supplementation could be very beneficial for the maintenance of anabolic/catabolic hormones, as shown by Sarfaraz et al. on testosterone levels [24]. However, short-term BR supplementation (maximum for 3 days) has not presented an improved

EIMD and anabolic/catabolic status after a damaging session of eccentric exercise [23] or high-intensity workouts [25], which opens the need for further research.

On the other hand, citrulline (CIT), a non-essential amino acid found primarily in watermelon and produced endogenously by recycling into arginine (ARG) and NO via argininosuccinate synthetase, increases NO availability and its effects [26]. In addition, CIT is an essential element of the urea cycle in the liver [27]. Therefore, it has been suggested that CIT supplementation may eliminate ammonia by urea production [28]. In the same way, CIT is an important activator of muscle protein synthesis in catabolic situations via activation of the mammalian target of rapamycin (mTOR) pathway due to its key role in the regulation of nitrogen homeostasis [29]. Based on these mechanisms, CIT supplementation may favor muscle performance and recovery in different ways, such as activating muscle protein synthesis, improving oxygen distribution to muscle, increasing oxidative ATP production during exercise and phosphocreatine (PCr) during exercise recovery and decreasing blood lactate and ammonium production [14,30,31], which could reduce fatigue and limit EIMD. However, although this proposal would be adequate for athletes, to the best of the authors' knowledge, there is little research on CIT supplementation in muscle recovery. In this regard, Da Silva et al. [27] did not observe improvements in functional (i.e., number of maximum repetitions, muscle pain and perceived effort), metabolic (i.e., CK and lactate), anabolic (i.e., testosterone and testosterone/cortisol (T/C) ratio) and physiological (electromyographic signal) outcomes of muscle recovery in untrained young adult males after CIT supplementation with 6 g at 60 min prior to the training session. These results of both CIT and BR supplementation on EIMD and anabolic/catabolic hormones may probably be due to the fact that the effects have only been investigated in the short term [28] and under isolated intakes [29,32], suggesting the need to investigate the effects of long-term combination of these two ergogenic aids. In this regard, it has been shown that the effects of some supplements can be synergic when combined over the long term [2,33]. Therefore, it could be considered that the combined effects of CIT (NO precursor and activator of muscle protein synthesis) and BR (NO precursor and antioxidant effect) could reduce EIMD and improve muscle recovery observed by anabolic/catabolic hormone profile [34,35]. This could favor some sporting performance variables [36]. In this sense, the supplementation of 6 g of CIT plus 520 mg of NO_3^- 6 h before the submaximal incremental cycling test has shown improvements in some cardiorespiratory variables, such as VO_2 [36].

Therefore, the main objective of this research was to assess the effect of the long-term (9 weeks) mixture of 3 g/day of CIT plus 2.1 g/day of BR (300 mg/day of NO_3^-) supplementation on recovery status, distance covered in the Cooper test, EIMD markers (urea, creatinine, AST, ALT, GGT, LDH and CK) and anabolic/catabolic hormones (testosterone, cortisol and T/C) in male trained triathletes. The hypothesis was that the combination of CIT plus BR could limit EIMD and improve endogenous recovery observed in lower cortisol and better testosterone and T/C than isolated CIT or BR supplementation.

2. Materials and Methods

2.1. Participants

Thirty-two male amateur triathletes from the same club (34.37 ± 7.08 years old and 58.79 ± 6.89 mL/min/kg of VO_{2max}) with at least 5 years of experience participated in this trial. All athletes rigorously performed the same training methodology, and thus, all of them were exposed to the same training load in terms of type, intensity and duration of exercise (Table 1): 15 h/week, 6 days/week during the 9 weeks. All participants completed a total of 135 h of training during the study.

Table 1. Type, intensity and duration of weekly training program.

	Monday	Tuesday	Wednesday	Thursday	Friday	Saturday	Sunday
1st session	- 20 warm-up - 15 min stretching - 45 min mindfulness	- 15 min run technique skills - 45 min strength training	- 120 m ride at 50–75% VO_{2max} - 15 min cooldown - 15 min core	- 15 min run technique skills - 45 min strength training	- 75 min run at 50–75% VO_{2max} - 20 min resistance training	75 min swim at 75–90% VO_{2max}	- 15 min run technique skills - 45 min strength training
2nd session	REST	- 30 min warm-up - 30 min run at 75–90% VO_{2max} - 15 min cooldown	REST	- 30 min warm-up - 30 min swim at 75–90% VO_{2max} - 15 min cooldown	REST	- 60 min walk	- 120 min ride at 50–75% VO_{2max} - 15 min core

Likewise, a certified nutritionist (CLR-0020) developed personalized diets for each participant. These diets were planned with the aim of ensuring adequate energy and macro and micronutrient intake considering the training load and the personal features of each triathlete following the international recommendations for an adequate sports performance [37].

All athletes also underwent a medical examination and completed a medical history questionnaire prior to the start of the study to find out whether they had any type of disease and/or injury [38]. The participants did not present any disease, and they did not drink alcohol, smoke or consume other drugs or stimulant substances during the study period which could alter the hormone response. Likewise, to eliminate the probable interference of other nutritional aids with the different outcomes measured in this research, a 2-week washout period was included [39–41].

All triathletes were completely informed of all actions of the study and signed a personal statement of informed consent, giving their individual agreement to take part in the proposed work. This trial was considered in accordance with the Declaration of Helsinki (2008) and the Fortaleza update (2013) and was approved by the Human Research Ethics Committee of the University of León, Spain (number: ULE-020-2020). Moreover, this study was registered in clinicaltrials.gov with NCT05143879 number.

2.2. Experimental Protocol and Evaluation Plan

This study was planned as a randomized, double-blind, placebo-controlled trial to assess the impact of a 9-week oral supplementation of the combination of CIT plus BR on recovery status by distance-covered performance test, EIMD markers and anabolic/catabolic hormones in this sport population. The proposed doses of CIT (3 g/day) and BR supplements were based on previous scientific studies that found favorable results with similar doses [42–44].

The 32 athletes were randomly assigned to four different groups of 8 participants (Table 2) by an independent statistician using the open-source software OxMaR (Oxford Minimization and Randomization, 2014): (I) placebo group (PLG); (II) CIT group (CITG); (III): nitrate-rich beetroot extract group (BRG); and (IV) CIT-BR group (CIT-BRG).

Table 2. Age and height of participants at the beginning of the study.

	PLG (n = 8)	CITG (n = 8)	BRG (n = 8)	CIT-BRG (n = 8)
Age (years)	34.01 ± 7.03	32.75 ± 7.01	32.67 ± 6.54	34.35 ± 7.95
Height (cm)	179 ± 8 cm	180 ± 9 cm	178 ± 8 cm	181 ± 6 cm

Data are presented as mean ± standard deviation.

CIT supplementation was included in 3 gelatin capsules of 1 g CIT by Hard Eight Nutrition LLC (7511 Eastgate Rd, Henderson, NV 89011). BR supplementation was included in 3 gelatin capsules of 700 mg (5:1 beetroot extract equivalent to 3500 mg of whole dried root, standardized to contain 0.3% betanin providing 100 mg of NO_3^-) by Lindens Health Nutrition (1 Calder Point, Monckton Road, Wakefield, WF2 7AL). The placebo (cellulose) capsules were made of both 1 g and 700 mg being of the same color and shape as the other two supplements to avoid the placebo effect [45]. All athletes took the same number of capsules per day (3 capsules of 1 g (BIG) and 3 capsules of 700 mg (SMALL)) based on their groups: PLG: 3 BIG of cellulose + 3 SMALL of cellulose; CIT: 3 BIG CIT and 3 SMALL of cellulose; BR: 3 BIG of cellulose + 3 SMALL BR; and CIT-BRG: 3 BIG CIT + 3 SMALL BR. In order to ensure blinding, all BIG capsules were white (CIT and placebo) and all SMALL capsules were red (BR and placebo).

All participants took 3 BIG and 3 SMALL capsules, either the placebo or aids, during the 7 days of the week after each of the 3 main meals (1-1-1) to eliminate any influence of circadian variation [46]. Athletes were informed that they should not brush their teeth or rinse their mouths for 2 h after the intake of the capsules, based on the effect of oral bacteria on the reduction of NO_2^- from NO_3^-. In addition, they were unaware of the contents of the capsules provided to them weekly by an independent nutritionist (LR003) who confirmed that all triathletes complied with the intake protocol.

2.3. Blood Collection

All triathletes arrived at the laboratory at 8:30 a.m. for blood extraction at two different moments during the trial: (T1) at baseline and (T2) after 9 weeks of supplementation. For the evaluation/assessment of EIMD and hormonal outcomes at T1 and T2, antecubital venous blood samples were collected. All samples were obtained after at least 12 h of fasting and 48 h without any previous exercise and after being at rest for 30 min.

The EIMD markers (urea, creatinine, AST, ALT, GGT, CK and LDH) were measured using the Hitachi 917® automatic autoanalyzer (Hitachi Ltd., Tokyo, Japan) [47]. Serum hormone outcomes (total testosterone and cortisol) were measured using an enzyme-linked fluorescent assay with the aid of a multiparametric analyzer (MINI VIDAS®, Biomerieux, Marcy l'Etoile, France) [3]. The substrate 4-methylumbelliferone was used, and fluorescence emission was performed at 450 nm and, after stimulation, at 370 nm [48]. The intra-assay CV was 5.7%, and the CV of the intermediate assay was 6.2%. Finally, T/C was considered by dividing testosterone by cortisol.

2.4. Cooper Test

After blood analysis and 2 h after the standardized breakfast (2 g of CHO/kg BM and consisting of rice, corn cereal with oat beverage, cooked fruit and biscuits with jam or sweet quince, cheese or paste) [37], the athletes performed a Cooper test. Athletes were familiar with this test given that they usually use this test throughout the season.

Before starting the test, a standardized 15 min warm-up was performed: 8 min incremental run; 3 min of core work; 2 min of trunk, hip and leg muscle exercises; and 2 min of different types of jumps. The Cooper 12 min run test was conducted under the observation of the research team on a 400 m synthetic sports track. The participants completed the traditional test protocol, which consisted of covering the farthest feasible distance in 12 min [49]. The total distance covered in this time was measured immediately after the test was completed using markers placed on the track at 50 m intervals [50].

2.5. Anthropometry

The same internationally certified anthropometrist (ISAK level 3 with certificate number: #636739292503670742) performed the anthropometric measurements for all triathletes based on the International Society for the Advancement of Kinanthropometry (ISAK) protocol [51]. Height (cm) was obtained by a SECA® measuring rod (Mod. 220; SECA Medical, Bradford, MA, USA), with 1 mm precision. Body mass (kg) was measured using a SECA® model scale (Mod. 220; SECA Medical, Bradford, MA, USA), with 0.1 kg precision. Body mass index (BMI) was considered by the equation body mass/height^2 (kg/m^2). Six skinfolds (mm) were assessed—triceps, subscapular, supraspinal, abdominal, front thigh and medial calf by a Harpenden® Skinfold Caliper (Harpenden Skinfold Caliber, British Indicators Ltd., London, UK) with 0.2 mm precision—and the sum of all of them was considered. Girths (cm) (relaxed arm, flexed arm, minimum waist, 1 cm below the buttock thigh, mid-thigh and calf girth) were measured with an inextensible metallic Lufkin® measuring tape model W606PM (Cooper Tools, Apex, NC, USA) with 1 mm precision. Fat mass (FM) and muscle mass (MM) were estimated by the Carter and Lee equations, respectively [52].

2.6. Dietary Assessment

The nutritionists participating in the study (J.B.-B. and J.M.-A.) informed all triathletes about proper food tracking. They tutored the participants regarding 2 validated methods of dietary recall [51]. The first method was a food frequency questionnaire (FFQ) previously used in other sport populations [53] which triathletes should complete at T2. The athletes should recall their average food "frequency" intake based on certain food groups over the previous 9 weeks. Food frequency was based on the number of times each food was consumed per day, week or month. The serving sizes consumed were estimated through the standard weight of food items or by determining the portion sizes by looking at a book containing over 500 photographs of food [54]. Energy (kcal) and macronutrient (g) consumption was determined by dividing the reported intake by the frequency in days using a validated software package (Easy diet©, online version 2020) [55]. The total energy and macronutrient intake per kilogram of body mass was calculated for each athlete. The second method was a seven-day dietary recall collected the week prior to T1 and during the week of T2. This method was used to check if the results of the FFQ were similar to those of this recall [52].

2.7. Statistical Analysis

The data are shown as means and standard deviations. The Shapiro–Wilk test ($n < 50$) was used to determine normality. Likewise, the homoscedasticity assumption was tested with the previous Levene test. Thereafter, differences from T1 to T2 in each group separately were assessed using Student's *t*-tests for parametric paired data. Then, a two-way repeated-measures analysis of variance (ANOVA) test was performed to assess the interaction effects (time × supplementation group).

On the other hand, the percentage changes of the outcomes studied between T1 and T2 in each study group were calculated as Δ (%): $((T2 - T1)/T1) \times 100)$. A one-way ANOVA test was performed to determine if there were significant differences between the means of the different outcomes analyzed among the 4 study groups. A Bonferroni post hoc test was applied for pairwise comparisons among supplemented groups to establish statistical significance levels.

Effect sizes as a qualitative measure were estimated by partial eta squared (η^2p). Given that this measure overestimates effect sizes, the values were interpreted based on Ferguson, who indicated no effect if $0 \leq \eta^2p < 0.05$, minimum effect if $0.05 \leq \eta^2p < 0.26$, moderate effect if $0.26 \leq \eta^2p < 0.64$ and strong effect if $\eta^2p \geq 0.64$ [56].

The analyses were completed by SPSS® software version 24.0 (SPSS, Inc., Chicago, IL, USA) and Microsoft Excel® version 24 (Microsoft Corporation, Redmond, WA, USA) and graphics using GraphPad Prism 6 software (GraphPad Software, Inc., San Diego, CA, USA). Statistical significance was designated when $p < 0.05$.

3. Results

During the trial, the triathletes did not present significant statistical differences ($p > 0.05$) in energy and macronutrient intake values among groups (Table 3).

Table 3. Energy and macronutrient intake of triathletes during 9 weeks of study.

	PLG	CITG	BRG	CIT + BRG
Energy (kcal/kg)	45 ± 6.4	45.2 ± 6.8	44.9 ± 6.5	45.3 ± 7.2
Protein (g/kg)	1.4 ± 0.5	1.5 ± 0.7	1.4 ± 0.8	1.4 ± 0.5
Fat (g/kg)	1.4 ± 0.4	1.5 ± 0.5	1.4 ± 0.6	1.5 ± 0.6
Carbohydrates (g/kg)	7.0 ± 1.0	7.1 ± 1.2	7.1 ± 1.4	7.0 ± 1.3

Data are shown as mean ± standard deviation. PLG: placebo group, CITG: citrulline group, BRG: NO_3^- group; CIT-BRG: citrulline plus NO_3^- supplemented group.

Body mass, BMI, muscle mass and fat mass percentage did not present significant differences ($p > 0.05$) in the interaction group-by-time (Table 4).

Table 4. Anthropometry and body composition outcomes of triathletes.

Group	T1	T2	p (T × G)	$\eta^2 p$
		Body mass (Kg)		
PLG	76.36 ± 7.03	76.31 ± 6.76	0.582	0.074
CITG	79.08 ± 7.36	77.70 ± 7.09		
BRG	74.11 ± 6.93	74.00 ± 6.90		
CIT + BRG	74.19 ± 11.26	74.29 ± 11.38		
		BMI (kg/m²)		
PLG	24.01 ± 1.89	23.98 ± 2.03	0.407	0.115
CITG	24.52 ± 2.53	23.99 ± 2.25		
BRG	23.25 ± 1.86	23.25 ± 1.85		
CIT + BRG	22.54 ± 1.63	22.53 ± 1.59		
		Muscle mass (kg)		
PLG	69.39 ± 5.42	69.65 ± 5.73	0.406	0.112
CITG	72.35 ± 6.21	66.05 ± 5.77		
BRG	67.38 ± 6.46	67.95 ± 6.31		
CIT + BRG	68.49 ± 9.49	68.49 ± 9.20		
		Fat mass (%)		
PLG	9.01 ± 2.05	8.66 ± 2.14	0.121	0.203
CITG	8.45 ± 1.44	7.77 ± 1.32		
BRG	9.07 ± 2.09	8.18 ± 0.96		
CIT + BRG	7.52 ± 1.66	7.58 ± 2.17		

Data are presented as mean ± standard deviation. p (T × G): interaction group-by-time ($p < 0.05$) by two-factor repeated-measures ANOVA.

Figure 1 shows the distance covered in the Cooper test at both T1 and T2. Significant differences can be seen in the group-by-time interaction in this parameter ($p = 0.002$; $\eta^2 p = 0.418$). In addition, significant increases ($p < 0.05$) were observed between study moments in distance covered (T1: 2953.1 ± 372.7 vs. T2: 3079.6 ± 423.5 m) in CIT-BRG.

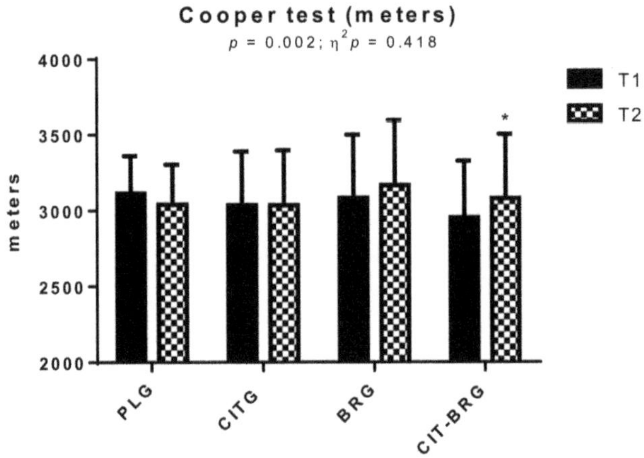

Figure 1. Distance covered in Cooper test by triathletes at T1 and T2 (after 9 weeks of supplementation). Data are presented as mean ± standard deviation. *: Significant differences with respect to T1. $p < 0.05$.

The EIMD markers did not present significant differences ($p > 0.05$) in the group-by-time interaction (Table 5). However, significant differences were observed between T1 and T2 for BR in creatinine (T1: 0.92 ± 0.11 vs. T2: 0.88 ± 0.09 mg/dL; η^2p: 0.063) and LDH (T1: 445.38 ± 247.59 vs. T2: 393.88 ± 63.37 UI/L; η^2p: 0.083).

Table 5. Serum EIMD markers of triathletes at T1 and T2 (after 9 weeks).

Group	T1	T2	p (T × G)	η^2p
		Urea (mg/dL)		
PLG	37.38 ± 6.63	38.00 ± 4.81	0.260	0.131
CITG	37.06 ± 6.92	34.58 ± 8.72		
BRG	38.38 ± 4.03	37.50 ± 2.83		
CIT + BRG	36.75 ± 8.55	41.13 ± 6.06		
		Creatinine (mg/dL)		
PLG	0.91 ± 0.10	0.92 ± 0.10	0.601	0.063
CITG	0.93 ± 0.57	0.92 ± 0.11		
BRG	0.92 ± 0.11	0.88 ± 0.09		
CIT + BRG	0.91 ± 0.09	0.91 ± 0.11		
		AST (UI/L)		
PLG	33.50 ± 9.06	37.13 ± 16.31	0.321	0.115
CITG	30.50 ± 9.09	24.38 ± 3.93		
BRG	39.38 ± 17.39	29.36 ± 4.79		
CIT + BRG	37.38 ± 12.02	29.75 ± 7.91		
		ALT (UI/L)		
PLG	28.00 ± 14.25	29.63 ± 13.76	0.327	0.114
CITG	25.88 ± 9.43	22.00 ± 5.40		
BRG	36.25 ± 29.36	23.38 ± 8.45		
CIT + BRG	33.25 ± 19.83	22.63 ± 7.65		
		GGT (UI/L)		
PLG	16.88 ± 4.55	18.50 ± 6.37	0.699	0.049
CITG	18.88 ± 8.04	19.88 ± 8.94		
BRG	15.50 ± 3.46	18.63 ± 6.41		
CIT + BRG	19.75 ± 5.92	20.25 ± 8.78		

Table 5. *Cont.*

Group	T1	T2	p (T × G)	$\eta^2 p$
		LDH (UI/L)		
PLG	438.86 ± 48.13	367.38 ± 77.77	0.498	0.083
CITG	330.88 ± 90.99 [a]	324.00 ± 71.97		
BRG	445.38 ± 247.59 [b]	393.88 ± 35.79 *		
CIT + BRG	431.00 ± 75.05	411.88 ± 63.37		
		CK (UI/L)		
PLG	319.50 ± 297.34	175.00 ± 51.77	0.238	0.138
CITG	327.63 ± 287.07	157.25 ± 60.78		
BRG	328.38 ± 247.59	208.88 ± 98.22		
CIT + BRG	379.38 ± 336.75	288.25 ± 209.86		

Data are presented as mean ± standard deviation. p (T × G): interaction group-by-time ($p < 0.05$) by two-way repeated-measures ANOVA. *: Significant differences between the two phases (T1 vs. T2) ($p < 0.05$). [a]: Significant differences with respect to PLG ($p < 0.05$). [b]: Significant differences with respect to CITG ($p < 0.05$).

Table 5 displays significant differences in the group-by-time interaction for cortisol ($p = 0.044$; $\eta^2 p = 0.247$) and T/C ($p = 0.005$; $\eta^2 p = 0.359$). In this sense, a significant difference was observed for T/C in CIT-BRG with respect to PLG at T2. On the other hand, a significant decrease in testosterone levels and T/C was observed in PLG, CITG and BRG after 9 weeks of supplementation (Table 6).

Table 6. Testosterone and cortisol status and testosterone/cortisol ratio of the triathletes at T1 and T2 (after 9 weeks).

Group	T1	T2	p (T × G)	$\eta^2 p$
		Testosterone (ng/mL)		
PLG	7.66 ± 2.26	4.51 ± 1.21 *	0.116	0.188
CITG	7.77 ± 1.10	5.50 ± 1.36 *		
BRG	7.11 ± 1.26	4.92 ± 1.16 *		
CIT + BRG	7.55 ± 1.06	6.69 ± 2.50		
		Cortisol (μg/dL)		
PLG	15.76 ± 1.34	20.37 ± 3.47 *	0.044	0.247
CITG	16.03 ± 2.48	18.20 ± 3.80		
BRG	15.89 ± 3.19	17.84 ± 3.76		
CIT + BRG	16.94 ± 2.33	15.30 ± 5.53		
		Testosterone/cortisol ratio		
PLG	49.07 ± 15.92	22.87 ± 8.12 *	0.005	0.359
CITG	49.13 ± 7.84	31.25 ± 11.02 *		
BRG	46.95 ± 13.83	28.63 ± 8.38 *		
CIT + BRG	45.97 ± 13.16	51.66 ± 30.00 [a]		

Data are presented as mean ± standard deviation. p (T × G): group-by-time interaction ($p < 0.05$) by two-way repeated-measures ANOVA. *: Significant differences between the two phases (T1 vs. T2) ($p < 0.05$). [a]: Significant differences with respect to PLG ($p < 0.05$).

Figure 2 shows the percentage change in distance covered in the Cooper test for each of the study groups. Significant differences can be observed in this parameter ($p = 0.002$; $\eta^2 p = 0.424$). Concretely, CIT-BRG presented a significantly higher value in the % change than PLG and CITG ($p < 0.05$).

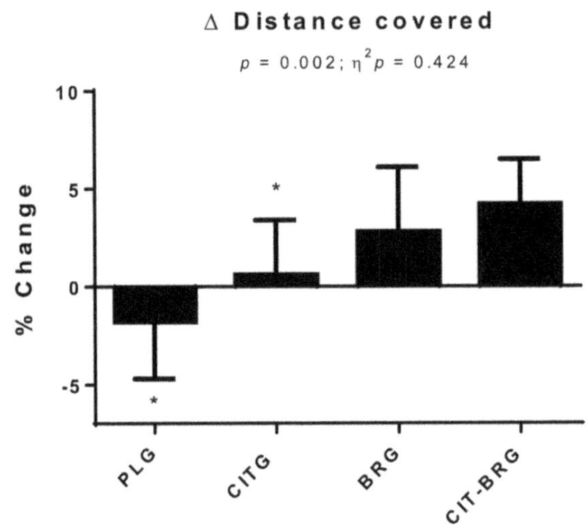

Figure 2. Percentage changes during the study in estimated distance covered in Cooper test in groups. Data are presented as mean ± standard deviation. Δ: ((T2 − T1)/T1) × 100. *: Significant differences with respect to CIT-BRG. $p < 0.05$.

Figure 3 indicates significant differences in cortisol percentage change ($p = 0.049$; $\eta^2p = 0.257$) between PLG and CIT-BRG. Moreover, T/C percentage change presented statistically significant differences ($p = 0.018$; $\eta^2p = 0.297$) between CIT-BRG and PLG. In the case of testosterone, there were no significant differences among groups in percentage change ($p = 0.149$).

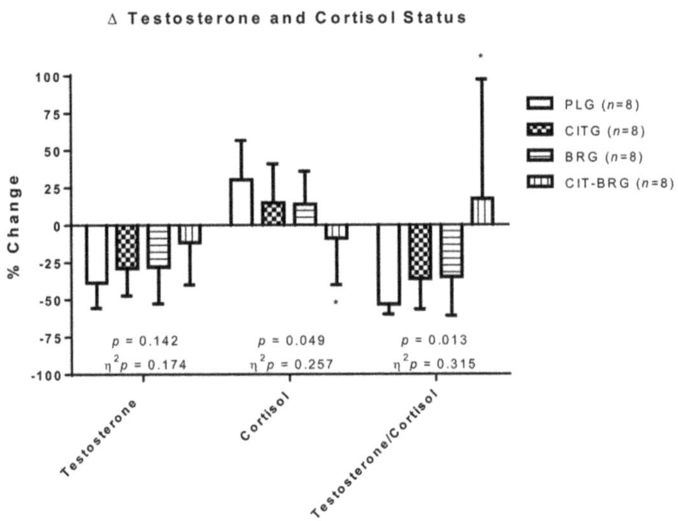

Figure 3. Percentage changes during the study in cortisol and testosterone hormone status and testosterone/cortisol ratio in the triathletes. Data are presented as mean ± standard deviation. Δ: ((T2 −T1)/T1) × 100. *: Significant differences with respect to PLG. $p < 0.05$.

4. Discussion

This study was planned to assess the effect of long-term (9 weeks) combination of 3 g/day of CIT plus 2.1 g/day of BR (300 mg/day of NO_3^-) supplementation on recovery status by distance covered in the Cooper test, serum EIMD markers and testosterone and cortisol in male triathletes. The EIMD markers (urea, creatinine, AST, ALT, GGT, LDH, CK) did not show any significant differences in the group-by-time interaction. However, triathletes showed a significantly better group-by-time interaction in distance covered in the Cooper test and anabolic/catabolic hormone status in CIT-BRG by preventing an increase in cortisol and a better T/C ratio. Furthermore, while CITG and BRG showed a significant decrease in testosterone levels, CIT + BR supplementation prevented a decline of this anabolic hormone. These significant results could be motivated by the synergistic effect that both supplements provided on the variables used to determine recovery status.

The balance between training loads and recovery are key factors in improving athletic performance [4]. To assess and control this balance, and with the intention of avoiding fatigue and maintaining an adequate performance, there are numerous parameters utilized, such as EIMD markers and anabolic/catabolic hormones [57,58]. Although there is an acute intensification of EIMD markers after exercise [2,59], long-term maintenance of high EIMD values could indicate a chronic fatigue status and inadequate adaptation to training [60]. In addition, it has been observed that anabolic/catabolic hormone status is changed after exercise due to an acute effect [58,61]. However, long-term variations in these hormones may be indicators of an adequate endogenous adaptation or, on the contrary, a fatigue status and, therefore, of an impaired sports performance [6]. Testosterone is an anabolic and androgenic hormone secreted by the hypothalamic–pituitary–testicular axis, and its increase specifies an overall anabolic state [62]. Nevertheless, cortisol, secreted by the hypothalamic–pituitary–adrenal axis, is a steroid hormone considered as a factor that indicates accumulated stress, and therefore, its increase suggests an accumulation of stress or catabolism [63]. Consequently, an increase in testosterone and/or a decrease in cortisol would lead to an increase in the testosterone/cortisol ratio, as an indicator of adaptation to training, thus indicating better endogenous recovery, while a decrease would indicate fatigue status [61,64]. In order to achieve these effects, some supplements that promote the NO pathway, such as CIT and BR, have been proposed [31].

It has been shown that NO can enhance recovery status through certain mechanisms [65], such as increasing protein synthesis through vasodilation of the arteries and veins of skeletal muscle that improve nutrient flow to the muscles, which in the long term favors muscle growth and repair [66]. In addition, it has been suggested that NO probably promotes angiogenesis in tissues by regulating the expression of the vascular endothelial growth factor [67]. Moreover, it has been demonstrated that skeletal muscle has the capacity to store, transport and metabolize NO_3^- and NO_2^- [68]. Therefore, chronic supplementation with NO precursor supplements (CIT and BR) would increase the levels of NO_3^- stored in skeletal muscle that is beneficial for NO production [69]. All these mentioned mechanisms could probably work in a complementary manner by enhancing endogenous recovery. A more efficient production of energy during exercise would reduce fatigue and thus decrease EIMD through an increase in protein synthesis [70]. This improved regeneration would lead to a decrease in stress and thus a reduced catabolic state, which would be reflected in anabolic/catabolic hormones [11].

In addition to the effect on NO, the CIT has been found to stimulate muscle protein synthesis by activating mTOR through the PI3K/MAPK/4E-BP1 pathway [71] and by increasing ARG production, which promotes growth hormone secretion [72]. Likewise, increased ARG production will improve intramuscular creatine levels, which will also allow an increase in phosphocreatine reserves, contributing to energy supply through a more efficient ATP regeneration and lowering fatigue, resulting in a decrease in EIMD after a high-demanding training [31]. Moreover, being part of the urea cycle, CIT facilitates the functioning of this cycle, helping to reduce the accumulation of ammonium and blood lactate concentration, improving the clearance capacity of these substances and, therefore, reducing the fatigue caused by their accumulation [73].

To the authors' knowledge, the effects of long-term combination of CIT plus BR supplementation on EIMD markers have not been studied in depth [74]. In this sense, the present trial did not present significant differences in the interaction between group and time in EIMD markers (urea, creatinine, AST, ALT, GGT, LDH and CK) ($p > 0.05$; Table 4). In the same line, some investigations that have evaluated the acute effects of these supplements individually have not found improvements in EIMD markers. Daab et al. did not find significant differences in CK and LDH before the Loughborough Intermittent Shuttle Test between the supplemented and the placebo groups after 7 days (3 days pre-exercise, test day and 3 days post-exercise) with 150 mL of BR juice (250 mg of NO_3^-) taken in two intakes per day (08:00 and 18:00 h) in soccer players [74]. Likewise, Martínez-Sanchez et al. did not present significant differences in the biochemical markers AST, ALT and CK between the supplemented group and placebo with CIT-enriched watermelon juice (3.45 g per 500 mL/day) taken two hours before a half-marathon race [66]. Therefore, considering that the results of this study did not offer any beneficial effects on the EIMD markers, the results obtained by the combination of CIT plus BR in the present study dismantle the original hypothesis in which it was predicted that both supplements could work in a complementary manner by reducing EIMD.

Although, to the authors' knowledge, the effects of long-term combination of CIT plus BR supplementation on anabolic/catabolic hormones have not been studied, in the current study, the combination of these supplements showed a better group-by-time interaction in distance covered in the Cooper test and anabolic/catabolic hormone status in CIT-BRG (Table 5 and Figure 3) by preventing an increase in cortisol ($p = 0.044$; $\eta^2p = 0.247$) and a better T/C ratio ($p = 0.005$; $\eta^2p = 0.359$). Furthermore, while CITG and BRG showed a significant decrease in the testosterone level after 9 weeks ($p < 0.05$), CIT + BR supplementation prevented a decline in this anabolic hormone. Nevertheless, some authors have shown the effect of both supplements individually on these hormones. In this way, Da Silva et al. did not observe improvements in the T/C ratio during the recovery period at 24, 48 and 72 h post-exercise in untrained young adult men after 6 g CIT supplementation before a 60 min workout [27]. These authors indicated that the inability to improve anabolic factors results in no beneficial effect of CIT supplementation on muscle regeneration during an acute recovery period. However, the chronic changes in cortisol and testosterone can be related to accumulated stress and body regeneration during the sports season [6]. In this way, Garnacho-Castaño et al. showed that BR supplementation did not appear to influence anabolic/catabolic status in response to acute high-intensity workouts after drinking 140 mL of BJ (~12.8 mmol NO_3^-) [25]. On the contrary, in this study, CITG and BRG presented a maintenance of distance covered in the Cooper test and a decrease in testosterone levels and T/C after 9 weeks of supplementation. These data could indicate an inadequate recovery status in these groups because after 9 weeks with adequate training, a better sports performance would be expected. Nevertheless, this study presented a significantly better recovery status in CIT-BRG represented as an increase in distance covered in the Cooper test and maintenance of testosterone and T/C ratio after 9 weeks of combined supplementation. These adaptations were obtained by preventing an increase in cortisol and/or a decline in testosterone in CIT-BRG with respect to other supplementation groups.

In this sense, CIT is a key activator of muscle protein synthesis in catabolic situations, such as high-intensity training periods, via activation of the mTOR pathway due to its key role in the regulation of nitrogen homeostasis [75]. Likewise, testosterone increases mTOR pathway [76] and cortisol inhibits mTOR pathway signaling [75]. Thus, increasing testosterone and controlling cortisol secretion could result in lower stress and adequate muscle regeneration [6]. In this sense, the long-term effect of CIT enhancing NO could increase blood flow in the testis promoting testosterone synthesis [19] and maintaining the testosterone level by vasodilator effect [77]. In addition, the enhancement of NO reduces ACTH-mediated cortisol production [18]. Consequently, although this hypothesis is speculative, increasing NO could be successful in maintaining an anabolic state, decreasing metabolic stress [2,9]. Therefore, the results obtained in the CIT-BRG group could show how independent pathways in muscle recovery (NO and mTOR) can be synergistically activated with both supplements to obtain better results.

4.1. Limitations, Strengths and Future Research

It should be noted that it is difficult to obtain larger samples in athletes as not many of them have the availability to comply with the training and supplementation instructions required by the study. In addition, the effects that both supplements used could have on the muscle were speculative because no evaluation was included in this regard. On the other hand, sampling using a convenient, non-probabilistic sampling procedure may produce results that are not representative of the rest of the population. These limitations may underrepresent the results and may affect study outcomes. For this reason, the results should be considered in the context of the study. However, the methodology used in this trial, a double-blind, placebo-controlled trial, is the most important strength. In addition, another strength was the control of the triathletes' diet, as well as the control of the body composition throughout the intervention process, so that these outcomes did not influence the final results. Another strength is the synergistic potential of the study.

Future research should continue to study the long-term effects of this combination on recovery, using different markers, such as sports performance, in order to expand the existing knowledge on this combination. It should also examine the effectiveness of these supplements in athletes who have already been diagnosed with an overtraining state to determine whether the use of these supplements as part of treatment would accelerate recovery. In addition, it should analyze how this potential combination affects the female population or anaerobic sports, given that this study only focused on males and measured aerobic performance.

4.2. Practical Application

This research could be of interest to physicians and nutritionists who want to achieve better post-exercise recovery for their athletes. Considering that 3 g/day of CIT plus 2.1 g/day of BR (300 mg/day of NO_3^-) for 9 weeks could advance muscle and endogenous recovery, supplementation phases could be considered in the intensive training phases.

5. Conclusions

In conclusion, although the combination of 3 g/day of CIT plus 2.1 g/day of BR (300 mg/day of NO_3^-) supplementation for 9 weeks did not present any benefit for EIMD, it prevented an increase in cortisol and a decline in T/C compared with placebo or isolated supplementation. Moreover, this combination promoted a better distance covered in the Cooper test after 9 weeks of supplementation. Therefore, the combined use of 3 g/day of CIT and BR (300 mg/day of NO_3^-) could promote a faster muscle recovery status but without preventing EIMD.

Author Contributions: All authors have read and agreed to the published version of the manuscript. J.B. and J.M.-A.: conception and design of research, analysis and interpretation of the data, drafting of the paper, critical review and approval of the final version submitted for publication. J.C.-G. and A.V.: analysis and interpretation of the data, drafting of the paper, critical review and approval of the final version submitted for publication. D.F.-L., J.O.-I. and J.S.-C.: drafting of the paper, critical review and approval of the final version submitted for publication.

Funding: The authors declare no funding sources.

Institutional Review Board Statement: This trial was designed in accordance with the Declaration of Helsinki (2008) and the Fortaleza update (2013) [47] and was approved by the Human Research Ethics Committee of the University of León, Spain (number: ULE-020-2020).

Informed Consent Statement: All triathletes were completely informed of all actions of the study and signed the personal statement of informed consent, giving their individual agreement to take part in the proposed work.

Data Availability Statement: No new data were created or analyzed in this study. Data sharing is not applicable to this article.

Acknowledgments: The authors want to thank the triathletes and other assistants involved in this investigation for their participation, enthusiasm and cooperation.

Conflicts of Interest: The authors affirm no conflict of interest.

References

1. Del Coso, J.; González, C.; Abian-Vicen, J.; Salinero Martín, J.J.; Soriano, L.; Areces, F.; Ruiz, D.; Gallo, C.; Lara, B.; Calleja-González, J. Relationship between physiological parameters and performance during a half-ironman triathlon in the heat. *J. Sports Sci.* **2014**, *32*, 1680–1687. [CrossRef]
2. Fernández-Landa, J.; Fernández-Lázaro, D.; Calleja-González, J.; Caballero-García, A.; Córdova, A.; León-Guereño, P.; Mielgo-Ayuso, J. Long-Term Effect of Combination of Creatine Monohydrate Plus β-Hydroxy β-Methylbutyrate (HMB) on Exercise-Induced Muscle Damage and Anabolic/Catabolic Hormones in Elite Male Endurance Athletes. *Biomolecules* **2020**, *10*, 140. [CrossRef]
3. Neubauer, O.; König, D.; Wagner, K.H. Recovery after an Ironman triathlon: Sustained inflammatory responses and muscular stress. *Eur. J. Appl. Physiol.* **2008**, *104*, 417–426. [CrossRef]
4. Meeusen, R.; Duclos, M.; Foster, C.; Fry, A.; Gleeson, M.; Nieman, D.; Raglin, J.; Rietjens, G.; Steinacker, J.; Urhausen, A. Prevention, diagnosis and treatment of the overtraining syndrome: Joint consensus statement of the European College of Sport Science (ECSS) and the American College of Sports Medicine (ACSM). *Eur. J. Sport Sci.* **2013**, *13*, 1–24. [CrossRef]
5. Del Coso, J.; Areces, F.; Salinero, J.J.; González-Millán, C.; Abián-Vicén, J.; Soriano, L.; Ruiz, D.; Gallo, C.; Lara, B.; Calleja-Gonzalez, J. Compression stockings do not improve muscular performance during a half-ironman triathlon race. *Eur. J. Appl. Physiol.* **2014**, *114*, 587–595. [CrossRef]
6. Mielgo-Ayuso, J.; Zourdos, M.C.; Urdampilleta, A.; Calleja-González, J.; Seco, J.; Córdova, A. Relationship of long-term macronutrients intake on anabolic-catabolic hormones in female elite volleyball players. *Nutr. Hosp.* **2017**, *34*, 1155–1162. [CrossRef]
7. Hotfiel, T.; Mayer, I.; Huettel, M.; Hoppe, M.W.; Engelhardt, M.; Lutter, C.; Pöttgen, K.; Heiss, R.; Kastner, T.; Grim, C. Accelerating Recovery from Exercise-Induced Muscle Injuries in Triathletes: Considerations for Olympic Distance Races. *Sports* **2019**, *7*, 143. [CrossRef]
8. Bescós, R.; Sureda, A.; Tur, J.A.; Pons, A. The effect of nitric-oxide-related supplements on human performance. *Sports Med.* **2012**, *42*, 99–117. [CrossRef]
9. Beck, K.L.; Thomson, J.S.; Swift, R.J.; von Hurst, P.R. Role of nutrition in performance enhancement and postexercise recovery. *Open Access J. Sport. Med.* **2015**, *6*, 259. [CrossRef]
10. Bryan, N.S. *Safe and Effective Use of Nitric Oxide-Based Supplements and Nutrition for Sports Performance*, 2nd ed.; Elsevier Inc.: Amsterdam, The Netherlands, 2018.
11. Gonzalez, A.M.; Trexler, E.T. Effects of Citrulline Supplementation on Exercise Performance in Humans: A Review of the Current Literature. *J. Strength Cond. Res.* **2020**, *34*, 1480–1495. [CrossRef]
12. Gilligan, D.M.; Panza, J.A.; Kilcoyne, C.M.; Waclawiw, M.A.; Casino, P.R.; Quyyumi, A.A. Contribution of endothelium-derived nitric oxide to exercise-induced vasodilation. *Circulation* **1994**, *90*, 2853–2858. [CrossRef]
13. Anderson, J.E. A Role for Nitric Oxide in Muscle Repair: Nitric Oxide-Mediated Activation of Muscle Satellite Cells. *Mol. Biol. Cell* **2000**, *11*, 1859–1874. [CrossRef]
14. Arazi, H.; Eghbali, E. Possible Effects of beetroot Supplementation on Physical Performance Through Metabolic, Neuroendocrine, and Antioxidant Mechanisms: A Narrative Review of the Literature. *Front. Nutr.* **2021**, *8*, 660150. [CrossRef]
15. Viribay, A.; Arribalzaga, S.; Mielgo-Ayuso, J.; Castañeda-Babarro, A.; Seco-Calvo, J.; Urdampilleta, A. Effects of 120 g/h of carbohydrates intake during a mountain marathon on exercise-induced muscle damage in elite runners. *Nutrients* **2020**, *12*, 1367. [CrossRef]
16. Urdampilleta, A.; Arribalzaga, S.; Viribay, A.; Castañeda-Babarro, A.; Seco-Calvo, J.; Mielgo-Ayuso, J. Effects of 120 vs. 60 and 90 g/h Carbohydrate Intake during a Trail Marathon on Neuromuscular Function and High Intensity Run Capacity Recovery. *Nutrients* **2020**, *12*, 94. [CrossRef]
17. Wang, J.M.; Gu, C.H.; Tao, L.; Wu, X.L. Effect of surgery and efferent duct ligation on testicular blood flow and testicular steroidogenesis in the rat. *J. Reprod. Fertil.* **1985**, *73*, 191–196. [CrossRef]
18. Monau, T.R.; Vargas, V.E.; Zhang, L.; Myers, D.A.; Ducsay, C.A. Nitric oxide inhibits ACTH-induced cortisol production in near-term, long-term hypoxic ovine fetal adrenocortical cells. *Reprod. Sci.* **2010**, *17*, 955–962. [CrossRef]
19. Banihani, S.A. Biomolecules Testosterone in Males as Enhanced by Onion (*Allium Cepa*, L.). *Biomolecules* **2019**, *9*, 75. [CrossRef]

20. Hernández, A.; Schiffer, T.A.; Ivarsson, N.; Cheng, A.J.; Bruton, J.D.; Lundberg, J.O.; Weitzberg, E.; Westerblad, H. Dietary nitrate increases tetanic [Ca 2+] i and contractile force in mouse fast-twitch muscle. *J. Physiol.* **2012**, *590*, 3575–3583. [CrossRef]
21. Milton-Laskibar, I.; Alfredo Martínez, J.; Portillo, M.P. Current knowledge on BRroot bioactive compounds: Role of nitrate and betalains in health and disease. *Foods* **2021**, *10*, 1314. [CrossRef]
22. Jones, L.; Bailey, S.J.; Rowland, S.N.; Alsharif, N.; Shannon, O.M.; Clifford, T. The Effect of Nitrate-Rich BRroot Juice on Markers of Exercise-Induced Muscle Damage: A Systematic Review and Meta-Analysis of Human Intervention Trials. *J. Diet. Suppl.* **2021**. [CrossRef]
23. Clifford, T.; Bell, O.; West, D.J.; Howatson, G.; Stevenson, E.J. The effects of BRroot juice supplementation on indices of muscle damage following eccentric exercise. *Eur. J. Appl. Physiol.* **2016**, *116*, 353–362. [CrossRef]
24. Sarfaraz, S.; Ikram, R.; Osama, M.; Anser, H. Effect of different doses of lyophilized BRroot on fertility and reproductive hormones. *Pak. J. Pharm. Sci.* **2020**, *33*, 2505–2510. [CrossRef]
25. Garnacho-Castaño, M.V.; Palau-Salvà, G.; Serra-Payá, N.; Ruiz-Hermosel, M.; Berbell, M.; Viñals, X.; Bataller, M.G.; Carbonell, T.; Vilches-Saez, S.; Cobo, E.P.; et al. Understanding the effects of BRroot juice intake on CrossFit performance by assessing hormonal, metabolic and mechanical response: A randomized, double-blind, crossover design. *J. Int. Soc. Sports Nutr.* **2020**, *17*. [CrossRef]
26. Larsen, F.J.; Schiffer, T.A.; Borniquel, S.; Sahlin, K.; Ekblom, B.; Lundberg, J.O.; Weitzberg, E. Dietary Inorganic Nitrate Improves Mitochondrial Efficiency in Humans. *Cell Metab.* **2011**, *13*, 149–159. [CrossRef]
27. Da Silva, D.K.; Jacinto, J.L.; de Andrade, W.B.; Roveratti, M.C.; Estoche, J.M.; Balvedi, M.C.W.; de Oliveira, D.B.; da Silva, R.A.; Aguiar, A.F. Citrulline malate does not improve muscle recovery after resistance exercise in untrained young adult men. *Nutrients* **2017**, *9*, 1132. [CrossRef]
28. Haider, G.; Folland, J.P. Nitrate supplementation enhances the contractile properties of human skeletal muscle. *Med. Sci. Sports Exerc.* **2014**, *46*, 2234–2243. [CrossRef]
29. Radak, Z.; Naito, H.; Taylor, A.W.; Goto, S. Nitric oxide: Is it the cause of muscle soreness? *Nitric Oxide Biol. Chem.* **2012**, *26*, 89–94. [CrossRef]
30. Bahri, S.; Zerrouk, N.; Aussel, C.; Moinard, C.; Crenn, P.; Curis, E.; Chaumeil, J.C.; Cynober, L.; Sfar, S. Citrulline: From metabolism to therapeutic use. *Nutrition* **2013**, *29*, 479–484. [CrossRef]
31. Rhim, H.C.; Kim, S.J.; Park, J.; Jang, K.M. Effect of citrulline on post-exercise rating of perceived exertion, muscle soreness, and blood lactate levels: A systematic review and meta-analysis. *J. Sport Health Sci.* **2020**, *9*, 553–561. [CrossRef]
32. Mathiowetz, V.; Weber, K.; Volland, G.; Kashman, N. Reliability and validity of grip and pinch strength evaluations. *J. Hand Surg.* **1984**, *9*, 222–226. [CrossRef]
33. Fernández-Landa, J.; Fernández-Lázaro, D.; Calleja-González, J.; Caballero-García, A.; Martínez, A.C.; León-Guereño, P.; Mielgo-Ayuso, J. Effect of ten weeks of creatine monohydrate plus HMB supplementation on athletic performance tests in elite male endurance athletes. *Nutrients* **2020**, *12*, 193. [CrossRef]
34. Allen, A.E.; Dupont, C.L.; Oborník, M.; Horák, A.; Nunes-Nesi, A.; McCrow, J.P.; Zheng, H.; Johnson, D.A.; Hu, H.; Fernie, A.R.; et al. Evolution and metabolic significance of the urea cycle in photosynthetic diatoms. *Nature* **2011**, *473*, 203–207. [CrossRef]
35. Alvares, T.S.; Conte-Junior, C.A.; Silva, J.T.; Paschoalin, V.M.F. L-arginine does not improve biochemical and hormonal response in trained runners after 4 weeks of supplementation. *Nutr. Res.* **2014**, *34*, 31–39. [CrossRef]
36. Le Roux-Mallouf, T.; Pelen, F.; Vallejo, A.; Halimaoui, I.; Doutreleau, S.; Verges, S. Effect of chronic nitrate and citrulline supplementation on vascular function and exercise performance in older individuals. *Aging* **2019**, *11*, 3315. [CrossRef]
37. Thomas, D.; Erdman, K.; Burke, L. American College of Sports Medicine Joint Position Statement. Nutrition and Athletic Performance. *Med. Sci. Sports Exerc.* **2016**, *48*, 543–568. [CrossRef]
38. Pruna, R.; Lizarraga, A.; Domínguez, D. Medical assessment in athletes. *Med. Clin.* **2018**, *150*, 268–274. [CrossRef]
39. Wax, B.; Kavazis, A.N.; Luckett, W. Effects of Supplemental Citrulline-Malate Ingestion on Blood Lactate, Cardiovascular Dynamics, and Resistance Exercise Performance in Trained Males. *J. Diet. Suppl.* **2016**, *13*, 269–282. [CrossRef]
40. Pérez-Guisado, J.; Jakeman, P.M. Citrulline malate enhances athletic anaerobic performance and relieves muscle soreness. *J. Strength Cond. Res.* **2010**, *24*, 1215–1222. [CrossRef]
41. Robineau, J.; Marrier, B.; Le Meur, Y.; Piscione, J.; Peeters, A.; Lacome, M. Road to Rio: A Case Study of Workload Periodization Strategy in Rugby-7s During an Olympic Season. *Front. Sports Act. Living* **2020**, *1*, 72. [CrossRef]
42. Anderson, O.K.; Martinez-Ferran, M.; Lorenzo-Calvo, J.; Jiménez, S.L.; Pareja-Galeano, H. Effects of Nitrate Supplementation on Muscle Strength and Mass: A Systematic Review. *J. Strength Cond. Res.* **2021**. [CrossRef]
43. Hoon, M.W.; Johnson, N.A.; Chapman, P.G.; Burke, L.M. The effect of nitrate supplementation on exercise performance in healthy individuals: A systematic review and meta-analysis. *Int. J. Sports Nutr. Exerc. Metab.* **2013**, *23*, 522–532. [CrossRef]
44. Trexler, E.T.; Persky, A.M.; Ryan, E.D.; Schwartz, T.A.; Stoner, L.; Smith-Ryan, A.E. Acute Effects of Citrulline Supplementation on High-Intensity Strength and Power Performance: A Systematic Review and Meta-Analysis. *Sports Med.* **2019**, *49*, 707–718. [CrossRef]
45. Jones, A.M.; Thompson, C.; Wylie, L.J.; Vanhatalo, A. Dietary nitrate and physical performance. *Ann. Rev. Nutr.* **2018**, *38*, 303–328. [CrossRef]
46. Wehrens, S.M.T.; Christou, S.; Isherwood, C.; Middleton, B.; Gibbs, M.A.; Archer, S.N.; Skene, D.J.; Johnston, J.D. Meal Timing Regulates the Human Circadian System. *Curr. Biol.* **2017**, *27*, 1768. [CrossRef]

47. Nowatzke, W.; Sarno, M.J.; Birch, N.C.; Stickle, D.F.; Eden, T.; Cole, T.G. Evaluation of an assay for serum 1,5-anhydroglucitol (GlycoMark^{TM}) and determination of reference intervals on the Hitachi 917 analyzer. *Clin. Chim. Acta* **2004**, *350*, 201–209. [CrossRef]
48. Mehta, P.H.; Welker, K.M.; Zilioli, S.; Carré, J.M. Testosterone and cortisol jointly modulate risk-taking. *Psychoneuroendocrinology* **2015**, *56*, 88–99. [CrossRef]
49. Penry, J.T.; Wilcox, A.R.; Yun, J. Validity and reliability analysis of Cooper's 12-minute run and the multistage shuttle run in healthy adults. *J. Strength Cond. Res.* **2011**, *25*, 597–605. [CrossRef]
50. Alvero-Cruz, J.R.; Carnero, E.A.; Giráldez García, M.A.; Alacid, F.; Rosemann, T.; Nikolaidis, P.T.; Knechtle, B. Cooper Test Provides Better Half-Marathon Performance Prediction in Recreational Runners Than Laboratory Tests. *Front. Physiol.* **2019**, *10*, 1349. [CrossRef]
51. Da Silva, V.S.; Vieira, M.F.S. International society for the advancement of kinanthropometry (Isak) global: International accreditation scheme of the competent anthropometrist. *Revista Brasileira de Cineantropometria Desempenho Humano* **2020**, *22*, 1–6. [CrossRef]
52. Mielgo-Ayuso, J.; Maroto-Sánchez, B.; Luzardo-Socorro, R.; Palacios, G.; Gil-Antuñano, N.P.; González-Gross, M. EXERNET Study Group Evaluation of nutritional status and energy expenditure in athletes. *Nutr. Hosp.* **2015**, *31*, 227–236. [CrossRef]
53. Mielgo-Ayuso, J.; Zourdos, M.C.; Calleja-González, J.; Urdampilleta, A.; Ostojic, S.M. Dietary intake habits and controlled training on body composition and strength in elite female volleyball players during the season. *Appl. Physiol. Nutr. Metab.* **2015**, *40*, 827–834. [CrossRef]
54. Russolillo Femenías, G.; Marques-Lopes, I. *Láminas de Porciones de Alimentos a Tamaño Real*; Librería Proteo: Málaga, Spain, 2011.
55. Valensise, H.; Andreoli, A.; Lello, S.; Magnani, F.; Romanini, C.; De Lorenzo, A. Total-body skeletal muscle mass: Development and cross-validation of anthropometric prediction models. *Am. J. Clin. Nutr.* **2000**, *72*, 796–803. [CrossRef]
56. Ferguson, C.J. An Effect Size Primer: A Guide for Clinicians and Researchers. *Prof. Psychol. Res. Pract.* **2009**, *40*, 532–538. [CrossRef]
57. Banfi, G.; Colombini, A.; Lombardi, G.; Lubkowska, A. *Metabolic Markers in Sports Medicine*, 1st ed.; Elsevier Inc.: Amsterdam, The Netherlands, 2012; Volume 56.
58. Baur, D.A.; Bach, C.W.; Hyder, W.J.; Ormsbee, M.J. Fluid retention, muscle damage, and altered body composition at the Ultraman triathlon. *Eur. J. Appl. Physiol.* **2016**, *116*, 447–458. [CrossRef]
59. Danielsson, T.; Rg Carlsson, J.; Schreyer, H.; Ahnesjö, J.; Ten Siethoff, L.; Ragnarsson, T.; Tugetam, Å.; Bergman, P. Blood biomarkers in male and female participants after an Ironman-distance triathlon. *PLoS ONE* **2017**, *12*, e0179324. [CrossRef]
60. Požgain, I.; Požgain, Z.; Degmečić, D. *Placebo and Nocebo Effect: A Mini-Review*; Medicinska naklada: Zagreb, Croatia, 2014; Volume 26.
61. Greenham, G.; Buckley, J.D.; Garrett, J.; Eston, R.; Norton, K. *Biomarkers of Physiological Responses to Periods of Intensified, Non-Resistance-Based Exercise Training in Well-Trained Male Athletes: A Systematic Review and Meta-Analysis*; Springer International Publishing: Berlin/Heidelberg, Germany, 2018; Volume 48.
62. Grossmann, M.; Wittert, G.A. Dysregulation of the Hypothalamic-Pituitary-Testicular Axis due to Energy Deficit. *J. Clin. Endocrinol. Metab.* **2021**, *106*, e4861–e4871. [CrossRef]
63. Casto, K.V.; Edwards, D.A. Testosterone, cortisol, and human competition. *Horm. Behav.* **2016**, *82*, 21–37. [CrossRef]
64. Hotta, Y.; Kataoka, T.; Kimura, K. Testosterone Deficiency and Endothelial Dysfunction: Nitric Oxide, Asymmetric Dimethylarginine, and Endothelial Progenitor Cells. *Sex. Med. Rev.* **2019**, *7*, 661–668. [CrossRef]
65. Habib, S.; Moinuddin; Ali, A. *Role of Nitric Oxide in Sports Nutrition*, 2nd ed.; Elsevier Inc.: Amsterdam, The Netherlands, 2018.
66. Rammos, C. Potential of dietary nitrate in angiogenesis. *World J. Cardiol.* **2015**, *7*, 652. [CrossRef]
67. Huerta Ojeda, Á. Efecto de la suplementación de L-arginina y L-citrulina sobre el rendimiento físico: Una revisión sistemática. *Nutr. Hosp.* **2019**, *36*, 1389–1402.
68. Nyakayiru, J.; van Loon, L.J.C.; Verdijk, L.B. Could intramuscular storage of dietary nitrate contribute to its ergogenic effect? A mini-review. *Free Radic. Biol. Med.* **2020**, *152*, 295–300. [CrossRef]
69. Brownlee, K.K.; Moore, A.W.; Hackney, A.C. Relationship between circulating cortisol and testosterone: Influence of physical exercise. *J. Sports Sci. Med.* **2005**, *4*, 76–83. [CrossRef]
70. Markus, I.; Constantini, K.; Hoffman, J.R.; Bartolomei, S.; Gepner, Y. Exercise-induced muscle damage: Mechanism, assessment and nutritional factors to accelerate recovery. *Eur. J. Appl. Physiol.* **2021**, *121*, 969–992. [CrossRef]
71. Le Plénier, S.; Goron, A.; Sotiropoulos, A.; Archambault, E.; Guihenneuc, C.; Walrand, S.; Salles, J.; Jourdan, M.; Neveux, N.; Cynober, L.; et al. Citrulline directly modulates muscle protein synthesis via the PI3K/MAPK/4E-BP1 pathway in a malnourished state: Evidence from in vivo, ex vivo, and in vitro studies. *Am. J. Physiol. Endocrinol. Metab.* **2017**, *312*, E27–E36. [CrossRef]
72. Botchlett, R.; Lawler, J.M.; Wu, G. *l-Arginine and l-Citrulline in Sports Nutrition and Health*, 2nd ed.; Elsevier Inc.: Amsterdam, The Netherlands, 2018.
73. Wylie, L.J.; Park, J.W.; Vanhatalo, A.; Kadach, S.; Black, M.I.; Stoyanov, Z.; Schechter, A.N.; Jones, A.M.; Piknova, B. Human skeletal muscle nitrate store: Influence of dietary nitrate supplementation and exercise. *J. Physiol.* **2019**, *597*, 5565–5576. [CrossRef]
74. Daab, W.; Bouzid, M.A.; Lajri, M.; Bouchiba, M.; Saafi, M.A.; Rebai, H. Chronic BRroot Juice Supplementation Accelerates Recovery Kinetics following Simulated Match Play in Soccer Players. *J. Am. Coll. Nutr.* **2021**, *40*, 61–69. [CrossRef]

75. Liao, Y.; Su, R.; Zhang, P.; Yuan, B.; Li, L. Cortisol inhibits mTOR signaling in avascular necrosis of the femoral head. *J. Orthop. Surg. Res.* **2017**, *12*, 1–7. [CrossRef]
76. Morley, J.E. The mTOR Conundrum: Essential for Muscle Function, but Dangerous for Survival. *J. Am. Med. Dir. Assoc.* **2016**, *17*, 963–966. [CrossRef]
77. Jones, R.D.; Jones, T.H.; Channer, K.S. The influence of testosterone upon vascular reactivity. *Eur. J. Endocrinol.* **2004**, *151*, 29–37. [CrossRef]

biology

MDPI

Article

Acid–Base Balance, Blood Gases Saturation, and Technical Tactical Skills in Kickboxing Bouts According to K1 Rules

Łukasz Rydzik [1,*], Mateusz Mardyła [2], Zbigniew Obmiński [3], Magdalena Więcek [2], Marcin Maciejczyk [2], Wojciech Czarny [4], Jarosław Jaszczur-Nowicki [5,*] and Tadeusz Ambroży [1]

1 Institute of Sports Sciences, University of Physical Education, 31-571 Cracow, Poland; tadek@ambrozy.pl
2 Department of Physiology and Biochemistry, Faculty of Physical Education and Sport, University of Physical Education, 31-571 Cracow, Poland; mmardyla93@gmail.com (M.M.); magdalena.wiecek@awf.krakow.pl (M.W.); marcin.maciejczyk@awf.krakow.pl (M.M.)
3 Department of Endocrinology, Institute of Sport—National Research Institute, 01-982 Warsaw, Poland; zbigniew.obminski@insp.waw.pl
4 College of Medical Sciences, Institute of Physical Culture Studies, University of Rzeszow, 35-959 Rzeszow, Poland; wojciechczarny@wp.pl
5 Department of Tourism, Recreation and Ecology, University of Warmia and Mazury in Olsztyn, 10-719 Olsztyn, Poland
* Correspondence: lukasz.gne@op.pl (Ł.R.); j.jaszczur-nowicki@uwm.edu.pl (J.J.-N.); Tel.: +48-730-696-377 (Ł.R.)

Simple Summary: The aim of our study was to analyze the changes in ABB after a three-round kickboxing fight and the level of technical and tactical skills presented during the fight. Fighting in kickboxing under K1 rules takes place with a high presence of anaerobic metabolism. Kickboxing athletes must have a good tolerance for metabolic acidosis and the ability to conduct an effective duel despite ABB disorders. Properly developed post-workout regeneration also plays an extremely important role.

Abstract: Background: Acid–base balance (ABB) is a major component of homeostasis, which is determined by the efficient functioning of many organs, including the lungs, kidneys, and liver, and the proper water and electrolyte exchange between these components. The efforts made during competitions by combat sports athletes such as kickboxers require a very good anaerobic capacity, which, as research has shown, can be improved by administering sodium bicarbonate. Combat sports are also characterized by an open task structure, which means that cognitive and executive functions must be maintained at an appropriate level during a fight. The aim of our study was to analyze the changes in ABB in capillary blood, measuring levels of H^+, pCO_2, pO_2, HCO_3^-, BE and total molar CO_2 concentration (TCO_2), which were recorded 3 and 20 min after a three-round kickboxing bout, and the level of technical and tactical skills presented during the fight. Methods: The study involved 14 kickboxers with the highest skill level (champion level). Statistical comparison of mentioned variables recorded prior to and after a bout was done with the use of Friedman's ANOVA. Results: 3 min after a bout, H^+ and pO_2 were higher by 41% and 11.9%, respectively, while pCO_2, HCO_3^-, BE and TO_2 were lower by 14.5%, 39.4%, 45.4% and 34.4%, respectively. Furthermore, 20 min after the bout all variables tended to normalization and they did not differ significantly compared to the baseline values. Scores in activeness of the attack significantly correlated ($r = 0.64$) with pre–post changes in TCO_2. Conclusions: The disturbances in ABB and changes in blood oxygen and carbon dioxide saturation observed immediately after a bout indicate that anaerobic metabolism plays a large part in kickboxing fights. Anaerobic training should be included in strength and conditioning programs for kickboxers to prepare the athletes for the physiological requirements of sports combat.

Keywords: acid–base balance; kickboxing; metabolic acidosis

Citation: Rydzik, Ł.; Mardyła, M.; Obmiński, Z.; Więcek, M.; Maciejczyk, M.; Czarny, W.; Jaszczur-Nowicki, J.; Ambroży, T. Acid–Base Balance, Blood Gases Saturation, and Technical Tactical Skills in Kickboxing Bouts According to K1 Rules. *Biology* **2022**, *11*, 65. https://doi.org/10.3390/biology11010065

Academic Editors: Filipe Manuel Clemente, Georgian Badicu, Eugenia Murawska-Ciałowicz and Paulo Gentil

Received: 6 December 2021
Accepted: 31 December 2021
Published: 2 January 2022

1. Introduction

Acid–base balance (ABB) is a major component of homeostasis, which is determined by the efficient functioning of many organs, including the lungs, kidneys, and liver [1,2], and the proper water and electrolyte exchange between these components [3]. At rest, an equilibrium is maintained between the pulmonary gas pressures (pO_2 and pCO_2) in the human body, and the buffering system in the blood consists of hydrogen ion (H^+) acceptors, which include bicarbonate ions (HCO_3^-), proteins, amino acids, hydrogen phosphate ions (HPO_4^{2-}), and hemoglobin contained in erythrocytes. The buffering system compensates for small fluctuations in acidity, but intense exercise disrupts ABB in the body. Studies using 31P nuclear magnetic resonance spectroscopy have demonstrated a significant intracellular increase in H^+ concentration and a concomitant decrease in phosphocreatine (PCr) concentration during some seconds of repeated supramaximal exercise [4]. With buffering, the post-exercise blood H^+ concentration is much lower than in the cytosol. The peripheral and central chemoreceptors responsible for the regulation of the pulmonary ventilation rate respond to post-exercise changes in pH and disturbances in gas parameters of ABB [5,6]. Studies have shown that the higher the intensity of anaerobic exercise, the higher the blood levels of lactate and hydrogen ions and the greater the decrease in HCO_3^- [7–13].

It is believed that the temporary high acidification of cytosol in muscle fiber cells is not the only cause of the decrease in maximal power [14]. Nevertheless, numerous studies have confirmed that an increase in blood buffering capacity after oral administration of sodium bicarbonate or other alkalizing fluids reduces post-exercise pH disturbances and increases exercise capacity [15–21].

Exercise during competitions in combat sports such as kickboxing, boxing, taekwondo, and wrestling requires very good anaerobic capacity, which, as demonstrated by studies, can be improved by administering sodium bicarbonate [22–24]. Combat sports are also characterized by an open task structure, which means that cognitive functions must be maintained at an appropriate level during a fight [25]. The following cognitive abilities have been most often studied in combat sport athletes because the levels of these features are in a great part related to athletic skills. The most frequently tested cognitive functions in athletes are visuo-motor coordination [26–28], information processing and planning [29,30], and accuracy of decision-making [31]. It should be noted that in hitting sports, such as boxing and kickboxing, testing of cognitive functions matters for assessment of brain micro-injuries among athletes [32,33]. In laboratory tests, measurements of the speed and accuracy of reactions to visual stimuli are used to evaluate the level of some of the above-mentioned skills, including simple reaction tests, choice reaction tests, GO/NOGO tasks, Stroop tests, and trail making tests. Many published results have indicated a relationship between the level of performance during these tests and the physiological responses induced by various laboratory efforts. Immediately after intense exercise-induced metabolic acidosis, the results of the reaction time test and Stroop test are worse than at rest, but after a 15 min rest, they partially normalize [34]. Combined with concurrent psychometric testing, exercise tests do not fully replicate the task structure of a typical real fight. In most cases, the scale of difficulty in completing a mental and physical task during official combat sports competitions is much greater compared to laboratory psychophysical tests.

The experiments conducted in these studies demonstrated bidirectional changes in the level of performance during psychometric tests in response to laboratory test exercises of increasing intensity. These are fundamental to understanding the relationships between exercise intensity and physiological responses with level of performance of psychometric tasks. During the first phase of low-intensity exercise, the choice reaction time progressively decreases until a blood lactate concentration of 5.5 mmol/L is reached, and progressively increases once this concentration is exceeded [35]. A similar biphasic pattern of choice reaction time has been recorded during running with increasing speed [36]. Furthermore, higher levels of cognitive function are presented by individuals with a higher physical capacity [37]. In the case of psychometric tests that examine simultaneously the speed and accuracy of reactions using the GO/NOGO test, there is a need to choose between two contradictory decisions, one favoring speed and the other oriented towards accuracy. It has been demonstrated that the choice between these options may depend on the task structure of the sport.

Studies have shown that karate athletes prefer a higher speed of response to a stimulus but make more mistakes, while rowers do the opposite [31]. In this study, both groups demonstrated improved performance on psychometric tests in subsequent attempts. This phenomenon is known as the effect of learning a response to the same repeated stimuli [38]. In the case of executive functions used during fighting with an opponent the athlete does not know yet, there is a very large variety of stimuli and many choices of responses to them, which minimizes the learning effect observed during repeated laboratory testing. For this reason, in addition to physical fitness, the outcome of the competition is determined by the level of technical and tactical skills.

There have been few attempts to numerically assess the level of specific executive functions as a technical skill. Three basic parameters that characterize the level of performance during a real kickboxing bout have been developed and implemented [32,33,39,40]. The literature to date lacks a comparison of the assessment of such technical skills with measurements of physiological responses during competitions in combat sports, with athletes using fist punches and/or kicks. Ass mentioned earlier, boxing, kickboxing and taekwondo athletes are exposed to head injuries during competitions, which can reduce cognitive and executive abilities and impair technical skills [41,42]. In addition, three repeated maximal physical efforts may contribute to physiological changes [43,44], which are, in part, responsible for accumulation of fatigue. The most visible symptom of increasing fatigue during successive bouts may be an increasing number of pauses and total time of pauses [45]. Although the analysis of the used types of offensive actions and their number during kickboxing matches have been presented in the literature, the novel skill parameters as a mirror of the levels of cognitive–executive function, together with complex physiological responses to the contest, are lacking. Thus, the aim of our study was to analyze the changes in ABB after a three-round kickboxing bout and the level of technical and tactical skills presented during the bout.

2. Materials and Methods

The study involved 14 kickboxers presenting the highest sports skill level (champion level).

The sports skill level was evaluated based on sporting achievements and having a kickboxing master's degree, and the coach's opinion. The minimum training experience of the subjects was between 8 and 10 years. The participants were aged 19 to 35 years. Details of the study group are presented in Table 1.

Table 1. Anthropometric measures of study participants.

Variables	No	M	Me	Min	Max	Q1	Q3	SD
Body mass	14	84.90	85.50	75.00	90.00	83.00	88.50	4.93
Body height	14	181.05	180.00	175.00	189.00	179.00	183.50	3.39
BMI	14	26.04	25.99	24.12	28.64	25.15	26.73	1.24

No—number, M—mean, Me—median, Min—minimum, Max—maximum, Q1—first quartile, Q3—third quartile, SD—standard deviation.

2.1. Analysis of the Fight

The competitors had one bout each according to K1 rules in the morning after two days of a training break. The fights were held according to the rules of the World Association of Kickboxing Organizations (WAKO) and consisted of three 2-min rounds separated by 1-min rests.

The fights were simulated, but took place on neutral ground and were refereed by a qualified referee. The athletes were matched in a manner consistent with their weight category.

The determination of the technical and tactical performance parameters was made based on video recordings of the bouts. Subsequent analysis was conducted using specialized formulas [40,46,47].

The efficiency of the attack indicates a number of scored points influencing the final result of the bout compared to the number of bouts observed.

Efficiency of the attack (S_a)

$$S_a = \frac{n}{N}$$

n—number of attacks scoring 1 point.
* In K1 rules, each clean hit of the opponent scores 1 point
N—sum of observed bouts ($N = 1$ in this study)

The effectiveness of the attack denotes the number of scoring techniques compared to all the offensive actions performed.

Effectiveness of the attack (E_a)

$$E_a = \frac{number\ of\ effective\ attacks}{number\ of\ all\ attacks} \times 100$$

An effective attack is a technical action awarded a point
Number of all attacks is a number of all offensive actions

The activeness of the attack describes the engagement of the athlete, indicating the number of offensive actions performed during the observed fights.

Activeness in the attack (A_a)

$$A_a = \frac{number\ of\ all\ registered\ offensive\ actions\ of\ a\ kickboxer}{number\ of\ fights\ fought\ by\ a\ kickboxer\ (1\ in\ this\ study)}$$

2.2. *Acid–Base Balance Analysis*

ABB parameters were analyzed using an EPOC gasometer (Siemens, Ottawa, ON, Canada) immediately after 95 μL of arterialized fingertip blood was drawn into glass capillaries containing calcium-balanced lithium heparin (65 IU/mL). The determinations were made 5 min before the bout (measurement I), and 3 min (measurement II) and 20 min after the bout (measurement III).

Hydrogen ion concentration (H^+), partial pressure of oxygen (pO_2), and partial pressure of carbon dioxide (pCO_2) were measured, and base excess in the extracellular fluid (BEecf) concentration of bicarbonate ions HCO_3^- and TCO_2 (total molar carbon dioxide concentration) were calculated.

2.3. *Bioethics Committee*

Prior to participation in the tests, the competitors were informed about the research procedures, which were in accordance with the ethical principles of the Declaration of Helsinki WMADH (2000). Obtaining the competitors' written consent was the condition for their participation in the project. The research was approved by the Bioethics Committee at the Regional Medical Chamber (No. 287/KBL/OIL/2020).

2.4. *Statistical Analysis*

Statistica 13.1 software (StatSoft, Cracow, Poland) was employed for statistical analysis. Friedman's ANOVA test was used to compare the results of repeated measures. The post-hoc test was Dunn's test. Correlation analysis between selected variables was performed using Pearson's linear correlation test. The Shapiro–Wilk test was used to test data for normal distribution. Effect size was calculated according to the formula:

Kendall's W = $Chi^{2}/N(K - 1)$, N = sample size, K = number of measurements. The level of statistical significance was set at $p < 0.05$.

3. Results

The biochemical indices studied changed significantly during the bout and recovery. The greatest changes were observed after the second measurement (3 min after the bout) (Table 2). For pCO_2, the difference was significant only between measurements I and II, for H^+, pO_2, HCO_3^- between measurements I and II and between measurements II and III, and for BE between all measurements (Table 2).

Table 2. The level of acid–base balance parameters in the tested group of athletes in three consecutive measurements.

Parameter	Measurement									Friedman's ANOVA		Post-Hoc (Dunn's Test)	Effect Size
	I (*n* = 14)			II (*n* = 14)			III (*n* = 14)						
	M	Me	SD	M	Me	SD	M	Me	SD	Chi[2]	*p*	I-II	I-II
H^+ (nmol/L)	37.9	37.0	3.3	54.0	49.0	9.8	41.1	40.0	3.9	22.29	<0.001	<0.05	0.80
pCO_2 (mmHg)	37.2	37.3	3.3	31.8	31.9	2.6	35.2	35.0	0.7	7.43	0.024	<0.05	0.27
pO_2 (mmHg)	77.2	75.2	6.0	85.6	85.1	8.5	73.9	75.8	4.5	16.15	<0.001	<0.05	0.58
HCO_3^- (mmol/L)	24.6	25.3	1.3	14.9	15.4	1.6	21.3	21.6	1.8	24.57	<0.001	<0.05	0.88
BE mmol/L	0.5	0.9	1.2	−11.9	−10.6	2.7	−3.7	−3.2	2.4	28.00	<0.001	<0.05	1.00
TCO_2(mmol/L)	24.1	25.1	1.3	15.8	16.1	1.4	21.5	21.7	1.1	24.50	<0.001	<0.05	0.88

M—mean, Me—median, SD—standard deviation. NS—not statistically significant, I—before exercise, II—3 min after exercise, III—20 min after exercise.

In most cases, the differences between measurements I and II were highest. They were also relatively high between measurements II and III, while the smallest (non-significant) differences were observed between measurements I and III (Table 2).

The recovery rate 20 min after the end of the bout was highest for H^+ and was $96.97 \pm 45.94\%$, whereas for the other ABB parameters the rate reached $68.57 \pm 18.44\%$ for HCO_3^- and $68.30 \pm 13.62\%$ for BE.

The activeness of the attack was evaluated at a mean level of 96.9 ± 43.6 and the range of scores was from 68 to 198. For the efficiency of the attack, it was a mean score of 50.1 ± 12.8, and ranged from 37 to 79, whereas for the effectiveness of the attack, it was a mean of 54.5 ± 7.9, ranging from 39.9 to 64.5 (Table 3).

Table 3. Value of activeness, efficiency, and effectiveness of the attack.

Variables	No	M	Me	Min	Max	Q1	Q3	SD
Activeness of the attack	14	96.9	79.0	68.0	198.0	76.0	96.0	43.6
Efficiency of the attack	14	50.1	47.0	37.0	79.0	45.0	49.0	12.8
Effectiveness of the attack	14	54.5	54.4	39.9	64.5	49.0	60.8	7.9

There were no significant correlations between the changes in the parameters induced by the bouts for $[H^+]$, pCO_2, pO_2, HCO_3^-, BE (ecf), or BE (b). The only positive correlation was found between molar concentrations of CO_2 (TCO_2) and the activity of the attack, which suggests that the greater the physical activity, the greater rise of CO_2 concentration in the blood (Table 4).

Table 4. Matrix of correlation coefficients between examined variables.

Variables		Activeness of the Attack		Efficiency of the Attack		Effectiveness of the Attack	
		R	*p*	R	*p*	R	*p*
Δ = I-II	$[H^+]$	0.11	0.62	0.07	0.81	0.07	0.808
	pCO_2 (mmHg)	0.14	0.14	−0.03	0.90	0.00	1.00
	pO_2 (mmHg)	0.14	0.62	−0.32	0.26	−0.03	0.90
	HCO_3^- (mmol/L)	−0.21	0.46	−0.25	0.38	0.32	0.26
	BE (ecf) mmol/L	−0.10	0.71	−0.01	0.95	−0.41	0.13
	TCO_2 mmol/L	0.64	0.01	−0.32	0.26	−0.17	0.54

Valuesin bold are statistically significant

4. Discussion

The significant changes in ABB parameters and blood oxygen and carbon dioxide saturation immediately after the bout indicate a large contribution of anaerobic metabolism in generating the physical work and gas exchange rate in kickboxers. The examinations conducted by other authors in kickboxers immediately after each of three rounds or after an entire bout showed a significant decline in muscle strength, and a progressive increase in blood lactate levels and heart rates [43–45,48]. In our study there were very weak correlations between activity in attacks and rise of hydrogen ion levels, but a significant positive link between activity and post bout increase of total CO_2 level. That relationship may suggest relatively high mechanical efficiency in tested contestants, according to the model describing relationships between mechanical work output and metabolism [49]. As was mentioned earlier, kickboxing belongs to a family of hitting martial arts similar to Muay Thai, karate and taekwondo, where upper and lower limbs are engaged in offensive actions [50]. Practitioners of all these sports use the same techniques of punches and kicks. One of the most effective kicking techniques is the roundhouse kick, when it is performed correctly. Biomechanical factors of this kick have been explored and described in detail [51]. To date, the literature lacks data on ABB and gasometric studies after kickboxing bouts. Few and fragmented data on ABB and gasometry have been published after boxing fights [13]. This sport differs from kickboxing in its task structure and variety of attacks, but the requirements for general physical fitness, technical skills, and the type of attacks using the upper limbs are very similar. Bout intensity assessed based on lactate and post-exercise changes in magnitude and direction for ABB and gas saturation in boxers [13,22] are also consistent with our findings. We also conducted our ABB and gasometry observations during the short-term recovery period. The results showed that there were no statistically significant differences between baseline and post-exercise recovery measured at 20 min after the bout. However, the results showed that the normalization of parameters was not fully achieved, which indicates a deep disruption of homeostasis caused by the bout. Glycolytic metabolism exercise significantly decreases the levels of bicarbonate as a main factor in neutralizing hydrogen ions in the blood. This process occurs according to the following equation: $H^+ + HCO_3^- \rightarrow H_2O + CO_2 \uparrow$. Despite the increased release of carbon dioxide into the blood, its saturation decreased during the bout due to increased gas exchange in the alveoli and intensifying hyperventilation [52,53]. Decreases in blood pCO_2 and bicarbonate levels during intense exercise have also been reported by previous researchers. Similar post-exercise changes in ABB and blood gas parameters were noted in non-athletes and athletes, but a slight increase in oxygen saturation (by 14%) was observed only in athletes [54]. We found a similar (although slightly smaller, ca. 10%) increase in oxygen saturation after a kickboxing bout. This may suggest that regular physical training induces such an adaptive mechanism. It is important to mention that under resting conditions, hyperventilation is responsible for blood alkalization, since in the case of negligible lactate levels and the associated source of hydrogen ions, the main H^+ donor is the reversible reaction of $CO_2 + H_2O \leftrightarrow HCO_3^- + H^+$. It has been shown that short-term resting hyperventilation at a mean lactate concentration of 1.9 mmol/L leads to a decrease in the partial pressure of CO_2 to a mean value of 21 mmHg and significant blood alkalization (pH = 7.6, H^+ = 23 nmol/L). During competitions in combat sports, it is not possible to quantify the level of hyperventilation. We also did not measure the ventilation immediately before the bout. Hyperventilation attenuates the post-exercise decrease in pH, reduces CO_2 saturation, and increases anaerobic power, especially at the end of a set of efforts [55]. A beneficial effect of pre-exercise hyperventilation before a competition on short-distance swimming performance has also been reported [56]. The reduction in CO_2 saturation and hydrogen ion concentration due to hyperventilation improves physical performance during repeated resistance efforts [55]. Therefore, as documented, the beneficial effect of exercise-induced hyperventilation results from physiological responses. Blood alkalization obtained pharmacologically prior to graded exercise has been found to delay the onset of hyperventilation during work, which, according to researchers, confirms that lactate acidosis increases respiratory rate [57].

The appropriate level of cognitive functions is of great importance in open skill sports. The results of psychometric measurements, reaction time, and decision-making in female and male kickboxers have shown differentiation depending on sex and rules of competition (light vs. full contact) [58], but, to date, there has been no comparison of the results of laboratory psychometric tests with the assessment of executive functions, i.e., the technical performance during a real bout. The few studies conducted in kickboxers have only been designed to assess activeness during the bout. A slightly higher number of higher- and lower-intensity actions was demonstrated in light-category athletes [45] and a higher ratio of activity time to rest time was found in another study [59]. Noticeable differences have been reported in activity and offensive fighting style in winners and losers based on the number of combined punches, kicks, and alternating hand and leg actions [60].

5. Conclusions

1. The disturbances in ABB and changes in blood oxygen and carbon dioxide saturation observed immediately after a bout indicate that anaerobic metabolism plays a large part in kickboxing fights. Anaerobic training should be included in strength and conditioning programs for kickboxers to prepare the athletes for the physiological requirements of sports combat.
2. K1 kickboxers must be characterized by good metabolic acidosis tolerance and the ability to fight effectively despite ABB disturbances, and show good post-exercise recovery.

Author Contributions: Conceptualization, Ł.R.,T.A., M.M. (Mateusz Mardyła); methodology, Ł.R., T.A., M.M. (Mateusz Mardyła), Z.O.; software, Ł.R., T.A.; validation, Ł.R., T.A.; formal analysis, Ł.R., T.A., J.J.-N.; investigation, Ł.R., T.A., W.C.; resources, Ł.R., T.A., M.M. (Mateusz Mardyła), M.M. (Marcin Maciejczyk); data curation, Ł.R., T.A., M.W.; writing—original draft preparation, Ł.R., T.A., M.M. (Mateusz Mardyła), J.J.-N.; writing—review and editing, Ł.R., T.A., Z.O., M.M. (Marcin Maciejczyk), M.W.; visualization, Ł.R., T.A., M.W., M.M. (Marcin Maciejczyk); supervision, Ł.R., T.A., W.C., J.J.-N.; project administration, Ł.R.; funding acquisition, Ł.R., T.A., M.M. (Marcin Maciejczyk), W.C. All authors have read and agreed to the published version of the manuscript.

Funding: Open Access financed by the programme of the Minister of Science and Higher Education entitled'Regional Initiative for Perfection' within the years 2019–2022, project No. 022/RID/2018/19, in the total of 11919908 PLN.

Institutional Review Board Statement: The research was approved by the Bioethics Committee at the Regional Medical Chamber (No. 287/KBL/OIL/2020).

Informed Consent Statement: Informed consent was obtained from all subjects involved in the study.

Data Availability Statement: The data presented in this study are available on request from the corresponding author.

Acknowledgments: Thank you to all the fighters who made the survey research.

Conflicts of Interest: The authors declare no conflict of interest.

References

1. Hamm, L.L.; Nakhoul, N.; Hering-Smith, K.S. Acid-Base Homeostasis. *Clin. J. Am. Soc. Nephrol.* **2015**, *10*, 2232–2242. [CrossRef]
2. Seifter, J.L.; Chang, H.-Y. Disorders of Acid-Base Balance: New Perspectives. *Kidney Dis.* **2016**, *2*, 170–186. [CrossRef] [PubMed]
3. Seifter, J.L.; Chang, H.-Y. Extracellular Acid-Base Balance and Ion Transport Between Body Fluid Compartments. *Physiology* **2017**, *32*, 367–379. [CrossRef]
4. Belfry, G.R.; Raymer, G.H.; Marsh, G.D.; Paterson, D.H.; Thompson, R.T.; Thomas, S.G. Muscle metabolic status and acid-base balance during 10-s work: 5-s recovery intermittent and continuous exercise. *J. Appl. Physiol.* **2012**, *113*, 410–417. [CrossRef]
5. Duffin, J. Role of acid-base balance in the chemoreflex control of breathing. *J. Appl. Physiol.* **2005**, *99*, 2255–2265. [CrossRef]
6. Langer, T.; Zanella, A.; Caironi, P. Understanding the role of the cerebrospinal fluid in acid–base disorders. *Intensive Care Med.* **2016**, *42*, 436–439. [CrossRef]
7. Wiecek, M.; Maciejczyk, M.; Szymura, J.; Szygula, Z. Changes in oxidative stress and acid-base balance in men and women following maximal-intensity physical exercise. *Physiol. Res.* **2015**, *64*, 93–102. [CrossRef]
8. Wiecek, M.; Szymura, J.; Maciejczyk, M.; Kantorowicz, M.; Szygula, Z. Anaerobic Exercise-Induced Activation of Antioxidant Enzymes in the Blood of Women and Men. *Front. Physiol.* **2018**, *9*, 1006. [CrossRef] [PubMed]
9. Russell, M.; Kingsley, M.I.C. Changes in Acid-Base Balance During Simulated Soccer Match Play. *J. Strength Cond. Res.* **2012**, *26*, 2593–2599. [CrossRef]
10. Vilmi, N.; AyrAma, S. Oxygen Uptake, Acid-Base Balance and Anaerobic Energy System Contribution in Maximal 300–400 M Running in Child, Adolescent and Adult Athletes. *J. Athl. Enhanc.* **2016**, *5*. [CrossRef]
11. Ratel, S.; Duche, P.; Hennegrave, A.; Van Praagh, E.; Bedu, M. Acid-base balance during repeated cycling sprints in boys and men. *J. Appl. Physiol.* **2002**, *92*, 479–485. [CrossRef]
12. Beneke, R.; Hütler, M.; Jung, M.; Leithäuser, R.M. Modeling the blood lactate kinetics at maximal short-term exercise conditions in children, adolescents, and adults. *J. Appl. Physiol.* **2005**, *99*, 499–504. [CrossRef]
13. Hanon, C.; Savarino, J.; Thomas, C. Blood Lactate and Acid-Base Balance of World-Class Amateur Boxers After Three 3-Minute Rounds in International Competition. *J. Strength Cond. Res.* **2015**, *29*, 942–946. [CrossRef]
14. Westerblad, H. Acidosis Is Not a Significant Cause of Skeletal Muscle Fatigue. *Med. Sci. Sport. Exerc.* **2016**, *48*, 2339–2342. [CrossRef]
15. Hietavala, E.-M.; Stout, J.R.; Hulmi, J.J.; Suominen, H.; Pitkänen, H.; Puurtinen, R.; Selänne, H.; Kainulainen, H.; Mero, A.A. Effect of diet composition on acid–base balance in adolescents, young adults and elderly at rest and during exercise. *Eur. J. Clin. Nutr.* **2015**, *69*, 399–404. [CrossRef]
16. Gough, L.A.; Brown, D.; Deb, S.K.; Sparks, S.A.; McNaughton, L.R. The influence of alkalosis on repeated high-intensity exercise performance and acid–base balance recovery in acute moderate hypoxic conditions. *Eur. J. Appl. Physiol.* **2018**, *118*, 2489–2498. [CrossRef]
17. Chycki, J.; Golas, A.; Halz, M.; Maszczyk, A.; Toborek, M.; Zajac, A. Chronic Ingestion of Sodium and Potassium Bicarbonate, with Potassium, Magnesium and Calcium Citrate Improves Anaerobic Performance in Elite Soccer Players. *Nutrients* **2018**, *10*, 1610. [CrossRef] [PubMed]
18. Chycki, J.; Kurylas, A.; Maszczyk, A.; Golas, A.; Zajac, A. Alkaline water improves exercise-induced metabolic acidosis and enhances anaerobic exercise performance in combat sport athletes. *PLoS ONE* **2018**, *13*, e0205708. [CrossRef] [PubMed]
19. Dalle, S.; De Smet, S.; Geuns, W.; Van Rompaye, B.; Hespel, P.; Koppo, K. Effect of Stacked Sodium Bicarbonate Loading on Repeated All-out Exercise. *Int. J. Sports Med.* **2019**, *40*, 711–716. [CrossRef] [PubMed]
20. Durkalec-Michalski, K.; Nowaczyk, P.M.; Adrian, J.; Kamińska, J.; Podgórski, T. The influence of progressive-chronic and acute sodium bicarbonate supplementation on anaerobic power and specific performance in team sports: A randomized, double-blind, placebo-controlled crossover study. *Nutr. Metab. (Lond.)* **2020**, *17*, 1–15. [CrossRef]
21. Kamińska, J.; Podgórski, T.; Rachwalski, K.; Pawlak, M. Does the Minerals Content and Osmolarity of the Fluids Taken during Exercise by Female Field Hockey Players Influence on the Indicators of Water-Electrolyte and Acid-Basic Balance? *Nutrients* **2021**, *13*, 505. [CrossRef]
22. Siegler, J.C.; Hirscher, K. Sodium Bicarbonate Ingestion and Boxing Performance. *J. Strength Cond. Res.* **2010**, *24*, 103–108. [CrossRef]
23. Lopes-Silva, J.P.; Da Silva Santos, J.F.; Artioli, G.G.; Loturco, I.; Abbiss, C.; Franchini, E. Sodium bicarbonate ingestion increases glycolytic contribution and improves performance during simulated taekwondo combat. *Eur. J. Sport Sci.* **2018**, *18*, 431–440. [CrossRef] [PubMed]
24. Durkalec–Michalski, K.; Zawieja, E.E.; Zawieja, B.E.; Michałowska, P.; Podgórski, T. The gender dependent influence of sodium bicarbonate supplementation on anaerobic power and specific performance in female and male wrestlers. *Sci. Rep.* **2020**, *10*, 1–12. [CrossRef]
25. Coco, M.; Buscemi, A.; Guerrera, C.S.; Di Corrado, D.; Cavallari, P.; Zappalà, A.; Di Nuovo, S.; Parenti, R.; Maci, T.; Razza, G.; et al. Effects of a Bout of Intense Exercise on Some Executive Functions. *Int. J. Environ. Res. Public Health* **2020**, *17*, 898. [CrossRef]
26. Obmiński, Z.; Mroczkowska, H.; Tomaszewski, W. Relationships between personality traits, resting serum hormones and visuomotor ability in male judokas. *Ann. Agric. Environ. Med.* **2016**, *23*, 79–83. [CrossRef] [PubMed]

27. Darby, D.; Moriarity, J.; Pietrzak, R.; Kutcher, J.; McAward, K.; McCrory, P. Prediction of winning amateur boxers using pretournament reaction times. *J. Sport. Med. Phys. Fit.* **2014**, *54*, 509–519. [CrossRef]
28. Gierczuk, D.; Bujak, Z.; Cieśliński, I.; Lyakh, V.; Sadowski, J. Response Time and Effectiveness in Elite Greco-Roman Wrestlers Under Simulated Fight Conditions. *J. Strength Cond. Res.* **2018**, *32*, 3433–3440. [CrossRef]
29. Supiński, J.; Obmiński, Z.; Kubacki, R.; Kosa, J.; Moska, W. Usefulness of the psychomotor tests for distinguishing the skill levels among older and younger judo athletes. *Arch. Budo* **2014**, *10*, 315–322.
30. Obminski, Z.; Litwiniuk, A.; Staniak, Z.; Zdanowicz, R.; Weimo, Z. Intensive specific maximal judo drills improve psycho-motor ability but may impair hand grip isometric strength. *Ido Mov. Cult.* **2015**, *15*, 52–58. [CrossRef]
31. Obmiński, Z.; Supiński, J.; Mroczkowska, H.; Borkowski, L.; Zdanowicz, R. The effect of aerobic fitness on acute changes in cognitive functions and blood hormones after an exhaustive effort. *Pol. J. Sport Med.* **2017**, *33*, 97–106.
32. Tanriverdi, F.; Suer, C.; Yapislar, H.; Kocyigit, I.; Selcuklu, A.; Unluhizarci, K.; Casanueva, F.F.; Kelestimur, F. Growth hormone deficiency due to sports-related head trauma is associated with impaired cognitive performance in amateur boxers and kickboxers as revealed by P300 auditory event-related potentials. *Clin. Endocrinol. (Oxf.)* **2013**, *78*, 730–737. [CrossRef]
33. Kim, G.H.; Kang, I.; Jeong, H.; Park, S.; Hong, H.; Kim, J.; Kim, J.Y.; Edden, R.A.E.; Lyoo, I.K.; Yoon, S. Low Prefrontal GABA Levels Are Associated With Poor Cognitive Functions in Professional Boxers. *Front. Hum. Neurosci.* **2019**, *13*. [CrossRef]
34. Russo, G.; Ottoboni, G. The perceptual–Cognitive skills of combat sports athletes: A systematic review. *Psychol. Sport Exerc.* **2019**, *44*, 60–78. [CrossRef]
35. Chmura, J.; Nazar, K.; Kaciuba-Uścilko, H. Choice Reaction Time During Graded Exercise in Relation to Blood Lactate and Plasma Catecholamine Thresholds. *Int. J. Sports Med.* **1994**, *15*, 172–176. [CrossRef] [PubMed]
36. Chmura, J.; Nazar, K. Parallel changes in the onset of blood lactate accumulation (OBLA) and threshold of psychomotor performance deterioration during incremental exercise after training in athletes. *Int. J. Psychophysiol.* **2010**, *75*, 287–290. [CrossRef] [PubMed]
37. Labelle, V.; Bosquet, L.; Mekary, S.; Bherer, L. Decline in executive control during acute bouts of exercise as a function of exercise intensity and fitness level. *Brain Cogn.* **2013**, *81*, 10–17. [CrossRef] [PubMed]
38. Lee, S.-Y.; Bae, S.-S.; Han, J.-T.; Byun, S.-D.; Chang, J.-S. The Effect of Motor Learning of Serial Reaction Time Task (SRTT) Through Action Observation on Mu Rhythm and Improvement of Behavior Abilities. *J. Clin. Med. Res.* **2012**, *4*, 114–118. [CrossRef] [PubMed]
39. Rydzik, Ł.; Niewczas, M.; Kędra, A.; Grymanowski, J.; Czarny, W.; Ambroży, T. Relation of indicators of technical and tactical training to demerits of kickboxers fighting in K1 formula. *Arch. Budo Sci. Martial Arts Extrem. Sport.* **2020**, *16*, 1–5.
40. Rydzik, Ł.; Ambroży, T. Physical fitness and the level of technical and tactical training of kickboxers. *Int. J. Environ. Res. Public Health* **2021**, *18*, 3088. [CrossRef]
41. Di Russo, F.; Spinelli, D. Sport is not always healthy: Executive brain dysfunction in professional boxers. *Psychophysiology* **2010**, *47*, 425–434. [CrossRef]
42. Di Virgilio, T.G.; Ietswaart, M.; Wilson, L.; Donaldson, D.I.; Hunter, A.M. Understanding the Consequences of Repetitive Subconcussive Head Impacts in Sport: Brain Changes and Dampened Motor Control Are Seen After Boxing Practice. *Front. Hum. Neurosci.* **2019**, *13*, 294. [CrossRef] [PubMed]
43. Ouergui, I.; Hammouda, O.; Chtourou, H.; Gmada, N.; Franchini, E. Effects of recovery type after a kickboxing match on blood lactate and performance in anaerobic tests. *Asian J. Sports Med.* **2014**, *5*, 99–107.
44. Karadağ, M. Compare the Values of Blood Lactate and Heart Rate of Kickboxers during Kickboxing Matches. *J. Educ. Train. Stud.* **2017**, *5*, 13–19. [CrossRef]
45. Ouergui, I.; Hssin, N.; Haddad, M.; Franchini, E.; Behm, D.G.; Wong, D.P.; Gmada, N.; Bouhlel, E. Time-Motion Analysis of Elite Male Kickboxing Competition. *J. Strength Cond. Res.* **2014**, *28*, 3537–3543. [CrossRef] [PubMed]
46. Ambroży, T.; Rydzik, Ł.; Obmiński, Z.; Klimek, A.T.; Serafin, N.; Litwiniuk, A.; Czaja, R.; Czarny, W. The Impact of Reduced Training Activity of Elite Kickboxers on Physical Fitness, Body Build, and Performance during Competitions. *Int. J. Environ. Res. Public Health* **2021**, *18*, 4342. [CrossRef] [PubMed]
47. Rydzik, Ł.; Maciejczyk, M.; Czarny, W.; Kędra, A.; Ambroży, T. Physiological Responses and Bout Analysis in Elite Kickboxers During International K1 Competitions. *Front. Physiol.* **2021**, *12*, 737–741. [CrossRef] [PubMed]
48. Salci, Y. The metabolic demands and ability to sustain work outputs during kickboxing competitions. *J. Perform. Anal. Sport* **2015**, *15*, 39–52. [CrossRef]
49. Peyré-Tartaruga, L.A.; Coertjens, M. Locomotion as a Powerful Model to Study Integrative Physiology: Efficiency, Economy, and Power Relationship. *Front. Physiol.* **2018**, *9*, 1789. [CrossRef]
50. Diniz, R.; Del Vecchio, F.B.; Schaun, G.Z.; Oliveira, H.B.; Portella, E.G.; da Silva, E.S.; Formalioni, A.; Campelo, P.C.C.; Peyré-Tartaruga, L.A.; Pinto, S.S. Kinematic Comparison of the Roundhouse Kick Between Taekwondo, Karate, and Muaythai. *J. Strength Cond. Res.* **2021**, *35*, 198–204. [CrossRef]
51. Gavagan, C.J.; Sayers, M.G.L. A biomechanical analysis of the roundhouse kicking technique of expert practitioners: A comparison between the martial arts disciplines of Muay Thai, Karate, and Taekwondo. *PLoS ONE* **2017**, *12*, e0182645. [CrossRef]
52. Péronnet, F.; Aguilaniu, B. Lactic acid buffering, nonmetabolic CO2 and exercise hyperventilation: A critical reappraisal. *Respir. Physiol. Neurobiol.* **2006**, *150*, 4–18. [CrossRef]

53. Stickland, M.K.; Lindinger, M.I.; Olfert, I.M.; Heigenhauser, G.J.F.; Hopkins, S.R. Pulmonary Gas Exchange and Acid-Base Balance During Exercise. In *Comprehensive Physiology*; 2013; pp. 693–739.
54. Chin, L.M.K.; Leigh, R.J.; Heigenhauser, G.J.F.; Rossiter, H.B.; Paterson, D.H.; Kowalchuk, J.M. Hyperventilation-induced hypocapnic alkalosis slows the adaptation of pulmonary O_2 uptake during the transition to moderate-intensity exercise. *J. Physiol.* **2007**, *583*, 351–364. [CrossRef]
55. Sakamoto, A.; Naito, H.; Chow, C.-M. Hyperventilation as a Strategy for Improved Repeated Sprint Performance. *J. Strength Cond. Res.* **2014**, *28*, 1119–1126. [CrossRef] [PubMed]
56. Jacob, C.; Keyrouz, C.; Bideau, N.; Nicolas, G.; El Hage, R.; Bideau, B.; Zouhal, H. Pre-exercise hyperventilation can significantly increase performance in the 50-meter front crawl. *Sci. Sports* **2015**, *30*, 173–176. [CrossRef]
57. Meyer, T. Is lactic acidosis a cause of exercise induced hyperventilation at the respiratory compensation point? *Br. J. Sports Med.* **2004**, *38*, 622–625. [CrossRef] [PubMed]
58. Çetin, M.Ç.; Taşğm, Ö.; Arslan, F. The relationship between reaction time and decision-making in elite kickboxing athletes. *World Appl. Sci. J.* **2011**, *12*, 1826–1831.
59. Slimani, M.; Chaabene, H.; Miarka, B.; Franchini, E.; Chamari, K.; Cheour, F. Kickboxing review: Anthropometric, psychophysiological and activity profiles and injury epidemiology. *Biol. Sport* **2017**, *34*, 185. [CrossRef] [PubMed]
60. Ouergui, I.; Hssin, N.; Franchini, E.; Gmada, N.; Bouhlel, E. Technical and tactical analysis of high level kickboxing matches. *Int. J. Perform. Anal. Sport* **2013**, *13*, 294–309. [CrossRef]

biology

MDPI

Article

High-Intensity Interval Training Improves Cardiac Autonomic Function in Patients with Type 2 Diabetes: A Randomized Controlled Trial

Lucas Raphael Bento Silva [1,2,*], Paulo Gentil [2,3,4], Camila Simões Seguro [5], Jordana Campos Martins de Oliveira [1,2], Maria Sebastiana Silva [3], Vitor Alves Marques [2], Thomas Beltrame [6,7] and Ana Cristina Silva Rebelo [2,3,8]

1 Department of Physical Education, Araguaia University Center, Goiania 74223-060, Brazil; jordana.oliveira@uniaraguaia.edu.br
2 Post-Graduate Program in Health Sciences, Faculty of Medicine, Federal University of Goias, Goiania 74605-050, Brazil; paulogentil@ufg.br (P.G.); vitor_alvesmarques@hotmail.com (V.A.M.); ana_rebelo@ufg.br (A.C.S.R.)
3 Faculty of Physical Education and Dance, Federal University of Goiás, Goiania 74690-900, Brazil; mssilva@ufg.br
4 Hypertension League, Federal University of Goias, Goiania 74605-020, Brazil
5 Post-Graduate Program in Nutrition and Health, Faculty of Nutrition, Federal University of Goias, Goiania 74605-080, Brazil; camila.simoes@discente.ufg.br
6 Institute of Computing, University of Campinas, Campinas 13083-852, Brazil; beltramethomas@gmail.com
7 Department of Physiotherapy, Federal University of Sao Carlos, Sao Carlos 13565-905, Brazil
8 Department of Morphology, Institute of Biological Sciences, Federal University of Goiás, Goiania 74690-900, Brazil
* Correspondence: lucas.silva@uniaraguaia.edu.br

Citation: Silva, L.R.B.; Gentil, P.; Seguro, C.S.; de Oliveira, J.C.M.; Silva, M.S.; Marques, V.A.; Beltrame, T.; Rebelo, A.C.S. High-Intensity Interval Training Improves Cardiac Autonomic Function in Patients with Type 2 Diabetes: A Randomized Controlled Trial. *Biology* **2022**, *11*, 66. https://doi.org/10.3390/biology11010066

Academic Editor: Johannes Vogel

Received: 15 November 2021
Accepted: 28 December 2021
Published: 2 January 2022

Simple Summary: Diabetes mellitus is a metabolic disorder characterized by an increased blood glucose concentration. The most common diabetes is type 2, corresponding to approximately 95% of the diagnosed cases. Chronic hyperglycemia can lead to many complications such as increased incidence of cardiovascular diseases as well as renal and ophthalmologic complications. Physical exercise is seen as an effective non-pharmacological strategy for managing the disease. In the present study, 44 middle-aged adults with type 2 diabetes were recruited and stratified into three exercise groups: HIIT-30:30, HIIT-2:2, and MICT. All patients were submitted to anamnesis, evaluation of cardiorespiratory fitness, and cardiac autonomic modulation, and were submitted to physical exercise programs for eight weeks. From the results found, it was possible to infer that high intensity physical training programs can be safe and effective for patients with type 2 diabetes and might be performed in different phases of a rehabilitation program. However, it is necessary to know how to work with the prescription of these exercises considering its cost effectiveness, because, in this study, the protocols HIIT-2:2 and HIIT-30:30 presented superior benefits to the MICT protocol.

Abstract: Different exercise models have been used in patients with type 2 diabetes mellitus (T2D), like moderate intensity continuous training (MICT) and high intensity interval training (HIIT); however, their effects on autonomic modulation are unknown. The present study aimed to compare the effects of different exercise modes on autonomic modulation in patients with T2D. In total, 44 adults with >5 years of T2D diagnosis were recruited and stratified into three groups: HIIT-30:30 ($n = 15$, age 59.13 ± 5.57 years) that performed 20 repetitions of 30 s at 100% of VO2peak with passive recovery, HIIT-2:2 ($n = 14$, age 61.20 ± 2.88) that performed 5 repetitions of 2 min at 100% of VO2peak with passive recovery, and MICT ($n = 15$, age 58.50 ± 5.26) that performed 14 min of continuous exercise at 70% of VO2peak. All participants underwent anamnesis and evaluation of cardiorespiratory fitness and cardiac autonomic modulation. All protocols were equated by total distance and were performed two times per week for 8 weeks. Group × time interactions were observed for resting heart rate (HRrest) [$F(2.82) = 3.641$; $p = 0.031$] and *SDNN* [$F(2.82) = 3.462$; $p = 0.036$]. Only the HIIT-30:30 group significantly reduced *SDNN* ($p = 0.002$ and 0.025, respectively). HRrest reduced more in the HIIT-30:30 group compared with the MICT group ($p = 0.038$). Group × time interactions were also observed for offTAU [$F(2.82) = 3.146$; $p = 0.048$] and offTMR [$F(2.82) = 4.424$; $p = 0.015$]. The MICT group presented increased values of offTAU compared with the HIIT-30:30 and HIIT-2:2 groups ($p = 0.001$ and 0.013, respectively), representing a slower HR response after eight weeks of intervention. HIIT, specially HIIT-30:30, represents a promising measure for improving autonomic modulation in patients with T2D.

Keywords: type 2 diabetes; physical exercise; high-intensity interval training; cardiac autonomic modulation; heart rate recovery; heart rate variability; aerobic training; health

---

1. Introduction

Projections estimate that, by 2045, the number of people diagnosed with diabetes will reach 693 million worldwide [1]. Type 2 diabetes (T2D) represents 90–95% of diagnoses, is more prevalent in adults over 40 years old, and can be caused by environmental (association with risk factors, such as obesity and sedentary lifestyle) or genetic factors [2]. T2D is associated with an increased risk of mortality, especially oeing to cardiovascular diseases [3].

Among the many problems in the cardiovascular system associated with T2D [4], there seems to be an association between chronic hyperglycemia and changes in cardiovascular autonomic nervous system, as verified by heart rate variability (HRV) indexes [5,6]. In previous studies, our research group showed attenuation of the response of the heart rate recovery (HRR) in patients with T2D and an association between blood glucose levels and the slowed response observed from the amplitude (Amp) indexes, which reflects the angulation of the heart rate (HR) curve after interruption of physical exercise and the tau time constant (τ), representing the time of HR decay in adults with T2D [7].

The association between cardiac autonomic modulation and cardiovascular events, as well as all-cause mortality, in the general population has been described in the current literature; individuals with low values for HRV and HRR present a higher risk of some severe cardiovascular outcome, which seems to associated with a deficiency in response to physiological stress [3,8]. Thus, it seems important to evaluate these variables in populations with different clinical conditions and under stress.

The importance of physical exercise as a non-pharmacological measure for the prevention of T2D [9] has been increasingly acknowledged, and physical exercise may decrease the risk of T2D and its cardiovascular complications [10,11]. Moderate aerobic physical training is related to numerous beneficial effects on glycemic control [12,13] and positive clinical outcomes in individuals with T2D, including a reduction in glycated hemoglobin (Hb1Ac) [14,15], increased oxygen consumption (VO2peak) [16], and an improvement in insulin sensitivity [17]. However, in recent decades, studies revealed that high-intensity interval training (HIIT) has many benefits and might be a good time-efficient approach for exercise [18–20]. However, HIIT might be performed in many different ways, with different combinations of effort and recovery, which might affect its acute and chronic responses [21–24].

Previous studies evaluated the acute effects of three exercise modes in healthy young people [25,26], the protocols tested were based on the intensity at which the VO2max was achieved (WLVO2max): HIIT-4:3 (3 repetitions of 4 min at 90% of WLVO2max and 3 min recovery at 60% of WLVO2max), HIIT-30:30 (29 repetitions of 30 s at 100% of WLVO2max and 30 s of passive recovery between), and MICT (continuous intensity for 21 min at 70% of WLVO2max). According the results, HIIT-4:3 resulted in a higher heart rate and increased ratings of perceived exertion. Moreover, HIIT-4:3 and MICT promoted slower HRR responses when compared with HIIT-30:30, and both also showed exacerbation of sympathetic modulation after physical exercise in HRV measurements. These results suggested that performing longer bouts during HIIT might impose greater cardiovascular stress, while HIIT with sorter bouts might be safer. However, interventions with long-term exercise, as well as their responses in individuals with T2D, require further elucidation.

It is important to compare different types of physical exercise to identify interventions that promote better physiological adaptations and have a better understanding of their cost–benefit. Therefore, the aim of this study was to evaluate the effects of different modes of physical exercise on cardiac autonomic modulation in T2D patients.

2. Materials and Methods

2.1. Study Design

The present study is characterized as an experimental, randomized clinical trial. Sixty patients were selected for the initial tests. The following inclusion criteria were employed: (1) age between 50 and 65 years, (2) diagnosed with T2D for at least five years, and (3) not being involved with regular physical activity over the last 6 months.

Tests were performed in four stages: cardiopulmonary exercise test (CPET) and cardiac autonomic modulation. After CPET, eight patients were excluded, including three with an exercise capacity <6 METS, two with uncontrolled arrhythmias during physical exertion, one with unstable angina, and two with a reduction in SBP less than that of SBP levels at rest and during exercise. After the exclusion of ineligible participants, participants were randomized into three groups using the website www.random.org (accessed on 1 November 2021).

The characteristics of the study participants who were stratified into three groups, including HIIT-30:30 (n = 15), HIIT-2:2 (n = 14), and MICT (n = 15), are shown in Table 1. Six participants did not reach minimal attendance of >85% and were excluded from the analysis.

Table 1. Characteristics, risk factors, and medications of the study volunteers.

Variables	HIIT-30:30 (n = 15)	HIIT-2:2 (n = 14)	MICT (n = 15)
Age, years	59.13 ± 5.57	61.20 ± 2.88	58.50 ± 5.26
Sex (M/F)	6/9	6/8	7/8
Fasting blood glucose	142.70 ± 63.01	145.47 ± 66.00	129.79 ± 57.12
HbA1c, %	9.62 ± 1.90	9.85 ± 2.79	8.50 ± 2.43
Diagnosis time	>5 years	>5 years	>5 years
Biguandas, %	80 (12)	73.33 (11)	71.42 (10)
Medicines for sulphonylurea, %	-	6.66 (1)	21.42 (3)
SGLT2 inhibitors, %	-	13.33 (2)	7.14 (1)
DPP-4 inhibitors, %	6.66 (1)	6.66 (1)	7.14 (1)
GLP-1 analogue, %	6.66 (1)	-	14.28 (2)
Pioglitazone, %	6.66 (1)	-	7.14 (1)
Insulin, %	46.66 (7)	53.33(8)	35.71 (5)
Hypertension, %	100 (15)	100 (15)	100 (14)
Beta-blockers, %	-	-	-
ACE inhibitors, %	46,66 (7)	35,71 (5)	28,57 (4)
Diuretics, %	53,33 (8)	64,28 (9)	46,66 (7)
Dyslipidemia, %	80 (12)	86.66 (13)	78.57 (11)

Percentage values are expressed as % (absolute number), M: male, F: female, HbA1c: glycated hemoglobin, SGLT-2: sodium glucose linked transporter 2, DPP-4: dipeptidil peptidase 4, GLP-1: glucagon-like peptide 1, ACE: angiotensin-stenin-sizing enzyme.

Two patients were excluded because they presented irregular data for the analyses of the determined variables, as shown in Figure 1. The project was approved by the research ethics committee (CEP) of the Institution under Opinion No. 2,667,732 and CAAE No. 54522016.6.0000.5083 and duly registered in the Brazilian Registry of Clinical Trials (ReBEC) under number TRIAL: RBR-4RJGC3.

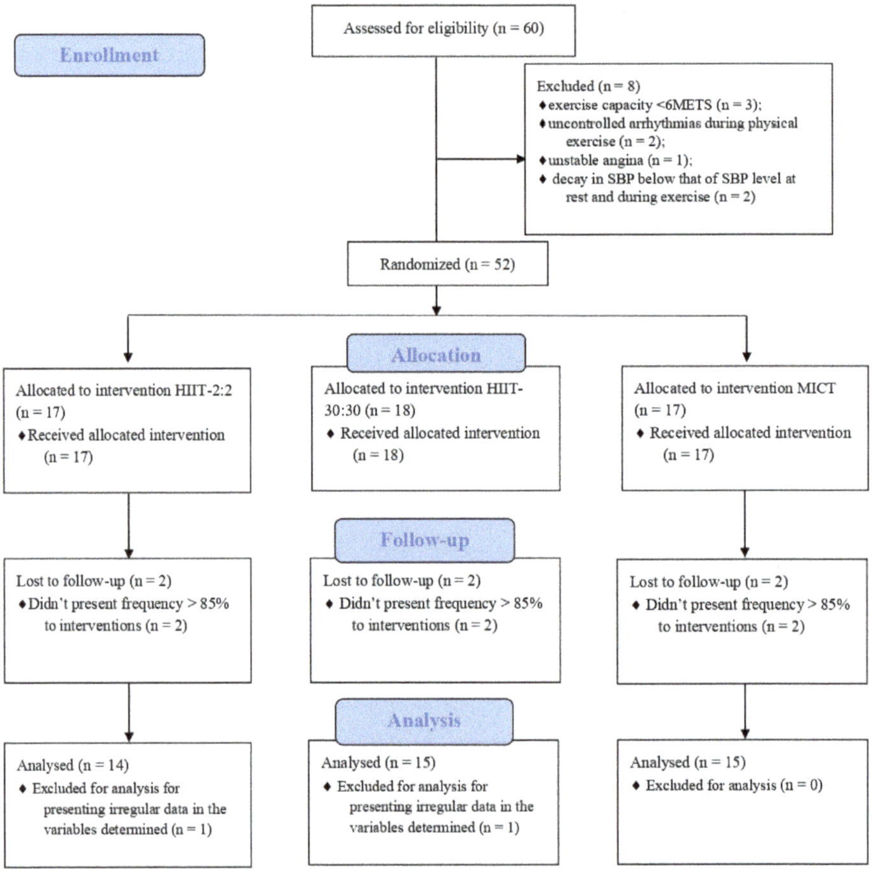

Figure 1. Study flow according to CONSORT recommendations. MICT: moderate intensity continuous training; HIIT: high-intensity interval training; SBP: sistolic blood pressure.

2.2. Test Protocols

The participants made four visits to the laboratories. The first visit occurred to explain the study and familiarise the participants with the procedures and equipment used. In the second visit, blood tests were conducted to confirm the diagnosis of T2D. In the third visit, cardiorespiratory fitness and cardiac autonomic modulation were evaluated. General reassessments were performed after eight weeks of intervention, as shown in Figure 2.

Postmenopausal women who reported drug use for hormone replacement therapy for at least 12 months were included in the study and evaluated during the placebo phase of the medication [6].

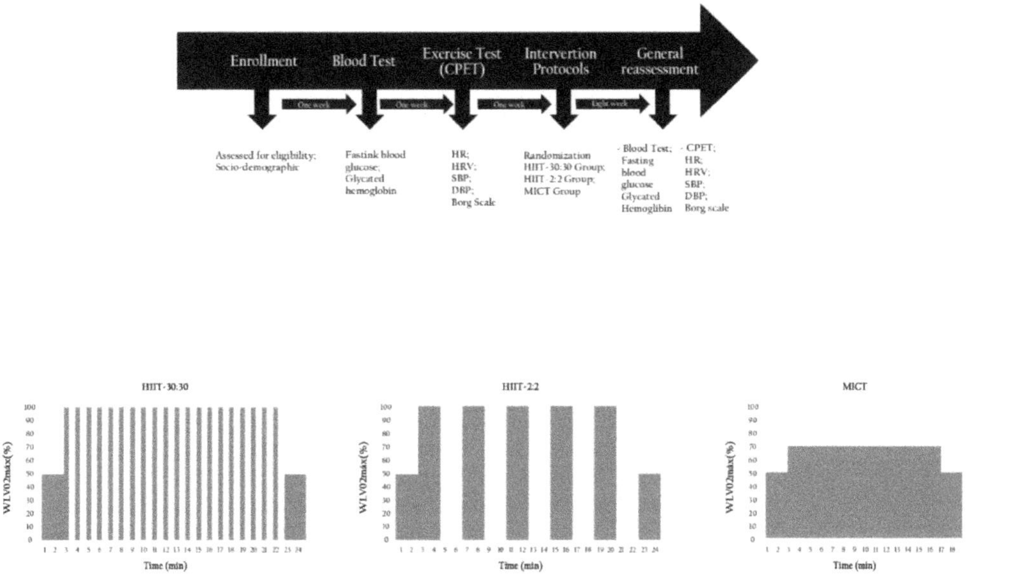

Figure 2. Study design. HR: heart rate; HRV: heart rate variability; SBP: systolic blood pressure; DBP: diastolic blood pressure.

2.2.1. Cardiopulmonary Exercise Test (CPET)

Volunteers were instructed to wear comfortable clothes and avoid vigorous physical exercise (for 24 h before the test), alcohol intake (12 h before the test), and caffeine intake (3 h before the test). All CPETs were performed in the morning to avoid the influence of circadian rhythm on the studied variables. Room temperature (22–24 °C), relative humidity (40–60%), and lighting were controlled according to the preliminary conditions to ensure consistent evaluations. The volunteers were informed regarding protocols, rating of perceived exertion (RPE) scale, and the criteria for interrupting the test.

The ramp-type load increment protocol was applied with a total duration of eight to 12 min. Each volunteer started the test with a warm-up and adaptation for two minutes at 50% speed of the initial values predicted for age and sex. The speed was increased by 0.5 km/h for every 15 s of warm-up.

During the test, the initial treadmill speeds were previously programmed according to age and gender, following the recommendations of the Brazilian Society of Cardiology (SBC). The velocity increased 0.1 km/h every 10, 20, or 30 s, and there was no increase in inclination, as the protocols were executed without inclination.

The active recovery period lasted four minutes at 0% inclination and 50% of the maximum speed reached, and the speed decreased by 10% every 30 s.

Cardiorespiratory fitness was directly assessed through the CPET. The treadmill (Micromed®, Centurion 200, Brasília, Brazil) was coupled to a computer for data processing. Gas analysis was performed using the Cortex analyser®(Metalyser II, Rome, Italy). The calibration of the equipment was performed for barometric pressure, ambient gas, gas mixture, flow, and volume according to the manufacturer's recommendations. During the CPET protocol, data on HR, blood pressure (SBP and DBP), subjective perception of exertion, and ventilatory parameters were collected.

HR was continuously monitored using a heart monitor (Polar v800, Oulu, Finland). Blood pressure was measured by Korotkoff auscultation with a mercury sphygmomanometer (WanMed, São Paulo, SP, Brazil) and a stethoscope (Littman, São Paulo, Minnesota,

USA) in the following positions: supine, sitting, and pre-CPET. To assess the subjective perception of exertion, the Borg scale was used [27].

The criteria for the interruption of CPET were recommended by the American College of Sports Medicine [28] as follows:

(1) discontinuity of stride during the treadmill running phase;
(2) reaching the predicted maximum heart rate (HRmax) for the patient's age;
(3) respiratory exchange ratio > 1.15.

2.2.2. Heart Rate Variability (HRV)

Volunteers were instructed to sit and rest for 10 min to obtain a reliable baseline R-Ri recording. Then, a heart rate monitor (Polar v800, Finland) attached to the chest was used to record the 10-min R–R interval between the rest and supine positions, which was transmitted to the heart monitor in real time.

Subsequently, the data were transferred to Microsoft Excel® (2016 version, Redmond, WA, USA) to remove artifacts. After recording the HRV at rest, the data were transferred to the portable microcomputer via a USB cable. The data obtained were processed by the Kubios HRV version 3.2 software (Kuopio, Finland). Average correction filters and visual data analysis were used for error detection and correction.

For time domain analysis, the *RMSSD* (square root of the square mean of the differences between adjacent normal *RR* intervals in a time interval, expressed in ms) was used as a marker of parasympathetic, and *SDNN* indices (standard deviation of all normal *RR* intervals recorded in a time interval, expressed in ms) were used as the global HRV index. Further, the variable PNN50 represents the percentage of R–R intervals with variation greater than 50 ms, which also corresponds to the vagal modulation of the individual. *SDDN* and rMSSD were calculated according to the following formulas:

$$SDNN = \sqrt{\frac{1}{N-1}\sum_{j=1}^{N}\left(RR_j - \overline{RR}\right)^2} \tag{1}$$

$$RMSSD = \sqrt{\frac{1}{N-1}\sum_{j=1}^{N-1}\left(RR_{j+1} - RR_j\right)^2} \tag{2}$$

where RR_j corresponds to the value of the j-th RR interval, N represents the total number of successive intervals, and $\overline{RR}$ represents the average of the RR intervals.

2.2.3. Symbolic Analysis and Shannon's Entropy

Symbolic analysis was performed by dividing all possible symbolic patterns into four categories: (I) patterns are the same and there is no change (0V, such as 2–2–2 or 4–4–4 or 5–5–5); (II) a single change in the pattern (1V, which is 2–2–3 or 4–2–2); (III) two similar forms of change in the pattern that form ascending or descending lines (2LV, which is 5–3–1 or 1–2–4); and (IV) two distinct variations, in which three symbols form a peak or valley (2UV, such as 1–4–2 or 5–2–4), being evaluated according to the frequency of occurrence (0V%, 1V%, 2LV%, and 2UV%). Previous studies have evaluated that the 0V family represents only sympathetic modulation, 1V represents parasympathetic and sympathetic modulations, and 2LV also represents both parasympathetic and sympathetic modulations being dominated by the valgus nerve. Finally, the 2UV family only represents parasympathetic modulation [29].

Shannon entropy (SE) reflects the complexity of distributing these symbolic patterns. If the distribution is flat (the pattern is evenly distributed), the SE will be high and, when a subset of the patterns is unlikely, nonexistent, or rare, it will be very low [30].

2.2.4. Monoexponential Analyses

The HRR data were obtained after the interruption of the CPET, and filtered and analyzed through a specific procedure developed in the OriginPro 8.0 software (OriginLab, Northampton, MA, USA) that applies an exponential model to the data for the entire recovery period (four minutes of active recovery and three minutes of passive recovery in the orthostatic position) [26,31].

To obtain the best parameters of this exponential curve, a nonlinear algorithm was used that adopts the minimization of the sum of square errors as a convergence criterion. Only the r > 0.95 function was included in the final analyses. The off kinetics was modulated using an exponential function of time as follows:

$$\text{HR (t)} = \text{HRend} - a * (1 - e^{-(t-TD)/t}) \quad (3)$$

where "t" is time; "HRend" is the heart rate at the end of the CPET; "Amp" is the amplitude of HR decrease after the end of the exercise; and "TD" is the delay time for the function. The inclusion of the term "TD" in this function was established owing to the possibility that the HRR is not immediately reduced after the load interruption. Because the parameter "τ" is a time constant in a negative decreasing exponential function, it can be inferred that, the lower its value, the faster the kinetics of the HRR [32].

2.3. *Exercise Protocols*

The protocols were customized with individualized monitoring of HR and their respective exercise intensities were adapted from previous studies [33,34].

The participants performed 2 min of warm-up at 50% of WLVO2max. In the cool down, the participants performed 2 min of recovery at 50% of WLVO2max. In the HIIT-30:30 protocol, participants performed 20 bouts of 30 s at 100% of WLVO2max with 30 s of passive intervals. In the HIIT-2:2 protocol, the participants performed 5 bouts of 2 min at 100% of WLVO2max with passive rest intervals. In the MICT protocol, the participants exercise continuously for 14 min at 70% of WLVO2max, as shown in Figure 2. In each training session, SBP and DBP, HR, and glycemia values were measured and recorded before and 10 min after the end of the session.

2.4. *Statistical Analyses*

We performed an a priori calculation to estimate sample size using GPower 3.1.9.2 (Düsseldorf University, Düsseldorf, Germany). The parameters were as follows: 0.5 effect size (medium), 0.05 alfa level, 0.8 power, 3 groups, 6 measures, and 0.5 correlation among repeated measures. The results suggested a total sample size of 27 for between-group comparisons. However, considering the possible attrition and large number of volunteers, we accepted considerably more participants.

The Shapiro–Wilk test was used to evaluate the normality of the data, and Levene's test was used to assess sample homogeneity. One-way ANOVA was used to verify the differences between the groups at baseline. Two-way ANOVA with repeated measurements was used to verify the inter-group differences. When these differences were found, the post hoc Ryan–Einot–Gabriel–Welsh Q (REGWQ) test was used.

The effect-size for the samples was based on the calculation of the rank-biserial correlation (rB). The rB values were classified according to the classification criteria for Pearson's correlation coefficient and, therefore, trivial ($rB < 0.10$), small ($0.10 \leq rB < 0.30$), medium ($0.30 \leq rB < 0.50$), and large ($rB \geq 0.50$) [35].

A statistical analysis was performed using the statistical program Statistical Package for the Social Sciences (SPSS; Armonk, NY, USA; IBM Corp.), version 21. $p < 0.05$ was considered significant in all analyses.

3. Results

Table 2 presents the results of HR and HRV responses for each group. The two-way ANOVA revealed the effect of the intervention in the variables HRpeak [F(2.82) = 5.091; $p = 0.009$] and pNN50 [F(2.82) = 5.071; $p = 0.008$], with a significant increase observed only in the HIIT 30:30 group (medium effect for variables). Time effect was observed in the variables HRrest [F(1.82) = 7,097; $p = 0.009$], with a reduction observed for HIIT-30:30 group ($r_B = 0.50$—medium effect), R-Ri [F(1.82) = 0.045; $p = 0.001$], pNN50 [F (1.81) = 10.413; $p = 0.002$], and 2UV% [F(1.82) = 9.285; $p = 0.003$], with the HIIT-30:30 group presenting a postintervention increase ($r_B = -0.24$,small effect; -0.39,medium effect; and -0.50, medium effect, respectively).

A group × time interaction was observed for the variables HRrest [F(2.82) = 3.641; $p = 0.031$]; the HIIT-30:30 group exhibited a greater reduction in HRrest than the MICT group ($p = 0.038$), R-Ri [F(2.82) = 4.420; $p = 0.015$]; the post-hoc analysis showed that there was significant diffusion when comparing the MICT group to groups HIIT-30:30 and HIIT-2:2 ($p = 0.007$ and 0.047 , respectively); *SDNN* [F(2.82) = 3.462; $p = 0.036$] showed a significant reduction only in the HIIT-30:30 ($p = 0.025$) and 2UV% [F(2.82) = 3.708; $p = 0.029$] groups; and two groups submitted to high intensity protocols, HIIT-30:30 and HIIT-2:2, showed a significant increase in the variable 2UV% ($p = 0.001$ and 0.025, respectively), which reflects vagal modulation, as observed in the post-hoc analysis.

Table 3 shows the results obtained from the analyses of HR kinetics (off parameters) in the groups in the pre- and post-assessments. A group effect was found for offTAU [F(2.82) = 4.710; $p = 0.012$] and offTMR [F(2.82) = 6.667; $p = 0.002$]. A time effect was found for offAMP [F(1.82) = 4.881; $p = 0.030$]. A group × time interaction was also observed for offTAU [F(2.82) = 3.146; $p = 0.048$] and offTMR [F(2.82) = 4.424; $p = 0.015$]. The MICT group presented higher values of offTAU compared with the HIIT-30:30 and HIIT-2:2 groups ($p = 0.001$—medium effect and 0.013—medium effect, respectively) in the post-intervention, representing increased slowness in the HR response after physical exertion. The MICT group showed a significant increase ($p = 0.012$) in offTMR as a function of time, with greater values than both the HIIT-30:30 and HIIT-2:2 groups ($p < 0.001$—medium effect and 0.008—trivial effect, respectively), indicating a slower HR response time after interruption of physical exercise.

Table 2. Linear and symbolic analysis of heart rate variability in groups submitted to the intervention program.

Variables	HIIT-30:30 (n = 15) Pre	Post	ES	HIIT-2:2 (n = 14) Pre	Post	ES	MICT (n = 15) Pre	Post	ES	Group	Time	Group*Time
HRrest, bpm	82.86 ± 9.64	72.13 ± 8.62	$0.50_{(medium)}$	84.85 ± 10.28	77.92 ± 8.96	$0.33_{(medium)}$	78.86 ± 8.56	80.73 ± 9.43	$-0.10_{(small)}$	0.278	0.009	0.031
HRpeak, bpm	152.46 ± 13.42	160.20 ± 14.06	$-0.56_{(medium)}$	152.00 ± 21.37	158.35 ± 19.26	$-0.31_{(medium)}$	146.40 ± 14.27	141.26 ± 17.46	$0.15_{(small)}$	0.009	0.408	0.274
R-Ri, ms	821.2 ± 147.55	905.0 ± 178.9	$-0.24_{(small)}$	822.7 ± 73.55	866.2 ± 85.7	$-0.26_{(small)}$	867.33 ± 105.0	718.0 ± 275.5	$0.33_{(medium)}$	0.224	0.832	0.015
Linear Anaysis												
SDNN, ms	20.13 ± 9.67	33.80 ± 22.04	$-0.37_{(medium)}$	23.45 ± 15.12	28.65 ± 18.08	$-0.15_{(small)}$	29.22 ± 20.92	21.85 ± 9.11	$0.22_{(small)}$	0.932	0.133	0.036
rMSSD, ms	23.63 ± 13.87	38.39 ± 27.76	$-0.31_{(medium)}$	18.51 ± 9.73	25.07 ± 10.33	$-0.31_{(medium)}$	29.22 ± 23.19	22.16 ± 13.89	$0.18_{(small)}$	0.150	0.217	0.065
pNN50, %	1.92 ± 3.21	9.70 ± 12.31	$-0.39_{(medium)}$	1.02 ± 1.97	3.76 ± 3.61	$-0.42_{(medium)}$	6.00 ± 8.59	12.85 ± 12.75	$-0.30_{(medium)}$	0.008	0.002	0.485
Nonlinear Analysis												
0V%	39.23 ± 25.05	28.20 ± 11.14	$0.27_{(small)}$	41.89 ± 25.50	31.09 ± 14.65	$0.25_{(small)}$	38.33 ± 20.98	47.59 ± 27.75	$-0.18_{(small)}$	0.226	0.355	0.111
1V%	33.67 ± 14.32	33.78 ± 13.97	$-0.00_{(trivial)}$	31.93 ± 16.75	36.41 ± 12.11	$-0.15_{(small)}$	35.20 ± 12.99	27.41 ± 12.18	$0.29_{(small)}$	0.692	0.718	0.234
2LV%	9.62 ± 9.82	9.54 ± 5.58	$0.00_{(trivial)}$	8.12 ± 5.16	8.37 ± 6.25	$-0.02_{(trivial)}$	7.71 ± 7.37	7.05 ± 7.15	$0.04_{(trivial)}$	0.482	0.918	0.970
2UV%	17.46 ± 10.72	28.47 ± 7.88	$-0.50_{(medium)}$	16.78 ± 9.64	24.57 ± 7.00	$-0.41_{(medium)}$	18.96 ± 8.21	17.73 ± 9.96	$0.06_{(trivial)}$	0.146	0.003	0.029
SE	2.90 ± 1.29	3.18 ± 1.14	$-0.11_{(small)}$	2.65 ± 1.38	2.94 ± 0.90	$-0.12_{(small)}$	3.14 ± 1.01	2.68 ± 1.02	$0.22_{(small)}$	0.724	0.888	0.380

HIIT: high intensity interval training, MICT: moderate continuous intensity training, Group*Time: corresponds to the group x time interaction, HRrest: resting heart rate in supine position, HRpeak: heart rate at peak physical exercise, iR–R: intervals R–R, *SDDN*: standard deviation of all normal *RR* intervals recorded in a time interval, expressed in milliseconds, rMSSD: square root of the mean square differences between adjacent normal *RR* intervals, in a time interval, expressed in milliseconds, PNN50: percentage of adjacent IRR with duration differences greater than 50 ms,0V%: percentage of pattern without variation, 1V%: percentage of pattern with one variation, 2LV%: percentage of pattern with two like variation, 2UV%: percentage of pattern with two unlike variations, SE: Shannon entropy, ES: effect size.

Table 3. Evaluation of HR kinetics at the beginning and end of the intervention program.

Variables	HIIT-30:30 (n = 15) Pre	Post	ES	HIIT-2:2 (n = 14) Pre	Post	ES	MICT (n = 15) Pre	Post	ES	Group	Time	Group*Time
*off*AMP	160.21 ± 16.77	179.73 ± 29.88	$-0.37_{(medium)}$	160.17 ± 25.00	175.36 ± 44.06	$-0.20_{(small)}$	153.21 ± 20.21	172.53 ± 67.22	$-0.18_{(small)}$	0.756	0.030	0.972
*off*TAU	123.19 ± 65.98	72.18 ± 29.20	$0.44_{(medium)}$	131.39 ± 71.05	92.05 ± 33.17	$0.33_{(medium)}$	135.08 ± 99.20	170.44 ± 98.46	$-0.17_{(small)}$	0.012	0.237	0.048
*off*TMR	140.36 ± 75.72	96.95 ± 23.61	$0.36_{(medium)}$	142.44 ± 67.87	136.24 ± 30.84	$0.05_{(trivial)}$	153.00 ± 100.63	224.82 ± 112.65	$-0.31_{(medium)}$	0.002	0.651	0.015

HIIT: high intensity interval training, MICT: moderate intensity continuous training, Group*Time: corresponds to the group × time interaction, Amp: amplitude, TAU: time constant, TMR: average heart rate response time, ES: effect size.

4. Discussion

The present study aimed to compare the effects of three different physical exercise programmes on cardiac autonomic modulation in patients with T2D. The main results of this study show that 8 weeks of HIIT (especially HIIT-30:30) presented better results in cardiac autonomic modulation compared with MICT. No significant differences were observed between the two HIIT programmes. It is difficult to compare our results with previous studies, as the vast majority of studies to date have chosen only the application of MICT in patients with T2D [36–38].

Previous studies have shown that increased HRrest is associated with all-cause mortality [39–41]. In particular, the study by Prasada et al [42] evaluated individuals with T2D and showed that an increase of one unit in the standard deviation of the HRrest is associated with a 20% increase in the risk of mortality from cardiovascular disease in patients with T2D. In this regard, HIIT-30:30 might be particularly interesting, as it promoted a significant reduction in this parameter, with a greater reduction than those observed in the MICT group. This corroborates previous studies that observed an improvement in HRrest in individuals undergoing HIIT protocols [43,44].

Regarding HRV, the HIIT-30:30 protocol exhibited a 69.7% improvement in the *SDNN* index, and this result was statistically superior to the other protocols. This is consistent with the previous study that followed individuals with metabolic syndrome and found a significant increase in *SDDN* index, especially in the group that realized the HIIT [45]. The HIIT-30:30 and HIIIT-2:2 groups showed increases of 63.05% and 46.42% in the 2UV% index, respectively, which were higher than those observed after MICT. Both HIIT groups showed a tendency to reduce the sympathetic modulation perceived through the 0V index. These suggest a positive autonomic cardiac adaptation after eight weeks of intervention with HIIT.

In the present study, HIIT-30:30 and HIIT-2:2 resulted in a faster decrease in HRR after the CPET when compared with the MICT protocol. No significant differences were noted between the two HIIT programmes. The benefit of HIIT on HRR has been suggested by Dall et al [46] when comparing the responses of 12 weeks of different exercise intensities in cardiac transplant recipients. The HIIT group performed 16 min interval training, with intervals of 4, 2, and 1 min duration and intensity above 80% of VO2peak with active recovery and duration of 2 min with intensity close to 60% of VO2peak. The MICT group performed 45 min of training on a bicycle with intensity between 60% to 70% of VO2peak. The authors observed an improvement in HRR in both groups, with a trend of greater improvement for the group that performed HIIT.

In agreement with our findings, Villelabeitia-Jaureguizar et al. [47] submitted 73 patients with coronary heart disease to eight weeks of physical training. The participants were divided into two groups. The HIIT performed 20 s at 50% of the maximum load and recovery of 40 s to 10% of the maximum load obtained in a steep ramp test, with progressions in the number of repetitions each week. The MICT group performed continuous exercise with the load of the first threshold with progression in time throughout the program. According to the results, the HIIT group presented better values in post-intervention HRR when compared with MICT.

There is already consensus in the literature that an attenuated HRR response is associated with the risk of cardiovascular events and all-cause mortality in the general population; people who have slowing HRR are 68% more likely to be affected by some cardiovascular event and have a 69% greater chance of mortality when compared with those who have a rapid response in HRR [48,49].

In our study, participants who performed eight weeks of HIIT protocols obtained better responses in autonomic modulation and HR parameters when compared with those who performed the MICT protocol. The reasons for these differences are not completely known. However, it might be possible that high intensity exercises influence autonomic nervous system adjustments to the heart and blood vessels to mediate the increased cardiovascular responses and metabolic demands [50]. In this regard, previous studies suggested that dynamic interactions between feed-forward and feedback circuits of central command and exercise pressor reflex are important in determining the long-term adjustments in the sympathetic and parasympathetic nervous systems [50,51]. Considering that these demands are intensity-dependent [52], HIIT might have promoted superior acute stress, resulting in higher adaptation than MICT.

Based on our results, we believe that HIIT should be included as a part of the patients' T2D routine in order to improve cardiac autonomic modulation. Future studies are suggested to evaluate the effects of other modes of HIIT, manipulating exercise types and volume and intensity of training, in order to define the optimal protocol.

Limitations

A limitation of our study was the fact that we did not evaluate HRV for longer periods, such as 12 or 24 h after the physical exercise sessions, in order to verify the cardiovascular stress generated by the intensity of the applied protocol. Another limitation is the absence of nutritional control in the patients.

5. Conclusions

HIIT, especially the HIIT-30:30 protocol, promoted an increase in R-Ri, *SDNN*, pNN50, and 2UV%, in addition to reducing HRrest, compared with continuous moderate intensity training. With these findings, we can suggest that HIIT is a feasible tool and that it can be implemented in cardiovascular rehabilitation programs.

Author Contributions: L.R.B.S. and A.C.S.R.: concept and study design, collected and analyzed data, wrote, reviewed, and edited the manuscript. C.S.S., J.C.M.d.O. and V.A.M.: collected data and edited the manuscript. P.G., M.S.S., T.B. and A.C.S.R.: contributed with data analysis, reviewed and edited the manuscript. All authors contributed to the article and approved. All authors have read and agreed to the published version of the manuscript.

Funding: This research received no external funding.

Institutional Review Board Statement: The studies involving human participants were reviewed and approved by Research Ethics Committee of the Federal University of Goias. The patients/participants provided their written informed consent to participate in this study.

Informed Consent Statement: Informed consent was obtained from all subjects involved in the study.

Data Availability Statement: The raw data supporting the conclusions of this article will be made available by the authors, without undue reservation.

Conflicts of Interest: The authors declare no conflict of interest.

References

1. Cho, N.H.; Shaw, J.E.; Karuranga, S.; Huang, Y.; da Rocha Fernandes, J.D.; Ohlrogge, A.W.; Malanda, B. IDF Diabetes Atlas: Global estimates of diabetes prevalence for 2017 and projections for 2045. *Diabetes Res. Clin. Pract.* **2018**, *138*, 271–281. [CrossRef]
2. Williams, R.; Karuranga, S.; Malanda, B.; Saeedi, P.; Basit, A.; Besançon, S.; Bommer, C.; Esteghamati, A.; Ogurtsova, K.; Zhang, P.; et al. Global and regional estimates and projections of diabetes-related health expenditure: Results from the International Diabetes Federation Diabetes Atlas, 9th edition. *Diabetes Res. Clin. Pract.* **2020**, *162*, 108072. [CrossRef]
3. Shaffer, F.; Ginsberg, J.P. An Overview of Heart Rate Variability Metrics and Norms. *Front. Public Health* **2017**, *5*, 258. [CrossRef]
4. Baldi, J.C.; Wilson, G.A.; Wilson, L.C.; Wilkins, G.T.; Lamberts, R.R. The Type 2 Diabetic Heart: Its Role in Exercise Intolerance and the Challenge to Find Effective Exercise Interventions. *Sports Med.* **2016**, *46*, 1605–1617. [CrossRef]
5. Stockhorst, U.; Huenig, A.; Ziegler, D.; Scherbaum, W.A. Unconditioned and conditioned effects of intravenous insulin and glucose on heart rate variability in healthy men. *Physiol. Behav.* **2011**, *103*, 31–38. [CrossRef]
6. Silva, L.R.B.E.; Zamunér, A.R.; Gentil, P.; Alves, F.M.; Leal, A.G.F.; Soares, V.; Silva, M.S.; Vieira, M.F.; Simões, K.; Pedrino, G.R.; et al. Cardiac Autonomic Modulation and the Kinetics of Heart Rate Responses in the on- and off-Transient during Exercise in Women with Metabolic Syndrome. *Front. Physiol.* **2017**, *8*, 542. [CrossRef]
7. Silva, L.R.B.; Gentil, P.; Seguro, C.S.; de Oliveira, G.T.; Silva, M.S.; Zamunér, A.R.; Beltrame, T.; Rebelo, A.C.S. High Fasting Glycemia Predicts Impairment of Cardiac Autonomic Control in Adults With Type 2 Diabetes: A Case-Control Study. *Front. Endocrinol.* **2021**, *12*, 760292. [CrossRef]

8. Dhoble, A.; Lahr, B.D.; Allison, T.G.; Kopecky, S.L. Cardiopulmonary fitness and heart rate recovery as predictors of mortality in a referral population. *J. Am. Heart Assoc.* **2014**, *3*, e000559. [CrossRef]
9. Pedersen, B.K.; Saltin, B. Exercise as medicine—Evidence for prescribing exercise as therapy in 26 different chronic diseases. *Scand. J. Med. Sci. Sport.* **2015**, *25*, 1–72. [CrossRef]
10. Rasmussen, M.G.; Grøntved, A.; Blond, K.; Overvad, K.; Tjønneland, A.; Jensen, M.K.; Østergaard, L. Associations between Recreational and Commuter Cycling, Changes in Cycling, and Type 2 Diabetes Risk: A Cohort Study of Danish Men and Women. *PLoS Med.* **2016**, *13*, e1002076. [CrossRef]
11. Abrignani, M.G. Physical exercise and risk of arterial hypertension and diabetes mellitus. Let's move, it is never too late. *Eur. J. Prev. Cardiol.* **2018**, *25*, 1063–1064. [CrossRef] [PubMed]
12. Winding, K.M.; Munch, G.W.; Iepsen, U.W.; Van Hall, G.; Pedersen, B.K.; Mortensen, S.P. The effect on glycaemic control of low-volume high-intensity interval training versus endurance training in individuals with type 2 diabetes. *Diabetes Obes. Metab.* **2018**, *20*, 1131–1139. [CrossRef]
13. Cassidy, S.; Vaidya, V.; Houghton, D.; Zalewski, P.; Seferovic, J.P.; Hallsworth, K.; MacGowan, G.A.; Trenell, M.I.; Jakovljevic, D.G. Unsupervised high-intensity interval training improves glycaemic control but not cardiovascular autonomic function in type 2 diabetes patients: A randomised controlled trial. *Diabetes Vasc. Dis. Res.* **2019**, *16*, 69–76. [CrossRef] [PubMed]
14. Helal, L.; Umpierre, D.; Moraes, R.S. High-intensity aerobic interval training improves aerobic fitness and HbA1c among persons diagnosed with type 2 diabetes: Considerations regarding HbA1c starting levels and intervention design. *Eur. J. Appl. Physiol.* **2017**, *117*, 2365–2366. [CrossRef]
15. Kawada, T. Effect of high-intensity aerobic exercise on aerobic fitness and HbA1c in patients with type 2 diabetes. *Eur. J. Appl. Physiol.* **2017**, *117*, 1519–1520. [CrossRef]
16. Støa, E.M.; Meling, S.; Nyhus, L.K.; Glenn, S.; Mangerud, K.M.; Helgerud, J.; Bratland-Sanda, S.; Støren, Ø. High-intensity aerobic interval training improves aerobic fitness and HbA1c among persons diagnosed with type 2 diabetes. *Eur. J. Appl. Physiol.* **2017**, *117*, 455–467. [CrossRef]
17. De Nardi, A.T.; Tolves, T.; Lenzi, T.L.; Signori, L.U.; Silva, A.M.V. da High-intensity interval training versus continuous training on physiological and metabolic variables in prediabetes and type 2 diabetes: A meta-analysis. *Diabetes Res. Clin. Pract.* **2018**, *137*, 149–159. [CrossRef]
18. Guiraud, T.; Nigam, A.; Gremeaux, V.; Meyer, P.; Juneau, M.; Bosquet, L. High-intensity interval training in cardiac rehabilitation. *Sport. Med.* **2012**, *42*, 587–605. [CrossRef]
19. Gillen, J.B.; Gibala, M.J. Is high-intensity interval training a time-efficient exercise strategy to improve health and fitness? *Appl. Physiol. Nutr. Metab.* **2014**, *39*, 409–412. [CrossRef]
20. Souza, D.; Coswig, V.; de Lira, C.A.B.; Gentil, P. H"IT"ting the Barriers for Exercising during Social Isolation. *Biology* **2020**, *9*, 245. [CrossRef]
21. Kilpatrick, M.W.; Martinez, N.; Little, J.P.; Jung, M.E.; Jones, A.M.; Price, N.W.; Lende, D.H. Impact of high-intensity interval duration on perceived exertion. *Med. Sci. Sport Exerc.* **2015**, *47*, 1038–1045. [CrossRef]
22. Martinez, N.; Kilpatrick, M.W.; Salomon, K.; Jung, M.E.; Little, J.P. Affective and Enjoyment Responses to High-Intensity Interval Training in Overweight-to-Obese and Insufficiently Active Adults. *J. Sport Exerc. Psychol.* **2015**, *37*, 138–149. [CrossRef]
23. Viana, R.B.; de Lira, C.A.B.; Naves, J.P.A.; Coswig, V.S.; Del Vecchio, F.B.; Ramirez-Campillo, R.; Vieira, C.A.; Gentil, P. Can We Draw General Conclusions from Interval Training Studies? *Sports Med.* **2018**, *48*, 2001–2009. [CrossRef]
24. Viana, R.B.; Naves, J.P.A.; Coswig, V.S.; de Lira, C.A.B.; Steele, J.; Fisher, J.P.; Gentil, P. Is interval training the magic bullet for fat loss? A systematic review and meta-analysis comparing moderate-intensity continuous training with high-intensity interval training (HIIT). *Br. J. Sports Med.* **2019**, *53*, bjsports-2018-099928. [CrossRef]
25. Naves, J.P.A.; Rebelo, A.C.S.; Silva, L.R.B.E.; Silva, M.S.; Ramirez-Campillo, R.; Ramírez-Vélez, R.; Gentil, P. Cardiorespiratory and perceptual responses of two interval training and a continuous training protocol in healthy young men. *Eur. J. Sport Sci.* **2019**, *19*, 653–660. [CrossRef]
26. Bento Silva, L.R.; Viana Gentil, P.R.; Beltrame, T.; Basso Filho, M.A.; Alves, F.M.; Silva, M.S.; Pedrino, G.R.; Ramirez-Campillo, R.; Coswig, V.; Silva Rebelo, A.C. Exponential model for analysis of heart rate responses and autonomic cardiac modulation during different intensities of physical exercise. *R. Soc. Open Sci.* **2019**, *6*, 190639. [CrossRef]
27. Borg, G.A. Psychophysical bases of perceived exertion. *Med. Sci. Sports Exerc.* **1982**, *14*, 377–381. [CrossRef] [PubMed]
28. Thompson, P.D.; Arena, R.; Riebe, D.; Pescatello, L.S. American College of Sports Medicine ACSM's New Preparticipation Health Screening Recommendations from ACSM's Guidelines for Exercise Testing and Prescription, Ninth Edition. *Curr. Sports Med. Rep.* **2013**, *12*, 215–217. [CrossRef] [PubMed]
29. Porta, A.; Faes, L.; Masé, M.; D'Addio, G.; Pinna, G.D.; Maestri, R.; Montano, N.; Furlan, R.; Guzzetti, S.; Nollo, G.; et al. An integrated approach based on uniform quantization for the evaluation of complexity of short-term heart period variability: Application to 24 h Holter recordings in healthy and heart failure humans. *Chaos* **2007**, *17*, 015117. [CrossRef] [PubMed]
30. Porta, A.; Guzzetti, S.; Montano, N.; Furlan, R.; Pagani, M.; Malliani, A.; Cerutti, S. Entropy, entropy rate, and pattern classification as tools to typify complexity in short heart period variability series. *IEEE Trans. Biomed. Eng.* **2001**, *48*, 1282–1291. [CrossRef] [PubMed]

31. Imai, K.; Sato, H.; Hori, M.; Kusuoka, H.; Ozaki, H.; Yokoyama, H.; Takeda, H.; Inoue, M.; Kamada, T. Vagally mediated heart rate recovery after exercise is accelerated in athletes but blunted in patients with chronic heart failure. *J. Am. Coll. Cardiol.* **1994**, *24*, 1529–1535. [CrossRef]
32. Bearden, S.E.; Moffatt, R.J. $\dot{V}O_2$ and heart rate kinetics in cycling: Transitions from an elevated baseline. *J. Appl. Physiol.* **2001**, *90*, 2081–2087. [CrossRef]
33. Wisløff, U.; Støylen, A.; Loennechen, J.P.; Bruvold, M.; Rognmo, Ø.; Haram, P.M.; Tjønna, A.E.; Helgerud, J.; Slørdahl, S.A.; Lee, S.J.; et al. Superior cardiovascular effect of aerobic interval training versus moderate continuous training in heart failure patients: A randomized study. *Circulation* **2007**, *115*, 3086–3094. [CrossRef]
34. Billat, L.V. Interval training for performance: A scientific and empirical practice. Special recommendations for middle- and long-distance running. Part II: Anaerobic interval training. *Sports Med.* **2001**, *31*, 75–90. [CrossRef]
35. Cohen, J. A power primer. *Psychol. Bull.* **1992**, *112*, 155–159. [CrossRef]
36. Zoppini, G.; Cacciatori, V.; Gemma, M.L.; Moghetti, P.; Targher, G.; Zamboni, C.; Thomaseth, K.; Bellavere, F.; Muggeo, M. Effect of moderate aerobic exercise on sympatho-vagal balance in Type 2 diabetic patients. *Diabet. Med.* **2007**, *24*, 370–376. [CrossRef]
37. Goulopoulou, S.; Baynard, T.; Franklin, R.M.; Fernhall, B.; Carhart, R.; Weinstock, R.; Kanaley, J.A. Exercise training improves cardiovascular autonomic modulation in response to glucose ingestion in obese adults with and without type 2 diabetes mellitus. *Metabolism* **2010**, *59*, 901–910. [CrossRef] [PubMed]
38. Goit, R.K.; Pant, B.N.; Shrewastwa, M.K. Moderate intensity exercise improves heart rate variability in obese adults with type 2 diabetes. *Indian Heart J.* **2018**, *70*, 486–491. [CrossRef] [PubMed]
39. Benetos, A.; Rudnichi, A.; Thomas, F.; Safar, M.; Guize, L. Influence of Heart Rate on Mortality in a French Population: Role of Age, Gender, and Blood Pressure. *Hypertension* **1999**, *33*, 44–52. [CrossRef] [PubMed]
40. Paul, L.; Hastie, C.E.; Li, W.S.; Harrow, C.; Muir, S.; John, J.M.; Dominiczak, A.F.; McInnes, G.T.; Padmanabhan, S. Resting heart rate pattern during follow-up and mortality in hypertensive patients. *Hypertension* **2010**, *55*, 67–74. [CrossRef] [PubMed]
41. Zhao, M.X.; Zhao, Q.; Zheng, M.; Liu, T.; Li, Y.; Wang, M.; Yao, S.; Wang, C.; Chen, Y.M.; Xue, H.; et al. Effect of resting heart rate on the risk of all-cause death in Chinese patients with hypertension: Analysis of the Kailuan follow-up study. *BMJ Open* **2020**, *10*, e032699. [CrossRef]
42. Prasada, S.; Oswalt, C.; Yeboah, P.; Saylor, G.; Bowden, D.; Yeboah, J. Heart rate is an independent predictor of all-cause mortality in individuals with type 2 diabetes: The diabetes heart study. *World J. Diabetes* **2018**, *9*, 33–39. [CrossRef]
43. Grace, F.; Herbert, P.; Elliott, A.D.; Richards, J.; Beaumont, A.; Sculthorpe, N.F. High intensity interval training (HIIT) improves resting blood pressure, metabolic (MET) capacity and heart rate reserve without compromising cardiac function in sedentary aging men. *Exp. Gerontol.* **2018**, *109*, 75–81. [CrossRef]
44. Bhatia, C.; Kayser, B. Preoperative high-intensity interval training is effective and safe in deconditioned patients with lung cancer: A randomized clinical trial. *J. Rehabil. Med.* **2019**, *51*, 712–718. [CrossRef]
45. Ramos, J.S.; Dalleck, L.C.; Borrani, F.; Beetham, K.S.; Mielke, G.I.; Dias, K.A.; Wallen, M.P.; Keating, S.E.; Fassett, R.G.; Coombes, J.S. High-intensity interval training and cardiac autonomic control in individuals with metabolic syndrome: A randomised trial. *Int. J. Cardiol.* **2017**, *245*, 245–252. [CrossRef]
46. Dall, C.H.; Snoer, M.; Christensen, S.; Monk-Hansen, T.; Frederiksen, M.; Gustafsson, F.; Langberg, H.; Prescott, E. Effect of high-intensity training versus moderate training on peak oxygen uptake and chronotropic response in heart transplant recipients: A randomized crossover trial. *Am. J. Transplant.* **2014**, *14*, 2391–2399. [CrossRef] [PubMed]
47. Villelabeitia-Jaureguizar, K.; Vicente-Campos, D.; Senen, A.B.; Jiménez, V.H.; Garrido-Lestache, M.E.B.; Chicharro, J.L. Effects of high-intensity interval versus continuous exercise training on post-exercise heart rate recovery in coronary heart-disease patients. *Int. J. Cardiol.* **2017**, *244*, 17–23. [CrossRef] [PubMed]
48. Qiu, S.; Cai, X.; Sun, Z.; Li, L.; Zuegel, M.; Steinacker, J.M.; Schumann, U. Heart rate recovery and risk of cardiovascular events and all-cause mortality: A meta-analysis of prospective cohort studies. *J. Am. Heart Assoc.* **2017**, *6*, e005505. [CrossRef] [PubMed]
49. Qiu, S.H.; Xue, C.; Sun, Z.L.; Steinacker, J.M.; Zügel, M.; Schumann, U. Attenuated heart rate recovery predicts risk of incident diabetes: Insights from a meta-analysis. *Diabet. Med.* **2017**, *34*, 1676–1683. [CrossRef] [PubMed]
50. Besnier, F.; Labrunée, M.; Pathak, A.; Pavy-Le Traon, A.; Galès, C.; Sénard, J.M.; Guiraud, T. Exercise training-induced modification in autonomic nervous system: An update for cardiac patients. *Ann. Phys. Rehabil. Med.* **2017**, *60*, 27–35. [CrossRef]
51. Fu, Q.; Levine, B.D. Exercise and the autonomic nervous system. *Handb. Clin. Neurol.* **2013**, *117*, 147–160. [CrossRef]
52. Fisher, J.P.; Young, C.N.; Fadel, P.J. Autonomic adjustments to exercise in humans. *Compr. Physiol.* **2015**, *5*, 475–512. [CrossRef]

biology

MDPI

Article

Effects of Resistance Training on Oxidative Stress Markers and Muscle Damage in Spinal Cord Injured Rats

Natalie de Almeida Barros [1], Felipe J. Aidar [2,3,4,5,*], Anderson Carlos Marçal [2,3], Jymmys L. Santos [2,3], Raphael Fabricio de Souza [2,3,4], Jainara Lima Menezes [5], Margarete Zanardo Gomes [6], Dihogo Gama de Matos [7], Eduardo Borba Neves [8], André Luiz Gomes Carneiro [9], Paulo Francisco de Almeida-Neto [10], Breno Guilherme de Araújo Tinoco Cabral [10], Reinaldo Viana Belo Neto [6], Beat Knechtle [11,12], Filipe Manuel Clemente [13,14] and Enilton Aparecido Camargo [1,15]

1 Graduate Program in Health Science, Federal University of Sergipe (UFS), São Cristovão 49100-000, Brazil; barrosnatalie@yahoo.com.br (N.d.A.B.); enilton.camargo@gmail.com (E.A.C.)
2 Graduate Program of Physical Education, Federal University of Sergipe (UFS), São Cristovão 49100-000, Brazil; acmarcal@yahoo.com.br (A.C.M.); jymmys.lopes@gmail.com (J.L.S.); raphaelctba20@hotmail.com (R.F.d.S.)
3 Group of Studies and Research of Performance, Sport, Health and Paralympic Sports (GEPEPS), Federal University of Sergipe (UFS), São Cristovão 49100-000, Brazil
4 Department of Physical Education, Federal University of Sergipe (UFS), São Cristovão 49100-000, Brazil
5 Graduate Program of Physiological Science, Federal University of Sergipe (UFS), São Cristovão 49100-000, Brazil; jhaynnara@hotmail.com
6 Laboratory of Morphology and Experimental Pathology, University Tiradentes (UNIT), Aracaju 49010-390, Brazil; guetezanardo@yahoo.com.br (M.Z.G.); reinaldo.viana@souunit.com.br (R.V.B.N.)
7 Cardiovascular & Physiology of Exercise Laboratory, University of Manitoba, Winnipeg, MB R3T 2N2, Canada; dihogogmc@hotmail.com
8 Graduate Program on Biomedical Engineering, Federal Technological University of Paraná (PPGEB/UTFPR), Curitiba 80230-901, Brazil; eduardoneves@utfpr.edu.br
9 Department of Physical Education, State University of Montes Claros, Montes Claros 39401-089, Brazil; algcarneiro@gmail.com
10 Department of Physical Education, Federal University of Rio Grande do Norte, Natal 59064-741, Brazil; paulo220911@hotmail.com (P.F.d.A.-N.); brenotcabral@gmail.com (B.G.d.A.T.C.)
11 Institute of Primary Care, University of Zurich, 8091 Zurich, Switzerland; beat.knechtle@hispeed.ch
12 Medbase St. Gallen Am Vadianplatz, 9001 St. Gallen, Switzerland
13 Escola Superior Desporto e Lazer, Instituto Politécnico de Viana do Castelo, Rua Escola Industrial e Comercial de Nun'Álvares, 4900-347 Viana do Castelo, Portugal; filipe.clemente5@gmail.com
14 Instituto de Telecomunicações, Delegação da Covilhã, 1049-001 Lisboa, Portugal
15 Department of Physiology, Federal University of Sergipe (UFS), São Cristóvão 49100-000, Brazil
* Correspondence: fjaidar@gmail.com; Tel.: +55-(79)-99685-7777

Citation: Barros, N.d.A.; Aidar, F.J.; Marçal, A.C.; Santos, J.L.; de Souza, R.F.; Menezes, J.L.; Gomes, M.Z.; de Matos, D.G.; Neves, E.B.; Carneiro, A.L.G.; et al. Effects of Resistance Training on Oxidative Stress Markers and Muscle Damage in Spinal Cord Injured Rats. *Biology* **2022**, *11*, 32. https://doi.org/10.3390/biology11010032

Academic Editor: Carlo Reggiani

Received: 22 October 2021
Accepted: 18 December 2021
Published: 27 December 2021

Simple Summary: Spinal Cord Injury is a devastating condition that compromises the individual's health, quality of life and functional independence. Rats submitted to Spinal Cord Injury were evaluated after four weeks of resistance training. Analyses of levels of muscle damage and oxidative stress surgery were performed. Resistance training demonstrated increase antioxidative activity while decreased oxidative damage in injured rats, in addition to having presented changes in the levels of muscle damage in that same group. The results highlight that resistance training promoted a decrease in oxidative stress and a significant response in muscle damage markers.

Abstract: Background: Spinal cord injury (SCI) is a condition that affects the central nervous system, is characterized by motor and sensory impairments, and impacts individuals' lives. The objective of this study was to evaluate the effects of resistance training on oxidative stress and muscle damage in spinal cord injured rats. Methodology: Forty Wistar rats were selected and divided equally into five groups: Healthy Control (CON), Sham (SHAM) SCI Untrained group (SCI-U), SCI Trained group (SCI- T), SCI Active Trained group (SCI- AT). Animals in the trained groups were submitted to an incomplete SCI at T9. Thereafter, they performed a protocol of resistance training for four weeks. Results: Significant differences in muscle damage markers and oxidative stress in the trained groups, mainly in SCI- AT, were found. On the other hand, SCI- U group presented higher levels of oxidative stress and biomarkers of LDH and AST. Conclusion: The results highlight that resistance training promoted a decrease in oxidative stress and a significative response in muscle damage markers.

Keywords: spinal cord injury; resistance training; oxidative stress; muscle damage

1. Introduction

Spinal cord injury (SCI) is a devastating condition that affects the central nervous system and is characterized by motor and sensory impairments. It also impacts individual quality of life including physical, psychological, economic and social aspects [1,2]. The practice of physical exercise for functional recovery has been widely used and its mechanisms have been studied by researchers around the world, such as Fu et al. [3].

The evidence has demonstrated the exercise benefits in the general physiological and psychosocial functions of SCI population [4–6], which include locomotor function, mood, pain, and life satisfaction improvement [3,7,8]. Exercise as a treatment strategy for people with spinal cord injury has shown that residual functions have been preserved or improved in individuals trained in conventional ways or with partial support of body weight [9,10].

A recently published physical activity guideline recommends that individuals with incomplete spinal cord injury perform at least three sets of muscle strengthening exercises since there is a correlation between the muscle strength of the hip flexors, extensors and abductors and better walking performance [11,12]. There is still no definitive treatment for spinal cord injury, but many studies have been carried out, highlighting here pre-clinical studies which allow the investigation of disease-related mechanisms, along with new treatment alternatives for this serious health condition.

Studies using animal models of exercise have been shown to be useful in understanding the central nervous system and the regulation of exercise performance, and biological and biochemical analyses provide a means of investigating these responses. [13–15] Understanding the evolution process of the primary lesion as well as the responses to exercise can contribute to the development of more effective therapeutic strategies [16,17].

The effects of resistance training (RT) after SCI were the main objective of this study. We used a rat model of SCI partial transection, enough to produce a deficit but to allow weight-bearing exercise.

2. Materials and Methods

2.1. Study Design

Weeks 1 and 2: The animals were trained during the first week, without apparatus, for approximately 30 min/day, and in the second week were trained with an empty apparatus in the tail;

Weeks from 3 to 6: Animals in the training groups were trained to climb a vertical staircase with the loads tied to their tails, and rested between the series for 2 min in a dark box at the top of the stairs. Study design is shown in Table 1.

2.2. Animal Care and Ethics Committee Approval

Forty adult male Wistar rats weighing 150–180 g were obtained from the Animal Center of the Federal University of Sergipe. The animals were maintained at 21 ± 2 °C with food and water ad libitum under a 12:12 h light/dark cycle. All experimental procedures were approved by the Ethics Committee for Animal Use in Research of the Federal University of Sergipe (protocol number 36/2017) and were conducted in compliance with the Guide for Care and Use of Laboratory Animals (National Institutes of Health) [18].

Table 1. Experimental drawing. Weekly training schedule.

Familiarization		Training
Week 1 1 × 30 min (3 × Week)	Week 2 1 × 30 min (3 × Week)	Week 3 to 6 (3 × Week) 3 × 30 min [Rest between sets: 2 min]
	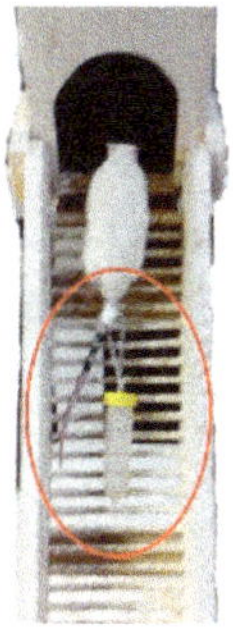	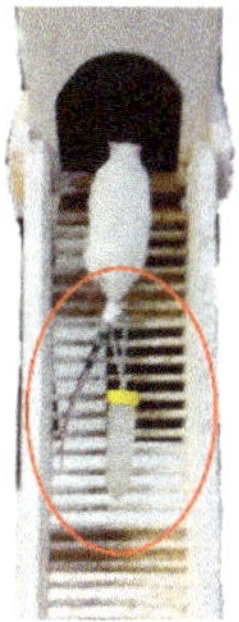

2.3. *Experimental Groups*

Animals were randomly divided into five groups, each one with eight animals (n = 8): Healthy Control (CON), constituted by rats not exposed to any surgical procedure or treatment, Sham (SHAM), surgical control animals submitted to laminectomy surgery without SCI, SCI Untrained group (SCI-U), animals that underwent partial transection but were untrained, SCI Trained group (SCI-T), animals that underwent partial transection and were trained, and SCI Active Trained group (SCI-AT), animals that trained previously, underwent partial transection and were trained again.

2.4. *Surgical Procedure*

The animals were anaesthetized with ketamine and xylazine at a dose of 75 and 14 mg/kg of body weight, respectively, and administrated intraperitonially as verified by the absence of the tail and paw withdrawal of the pinching reflex. After the trichotomy and aseptic measures, the animals were submitted to surgery to remove spinous process and expose spinal cord at the thoracic vertebra 9 (T9). A laminectomy and then a partial transection at the right side using micro scissors and a scalpel was performed and then the muscles and skin were sutured in layers. Postoperative treatment included antibiotic therapy (Pencivet®) intramuscularly to small animals [19]. The animals were placed in an appropriate box, isolated from other animals and accommodated in an air-conditioned room for post-surgical recovery. The general health of all animals was verified, and the bladder and intestine manually emptied until the restoration of normal conditions of urination and defection.

2.5. *Training Protocol*

Two weeks before surgery, all rats, except control group (CG) trained on the staircase for adaptation. On the first week, animals trained without apparatus in the tail, for approximately 30 min/day, and in the second week they trained with an empty apparatus in the tail. After the period of adaptation to the device, rats in the training groups were trained to climb a 1.1-m vertical staircase (60° incline) with the loads tied to their tails. At the top of the stairs there was a dark box (20 × 20 × 20 cm) that allowed the animals to rest between the series, and the adopted rest interval was 2 min. Protocol consisted of three sessions a week, 30 min per day (time corresponding to three sets of eight repetitions each) with a total training time period of four weeks.

Before each training, animals of SCI-T and SCI-AT groups were individually weighed to adjust loads. The weight lifted was fixed at 25% of the animal's weight in the first week and increased to 75% in the last week.

2.6. Euthanasia and Collection of Biological Material

After one hour of the last training, the animals were weighted and anesthetized via intraperitoneal injection of ketamine and xylazine (10 and 85 mg/kg, respectively). Already under the effect of the anesthetic, the animals were sacrificed by decapitation with a guillotine. Then, brain, heart, liver, right gastrocnemius muscle and triceps were collected, in addition to blood for further analysis.

After the collection, the blood sample was immediately centrifuged at 4000× g for 15 min at 4 °C and the supernatant was stored at −80 °C. The other organs were washed three times with 1.15% potassium chloride (KCl) solution, dried, and weighed. Thus, they were homogenized, wherein each gram of the tissue was mixed with 5 mL KCl + 10 µL phenylmethyl sulfonyl fluoride (100 mmol/L) + 15 µL 10% Triton solution and then centrifuged at 3000× g for 10 min at 4 °C and the supernatant stored at −80 °C.

2.7. Tissue Damage Analysis

Quantification of muscle damage was evaluated by enzymatic tissue damage markers such as Creatine Kinase (CK), Lactate Dehydrogenase (LDH), Alanine Aminotransferase (ALT), and Aspartate Aminotransferase (AST). A commercial kit (Labtest®, Lagoa Santa, Minas Gerais, Brazil) was used, 20 µL of each animal homogenized in specific reagents at 37 ± 0.2 °C, and readings taken using a spectrophotometer (Biospectro Model SP-22 UV/Visible, Curitiba, Brazil) at a wavelength of 340 nm.

2.8. Oxidative Stress (OS) Analysis

The lipid oxidation was determined by measuring Thio-barbituric Acid Reactive Substances (TBARS), according to the method described by Lapenna [20]. We used aliquots of 200 µL of the samples (blood and tissues) added to a 400 µL solution of trichloroacetic acid (TCA; 15%), HCl (0.25 N), TBA (0.375%), and butylated hydroxytoluene (BHT; 2.5 mM), 4 µL of sodium dodecyl sulfate (8.1%), heated for 30 min at 95 °C in an oven. The pH of the solution was adjusted to 0.9 with concentrated HCl. To prevent lipid peroxidation during heating, BHT was used. After cooling the solution to room temperature, 4 mL butanol was added, followed by centrifugation at 800× g for 15 min at 4 °C. Next, the absorbance of the supernatant was measured at 532 nm. A molar extinction coefficient of 1.54 10^5/M/cm was used. The TBARS results are expressed malondialdehyde (MDA) equivalents (nmol MDA eq/mL) for plasma and tissue samples.

Following the methodology described by Faure and Lafond [21], the antioxidant activity in the plasma and tissues was quantified through the sulfhydryl (SH) groups. Briefly, aliquots of 50 µL of samples were mixed with 1 mL of tris-EDTA buffer (pH 8.2). The absorbance (A1) was measured at 412 nm. The samples were then transferred to test tubes containing 20 µL DTNB (10 mM), diluted in methanol (4 mg/mL), and left undisturbed in a dark room. After 15 min, the absorbance (A2) was measured. The SH concentration was calculated using the following equation: $(A2 - A1) - B \times 1.57\ mM \times 1000$, and the result was expressed in nmol/mg tissue.

2.9. Statistical Analysis

After confirmation of normality through the test of Shapiro Wilk and homogeneity assumptions, one-way ANOVA with Bonferroni's post hoc was performed to compare the measurements groups. To check the effect size, partial Eta squared ($\eta^2 p$) was used, adopting values of low effect (≤0.05), medium effect (0.05 to 0.25), high effect (0.25 to 0.50), and very high effect (>0.50) for ANOVA [22]. The level of significance was set at $p < 0.05$. Data are presented as means (X) ± standard deviation (SD). All statistical analyses were performed using the computerized package Statistical Package for the Social Science (SPSS), version 22.0 (IBM Corp, Armonk, NY, USA).

3. Results

The results of muscle damage (CK, LDH, ALT and AST) are shown in Figure 1.

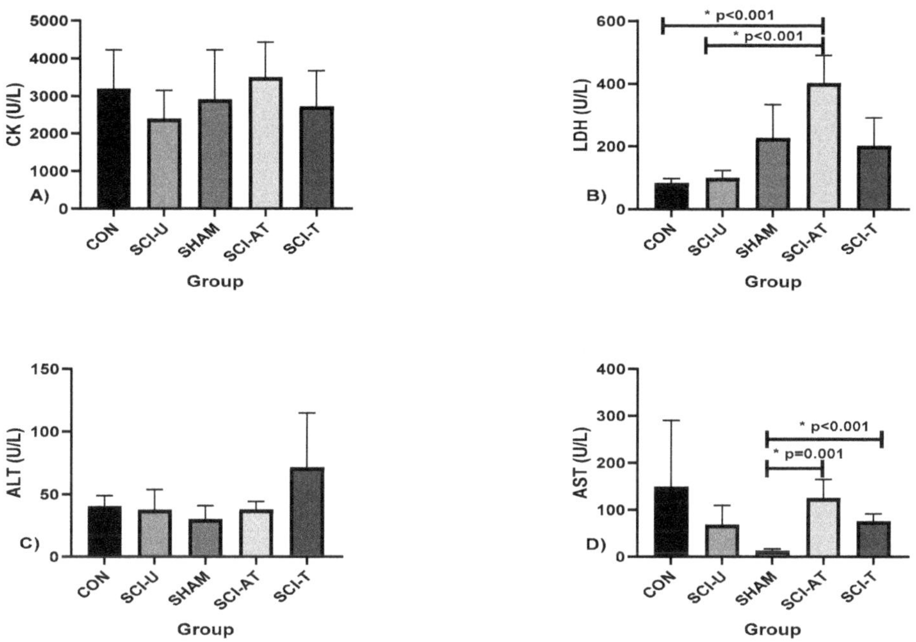

Figure 1. Muscle Damage levels presented by (**A**) Creatine Kinase (CK), (**B**) Lactate Dehydrogenase (LDH), (**C**) Alanine Aminotransferase (ALT), (**D**) Aspartate Aminotransferase (AST) after interventions in the different groups: Control (CON), constituted by rats not exposed to any surgical procedure or treatment, Sham (SHAM), surgical control animals submitted to laminectomy surgery without SCI, SCI Untrained group (SCI-U), animals that underwent partial transection but were untrained, SCI Trained group (SCI- T), animals that underwent partial transection and were trained, SCI Active Trained group (SCI- AT), animals that trained previously, underwent partial transection and were trained again. * $p < 0.05$ ANOVA (one-way), and Bonferroni post hoc.

There was no difference in CK and ALT among the groups. In relation to the LDH there was difference between CON and SCI-U, and in relation to SCI-AT groups ($p < 0.001$, $F(4,28) = 88.118$; $\eta^2p = 0.926$, very high effect). Regarding AST levels, there were differences between SHAM and SCI- AT groups ($p = 0.001$) and SCI-T ($p < 0.001$; $F(4,28) = 5.275$; $\eta^2p = 0.433$, high effect).

Sulfhydryl levels (SH) were analyzed in the blood and in the following tissues: brain, heart, liver, gastrocnemius and triceps muscles (Figure 2).

For blood concentration of SH, there were differences between SCI- AT and SCI- U groups ($p = 0.021$) and between SCI- AT and CON groups ($p = 0.036$; $F(4,28) = 8.452$; $\eta^2p = 0.547$) (very high effect). There were differences in SH brain levels between SCI-T and SCI-U groups ($p = 0.001$) and SCI- T and SHAM ($p = 0.015$, $F(4,28) = 7.150$; $\eta^2p = 0.505$, very high effect). In heart SH analysis, there were differences between SHAM and SCI-AT groups ($p = 0.019$; $F(4,28) = 7.781$; $\eta^2p = 0.526$ (high effect). There were differences in SH liver levels between CON and SHAM groups ($p < 0.001$) and CON and SCI-T ($p = 0.006$, $F(4,28) = 9.813$; $\eta^2p = 0.584$, very high effect). In liver SH analysis, there were differences between SHAM and SCI- AT groups ($p = 0.019$; $F(4,28) = 7.781$; $\eta^2p = 0.526$ (high effect). There were no differences between the groups in relation to gastrocnemius and triceps muscles.

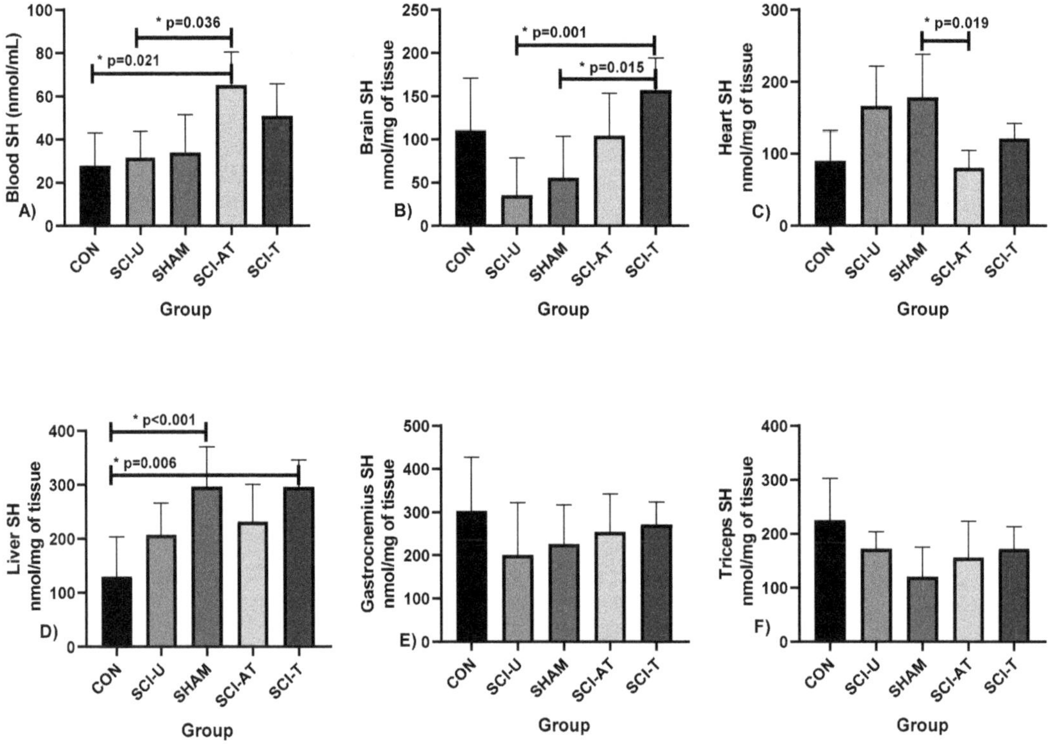

Figure 2. Oxidative Stress measured by Sulfhydryl (SH) levels measured in: (**A**) Blood, (**B**) Brain, (**C**) Heart, (**D**) Liver, (**E**) Gastrocnemius, (**F**) Triceps after interventions in the different groups: Control (CON), constituted by rats not exposed to any surgical procedure or treatment, Sham (SHAM), surgical control animals submitted to laminectomy surgery without SCI, SCI Untrained group (SCI-U), animals that underwent partial transection but were untrained, SCI Trained group (SCI- T), animals that underwent partial transection and were trained, SCI Active Trained group (SCI- AT), animals that trained previously, underwent partial transection and were trained again. * $p < 0.05$ ANOVA (one-way), and Bonferroni post hoc.

Malondialdehyde (MDA) levels analysis in the blood and in the following tissues: brain, heart, liver, gastrocnemius and triceps muscles, are shown in Figure 3.

Malondialdehyde (MDA) levels analysis showed differences between CON and SCI-U groups ($p = 0.031$; $F(4,28) = 7.194$; $\eta^2p = 0.507$) (very high effect). Regarding MDA levels in the brain, there were differences between CON and SCI-U groups ($p = 0.028$) and CON and SHAM groups ($p = 0.035$, $F(4,28) = 3.685$; $\eta^2p = 0.343$, high effect). In heart MDA levels, there were differences between CON and SCI-U ($p = 0.005$), CON and SHAM ($p = 0.008$) and CON and SCI-T ($p = 0.045$; $F(4,28) = 10.511$; $\eta^2p = 0.600$ (very high effect). Regarding liver MDA levels, there were differences between CON and SCI-U groups ($p = 0.048$), CON and SHAM ($p = 0.023$) and CON and SCI-T groups ($p = 0.018$; $F(4,28) = 9.814$; $\eta^2p = 0.584$ (very high effect). In Tríceps MDA levels, there were differences between the groups CON and SHAM ($p = 0.020$) and CON and SCI-T ($p = 0.033$; $F(4,28) = 6.973$; $\eta^2p = 0.499$ (high effect), while in gastrocnemius there was no difference between groups.

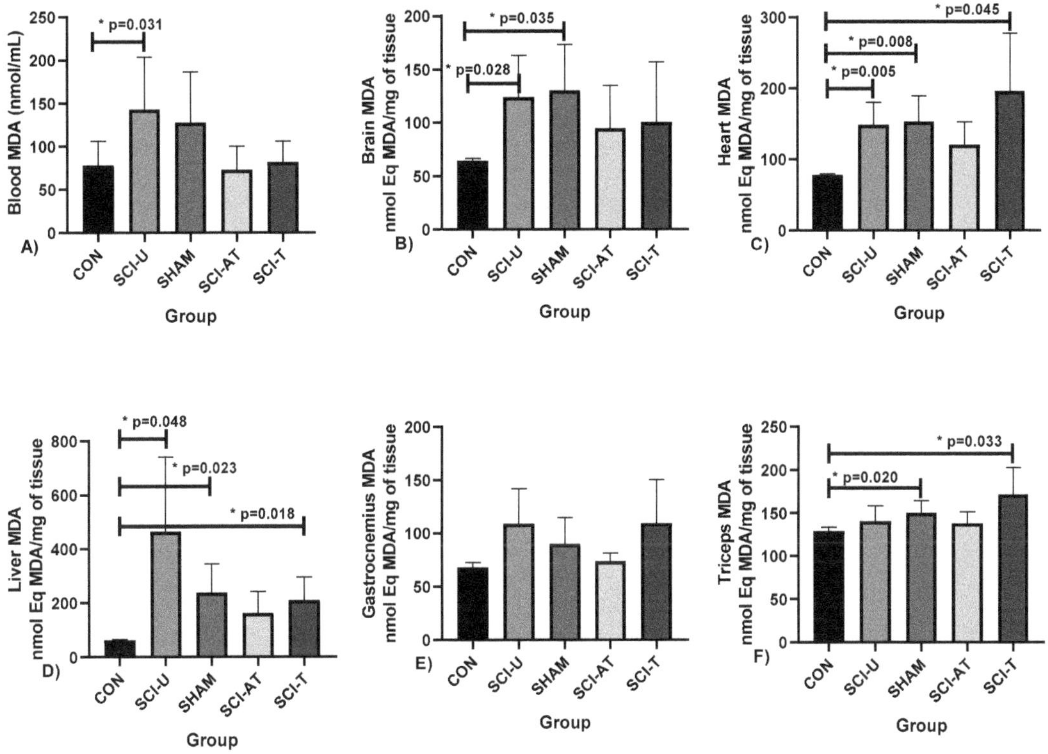

Figure 3. Malondialdehyde (MDA) levels measured in: (**A**) Blood, (**B**) Brain, (**C**) Heart, (**D**) Liver, (**E**) Gastrocnemius, (**F**) Triceps after interventions in the different groups: Control (CON), constituted by rats not exposed to any surgical procedure or treatment, Sham (SHAM), surgical control animals submitted to laminectomy surgery without SCI, SCI Untrained group (SCI-U), animals that underwent partial transection but were untrained, SCI Trained group (SCI- T), animals that underwent partial transection and were trained, SCI Active Trained group (SCI- AT), animals that trained previously, underwent partial transection and were trained again. * $p < 0.05$ ANOVA (one-way), and Bonferroni post hoc.

4. Discussion

This study aimed to analyze the effect of resistance training on muscle damage and oxidative stress in spinal cord injured rats, by biochemical indicators in the blood, brain, heart, liver, gastrocnemius and triceps tissues. The results highlighted that there was a decrease in oxidative stress and a significant response in muscle damage markers among the groups.

Significant differences in muscle damage and OS were found in the SCI-AT group compared with the SCI-T group. This suggests that four weeks of resistance training induces positive responses following spinal cord injury and ameliorates deleterious effects and so should be considered a potential strategy for SCI treatment.

The link between exercise, muscle damage and oxidative stress in different populations has been widely studied [23,24]. However, the impact of exercise on biomarkers in people with SCI was only recently brought to light. The importance of studying this relation is due to the fact that muscle wasting after SCI is associated with changes in body composition which contributes to increased risk of cardiovascular disease and type 2 diabetes, interfering in individual health and quality of life [25].

It is already known that the release of inflammatory mediators after SCI alters neural function, which impairs ionic conduction and synaptic transmission. Thus, knowing the possible mechanisms with which to control this inflammatory response can contribute to the reduction of damage caused by SCI. One of the ways to carry out this analysis is through biomarkers, which determine the evolution of the lesion and also point out the results of applied therapeutic strategies [26].

In the analysis of tissue damage, there are methods that use analysis of biomarkers such as creatine kinase (CK), lactate dehydrogenase (LDH), alanine aminotransferase (ALT), and aspartate aminotransferase (AST), which together estimate the potential exercise injury. The results revealed no differences in Creatine Kinase (CK) levels among the groups after 4 weeks of training. Different from the findings here, in the study of Dos Santos et al. rats submitted to squat exercise presented a significant increase in CK after four weeks of training and this may be related to the intense activity of healthy muscles of animals. On the other hand, the study of Mohammadi et al. [27] found no difference in levels of CK, in groups of healthy subjects submitted to different volumes of eccentric exercise.

Creatine Kinase is an enzyme located in skeletal muscle cells, considered an indirect, highly sensitive and specific indicator of muscle damage with a direct relationship between this and muscle activity, and is impaired in people with SCI, suggesting that the protocol training did not generate enough muscular contractions to generate damage in trained animals [28–31].

The results for Alanine Aminotransferase (ALT) and Aspartate Aminotransferase (AST) were different. While ALT did not present differences among the groups, AST presented differences between SHAM and SCI-AT groups. Both are important liver enzymes, ALT found in higher concentration in cytoplasm whose elevation is acute and directly proportional to the injury. The absence of significative difference may suggest that there was no liver damage after the protocol of RT. Concerning AST, this is an enzyme found in other tissues besides liver such as kidney, brain and skeletal muscle, therefore intense exercise to which the animals were submitted may explain the changes in levels. [32–34].

Partially supporting our results, studies performed with other exercise protocols with healthy animals showed no significant differences in ALT and AST levels. In our study, while ALT presented no differences among the groups, AST levels presented differences between CON and SCI- AT groups. In the study of Motta et al. [35], a protocol of swimming with weights ALT and AST levels presented a decrease, while another showed no differences in these markers. Low levels of ALT and AST can provide evidence that exercise did not cause hepatic damage [31,36].

In addition to muscle damage analysis, LDH levels presented differences in SCI-AT group compared with CON and SHAM groups. The increase in LDH levels is related to cellular damage, being better used as a damage confirmation marker rather than a damage extension marker [37]. Its levels increase more slowly than CK, but stay increased longer. LDH is a marker of skeletal muscle injuries due to micro-rupture of muscle fibers and shows the degree of metabolic adaptation of skeletal muscles [38]. Considering that the higher levels occurred in a group that was submitted to the resistance training protocol before the injury (SCI- AT) was different from the others, we can suggest that this group had evidence of muscle damage [39].

Since the discovery that prolonged exercise promotes oxidative stress, many studies have been conducted and have shown that exercise increases reactive oxygen species (ROS) which results in oxidative stress and can be measured by the analysis of oxidative damage in various tissues, including blood and skeletal muscle [40]. However, it must be remembered that regular physical exercise increases the regulation of the enzymatic antioxidant system and the modulation of oxidative damage through the regulation of the cellular redox state by modulating the metabolism in an intensity-dependent manner and/or by direct activation of ROS-generating enzymes [41]. Regular exercise can even promote an increase in the brain's antioxidant capacity [42,43] and the increase in hippocampal neurogenesis [44], among other effects.

Resulting from the SCI, rapid skeletal muscle atrophy occurs in the following six weeks, this period marked by an increase of oxidative stress (OS), while plasma antioxidant levels are decreased. Therefore, OS decrease or inhibition can reduce the deleterious effects of SCI [45].

Concerning the analysis of OS parameters, sulfhydryl (SH) and Malondialdehyde (MDA) levels were used. We observed that RT promoted a significant difference in SH levels in different tissues in the trained groups (SCI- AT and SCI-T) compared with the others (SCI-U, SHAM and CON groups), results that evidence the role of exercise, corroborating studies that demonstrated that exercise training can promote a decrease in oxidative stress, and also increase the antioxidant system [45,46].

Being considered an indirect antioxidant biomarker, SH is found in the GSH Cys residue and other antioxidants, whereby molecules stabilize free radicals by receiving their unpaired electron. The results in this study are divergent from de Araujo [47], in which Wistar rats were submitted to a protocol of high intensity interval training (HIIT) but did not find significant differences in SH levels [48,49].

The results showed that SCI induction caused a significant increase in blood, brain, heart, liver and triceps levels of Malondialdehyde (MDA), mainly SCI-U group. Considered the primary biomarker of oxidative damage, this is the result of many chemical reactions that occur in lipidic peroxidation, in short, the degradation of cell membrane by the action of reactive species, whose measurement is made by the formation of MDA through its complex action with thio-barbituric acid [50,51].

The lower levels of MDA in trained animals suggests a reduction of lipidic peroxidation. On the other hand, if we consider the higher levels of SH in these groups, we can deduce that there is a trend towards an increase of antioxidant activity. It is already known that exercise improves the capacity of the cell antioxidant defense system in order to neutralize ROS increases, in addition to improving the metabolic state, resulting in a redox balance, although the intensity necessary to produce this balance has not yet been found [52–54].

An important aspect that should be mentioned is that different responses of muscle damage and oxidative stress markers were found in the different tissues analyzed and this is due to the fact that each organ responds differently to exercise [55].

Nevertheless, despite these important findings, a limitation of the present study was the absence of an apparatus to accurately control the extent of the lesion, although the transection model is widely used in the literature. The Basso-Beatie-Bresnahan (BBB) functionality assessment scale was used in the postoperative period to ensure sample homogeneity and minimize these differences between groups.

It is important to highlight another limitation, related to the results of the SHAM group, which was exposed to the surgical procedure and laminectomy but did not suffer spinal cord injury. Based on the results found in this study, there was an alteration in the levels of several markers, which may be related to the stress of surgery, as well as the handling of animals. We can also infer that the invasive procedure, even without injury, may have triggered some physiological response that raised these parameters.

Furthermore, concerning spinal cord injury and the surgical procedure to which the animals were submitted, some aspects should be considered. In this sense, the different segments of the spinal cord end up by interfering in specific disease outcomes [56,57]. The spinal cord is a complex and dynamic neural structure, whose neurons, when injured, tend to interfere with the generation of sympathetic activity in many autonomic targets, including the heart and blood vessels [58,59], with evidence of the interaction between the spinal cord. and the heart, [60,61], and it can interfere in several activities, which were not the target of our study. Despite these possible influences that spinal cord injury tends to promote, we consider these points as secondary to our study, and the focus of research is only in terms of exercise-related spinal cord injury.

5. Conclusions

To the best of our knowledge, there are no similar studies involving SCI, resistance training, oxidative stress and muscle damage.

Spinal Cord Injury is known to affect all aspects of an individual's life as a result of decreased function associated with loss of skeletal muscle mass. Resistance training groups demonstrated reduce MDA levels compared with non-trained animals while increasing SH levels in the same groups. Thus, resistance training provides a potential strategy for reducing the deleterious effects of muscle damage and oxidative stress in individuals with SCI.

Author Contributions: Conceptualization, N.d.A.B., F.J.A. and E.A.C.; methodology, A.C.M., J.L.S. and J.L.M., software, R.F.d.S., J.L.M. and M.Z.G.; validation, D.G.d.M., E.B.N. and P.F.d.A.-N.; formal analysis, F.J.A., A.C.M. and R.F.d.S.; investigation, N.d.A.B., M.Z.G. and E.A.C.; resources, M.Z.G. and A.L.G.C.; data curation, E.B.N., B.G.d.A.T.C. and F.M.C.; writing—original draft preparation, P.F.d.A.-N. and B.G.d.A.T.C.; writing—review and editing, R.V.B.N., B.K. and F.M.C.; visualization, R.V.B.N. and B.K.; supervision, F.J.A. and A.L.G.C.; project administration, E.A.C. All authors have read and agreed to the published version of the manuscript.

Funding: We are grateful to the Social Demand Program of the Office to Coordinate Improvement of Higher Education Personnel (CAPES) and the Brazilian National Council for Scientific and Technological Development (CNPq/Brazil) for financial support. N.d.A.B. was a beneficiary of scholarships from the Social Demand Program of CAPES and EAC is a recipient of a scientific productivity grant from CNPq (no. 311671/2018-7).

Institutional Review Board Statement: The study was conducted according to the precepts of the Law. N°. 11.794, of 8 October 2008, of Decree N°. 6.899, of 15 July 2009, and with the rules edited by the National Council for the Control of Animal Experiments (CONCEA) approved by the Ethics in the Use of Animals Committee (CEUA) of the Federal University of Sergipe, registered with n° 36/2017.

Informed Consent Statement: Not applicable.

Data Availability Statement: The data that support this study can be obtained from the address: www.ufs.br/Department of Physical Education. Accessed on 16 October 2021.

Acknowledgments: The authors would like to acknowledge CAPES and CNPq for their funding support for this study.

Conflicts of Interest: The authors declare no conflict of interest.

References

1. Middleton, J.; Tran, Y.; Craig, A. Relationship between Quality of Life and Self-Efficacy in Persons with Spinal Cord Injuries. *Arch. Phys. Med. Rehabil.* **2007**, *88*, 1643–1648. [CrossRef] [PubMed]
2. Furlan, J.C.; Noonan, V.; Singh, A.; Fehlings, M.G. Assessment of impairment in patients with acute traumatic spinal cord injury: A systematic review of the literature. *J. Neurotrauma* **2011**, *28*, 1445–1477. [CrossRef] [PubMed]
3. Fu, J.; Wang, H.; Deng, L.; Li, J. Exercise Training Promotes Functional Recovery after Spinal Cord Injury. *Neural Plast.* **2016**, *2016*, 7. [CrossRef] [PubMed]
4. Hicks, A.L.; Martin, K.A.; Ditor, D.S.; Latimer, A.E.; Craven, C.; Bugaresti, J.; McCartney, N. Long-term exercise training in persons with spinal cord injury: Effects on strength, arm ergometry performance and psychological well-being. *Spinal Cord* **2003**, *41*, 34–43. [CrossRef] [PubMed]
5. Hicks, A.L.; Ginis, K.A.M.; Pelletier, C.A.; Ditor, D.S.; Foulon, B.; Wolfe, D.L. The effects of exercise training on physical capacity, strength, body composition and functional performance among adults with spinal cord injury: A systematic review. *Spinal Cord* **2011**, *49*, 1103–1127. [CrossRef] [PubMed]
6. Ditor, D.S.; Latimer, A.E.; Ginis, K.A.M.; Arbour, K.P.; McCartney, N.; Hicks, A.L. Maintenance of exercise participation in individuals with spinal cord injury: Effects on quality of life, stress and pain. *Spinal Cord* **2003**, *41*, 446–450. [CrossRef] [PubMed]
7. Van Koppenhagen, C.F.; Post, M.; De Groot, S.; Van Leeuwen, C.; Van Asbeck, F.; Stolwijk-Swüste, J.; van der Woude, L.; Lindeman, E. Longitudinal relationship between wheelchair exercise capacity and life satisfaction in patients with spinal cord injury: A cohort study in the Netherlands. *J. Spinal Cord Med.* **2014**, *37*, 328–337. [CrossRef] [PubMed]
8. Crane, D.A.; Hoffman, J.M.; Reyes, M.R. Benefits of an exercise wellness program after spinal cord injury. *J. Spinal Cord Med.* **2017**, *40*, 154–158. [CrossRef] [PubMed]

9. Barbeau, H. Locomotor training in neurorehabilitation: Emerging rehabilitation concepts. *Neurorehabil. Neural Repair.* **2003**, *17*, 3–11. [CrossRef] [PubMed]
10. Singh, A.; Balasubramanian, S.; Murray, M.; Lemay, M.; Houle, J. Role of spared pathways in locomotor recovery after body-weight-supported treadmill training in contused rats. *J. Neurotrauma* **2011**, *28*, 2405–2416. [CrossRef]
11. Kim, C.M.; Eng, J.J.; Whittaker, M.W. Level walking and ambulatory capacity in persons with incomplete spinal cord injury: Relationship with muscle strength. *Spinal Cord* **2004**, *42*, 156–162. [CrossRef]
12. Ginis, K.A.M.; Van Der Scheer, J.W.; Latimer-Cheung, A.E.; Barrow, A.; Bourne, C.; Carruthers, P.; Bernanrdi, M.; Ditor, D.S.; Gaudet, S.; de Groot, S.; et al. Evidence-based scientific exercise guidelines for adults with spinal cord injury: An update and a new guideline. *Spinal Cord* **2018**, *56*, 308–321. [CrossRef] [PubMed]
13. Seo, D.Y.; Lee, S.R.; Kim, N.; Ko, K.S.; Rhee, B.D.; Han, J. Humanized animal exercise model for clinical implication. *Pflug. Arch. Eur. J. Physiol.* **2014**, *466*, 1673–1687. [CrossRef] [PubMed]
14. Voss, M.W.; Vivar, C.; Kramer, A.F.; van Praag, H. Bridging animal and human models of exercise-induced brain plasticity. *Trends Cogn. Sci.* **2013**, *17*, 525–544. [CrossRef] [PubMed]
15. Cholewa, J.; Guimarães-Ferreira, L.; da Silva Teixeira, T.; Naimo, M.A.; Zhi, X.; de Sá, R.B.D.P.; Lodetti, A.; Cardozo, M.Q.; Zanchi, N.E. Basic models modeling resistance training: An update for basic scientists interested in study skeletal muscle hypertrophy. *J. Cell. Physiol.* **2014**, *229*, 1148–1156. [CrossRef]
16. Strickland, J.C.; Smith, M.A. Animal models of resistance exercise and their application to neuroscience research. *J. Neurosci. Methods* **2016**, *273*, 191–200. [CrossRef] [PubMed]
17. Sharif-Alhoseini, M.; Khormali, M.; Rezaei, M.; Safdarian, M.; Hajighadery, A.; Khalatbari, M.M.; Safdarian, M.; Meknatkhah, S.; Rezvan, M.; Chalangari, M.; et al. Animal models of spinal cord injury: A systematic review. *Spinal Cord* **2017**, *55*, 714–721. [CrossRef] [PubMed]
18. Albus, U. Guide for the Care and Use of Laboratory Animals. *Lab. Anim.* **2012**, *46*, 267–268. [CrossRef]
19. Yin, H.; Jiang, T.; Deng, X.; Yu, M.; Xing, H.; Ren, X. A cellular spinal cord scaffold seeded with rat adipose-derived stem cells facilitates functional recovery via enhancing axon regeneration in spinal cord injured rats. *Mol. Med. Rep.* **2018**, *17*, 2998–3004. [CrossRef]
20. Lapenna, D.; Ciofani, G.; Pierdomenico, S.D.; Giamberardino, M.A.; Cuccurullo, F. Reaction conditions affecting the relationship between thiobarbituric acid reactivity and lipid peroxidesin human plasma. *Free Radic. Biol. Med.* **2001**, *31*, 331–335. [CrossRef]
21. Faure, P.; Lafond, J. Measurement of plasma sulfhydryl and carbonyl. In *Analysis of Free Radicals in Biological Systems*; Birkhäuser: Basel, Switzerland, 1995; pp. 237–248.
22. Cohen, J. A power primer. *Psychol. Bull.* **1992**, *112*, 155–159. [CrossRef] [PubMed]
23. Thirupathi, A.; Pinho, R.A.; Ugbolue, U.C.; He, Y.; Meng, Y.; Gu, Y. Effect of running exercise on oxidative stress biomarkers: A systematic review. *Front. Physiol.* **2020**, *11*, 1789. [CrossRef] [PubMed]
24. Bethany, N.; Clifford, T. The Effect of Creatine Supplementation on Markers of Exercise-Induced Muscle Damage: A Systematic Review and Meta-Analysis of Human Intervention Trials. *Int. J. Sport Nutr. Exerc. Metab.* **2021**, *31*, 276–291. [CrossRef]
25. Wattananit, S.; Tornero, D.; Graubardt, N.; Memanishvili, T.; Monni, E.; Tatarishvili, J.; Miskinyte, G.; Ge, R.; Ahlenius, H.; Lindvall, O.; et al. Monocyte-derived macrophages contribute to spontaneous long-term functional recovery after stroke in mice. *J. Neurosci.* **2016**, *36*, 4182–4195. [CrossRef]
26. Savikj, M.; Kostovski, E.; Lundell, L.S.; Iversen, P.O.; Massart, J.; Widegren, U. Altered oxidative stress and antioxidant defence in skeletal muscle during the first year following spinal cord injury. *Physiol. Rep.* **2019**, *7*, e14218. [CrossRef]
27. Mohammadi, H.; Afzalpour, M.E.; Abtahi, S.H. Response of creatine kinase and lactate dehydrogenase enzymes to rest interval between sets and set-repetition configuration during bouts of eccentric exercise. *Interv. Med. Appl. Sci.* **2018**, *10*, 83–86. [CrossRef] [PubMed]
28. Isaacs, A.W.; Macaluso, F.; Smith, C.; Myburgh, K.H. C-reactive protein is elevated only in high creatine kinase responders to muscle damaging exercise. *Front. Physiol.* **2019**, *10*, 86. [CrossRef] [PubMed]
29. Del Coso, J.; Valero, M.; Salinero, J.J.; Lara, B.; Gallo-Salazar, C.; Areces, F. Optimum polygenic profile to resist exertional rhabdomyolysis during a marathon. *PLoS ONE* **2017**, *12*, e0172965. [CrossRef]
30. Bekkelund, S.I. Leisure physical exercise and creatine kinase activity. The Tromsø study. *Scand. J. Med. Sci. Sports* **2020**, *30*, 2437–2444. [CrossRef] [PubMed]
31. Dos Santos, J.L.; Dantas, R.E.A.; Lima, C.A.; de Araújo, S.S.; de Almeida, E.C.V.; Marçal, A.C.; Estevam, C.S. Protective effect of a hydroethanolic extract from Bowdichia virgilioides on muscular damage and oxidative stress caused by strenuous resistance training in rats. *J. Int. Soc. Sports Nutr.* **2014**, *11*, 1–10. [CrossRef] [PubMed]
32. Souza, L.M.V.; Aidar, F.J.; de Matos, D.G.; Marçal, A.C.; de Souza, R.F.; dos Santos, J.L.; Wartha, E.R.S.A.; da Silva, A.N.; Estevam, C.S.; de Araújo, S.S. Analysis of oxidative stress in Wistar rats submitted to high-intensity interval training. *Motricidade* **2020**, *16*, 274–281. [CrossRef]
33. Altinoz, E.; Ozmen, T.; Oner, Z.; Elbe, H.; Erdemli, M.E.; Bag, H.G. Effect of crocin on oxidative stress in recovery from single bout of swimming exercise in rats. *Gen. Physiol. Biophys.* **2016**, *35*, 87–94. [CrossRef] [PubMed]
34. Chung, S.M.; Moon, J.S.; Yoon, J.S.; Won, K.C.; Lee, H.W. Low alanine aminotransferase levels predict low muscle strength in older patients with diabetes: A nationwide cross-sectional study in Korea. *Geriatr. Gerontol. Int.* **2020**, *20*, 271–276. [CrossRef] [PubMed]

35. Motta, V.F.; Aguila, M.B.; Mandarim-De-Lacerda, C.A. High-intensity interval training (swimming) significantly improves the adverse metabolism and comorbidities in diet-induced obese mice. *J. Sports Med. Phys. Fit.* **2016**, *56*, 655–663.
36. Righi, T.; Carvalho, C.A.D.; Ribeiro, L.M.; Cunha, D.N.Q.D.; Paiva, A.C.S.; Natali, A.J.; Pereira, E.T.; Lima, L.M. Alcohol consumption and the influence of physical exercise on enzyme activity of Wistar rats. *Rev. Bras. Med. Esp.* **2016**, *22*, 40–44. [CrossRef]
37. Owens, D.J.; Twist, C.; Cobley, J.N.; Howatson, G.; Close, G.L. Exercise-induced muscle damage: What is it, what causes it and what are the nutritional solutions? *Eur. J. Sport Sci.* **2018**, *19*, 71–85. [CrossRef] [PubMed]
38. Isik, O.; Yildirim, I.; Ersoz, Y.; Koca, H.B.; Dogan, I.; Ulutas, E. Monitoring of pre-competition dehydration-induced skeletal muscle damage and inflammation levels among elite wrestlers. *J. Back Musculoskelet. Rehabil.* **2018**, *31*, 533–540. [CrossRef] [PubMed]
39. Abe, T.; Kitaoka, Y.; Kikuchi, D.M.; Takeda, K.; Numata, O.; Takemasa, T. High-intensity interval training-induced metabolic adaptation coupled with an increase in Hif-1α and glycolytic protein expression. *J. Appl. Physiol.* **2015**, *119*, 1297–1302. [CrossRef]
40. Powers, S.K.; Deminice, R.; Ozdemir, M.; Yoshihara, T.; Bomkamp, M.P.; Hyatt, H. Exercise-induced oxidative stress: Friend or foe? *J. Sport Health Sci.* **2020**, *9*, 415–425. [CrossRef] [PubMed]
41. Radak, Z.; Suzuki, K.; Higuchi, M.; Balogh, L.; Boldogh, I.; Koltai, E. Physical exercise, reactive oxygen species and neuroprotection. *Free Radic. Biol. Med.* **2016**, *98*, 187–196. [CrossRef] [PubMed]
42. So, J.H.; Huang, C.; Ge, M.; Cai, G.; Zhang, L.; Lu, Y.; Mu, Y. Intense exercise promotes adult hippocampal neurogenesis but not spatial discrimination. *Front. Cell. Neurosci.* **2017**, *11*, 13. [CrossRef] [PubMed]
43. Freitas, D.A.; Rocha-Vieira, E.; De Sousa, R.A.L.; Soares, B.A.; Rocha-Gomes, A.; Garcia, B.C.C.; Casilhas, R.C.; Mendonça, V.A.; Camargos, A.C.R.; De Gregorio, J.A.M.; et al. High-intensity interval training improves cerebellar antioxidant capacity without affecting cognitive functions in rats. *Behav. Brain Res.* **2019**, *376*, 112181. [CrossRef] [PubMed]
44. Freitas, D.A.; Rocha-Vieira, E.; Soares, B.A.; Nonato, L.F.; Fonseca, S.R.; Martins, J.B.; Mendonça, V.A.; Lacerda, A.C.R.; Massensini, A.R.; Poortams, J.R.; et al. High intensity interval training modulates hippocampal oxidative stress, BDNF and inflammatory mediators in rats. *Physiol. Behav.* **2018**, *184*, 6–11. [CrossRef] [PubMed]
45. Villanueva, C.; Kross, R.D. Antioxidant-induced stress. *Int. J. Mol. Sci.* **2012**, *13*, 2091–2109. [CrossRef] [PubMed]
46. Powers, S.K.; Talbert, E.E.; Adhihetty, P.J. Reactive oxygen and nitrogen species as intracellular signals in skeletal muscle. *J. Physiol.* **2011**, *589*, 2129–2138. [CrossRef] [PubMed]
47. de Araujo, G.G.; Papoti, M.; dos Reis, I.G.M.; de Mello, M.A.R.; Gobatto, C.A. Short and long term effects of high-intensity interval training on hormones, metabolites, antioxidant system, glycogen concentration, and aerobic performance adaptations in rats. *Front. Physiol.* **2016**, *7*, 505. [CrossRef]
48. McLeay, Y.; Stannard, S.; Houltham, S.; Starck, C. Dietary thiols in exercise: Oxidative stress defence, exercise performance, and adaptation. *J. Int. Soc. Sports Nutr.* **2017**, *14*, 1–8. [CrossRef]
49. Zhang, L.; Wang, X.; Cueto, R.; Effi, C.; Zhang, Y.; Tan, H.; Quin, X.; Ji, Y.; Yang, X.; Wang, H. Biochemical basis and metabolic interplay of redox regulation. *Redox Biol.* **2019**, *26*, 1–16. [CrossRef]
50. Tsikas, D. Assessment of lipid peroxidation by measuring malondialdehyde (MDA) and relatives in biological samples: Analytical and biological challenges. *Anal. Biochem.* **2017**, *524*, 13–30. [CrossRef]
51. Garcia, S.C.; Grotto, D.; Bulcão, R.P.; Moro, A.M.; Roehrs, M.; Valentini, J.; de Freitas, F.A.; Paniz, C.; Bubols, G.B.; Charão, M.F. Evaluation of lipid damage related to pathological and physiological conditions. *Drug Chem. Toxicol.* **2013**, *36*, 306–312. [CrossRef] [PubMed]
52. Parker, L.; Trewin, A.; Levinger, I.; Shaw, C.S.; Stepto, N.K. Exercise-intensity dependent alterations in plasma redox status do not reflect skeletal muscle redox-sensitive protein signaling. *J. Sci. Med. Sport* **2018**, *21*, 416–421. [CrossRef] [PubMed]
53. Bouzid, M.A.; Hammouda, O.; Matran, R.; Robin, S.; Fabre, C. Low intensity aerobic exercise and oxidative stress markers in older adults. *J. Aging Phys. Act.* **2014**, *22*, 536–542. [CrossRef] [PubMed]
54. Vezzoli, A.; Pugliese, L.; Marzorati, M.; Serpiello, F.R.; La Torre, A.; Porcelli, S. Time-course changes of oxidative stress response to high-intensity discontinuous training versus moderate-intensity continuous training in masters runners. *PLoS ONE* **2014**, *9*, e87506. [CrossRef] [PubMed]
55. Kruk, J.; Kotarska, K.; Aboul- Enein, B.H. Physical exercise and catecholamines response: Benefits and health risk: Possible mechanisms. *Free Radic. Res.* **2020**, *54*, 105–125. [CrossRef] [PubMed]
56. Wang, Q.; Li, Z.X.; Li, Y.J.; He, Z.G.; Chen, Y.L.; Feng, M.H.; Li, S.Y.; Wu, D.Z.; Xiang, H.B. Identification of lncRNA and mRNA expression profiles in rat spinal cords at various time-points following cardiac ischemia/reperfusion. *Int. J. Mol. Med.* **2019**, *43*, 2361–2375. [CrossRef] [PubMed]
57. Li, Z.X.; Li, Y.J.; Wang, Q.; He, Z.G.; Feng, M.H.; Xiang, H.B. Characterization of novel lncRNAs in upper thoracic spinal cords of rats with myocardial ischemia-reperfusion injury. *Exp. Ther. Med.* **2021**, *21*, 15. [CrossRef]
58. Lujan, H.L.; Palani, G.; DiCarlo, S.E. Structural neuroplasticity following T5 spinal cord transection: Increased cardiac sympathetic innervation density and SPN arborization. *Am. J. Physiol.-Regul. Integr. Comp. Physiol.* **2010**, *299*, 985–995. [CrossRef] [PubMed]
59. Cowley, K.C. A new conceptual framework for the integrated neural control of locomotor and sympathetic function: Implications for exercise after spinal cord injury. *Appl. Physiol. Nutr. Metab.* **2018**, *43*, 1140–1150. [CrossRef]

60. Waldron, N.H.; Fudim, M.; Mathew, J.P.; Piccini, J.P. Neuromodulation for the treatment of heart rhythm disorders. *JACC Basic Transl. Sci.* **2019**, *4*, 546–562. [CrossRef]
61. Zhu, C.; Hanna, P.; Rajendran, P.S.; Shivkumar, K. Neuromodulation for ventricular tachycardia and atrial fibrillation: A clinical scenario-based review. *JACC Clin. Electrophysiol.* **2019**, *5*, 881–896. [CrossRef] [PubMed]

biology

MDPI

Systematic Review

Effects of Exercise during Pregnancy on Postpartum Depression: A Systematic Review of Meta-Analyses

Priscila Marconcin [1], Miguel Peralta [2,3], Élvio R. Gouveia [4,5], Gerson Ferrari [6], Eliana Carraça [7], Andreas Ihle [8,9,10] and Adilson Marques [2,3,*]

1. Faculdade de Motricidade Humana, Universidade de Lisboa, 1499-002 Cruz Quebrada, Portugal; priscilamarconcin@fmh.ulisboa.pt
2. Centro Interdisciplinar de Estudo da Performance Humana (CIPER), Faculdade de Motricidade Humana, Universidade de Lisboa, 1499-002 Cruz Quebrada, Portugal; mperalta@fmh.ulisboa.pt
3. Instituto de Saúde Ambiental (ISAMB), Universidade de Lisboa, 1649-004 Lisbon, Portugal
4. Departamento de Educação Física e Desporto, Universidade da Madeira, 9000-390 Funchal, Portugal; erubiog@staff.uma.pt
5. Interactive Technologies Institute, LARSyS, 9000-390 Funchal, Portugal
6. Escuela de Ciencias de la Actividad Física, el Deporte y la Salud, Universidad de Santiago de Chile (USACH), Santiago 7500618, Chile; gerson.demoraes@usach.cl
7. Centro de Investigação em Desporto, Educação Física, Exercício e Saúd (CIDEFES), Faculdade de Educação Física e Desporto, Universidade Lusófona de Humanidades e Tecnologias, 1749-024 Lisbon, Portugal; elianacarraca@gmail.com
8. Center for the Interdisciplinary Study of Gerontology and Vulnerability, University of Geneva, 1205 Geneva, Switzerland; andreas.ihle@unige.ch
9. Swiss National Centre of Competence in Research LIVES—Overcoming Vulnerability: Life Course Perspectives, 1015 Lausanne, Switzerland
10. Department of Psychology, University of Geneva, 1205 Geneva, Switzerland

* Correspondence: amarques@fmh.ulisboa.pt

Citation: Marconcin, P.; Peralta, M.; Gouveia, É.R.; Ferrari, G.; Carraça, E.; Ihle, A.; Marques, A. Effects of Exercise during Pregnancy on Postpartum Depression: A Systematic Review of Meta-Analyses. *Biology* **2021**, *10*, 1331. https://doi.org/10.3390/biology10121331

Academic Editor: Johannes Vogel

Received: 24 November 2021
Accepted: 12 December 2021
Published: 15 December 2021

Simple Summary: Postpartum depression (PPD) is a public health problem. Exercise is a nonpharmacologic alternative to deal with PPD. This study conducted a systematic review of previous meta-analyses and an exploratory pooled analysis regarding the effects of exercise on depressive symptoms among women during the postpartum period. We searched for previous meta-analyses of experimental studies. Of the 52 records selected, we included five in the analyses, because they were focused on PPD. From the results, it was clear that exercise had a significant but small effect on depressive symptoms. This study shows that exercise is effective in reducing PPD symptoms.

Abstract: Postpartum depression (PPD) is a public health issue. Exercise is a nonpharmacologic alternative to deal with PPD. This study conducted a systematic review of previous meta-analyses and an exploratory pooled analysis regarding the effects of exercise on depressive symptoms among women during the postpartum period. We searched for previous meta-analyses of randomised controlled trials on PubMed, Web of Science and Scopus, date of inception to 31 May 2021. The methodological quality was assessed using the Assessment of Multiple Systematic Reviews 2 (AMSTAR2) instrument. We pooled the standardised mean differences from the selected studies. Of the 52 records screened, five were included. The results revealed a significant moderate effect of exercise on depressive symptoms among women during the postpartum period (SMD = −0.53; 95% CI: −0.80 to −0.27, $p < 0.001$). The pooled effect of the five meta-analyses established that exercise had a significant, small effect on depressive symptoms (SMD = −0.41; 95% CI: −0.50 to −0.32, $p < 0.001$). Our study indicates that exercise is effective in reducing PPD symptoms. Compared with traditional control approaches (psychosocial and psychological interventions), exercise seems have a superior effect on PPD symptoms. The implications of the present synthesis of past meta-analytical findings to guide health policies and research are discussed.

Keywords: mental health; physical activity; sports

1. Introduction

The prevalence of postpartum depression (PPD) varies between 11.9% and 19.2% during the perinatal period [1,2]. PPD refers to minor and major depression incidents that occur during pregnancy or shortly after (up until 12 months after birth) [3]. Due to its similarities to pregnancy and puerperium discomforts symptoms, PPD frequently goes undetected [4]. The symptoms of PPD embrace feeling sad or having a depressed mood, being uninterested in the new-born, unreasonable crying and fear of injuring or harming the baby [5]. Consequently, PPD can negatively impact the mother's well-being and the baby's development [6]. The impact on a child can be short for cognitive and motor development [7]. It can also be expressed in the long term, namely on psychological outcomes during adolescence [8]. For women, the impact can also be in the long term, as those reporting thoughts of self-harm after giving birth are known to have an increased risk of morbidity for the next seven years [9]. In addition, frequently, women with PPD also experience anxiety disorders [10]. Regarding risk factors, a history of major depression, lifetime anxiety disorder diagnosis and adverse life events were all considered important predictors of PPD [11].

PPD can be managed with psychotherapy, medication, lifestyle changes, a supportive environment or a combination of these [12]. Although medication is a feasible alternative, many women have constraints due to continuing breastfeeding [13]. Therefore, exercise can be an alternative that could help to deal with PPD. Exercise can be used as a preventive or treatment of mild depression at an early stage and as an addition to a treatment plan for major depressive disorder [12]. Exercising during pregnancy and postpartum improves psychological health and also benefits physical fitness [14,15], weight gain control [16,17] and the prevention or reduction of musculoskeletal discomfort and pain [18,19]. Therefore, the American College of Obstetrics and Gynaecologists has recommended that women during pregnancy and postpartum engage in moderate-intensity physical activity almost every day for 30 min a day [20].

Several meta-analyses have examined the exercise interventions' efficacy in preventing or reducing PPD symptoms [21–23]. These meta-analyses vary in the aim, the type of exercise intervention and study quality, hampering the overarching conclusions. However, no study has been done to synthesise these findings. Therefore, an overview of the existing systematic reviews is an efficient way to gather and summarise the best available evidence on the effectiveness of an intervention [24]. Additionally, it can provide useful and update evidence for practitioners and clinicians. Thus, this systematic review and meta-analysis aimed to (1) appraise past meta-analyses regarding the effects of exercise on PPD and (2) synthesise past meta-analytical findings to guide health policies and research.

2. Materials and Methods

2.1. Literature Search

This systematic review protocol was registered at PROSPERO (CRD42021254814), and the systematic review itself followed the PRISMA 2020 guidelines [25].

For selecting manuscripts, first, a screening based on titles and abstracts was performed, followed by full-text reads to establish the final selection. In both the screening and full-text read stages, two researchers performed the analysis (PM and AM). In case of disagreement, another researcher (MP) was asked to mediate.

We conducted a broad search on meta-analyses published until 31 May 2021 using PubMed, Web of Science and Scopus. The search terms and strategy were: "physical activ*" OR "physical inactiv" OR exercise OR training OR sport* OR fitness OR "movement behavio*" OR walking OR running OR yoga OR jogging OR swimming OR cycl* AND depress* OR "mental health" OR mood OR "psychological health" OR "psychological function*" OR "mental function*" OR worries OR worry OR "depressive disorder*" OR "baby blues" AND postpartum OR postpartum OR postnatal OR post-natal NOT Rats. They were limited to English articles.

2.2. Eligibility Criteria

The included articles met the following PICOS (participants, intervention, comparison, outcome and study design) guidelines [26]: (1) Population: women during the first year postpartum, (2) Intervention: any exercise intervention, (3) Comparison: any comparison condition, (4) Outcome: PPD or depressive symptoms and (5) Study design: meta-analyses of RCTs published from data inception to 31 May 2021. Meta-analyses involving animals were excluded.

2.3. Quality Assessment

The methodological quality of the studies was assessed by two independent authors (PM and AM) using the Assessment of Multiple Systematic Reviews (AMSTAR). This instrument uses dichotomous scoring (0 or 1) for assessing systematic reviews and meta-analyses' rigour. The scores range from 0 to 11, and studies are graded as high-quality (score between 8 and 11), medium-quality (score between 4 and 7) and low-quality (score between 0 and 3) [27]. The authors discussed the discrepancies in grading and reached a consensus.

2.4. Data Extraction

The study characteristics were extracted based on PICO criteria (population, intervention, comparison and outcomes) [26] by one author, as well as standardised mean differences (SMD) from meta-analytic comparisons. The SMD was classified as trivial if <0.20, small between 0.20 and 0.49, medium between 0.50 and 0.79 and large if >0.80 [28]. The following information was extracted: number of randomised controlled trials (RCTs) and sample included in each comparison.

2.5. Statistical Analysis

Heterogeneity data of the meta-analytic comparisons was assessed using the I^2 statistics, where reported values of 0–25%, 2550%, 50–75% and >75% indicated, respectively, low, moderate, large and very large inconsistencies [29]. Fixed effects were used in the meta-analysis of the meta-analyses. Pooled effect sizes were expressed by the standardised mean differences (SMD) of the effects of exercise on postpartum depressive symptoms. In addition, random effects were used to obtain the pooled effect size (i.e., SMD) derived from the RCTs included in the identified meta-analysis studies, excluding the overlapped studies/RCTs. The heterogeneity using the I^2 statistic, tau-square and Z-test for the overall effect were assessed. All statistical analyses and calculations were performed by Comprehensive Meta-Analysis version 3.0 software. In cases of information unavailability in the study, the authors were contacted and asked to complement the data extraction.

3. Results

3.1. Included Meta-Analyses

Figure 1 presents the PRISMA flow diagram of the selection process. The database search included 103 publications. After removing duplicates, 52 articles remained. During screening (title and abstract), 42 articles were excluded. Some interventions did not consist of exercise or physical activity; others did not include women in the postpartum period, and some were not meta-analyses. The remaining 10 articles were designated for a full-text read. Five articles were excluded, because the focus was on antenatal depression, including any study beyond RCTs. Finally, five meta-analyses were included in the present study [30–34].

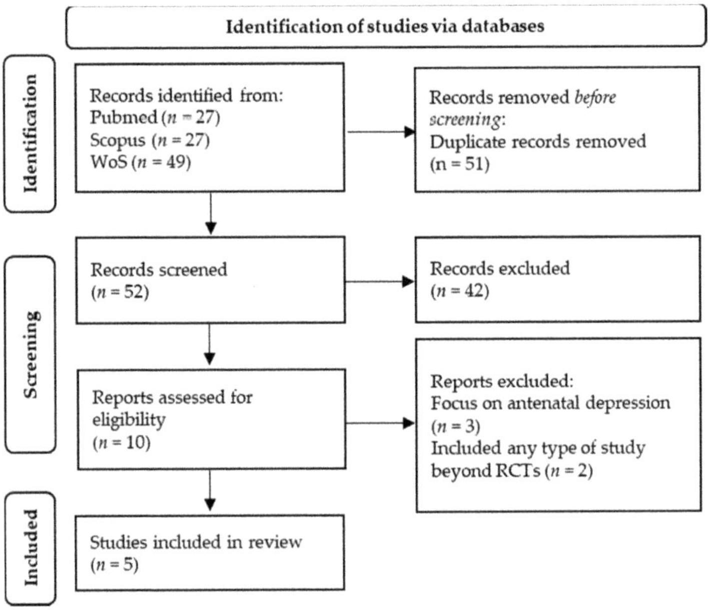

Figure 1. Preferred Reporting Items for Systematic Reviews and Meta-analysis (PRISMA) flow diagram of the search and selection process for the meta-analyses included in the current study.

3.2. Characteristics of Meta-Analyses

Table 1 presents a summary of the included meta-analyses following the PICO search strategy. All five included meta-analyses reported the effects of exercise interventions on postpartum depressive symptoms.

Table 1. Summary of the meta-analyses included in the current study following the PICO search strategy.

Reference	Included RCT's	Population	Interventions Characteristics	Comparison	Outcomes Measures
Carter et al. [30]	17	1927 primiparous or multiparous postnatal women.	Exercise-based (supervised, unsupervised, coaching-based, motivational, behavioural-oriented, universal, targeted or treatment based, in a community or clinical context).	Any control condition (including exercise).	Depression symptoms using a validated assessment tool (e.g., EPDS and PHQ).
Daley et al. [31]	5	221 women who were between 4 weeks and 18 months postpartum.	The exercise was defined as any planned, structured and repetitive bodily movement. Trials involving exercise with additional interventions (co-interventions) were eligible.	Social support intervention and standard care.	Clinical interview screened for probable depression using a recognised or diagnosed according to the clinical judgment of a health professional.
Mc Curdy et al. [32]	16	1327 postpartum women with and without depression.	Postpartum exercise (supervised or unsupervised exercise interventions).	Standard care.	Depressive symptoms or depressive episodes assessed by a validated questionnaire (e.g., EPDS, CES-D and HDRS).

Table 1. *Cont.*

Reference	Included RCT's	Population	Interventions Characteristics	Comparison	Outcomes Measures
Pritchett et al. [33]	13	1734 women up to 1 year postpartum.	Aerobic exercise, counselling exercise and group exercise.	Standard care.	Depressive symptoms measured by questionnaire or diagnostic interview.
Poyatos-León et al. [34]	12	932 pregnant women with a single foetus and an uncomplicated pregnancy or women who had a child aged between 6 weeks and 18 months: 471 in the intervention group and 461 in the control group.	Stretching and breathing exercises, a walking program, cardiovascular exercises, mixed cardiovascular and strength exercises, Pilates and yoga exercises and home-based programs.	Any intervention during pregnancy and postpartum period.	Depression scale: the EPDS or the BDI.

Abbreviations: BDI, Beck Depression Inventory; CESD-D, Center for Epidemiological Studies Depression Scale; EPDS, Edinburgh Postnatal Depression Score; HDRS, Hamilton Depression Rating Scale; PHQ, Patient Health Questionnaire.

3.2.1. Population

In total, 63 RCTs involving 6141 participants were included. The eligibility criteria for included participants varied between studies, including:

(a) primiparous or multiparous postnatal women [30].
(b) women who were between 4 weeks and 18 months postpartum [35].
(c) postpartum women with and without depression [32].
(d) women up to 1 year postpartum [33].
(e) pregnant women with a single foetus and an uncomplicated pregnancy, or women who had a child aged between 6 weeks and 18 months [34].

It excluded pregnant women and women with psychiatric diagnoses other than depression.

3.2.2. Intervention

Regarding exercise interventions, the studies included different types of interventions: coaching-based, motivational and counselling interventions [30,33]; supervised, planned exercise [30,32,35]; home-based program [34]; aerobic exercise [33,34]; stretching and breathing exercises; a walking program; mixed cardiovascular and strength exercises and Pilates and yoga [34]. Regarding the duration of the intervention, in the Carter study [30], 76% continued up to 12 weeks, and the duration of the supervised delivered sessions ranged from 30 to 90 min. Most sessions were delivered at a moderate intensity, and the frequency ranged from 1 to 4. In the Daley study [31], all trials included interventions of 12 weeks, and the duration of the sessions ranged from 30 to 45 min. In the McCurdy study [32], the frequency of the exercise ranged from one to five times per week for 30–60 min and lasted between 6 weeks and 12 months. In the Pritchett study [33], the interventions ranged from 4 weeks to 6 months in duration, and most of the included studies aimed to achieve 30 min of moderate activity three to five times weekly. In the León study [34], the session frequency varied from 1 to 5 days per week, and the intensity was measured as low, moderate and moderate to high.

3.2.3. Comparison

Exercising was compared to any control condition (including exercise) [30], usual care [32,33,35], social support intervention [35] and any intervention during pregnancy and the postpartum period [34].

3.2.4. Outcomes Measure

The main outcome measure across the studies was depressive symptoms, using the Edinburgh Postnatal Depression Scale (EPDS) [30,32,34,35], Patient Health Questionnaire [30],

Beck Depression Inventory [34,35], clinical diagnoses [33,35], Center for Epidemiological Studies Depression Scale (CES-D) and Hamilton Depression Rating Scale (HDRS) [32].

3.2.5. Adverse Events

Little information was disclosed about potential adverse events. Poyatos-Leon et al. [34] mentioned that most studies did not disclose adverse effects attributable to interventions. The other four meta-analyses did not report information about adverse events from the included studies.

3.3. Methodological Quality

Table 2 presents an item-by-item quality assessment for each study using the AMSTAR 2 instrument. The meta-analyses by Carter et al. [30], Pritchett et al. [33] and Poyatos-León et al. [34] were classified as moderate quality, and Daley et al. [35] and McCurdy et al. [32] were classified as low quality essentially, because both studies did not discuss how the publication biases impacted the results of their reviews.

Table 2. Quality of the meta-analyses according to the AMSTAR 2 criteria.

AMSTAR 2 Criteria	Carter et al. [30]	Daley et al. [31]	McCurdy et al. [32]	Pritchett et al. [33]	Poyatos-León et al. [34]
1. Did the research questions and inclusion criteria for the review include the components of PICO?	V	-	-	V	-
2. Did the report of the review contain an explicit statement that the review methods were established before the conduct of the review, and did the report justify any significant deviations from the protocol?	V	-	-	V	V
3. Did the review authors explain their selection of the study designs for inclusion in the review?	V	V	V	V	V
4. Did the review authors use a comprehensive literature search strategy?	V	V	V	V	V
5. Did the review authors perform study selection in duplicate?	V	X	V	V	V
6. Did the review authors perform data extraction in duplicate?	V	V	V	V	V
7. Did the review authors provide a list of excluded studies and justify the exclusions?	V	V	V	V	V
8. Did the review authors describe the included studies in adequate detail?	V	V	V	V	V
9. Did the review authors use a satisfactory technique for assessing the risk of bias (RoB) in individual studies included in the review?	V	V	V	V	V
10. Did the review authors report on funding sources for the studies included in the review?	-	-	-	-	-
11. If meta-analysis was performed, did the review authors use appropriate methods for statistical combination of results?	V	V	V	V	V
12. If meta-analysis was performed, did the review authors assess the potential impact of RoB in individual studies on the results of the meta-analysis or other evidence synthesis?	V	-	V	V	V
13. Did the review authors account for RoB in individual studies when interpreting/discussing the results of the review?	V	-	-	V	V
14. Did the review authors provide a satisfactory explanation for, and discussion of, any heterogeneity observed in the results of the review?	V	V	V	V	V

Table 2. *Cont.*

AMSTAR 2 Criteria	Carter et al. [30]	Daley et al. [31]	McCurdy et al. [32]	Pritchett et al. [33]	Poyatos-León et al. [34]
15. If they performed quantitative synthesis, did the review authors carry out an adequate investigation of publication bias (small study bias) and discuss its likely impact on the review results?	V	-	-	-	V
16. Did the review authors report any potential sources of conflict of interest, including any funding they received for conducting the review?	V	V	V	V	V
	MQR	LQR	LQR	MQR	MQR

Abbreviation: MQR, moderate quality review; LQR, low-quality review; V, meets the criteria; X, does not meet the criteria.

3.4. Quality of Evidence

Only Carter's study [30] reported a quality of evidence and presented that the overall quality of evidence for exercise on depression symptoms was low and with a small treatment effect.

3.5. Results of Individual Meta-Analyses

The review findings of the individual meta-analysis are summarised in Table 3. All five meta-analyses reported the effects of different types of exercise interventions (e.g., coaching-based, aerobic and yoga) on depressive symptoms among women during the postpartum period. Similar methods to calculate aggregated effects were used, namely SMD using a random effects model. All the included meta-analyses weighted the studies to give larger samples more influence. As can be seen, a SMD reduction in depression symptoms was observed for all the meta-analyses, favouring the exercise interventions group. Carter et al. [30] found a medium, significant SMD, although the I^2 was 85%. The results of a post hoc sensitivity analysis presented small, significant results and a moderate heterogeneity (SMD = −0.25, 95% CI: −0.39 to −0.11, $p < 0.001$, $I^2 = 29\%$). Daley et al. [31] found a large, significant SMD. However, a significant heterogeneity between studies was found when one trial (included exercise as a cointervention with social support) was eliminated. The SMD was reduced to a small effect size, but the heterogeneity was zero. McCurdy et al. [32], Pritchett et al. [33] and Poyatos-Leon et al. [34] found small, significant SMD. The heterogeneity was 37% in the McCurdy study [32], 85% in the Pritchett study [33] and 33% in the Poyatos-Leon study [34].

Table 3. Review findings.

Reference	SMD (95% CI)	I^2 (%)	Conclusions
Carter et al. [30]	−0.64 (−0.96 to −0.33)	86.0%	Statistically significant medium treatment effect of exercise over control conditions for depression symptoms in postpartum women up to 52 weeks after childbirth.
Daley et al. [31]	−0.81 (−1.53 to −0.10)	81.7%	Exercise can reduce postpartum depression, but this finding is contingent on one trial that included exercise as a co-intervention.
McCurdy et al. [32]	−0.34 (−0.50 to 0.19)	37%	Post-intervention depressive symptoms were lower in the exercise compared with the control group. In women with depression, exercise improved the odds of resolving depression post-intervention by 54%.
Pritchett et al. [33]	−0.44 (−0.75 to −0.12)	85.0%	Exercise interventions significantly reduced depressive symptoms
Poyatos-León et al. [34]	−0.41 (−0.28 to −0.54)	33.1%	Decrease in postpartum depressive symptom scores favour of the physical activity group.

Abbreviations: I^2, I-squared statistic for heterogeneity; SMD, standardised mean difference.

3.6. Subgroup Analyses

The study of Carter et al. [30] performed four subgroup analyses. First, this study found that targeted prevention or treatment interventions yielded a greater effect size (SMD = −0.75, 95% CI: −1.22 to −0.28, p = 0.002) compared to the universal prevention interventions (SMD = −0.52, 95% CI: −0.99 to −0.05, p = 0.030). Second, it showed that interventions with active exercise-oriented components yielded larger effects (SMD = −1.19, 95% CI: −1.84 to −0.53, p <0.001) than those with coaching/motivational components (SMD = −0.21, 95% CI: −0.37 to −0.05, p = 0.009. Third, it showed that, when tested against an active control, the exercise-based interventions yielded a smaller effect (SMD = −0.46, 95% CI: −0.86 to −0.05, p = 0.03) than those tested against nonactive control groups (SMD = −0.70, 95% CI: −1.09 to −0.32, p < 0.001). Fourth, interventions with a shorter duration (SMD = −1.72, 95% CI; −3.05 to −0.39, p = 0.010) yielded larger effect sizes than those of longer durations (SMD = −0.52, 95% CI: −0.84 to −0.19, p = 0.002).

The study of Daley et al. [31] excluded one trial that included exercise as a cointervention with social support. The SMD was reduced to −0.42 (95% CI: −0.90 to 0.05). The weighted mean difference in the size of the effect of the Edinburgh Postnatal Depression Score was −2.03 (95% CI: −4.34 to 0.29).

The study of McCurdy et al. [32] conducted sub-analyses and showed no difference between supervised and unsupervised exercises. Although analyses with only women with postpartum depression, before starting the intervention (10 trials), exercise had a moderate effect in treating depressive symptoms (SMD = 20.48, 95% CI: 20.22–20.73, I^2 = 42%). Comparing supervised and unsupervised exercises, only supervised exercise improved depressive symptoms (SMD = 20.74, 95% CI: 20.40–21.07, I^2 = 27%).

In the study of Poyatos-Leon et al. [34], the PPD status subgroup analysis revealed an effect size of 0.67 (95% CI: 0.44–0.90) for mothers with PPD and 0.29 (95% CI: 0.14–0.45) for mothers without PPD. In the study of Pritchett et al. [33], no differences were observed comparing "depressed" postpartum populations (SMD = −0.32, 95% CI: −0.63 to −0.00, I^2 = 55%) and general postpartum populations (SMD = −0.57, 95% CI: −1.12 to −0.02, I^2 = 92). Additionally, no differences were observed comparing exercise-only interventions (SMD = −0.56, 95% CI −1.13 to 0.01, I^2 = 89%) and exercise with cointerventions (−0.35, 95% CI = −0.66 to −0.04, I^2 = 72%). In addition, no differences were observed comparing group exercise interventions (SMD = −1.10, 95% CI: −1.99 to −0.21, I^2 = 93%) and participant choice interventions such as exercise counselling with a personal choice of exercise (often exercising alone) (SMD = −0.20, 95% CI: −0.33 to −0.06, I^2 = 0%).

3.7. Pooled Summary SMD across Meta-Analyses

It was possible to pool the SMD from all the meta-analyses (n = 5), with 2564 women in the exercise group and 2626 in the control group, from 63 trials, displayed in Figure 2. This established that exercise had a significant, small effect on depressive symptoms among women during the postpartum period (SMD = −0.41, 95% CI −0.50 to −0.32, p< 0.001) using fixed models. There was a low heterogeneity (I^2 = 2%).

Study name	Std diff in means	Standard error	Variance	Lower limit	Upper limit	Z-value	*p*-value
Armstrong 2003	-1.640	0.497	0.247	-2.615	-0.665	-3.297	0.001
Armstrong 2004	-1.090	0.464	0.216	-2.000	-0.180	-2.347	0.019
Boath 2015	-1.070	0.434	0.188	-1.920	-0.220	-2.466	0.014
Buttner 2015	-0.460	0.266	0.071	-0.982	0.062	-1.729	0.084
Da Costa 2009	-0.080	0.211	0.045	-0.494	0.334	-0.379	0.705
Daley 2008	-0.220	0.347	0.121	-0.900	0.460	-0.634	0.526
Daley 2015	-0.410	0.216	0.047	-0.833	0.013	-1.899	0.058
Forsythet 2017	-0.170	0.400	0.160	-0.955	0.615	-0.425	0.671
Haruna 2013	-0.130	0.201	0.041	-0.525	0.265	-0.645	0.519
Heh 2008	-0.670	0.255	0.065	-1.170	-0.170	-2.627	0.009
Huang 2011	-0.230	0.177	0.031	-0.577	0.117	-1.301	0.193
Keller 2014	-0.140	0.209	0.044	-0.549	0.269	-0.670	0.503
Ko 2008	0.340	0.252	0.064	-0.155	0.835	1.347	0.178
Lewis 2014	-0.540	0.182	0.033	-0.896	-0.184	-2.970	0.003
Mohammad 2015	0.410	0.236	0.056	-0.052	0.872	1.740	0.082
Norman 2010	-0.240	0.172	0.030	-0.577	0.097	-1.396	0.163
Robichaud 2008	-0.090	0.254	0.064	-0.587	0.407	-0.355	0.723
Saeedi 2015	-4.410	0.588	0.346	-5.562	-3.258	-7.502	0.000
Shelton 2015	-0.930	0.663	0.439	-2.229	0.369	-1.403	0.161
Surkan 2012	-0.160	0.099	0.010	-0.354	0.034	-1.613	0.107
Thiruppathi 2014	-4.210	0.564	0.318	-5.315	-3.105	-7.470	0.000
Yang 2018	0.090	0.179	0.032	-0.262	0.442	0.502	0.616
	-0.538	0.135	0.018	-0.803	-0.274	-3.992	0.000

Statistics for each study

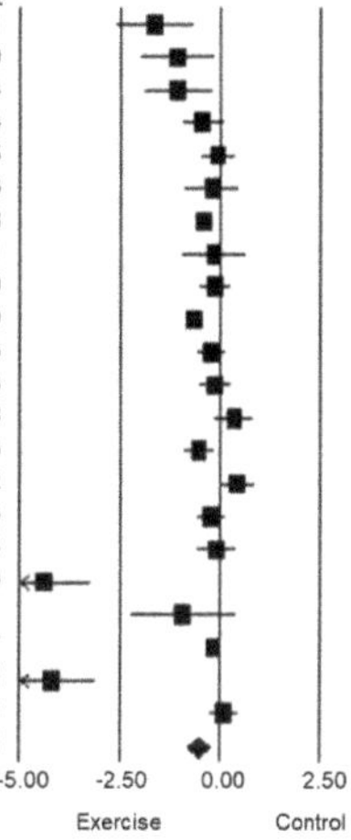

Heterogeneity: Tau^2 = 0.30; Q-value = 140.92, df = 21; p < 0.001; I^2 = 85%
Test for overall effect: Z = -3.99 (p < 0.001)

Figure 2. Forest plot of the standard mean differences of exercise intervention on postpartum depressive symptoms by the study.

3.8. Pooled Summary SMD across Studies without Overlap

The overlap of single studies within the five included meta-analyses caused a reduction in the final number of studies to 22 original studies. This analysis revealed that exercise had a significant, moderate effect on depressive symptoms among women in the postpartum period (SD = −0.53, 95% CI −0.80 to −0.27, $p < 0.001$, $I^2 = 85.9$) using a random-effects model (Figure 3).

Statistics for each study

Study	Std diff in means	Standard error	Variance	Lower limit	Upper limit	Z-value	*p*-value
Carter et al. [30]	-0.640	0.161	0.026	-0.955	-0.325	-3.986	0.000
Daley et al. [31]	-0.810	0.363	0.132	-1.521	-0.099	-2.233	0.026
McCurdy et al. [32]	-0.340	0.079	0.006	-0.495	-0.185	-4.303	0.000
Pritchett et al. [33]	-0.440	0.161	0.026	-0.755	-0.125	-2.740	0.006
Poyatos-León et al. [34]	-0.410	0.066	0.004	-0.540	-0.280	-6.190	0.000
	-0.414	0.046	0.002	-0.504	-0.324	-9.007	0.000

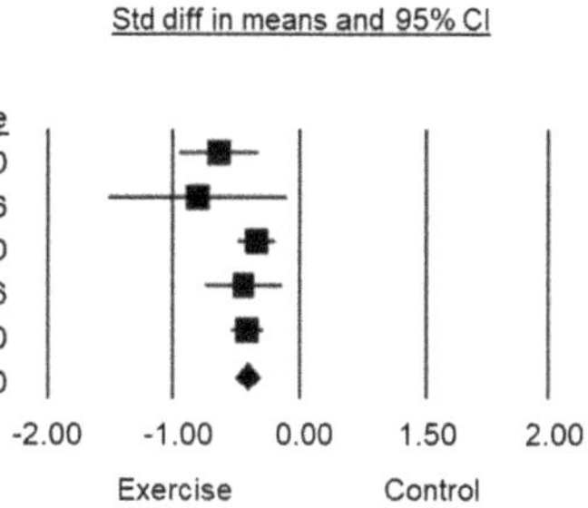

Heterogeneity: Tau^2 = 0.000; Q-value = 4.07, df = 4; p < 0.39; I^2 = 2%
Test for overall effect: Z = -9.00 (p < 0.001)

Figure 3. Forest plot of the standard mean differences of exercise intervention on postpartum depressive symptoms by the included meta-analyses.

4. Discussion

The purpose of this study was to conduct a systematic review of previous meta-analyses addressing the effects of exercise on PPD symptoms. This systematic review included five meta-analyses that comprised 6141 participants. Excluding the overlapping studies, the sample was reduced to a total of 2419 participants. We found that exercise significantly reduces PPD symptoms, with a small effect. We processed the analyses with only studies that were not overlapping, and the results remained the same: exercise significantly reduces PPD symptoms, with a moderate magnitude. Compared with traditional approaches that present a pooled effect size lower than 0.3 for any depression-related outcomes [36], exercise seems to be a feasible alternative. In addition, our results can be compared with a meta-analysis that found a significant small effect of low-intensity psychological interventions (e.g., as online cognitive behavioural therapy and self-help books) versus the usual care for depression in the general population [37]. Exercise has recognised benefits for women during the postpartum period (e.g., weight loss and pelvic floor strengthening) [38–40].

Moreover, no detrimental effects, except temporary changes in the composition of breast milk following maximal exercise [41], were found. In this sense, we might recommend exercise as a feasible alternative to control PPD symptoms. Since no serious adverse events were reported, exercise might be considered a safe intervention for this target population. However, we would like to highlight the need for more RCTs exploring the physiological and medical (after) effects of exercise in women with PPD symptoms. For example, the dose–response of exercise remains unknown so far.

Comparing the results of the included meta-analyses, the study of Daley et al. [31] presented the largest effect of exercise on PPD symptoms. However, this result considered one trial that included exercise as a cointervention. In addition, the AMSTAR quality of that particular meta-analysis was low, mainly because they did not account for the possible risk of bias in the individual studies. In second place, the study of Carter et al. [30] found a significant, moderate effect of exercise on PPD symptoms. However, when a sensitivity analysis was conducted, eliminating the studies with a high risk of bias, the magnitude of the effect became small. A small effect of exercise on PPD symptoms was found in the studies from McCurdy et al. [32], Pritchett et al. [33] and Poyatos-Leon et al. [34]. The McCurdy study's subgroup analyses showed that, for women with depression, exercise improved the odds of resolving depression post-intervention by 54%. However, we must consider that the McCurdy meta-analysis was classified as low-quality, mainly because it did not account for possible biases in the individual studies.

One point that must be highlighted is that the exercise was not tested as an exclusive treatment. PPD is a serious problem that can put both the mother and the baby at risk. In most cases, medication is used as the first option or a combination of treatments such as counselling or therapies [42]. Thus, many times, exercise appears as an adjunctive treatment. In Daley's meta-analysis [31], three out of five trials reported that the participants received other standard treatments. In Pritchett´s meta-analysis [33], five included studies presented cointerventions, such as dietary, an educational section on postpartum issues or social support. In the Carter [30], McCurdy [32] and Poyatos-Leon [34] meta-analyses, no information about adjunctive treatments was available.

We speculate that the source of the large heterogeneity between the studies came from interventions and outcomes. A variable range of exercise types was used in the trials included in the meta-analyses. The different measures and classifications were used to classify women with PDD.

5. Conclusions

As practical implications, this study provided a synthesis of reviews that practitioners and policymakers can use as an evidence map of the effectiveness of exercise on PPD symptoms. Only one study reported the level of evidence, and it was low. For future research, it is important to evaluate the preventive role of exercise during gestation on PPD symptoms, evaluate the dose–response of exercise and clarify the effects of different intervention modalities (e.g., frequency, intensity, time and type—FITT principles). Additionally, future studies can explore meta-regression analyses.

The strength of our study is that we conducted a comprehensive search including only the highest level of evidence (meta-analyses of RCT). Moreover, we provided a pooled effect size across the included studies to demonstrate the beneficial effects of exercise on PDD symptoms. Our results are of high interest to clinicians and researchers in the area considering exercise as an effective way to reduce depressive symptoms among postpartum women. However, several limitations should be acknowledged, mostly reflected by limitations in the original studies. There was a limited number of eligible meta-analyses.

The meta-analyses varied in their aims and, mostly, in the type of exercise interventions. For future studies, it is important to understand the exercise characteristics (e.g., frequency, intensity, time and type of exercise) that are most effective.

Author Contributions: Conceptualisation, formal analysis, writing—original draft and writing—review and editing, P.M. and E.C.; supervision, formal analysis and writing—review and editing, A.M.; writing—review and editing, G.F. and É.R.G.; formal analysis and writing—review and editing, M.P. and supervision and writing—review and editing, A.I. All authors have read and agreed to the published version of the manuscript.

Funding: This research received no external funding.

Institutional Review Board Statement: Not applicable.

Informed Consent Statement: Not applicable.

Data Availability Statement: Please contact the corresponding author to discuss the availability of the data and materials.

Conflicts of Interest: The authors declare no conflict of interest.

References

1. Gavin, N.I.; Gaynes, B.N.; Lohr, K.N.; Meltzer-Brody, S.; Gartlehner, G.; Swinson, T. Perinatal depression: A systematic review of prevalence and incidence. *Obstet. Gynecol.* **2005**, *106*, 1071–1083. [CrossRef]
2. Woody, C.A.; Ferrari, A.J.; Siskind, D.J.; Whiteford, H.A.; Harris, M.G. A systematic review and meta-regression of the prevalence and incidence of perinatal depression. *J. Affect. Disord.* **2017**, *219*, 86–92. [CrossRef]
3. Gaynes, B.N.; Gavin, N.; Meltzer-Brody, S.; Lohr, K.N.; Swinson, T.; Gartlehner, G.; Brody, S.; Miller, W.C. Perinatal Depression: Prevalence, Screening Accuracy, and Screening Outcomes: Evidence Report/Technology Assessment, Number 119. *PsycEXTRA Dataset* **2005**, *2005*, 1–8. [CrossRef]
4. Klein, M.H.; Essex, M.J. Pregnant or depressed? The effect of overlap between symptoms of depression and somatic complaints of pregnancy on rates of major depression in the second trimester. *Depression* **1994**, *2*, 308–314. [CrossRef]
5. American Psychiatric Association. *Diagnostic and Statistical Manual of Mental Disorders (DSM-5®)*; American Psychiatric Association: Washington, DC, USA, 2013.
6. Logsdon, M.C.; Wisner, K.L.; Pinto-Foltz, M.D. The Impact of Postpartum Depression on Mothering. *J. Obstet. Gynecol. Neonatal Nurs.* **2006**, *35*, 652–658. [CrossRef]
7. Cornish, A.; McMahon, C.; Ungerer, J.; Barnett, B.; Kowalenko, N.; Tennant, C. Postnatal depression and infant cognitive and motor development in the second postnatal year: The impact of depression chronicity and infant gender. *Infant Behav. Dev.* **2005**, *28*, 407–417. [CrossRef]
8. Sanger, C.; Iles, J.; Andrew, C.S.; Ramchandani, P. Associations between postnatal maternal depression and psychological outcomes in adolescent offspring: A systematic review. *Arch. Women's Ment. Health* **2015**, *18*, 147–162. [CrossRef] [PubMed]
9. Iliadis, S.I.; Skalkidou, A.; Ranstrand, H.; Georgakis, M.K.; Axfors, C.; Papadopoulos, F.C. Self-Harm Thoughts Postpartum as a Marker for Long-Term Morbidity. *Front. Public Health* **2018**, *6*, 34. [CrossRef] [PubMed]
10. Wisner, K.L.; Sit, D.K.Y.; McShea, M.C.; Rizzo, D.M.; Zoretich, R.A.; Hughes, C.L.; Eng, H.F.; Luther, J.F.; Wisniewski, S.; Costantino, M.L.; et al. Onset Timing, Thoughts of Self-harm, and Diagnoses in Postpartum Women with Screen-Positive Depression Findings. *JAMA Psychiatry* **2013**, *70*, 490–498. [CrossRef] [PubMed]
11. Guintivano, J.; Sullivan, P.F.; Stuebe, A.M.; Penders, T.; Thorp, J.; Rubinow, D.R.; Meltzer-Brody, S. Adverse life events, psychiatric history, and biological predictors of postpartum depression in an ethnically diverse sample of postpartum women. *Psychol. Med.* **2018**, *48*, 1190–1200. [CrossRef]
12. American Psychiatric Association. *Practice Guideline for the Treatment of Patients with Major Depressive Disorder*; American Psychiatric Association: Washington, DC, USA, 2010.
13. Turner, K.M.; Sharp, D.; Folkes, L.; Chew-Graham, C. Women's views and experiences of antidepressants as a treatment for postnatal depression: A qualitative study. *Fam. Pract.* **2008**, *25*, 450–455. [CrossRef] [PubMed]
14. de Oliveria Melo, A.S.; Silva, J.L.; Tavares, J.S.; Barros, V.O.; Leite, D.F.; Amorim, M.M. Effect of a physical exercise program during pregnancy on uteroplacental and fetal blood flow and fetal growth: A randomized controlled trial. *Obstet. Gynecol.* **2012**, *120*, 302–310. [CrossRef] [PubMed]
15. Price, B.B.; Amini, S.B.; Kappeler, K. Exercise in pregnancy: Effect on fitness and obstetric outcomes-a randomized trial. *Med. Sci. Sports Exerc.* **2012**, *44*, 2263–2269. [CrossRef] [PubMed]
16. Muktabhant, B.; Lumbiganon, P.; Ngamjarus, C.; Dowswell, T. Interventions for preventing excessive weight gain during pregnancy. *Cochrane Database Syst. Rev.* **2012**, *4*, Cd007145.

17. Elliott-Sale, K.J.; Barnett, C.T.; Sale, C. Exercise interventions for weight management during pregnancy and up to 1 year postpartum among normal weight, overweight and obese women: A systematic review and meta-analysis. *Br. J. Sports Med.* **2015**, *49*, 1336–1342. [CrossRef]
18. Liddle, S.D.; Pennick, V. Interventions for preventing and treating low-back and pelvic pain during pregnancy. *Cochrane Database Syst. Rev.* **2015**, *2015*, CD001139. [CrossRef]
19. Owe, K.M.; Bjelland, E.K.; Stuge, B.; Orsini, N.; Eberhard-Gran, M.; Vangen, S. Exercise level before pregnancy and engaging in high-impact sports reduce the risk of pelvic girdle pain: A population-based cohort study of 39 184 women. *Br. J. Sports Med.* **2016**, *50*, 817–822. [CrossRef]
20. Artal, R.; O'Toole, M.; White, S. Guidelines of the American College of Obstetricians and Gynecologists for exercise during pregnancy and the postpartum period. *Br. J. Sports Med.* **2003**, *37*, 6–12. [CrossRef]
21. Nakamura, A.; van der Waerden, J.; Melchior, M.; Bolze, C.; El-Khoury, F.; Pryor, L. Physical activity during pregnancy and postpartum depression: Systematic review and meta-analysis. *J. Affect. Disord.* **2019**, *246*, 29–41. [CrossRef] [PubMed]
22. Davenport, M.H.; McCurdy, A.P.; Mottola, M.F.; Skow, R.J.; Meah, V.L.; Poitras, V.J.; Garcia, A.J.; Gray, C.E.; Barrowman, N.; Riske, L.; et al. Impact of prenatal exercise on both prenatal and postnatal anxiety and depressive symptoms: A systematic review and meta-analysis. *Br. J. Sports Med.* **2018**, *52*, 1376–1385. [CrossRef]
23. Kwon, R.; Kasper, K.; London, S.; Haas, D.M. A systematic review: The effects of yoga on pregnancy. *Eur. J. Obstet. Gynecol. Reprod. Biol.* **2020**, *250*, 171–177. [CrossRef]
24. Ioannidis, J.P. Integration of evidence from multiple meta-analyses: A primer on umbrella reviews, treatment networks and multiple treatments meta-analyses. *Can. Med. Assoc. J.* **2009**, *181*, 488–493. [CrossRef]
25. Page, M.J.; McKenzie, J.E.; Bossuyt, P.M.; Boutron, I.; Hoffmann, T.C.; Mulrow, C.D.; Shamseer, L.; Tetzlaff, J.M.; Akl, E.A.; Brennan, S.E.; et al. The PRISMA 2020 statement: An updated guideline for reporting systematic reviews. *BMJ* **2021**, *372*, n71. [CrossRef]
26. Smith, V.; DeVane, D.; Begley, C.M.; Clarke, M. Methodology in conducting a systematic review of systematic reviews of healthcare interventions. *BMC Med. Res. Methodol.* **2011**, *11*, 15. [CrossRef]
27. Shea, B.J.; Grimshaw, J.M.; Wells, G.A.; Boers, M.; Andersson, N.; Hamel, C.; Porter, A.C.; Tugwell, P.; Moher, D.; Bouter, L.M. Development of AMSTAR: A measurement tool to assess the methodological quality of systematic reviews. *BMC Med. Res. Methodol.* **2007**, *7*, 10. [CrossRef] [PubMed]
28. Cohen, J. A power primer. *Psychol. Bull.* **1992**, *112*, 155–159. [CrossRef] [PubMed]
29. Higgins, J.P.T.; Thompson, S.G.; Deeks, J.J.; Altman, D.G. Measuring inconsistency in meta-analyses. *BMJ* **2003**, *327*, 557–560. [CrossRef]
30. Carter, T.; Bastounis, A.; Guo, B.; Morrell, C.J. The effectiveness of exercise-based interventions for preventing or treating postpartum depression: A systematic review and meta-analysis. *Arch. Women's Ment. Health* **2019**, *22*, 37–53. [CrossRef] [PubMed]
31. Daley, A.; Jolly, K.; MacArthur, C. The effectiveness of exercise in the management of post-natal depression: Systematic review and meta-analysis. *Fam. Pract.* **2009**, *26*, 154–162. [CrossRef]
32. McCurdy, A.P.; Boulé, N.G.; Sivak, A.; Davenport, M.H. Effects of Exercise on Mild-to-Moderate Depressive Symptoms in the Postpartum Period: A Meta-analysis. *Obstet. Gynecol.* **2017**, *129*, 1087–1097. [CrossRef]
33. Pritchett, R.V.; Daley, A.J.; Jolly, K. Does aerobic exercise reduce postpartum depressive symptoms? A systematic review and meta-analysis. *Br. J. Gen. Pract.* **2017**, *67*, e684–e691. [CrossRef] [PubMed]
34. Poyatos-León, R.; García-Hermoso, A.; Sanabria-Martínez, G.; Alvarez-Bueno, C.; Cavero-Redondo, I.; Martínez-Vizcaíno, V. Effects of exercise-based interventions on postpartum depression: A meta-analysis of randomized controlled trials. *Birth* **2017**, *44*, 200–208. [CrossRef]
35. Daley, A.J.; Jolly, K.; Sharp, D.J.; Turner, K.M.; Blamey, R.V.; Coleman, S.; McGuinness, M.; Roalfe, A.K.; Jones, I.; MacArthur, C. The effectiveness of exercise as a treatment for postnatal depression: Study protocol. *BMC Pregnancy Childbirth* **2012**, *12*, 45. [CrossRef] [PubMed]
36. Dennis, C.-L.; Dowswell, T. Psychosocial and psychological interventions for preventing postpartum depression. *Cochrane Database Syst. Rev.* **2013**, *2013*, 001134. [CrossRef] [PubMed]
37. Bower, P.; Kontopantelis, E.; Sutton, A.; Kendrick, T.; Richards, D.; Gilbody, S.; Knowles, S.; Cuijpers, P.; Andersson, G.; Christensen, H.; et al. Influence of initial severity of depression on effectiveness of low intensity interventions: Meta-analysis of individual patient data. *BMJ* **2013**, *346*, f540. [CrossRef]
38. Nascimento, S.L.; Pudwell, J.; Surita, F.; Adamo, K.; Smith, G. The effect of physical exercise strategies on weight loss in postpartum women: A systematic review and meta-analysis. *Int. J. Obes.* **2013**, *38*, 626–635. [CrossRef]
39. Sobhgol, S.S.; Priddis, H.; Smith, C.A.; Dahlen, H. The Effect of Pelvic Floor Muscle Exercise on Female Sexual Function During Pregnancy and Postpartum: A Systematic Review. *Sex. Med. Rev.* **2019**, *7*, 13–28. [CrossRef]
40. Davenport, M.H.; Nagpal, T.S.; Mottola, M.F.; Skow, R.J.; Riske, L.; Poitras, V.J.; Garcia, A.J.; Gray, C.; Barrowman, N.; Meah, V.L.; et al. Prenatal exercise (including but not limited to pelvic floor muscle training) and urinary incontinence during and following pregnancy: A systematic review and meta-analysis. *Br. J. Sports Med.* **2018**, *52*, 1397–1404. [CrossRef]
41. Kate, M.; Renee, K. Exercise and breastfeeding. *Breastfeed. Rev.* **2012**, *20*, 39–42.
42. Howard, L.M.; Megnin-Viggars, O.; Symington, I.; Pilling, S.; On behalf of the Guideline Development Group. Antenatal and postnatal mental health: Summary of updated NICE guidance. *BMJ* **2014**, *349*, g7394. [CrossRef]

biology MDPI

Article

Physical Training Increases Erythroferrone Levels in Men

Inga Dziembowska [1,*], Małgorzata Wójcik [2], Jakub Bukowski [1] and Ewa Żekanowska [1]

1 Department of Pathophysiology, Faculty of Pharmacy, Collegium Medicum in Bydgoszcz, Nicolaus Copernicus University in Toruń, Curie-Skłodowskiej 9, 85-094 Bydgoszcz, Poland; j_bukowski@o2.pl (J.B.); zhemostazy@cm.umk.pl (E.Ż.)

2 Institute of Health Sciences, Hipolit Cegielski State University of Applied Sciences in Gniezno, Ks. Kard. Stefana Wyszyńskiego 38, 62-200 Gniezno, Poland; malgo_wojcik@interia.pl

* Correspondence: i.dziembowska@cm.umk.pl

Simple Summary: Intense physical activity contributes to an increased consumption of oxygen transported by red blood cells. The red blood cells' differentiation and proliferation process is mainly stimulated by erythropoietin (EPO) and erythroferrone (ERFE), which are novel markers of erythroid activity. The purpose of this study was to assess the level of concentration of these hormones in athletes' blood. Seventy-three clinically healthy men took part in this study. Participants were divided into groups according to their physical activity, as assessed by the questionnaire survey. The first group included 39 athletes, the second group included 18 men with moderate physical activity, and the third—16 men with a sedentary lifestyle. Men with a high level of weekly physical activity had significantly different concentrations of ERFE and EPO than men with insufficient weekly physical activity. Higher endogenic ERFE and EPO levels are indicators of increased erythropoiesis in the period of intensified physical activity. The results obtained suggest the important role of endogenic EPO in the process of adaptation to intense physical activity.

Abstract: Intense physical activity contributes to an increased demand for red blood cells, which transport oxygen to working muscles. The purpose of this study was to assess the concentration of erythroferrone (ERFE), the novel marker of erythroid activity in athletes, during the beginning of their training season. The study group consisted of 39 athletes aged 23.24 ± 3.77 years. The study was carried out during the athletes' preparatory period of the training cycle. The control group consisted of 34 healthy men aged 22.33 ± 2.77 years. The erythropoietic activity was evaluated by determining athletes' concentrations of erythropoietin (EPO) and erythroferrone (ERFE). The level of physical activity was assessed using the International Physical Activity Questionnaire (IPAQ). In the athletes' group, we observed higher concentrations of EPO (Me = 12.65 mIU/mL) and ERFE (40.00 pg/mL) compared to the control group (EPO: Me = 5.74 mIU/ml, p = 0.001; ERFE: Me = 25.50 pg/mL, p = 0.0034). The average intensity of physical exercise significantly differentiated the participants as far as EPO and ERFE concentrations. These results suggest that intense physical activity, at least at the beginning of the training season, may stimulate EPO production, which increases ERFE release. This seems to be an adaptative mechanism that provides adequate iron for enhanced erythropoiesis.

Keywords: erythropoietin; erythroferrone; physical activity

Citation: Dziembowska, I.; Wójcik, M.; Bukowski, J.; Żekanowska, E. Physical Training Increases Erythroferrone Levels in Men. *Biology* **2021**, *10*, 1215. https://doi.org/10.3390/biology10111215

Academic Editors: Filipe Manuel Clemente, Georgian Badicu and Eugenia Murawska-Ciałowicz

Received: 9 September 2021
Accepted: 17 November 2021
Published: 21 November 2021

1. Introduction

Research has shown that erythropoiesis responds to physical activity. Exercise-induced erythropoiesis is reflected in the elevation of reticulocytes (premature erythrocytes) following physical training [1]. However, the maturation of reticulocytes in bone marrow takes 1–2 days; therefore, it is an inaccurate marker for determining immediate erythropoietic stimulation [1].

Based on established knowledge of the regulation of erythropoiesis, the primary focus of attention has been directed toward tracing circulating EPO levels. EPO is a glycoprotein hormone produced in the kidneys that stimulates the proliferation, differentiation, and maturation of erythroid progenitor cells (EPCs) in the bone marrow [2]. Previous studies have indicated that physical training leads to a mild, but temporary, increase in EPO concentrations [3,4]. The synthesis and secretion of EPO is primarily the result of hypoxia, inflammation, and endocrine system stimulation [5]. Research suggests that exercise can be employed as a model of temporary immunosuppression, which occurs during physical stress such as hypoxia [6]. Acute bouts of physical exercise also regulate the immune response, i.e., by transiently redistributing immune cells to peripheral tissues, resulting in a heightened state of immunocompetence [7]. However, it has already been proven that regular physical exercise enhances the immune function response, reinforces the antioxidative capacity, and reduces oxidative stress [8,9]. Furthermore, exercise leads to an increased destruction of RBCs with exercise, leading to compensatory erythropoiesis [10].

Numerous studies involving adult athletes practising altitude training have shown that the concentration of EPO in the blood can increase after several hours, whereas its peak occurs between 1 and 3 days from the beginning of such a taining. In the following days of a continuous reduction in oxygen partial pressure, a gradual decrease in EPO concentration is observed [11].

In 2014, Kautz et al. described a protein derived from erythroid precursor cells—erythroferrone (ERFE) [12]. The authors indicated that increased erythropoietic activity results in ERFE secretion by erythroblasts [12,13]. ERFE suppresses hepatic synthesis of the master iron-regulatory hormone, hepcidin, leading to increased availability of body iron resources for erythropoiesis [14].

Based on previous research, it can be concluded that the EPO response observed after a period of intense physical exercise may stimulate ERFE synthesis. It may be worthwhile to determine whether the level of physical activity influences EPO and ERFE concentrations. This would be particularly relevant to propose guidelines for people at risk of insufficient erythropoiesis.

Thus, the aim of the current study was the assessment of erythropoietic activity based on the erythropoietin and erythroferrone levels among athletes as compared to sedentary men.

2. Materials and Methods

2.1. Participants

The study was carried out on a total of 73 clinically healthy men (average age 22.97 $\pm$ 3.51 years). All participants were informed about the purpose of the research and familiarised with the research procedure, to which they gave their informed written consent. All participants performed biochemical tests of parameters characterising erythropoiesis and completed the International Physical Activity Questionnaire (IPAQ) to determine their physical activity characteristics. The study group consisted of 39 men with an average age of 23.24 $\pm$ 3.77 years, who regularly practised sports at least four times a week for a minimum of 1.5 h for the previous three years. Of the 39 athletes, 23 (61%) practised martial arts, including mixed martial arts (MMA), boxing, kickboxing, judo, taekwondo, and karate. The remaining 16 athletes (39%) practised team sports: football and basketball. All athletes had valid sport medical examinations.

The control group consisted of 34 healthy men: non-athletes aged 22.33 $\pm$ 2.77 years. The subjects were recruited through posters and information leaflets placed in public places (universities, clinics, and board game clubs). According to IPAQ, the inclusion criteria involved no abnormalities in the basic peripheral blood counts and, for the control group only, an insufficient level of physical activity. The exclusion criteria were as follows: smoking, chronic diseases, cancer, and taking medications or iron supplements. The study protocol was approved by the Bioethics Committee of the Nicolaus Copernicus University in Toruń functioning at Collegium Medicum in Bydgoszcz (permit No. KB/247/2014) and was performed following the ethical standards set forth in the 1964 Declaration of Helsinki and its later amendments.

2.2. Physical Activity Assessment

An IPAQ questionnaire was used to assess physical activity. It expresses physical activity in Metabolic Equivalent of the Task (MET)× min/week units, which permits the easy classification of respondents into one of three categories of activity: insufficient, moderate, and high [15]. The median physical activity for each group is presented in Table 1. In the group of sedentary men, 16 (47%) showed insufficient physical activity, and 18 (53%) showed a moderate level. All athletes (*N* = 39, 100%) declared physical activity at a high level.

Table 1. Group characteristics and biochemical parameters for the test groups.

Variable	Athletes N = 39 Me (Min–Max)	Non-Athletes N = 34 Me (Min–Max)	p	η^2
Age (years)	21.00 (18.00–32.00)	21.50 (18.00–29.00)	0.5321	0.0031
BMI (kg/m^2)	23.12 (17.28–27.15)	23.61 (17.94–28.12)	0.2431	0.0000
Physical activity (MET × min/week)	5120.00 (3100.00–6300.00)	1230.00 (340.00–2100.00)	<0.0001	0.5706
Years of training	9.00 (4.00–14.00)	N/A	N/A	N/A
Serum EPO levels (mIU/mL)	12.35 (0.90–32.25)	5.68 (0.38–32.15)	0.0001	0.1612
Serum ERFE levels (pg/mL)	40.00 (5.00–190.00)	25.5 (8.00–180.00)	0.0042	0.1861
Serum hepcidin levels (pg/mL)	8.43 (5.81–12.54)	8.21 (5.92–12.32)	0.5431	0.0001
Serum ferritin levels (ng/mL)	57.36 (20.58–158.32)	45.28 (21.52–162.45)	0.3412	0.0013
Serum sTfR levels (µg/mL)	1.28 (0.81–3.43)	1.83 (1.47–2.78)	0.0002	0.1014

EPO—erythropoietin; ERFE—erythroferrone; sTfR—serum transferrin receptor.

2.3. Blood Analysis

The test material was peripheral blood collected in the morning (7.00–9.00 a.m.) after a half-hour of rest via venipuncture from an arm. The athletes were asked not to perform any physical exercise on the day of the examination and the day before the examination to exclude the potential confounding effect of an acute bout of physical exercise on results.

Serum erythropoietin concentration was measured by an enzyme immunoassay (ELISA) using the EPO ELISA assay from Roche Diagnostics GmbH, Germany. The test is used to quantify erythropoietin in human serum and plasma. Serum erythroferrone concentration level was determined by enzyme immunoassay (ELISA) using the ELISA Kit for Erythroferrone (SEU-540Hu) from Cloud Clone, Houston, USA. The test is intended for the quantitative determination of erythroferrone concentration in human serum; however, it is not validated. To confirm that this assay detects the expected increase in ERFE, we compared the results of ERFE of blood donors (32 men, aged 21–56) before blood donation and a week after the donation, as the maximum ERFE peak is expected in 9 ± 4 days [16].

Furthermore, the following iron metabolism parameters were determined using the enzyme-linked immunosorbent method (ELISA): ferritin concentration (DRG Ferritin Kit reagent kit EIA-187) from DiaMetra, Spello, Perugia, Italy) soluble transferrin receptor concentration (Human sTfR ELISA from BioVendor Laboratory Medicine Inc. Brno, Czech Republic), and hepcidin concentration (ELISA Kit for Hepcidin, CEB-979Hu from Cloud Clone, Houston, TX, USA).

2.4. Statistical Analysis

Statistical analyses were performed using STATISTICA v. 13.1 (Statsoft, Cracov, Poland). The compliance of the distribution of individual features with the normal distribution was tested using the Shapiro–Wilk test. Linear variables were presented as median (Me), minimum (Min), and maximum (Max). The values of categorised variables are presented by quantity (N) and percentage values. To analyse differences in individual subgroups, the Mann–Whitney U test (for the comparison of 2 groups) and the Kruskal–Wallis rank test (for the comparison of groups) were used. A linearised non-linear regression model was performed to assess the relationship between EPO and ERFE in the studied groups. Differences with a p-value < 0.05 were considered statistically significant.

3. Results

The athletes showed higher ERFE and EPO concentrations in comparison to the non-athletes (Table 1).

Serum ERFE and EPO concentrations among highly physically active men were significantly higher than those declaring physical activity at an insufficient level. Furthermore, we observed that, although there were no between-group differences in hepcidin and ferritin levels, athletes presented lower serum transferrin receptor (sTfR) levels (Me 1.18 µg/mL) when compared to non-athletes (Me 1.83 µg/mL; $p = 0.0001$).

Athletes showed higher (Me = 16.32 mIU/ml) serum EPO concentration levels when compared to participants presenting moderate and insufficient physical activity (Me: 8.99 mIU/mL and 5.74 mIU/mL, respectively; $p = 0.0001$; $\eta^2 = 0.1521$ and 0.2132, respectively). The comparison of EPO concentration distributions in groups of different physical activity levels is presented in Figure 1.

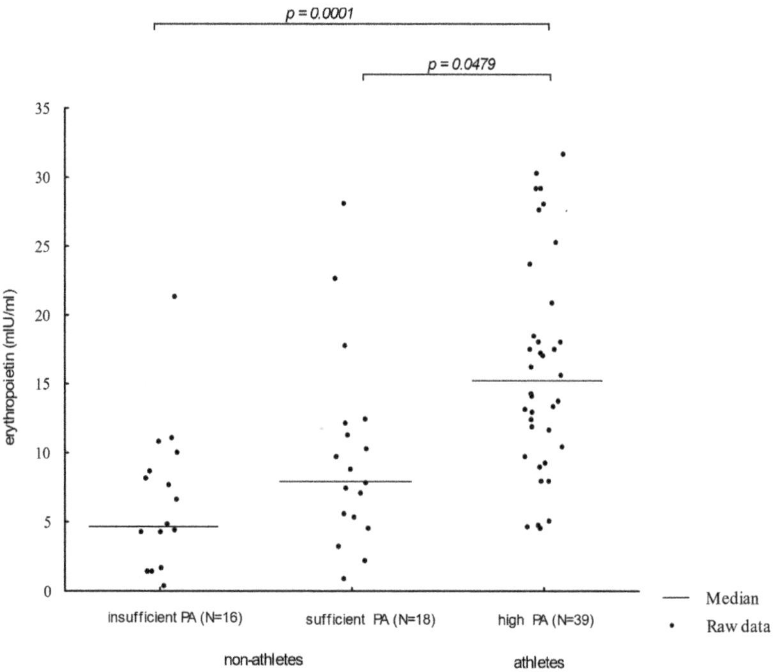

Figure 1. Erythropoietin levels in participants showing an insufficient, moderate, and high level of physical activity. PA—physical activity ($\eta^2 = 0.0811$).

Athletes showed a higher (Me = 52 pg/mL) serum ERFE concentration when compared to participants presenting moderate and insufficient physical activity (Me: 30 pg/mL and 25.50 pg/mL, respectively, $p = 0.0004$; $\eta^2 = 0.0811$ and 0.1232, respectively). The comparison of ERFE concentration distributions in groups of different physical activity levels is presented in Figure 2.

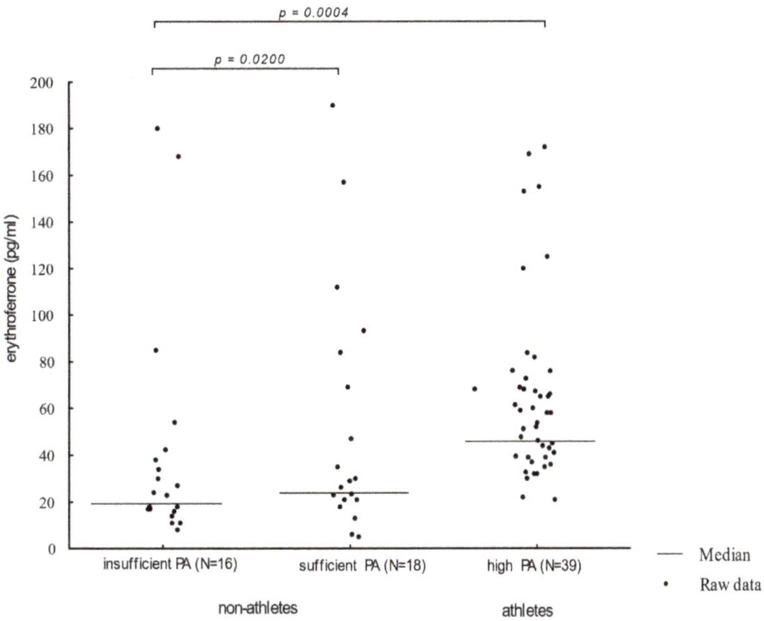

Figure 2. Erythroferrone levels in participants showing an insufficient, moderate, and high level of physical activity. PA—physical activity ($\eta^2 = 0.1230$).

Using a linearised logarithmic regression analysis, it was possible to set up a predictive model using an individual's EPO level that explained 72% of the variance in ERFE levels among athletes ($\beta_{stand} = 0.85$, $SE\ \beta_{stand} = 0.0691$, $p < 0.0001$) and 39% of the variance in ERFE levels among non-athletes ($\beta_{stand} = 0.64$, SE $\beta_{stand} = 0.1625$, $p = 0.0006$). The results of the logarithmic regression are presented in Figure 3.

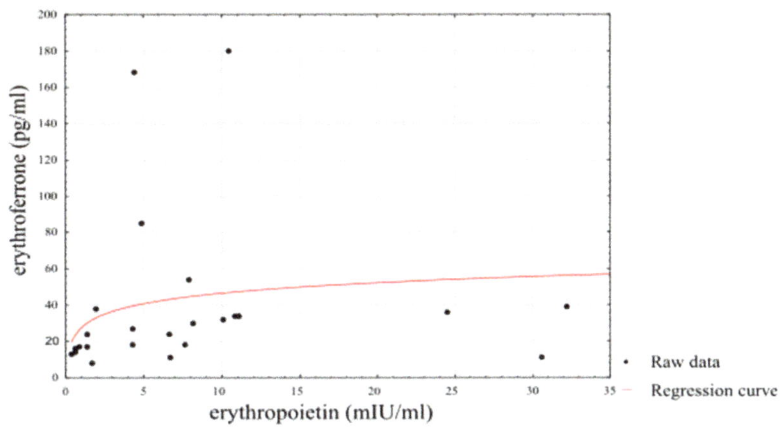

Figure 3. Linearised logarithmic regression analysis of ERFE concentrations in athletes.

4. Discussion

The aim of the current study was to compare EPO and ERFE levels between athletes and sedentary men. ERFE is produced by erythroblasts in response to EPO stimulation and mediates the inhibition of hepcidin synthesis [17]. A mouse model study showed that animals lacking the ERFE gene (*Erfe*−/−) developed anaemia; however, this phenomenon only affected the period of intensive growth [18]. Other studies in animal models have shown that bacterial pathogen-induced inflammatory anaemia, as well as increased erythropoiesis after significant blood loss, lead to increased ERFE synthesis [19]. Suppressing hepcidin expression with ERFE would exaggerate the availability of circulating iron for an increased iron demand for red blood cell haemoglobinisation [13]. In the present study, we observed greater ERFE levels among an examined group of athletes, which may suggest that intense physical effort following a recovery period enhances the erythropoietic activity, and thus ERFE synthesis. As the major factor determining ERFE production in erythroblasts is EPO, it is, therefore, possible that increased ERFE levels are connected with enhanced EPO synthesis in athletes. EPO is a hormone produced primarily in the kidneys and liver, and its most important role is to stimulate the proliferation, differentiation, and maturation of erythroid progenitor cells in the bone marrow [20]. During the last decade, researchers have also proven the local effects of EPO on blood vessel endothelium, nervous tissue, skeletal muscle, and heart muscle in response to physical or metabolic stress [21–25]. Based on these results, the role of EPO in the protection of nerve cells against oxidative stress during the process of neovascularisation is suggested [21–23]. EPO may also positively affect skeletal muscle regeneration, growth, and angiogenesis [21–25].

Expression of EPO mRNA is closely related to renal oxygen concentration, and hypoxia is considered the most important stimulus for EPO release [5]. Indeed, recent research showed that in elite athletes, EPO concentrations reached the highest value after 6 days of intermittent hypoxic exposure [24]. Previous studies also demonstrated the associations between muscle mass and erythropoiesis [25]. There are several lines of evidence supporting the concept that EPO augments exercise performance by activating an EPO receptor subtype in non-haematopoietic tissues, including skeletal muscle [26]. Recent research also draws attention to the small EPO fraction synthesised in skeletal muscle in response to intense physical exercise. Animal studies with mouse models have shown a 4–7 fold increase in EPO mRNA expression in skeletal muscle following exercise, and the increase was higher in glycolytic muscles and for trained mice [26]. The most recent data indicate that the endogenous EPO, through the EPO receptor in myocytes, controls mitochondrial biogenesis in skeletal muscle [27]. Nijholt et al. showed that a lack of EPO receptors in non-haematopoietic tissues lead to low mitochondrial content in skeletal muscles, as well as reduced myocyte growth and exercise capacity in response to voluntary exercise in mice [27]. Thus, it may be suggested that the source of erythropoietin released into circulation as a result of physical exertion may not only be from the kidneys but—at least in the early stages of exercise—from the working skeletal muscles as well [27].

Each bout of physical exercise contributes to the release of noradrenaline, cortisol, and androgens [28–31], which are hormones that stimulate the secretion of EPO, which, in turn, leads to the increased erythropoietic activity of the bone marrow [5]. Testosterone can also directly affect erythroblasts by increasing the number of receptors for EPO [28–31]. Current data have also revealed the complex interaction of noradrenaline within the bone marrow, and thus the dependence of the erythropoietic response on the intensity and duration of exercise and the associated stress response [28–31]. Review studies indicate that short-term exercise does not significantly affect EPO levels, while prolonged, intense physical effort (such as ultramarathon runs) leads to a temporary increase in EPO levels, both in trained and untrained men [4,31]. Therefore, a transient increase in EPO concentrations appears to have an adaptive significance to a substantial increase in physical effort compared to an individual's usual physical activity.

Another potential cause of higher EPO levels could be increased hemolysis and/or iron deficiency as a result of increased physical activity. However, ferritin levels, which reflect the iron pool, were similar in both groups. Recent work focusing on the characteristics of iron metabolism proves that sTfR may be a more sensitive marker for assessing iron status in athletes than ferritin [32]. An increased demand for iron in the group of athletes would be reflected by higher sTfR values. Interestingly, the results of our analysis showed that the concentration of sTfR in athletes was significantly lower than in the group of non-athletes, which reflect even lower iron demand. To complete the analysis of the endocrine system that regulates iron homeostasis, we analysed the serum hepcidin concertation in both groups. A review of 21 studies that analysed hepcidin concentration in response to endurance exercise (running, cycling, rowing, and walking) showed that post-workout increases in hepcidin concentrations typically peaked between 3 and 6 h after training and lasted up to 24 h [33,34]. In the presented study, athletes were asked not to engage in physical training for about 24 h before blood collection. Therefore, it seems that the increase in hepcidin concentrations caused by physical exertion is a temporary state, and surprisingly, ERFE may play a role in returning hepcidin to normal values.

It is worth noting that the current study was conducted during the preparatory period of the training cycle, which is characterised by intensified physical activity following the recovery period. Thus, increased physical activity may stimulate erythropoiesis and greater ERFE and EPO levels in athletes in comparison to sedentary and moderately active men. This is in line with a previous review showing a significant variation in reticulocyte counts, which represents erythropoietic activity in athletes throughout the year [1]. The authors reported generally higher reticulocyte counts at the beginning of the season but lower values after intensive training sessions, competitions, and at the end of the season [1]. The elevated EPO levels in athletes observed in the current research reflect a temporary state associated with the adaptation of the musculoskeletal system to the change in the intensity and type of physical training. Mathematical models created in this work showed that EPO concentration could be used to predict ERFE concentration within the group of athletes, explaining more variance (R^2 = 0.72) than within the group of sedentary men (R^2 = 0.39). These results may indicate the role of ERFE as a protein that provides optimal iron reserves during exercise-induced, enhanced erythropoiesis in athletes.

Despite the undoubted value in using ERFE to attempt to explain the physiological mechanism of exercise-induced erythropoiesis, there were limitations of this study introduced by the homogeneity of the study groups with respect to age, sex, and eating habits. In addition, the analysed group of athletes is not representative of all physically active men or the broadly understood population of athletes, as it was limited to only a few sport disciplines. Furthermore, the athletes are not only "highly active" people, but they are also competitors and have some genetic peculiarities. In addition, we observed increased EPO and ERFE levels in athletes compared to non-athletic individuals at one time-point, whereas it would be interesting to investigate the time-course changes in these parameters over the entire training cycle.

5. Conclusions

Based on the conducted analyses and the explained directional variability, one can assume the possible role of erythropoietin in the activation of erythroferrone that is in line with previous literature reports. Higher concentrations of EPO and ERFE observed in athletes indicate that regular, intense physical activity may stimulate the erythropoiesi process.

Author Contributions: Conceptualisation, I.D., M.W., and E.Ż.; methodology, I.D., M.W., and E.Ż.; formal analysis, I.D.; investigation I.D., M.W., and J.B.; data curation, I.D.; writing—original draft preparation, I.D., M.W., and J.B.; writing—review and editing, I.D., M.W., J.B., and E.Ż.; visualisation, I.D.; supervision, E.Ż.; project administration, I.D., M.W., J.B., and E.Ż.; funding acquisition, I.D., M.W., and E Ż. All authors have read and agreed to the published version of the manuscript.

Funding: This research received no external funding.

Institutional Review Board Statement: This study was conducted according to the guidelines of the Declaration of Helsinki and approved by the Bioethics Committee of the Nicolaus Copernicus University in Toruń functioning at Collegium Medicum in Bydgoszcz (permit No. KB/247/2014).

Informed Consent Statement: Informed consent was obtained from all subjects involved in the study.

Data Availability Statement: The data presented in this study are available on request from the corresponding author.

Acknowledgments: In this section, you can acknowledge any support given which is not covered by the author contribution or funding sections. This may include administrative and technical support, or donations in kind (e.g., materials used for experiments).

Conflicts of Interest: The authors declare no conflict of interest.

References

1. Lombardi, G.; Colombini, A.; Lanteri, P.; Banfi, G. Reticulocytes in sports medicine: An update. *Adv. Clin. Chem.* **2013**, *59*, 125–153.
2. Rybinska, I.; Cairo, G. Mutual Cross Talk Between Iron Homeostasis and Erythropoiesis. *Vitam. Horm.* **2017**, *105*, 143–160. [CrossRef] [PubMed]
3. Park, H.Y.; Park, W.; Lim, K. Living High-Training Low for 21 Days Enhances Exercise Economy, Hemodynamic Function, and Exercise Performance of Competitive Runners. *J. Sports Sci. Med.* **2019**, *18*, 427–437.
4. Montero, D.; Breenfeldt-Andersen, A.; Oberholzer, L.; Haider, T.; Goetze, J.P.; Meinild-Lundby, A.-K.; Lundby, C. Erythropoiesis with endurance training: Dynamics and mechanisms. *Am. J. Physiol. Regul. Integr. Comp. Physiol.* **2017**, *312*, R894–R902. [CrossRef] [PubMed]
5. Lobigs, L.M.; Sharpe, K.; Garvican-Lewis, L.A.; Gore, C.J.; Peeling, P.; Dawson, B.; Schumacher, Y.O. The athlete's hematological response to hypoxia: A meta-analysis on the influence of altitude exposure on key biomarkers of erythropoiesis. *Am. J. Hematol.* **2018**, *93*, 74–83. [CrossRef]
6. Campbell, J.P.; Turner, J.E. Debunking the myth of exercise-induced immune suppression: Redefining the impact of exercise on immunological health across the lifespan. *Front. Immunol.* **2018**, *9*, 648. [CrossRef]
7. Domínguez, R.; Sánchez-Oliver, A.J.; Mata-Ordoñez, F.; Feria-Madueño, A.; Grimaldi-Puyana, M.; López-Samanes, Á.; Pérez-López, A. Effects of an Acute Exercise Bout on Serum Hepcidin Levels. *Nutrients* **2018**, *10*, 209. [CrossRef]
8. Scheffer, D.D.L.; Latini, A. Exercise-induced immune system response: Anti-inflammatory status on peripheral and central organs. *Biochim. Biophys. Acta Mol. Basis Dis.* **2020**, *1866*, 165823. [CrossRef] [PubMed]
9. Kawamura, T.; Muraoka, I. Exercise-Induced Oxidative Stress and the Effects of Antioxidant Intake from a Physiological Viewpoint. *Antioxidants* **2018**, *7*, 119. [CrossRef]
10. Yusof, A.; Leithauser, R.M.; Roth, H.J.; Finkernagel, H.; Wilson, M.T.; Beneke, R. Exercise-induced hemolysis is caused by protein modification and most evident during the early phase of an ultraendurance race. *J. Appl. Physiol.* **2007**, *102*, 582–586. [CrossRef]
11. Płoszczyca, K.; Langfort, J.; Czuba, M. The Effects of Altitude Training on Erythropoietic esponse and Hematological Variables in Adult Athletes: A Narrative Review. *Front. Physiol.* **2018**, *9*, 375. [CrossRef]
12. Kautz, L.; Jung, G.; Valore, E.V.; Rivella, S.; Nemeth, E.; Ganz, T. Identification of erythroferrone as an erythroid regulator of iron metabolism. *Nat. Genet.* **2014**, *46*, 678–684. [CrossRef]
13. Ramirez Cuevas, K.; Schobinger, C.; Gottardo, E.; Voss, S.C.; Kuuranne, T.; Tissot, J.-D.; Favrat, B.; Townsend, N.; Leuenberger, N. Erythroferrone as a sensitive biomarker to detect stimulation of erythropoiesis. *Drug Test. Anal.* **2019**, *12*, 261–267. [CrossRef] [PubMed]
14. Ganz, T. Erythropoietic regulators of iron metabolism. *Free Radic. Biol. Med.* **2019**, *133*, 69–74. [CrossRef]
15. Bergier, J.; Kapka-Skrzypczak, L.; Biliński, P.; Paprzycki, P.; Wojtyla, A. Physical activity of Polish adolescents and young adults according to IPAQ: A population based study. *Ann. Agric. Environ. Med.* **2012**, *19*, 109–115. [PubMed]
16. Ganz, T.; Jung, G.; Naeim, A.; Ginzburg, Y.Z.; Pakbaz, Z.; Walter, P.B.; Kautz, L.; Nemeth, E. Immunoassay for human serum erythroferrone. *Blood* **2017**, *130*, 1243–1246. [CrossRef]
17. Kautz, L.; Jung, G.; Du, X.; Gabayan, V.; Chapman, J.; Nasoff, M.; Nemeth, E.; Ganz, T. Erythroferrone contributes to hepcidin suppression and ironoverload in a mouse model of β-thalassemia. *Blood* **2015**, *126*, 2031–2037. [CrossRef] [PubMed]
18. Kautz, L.; Jung, G.; Nemeth, E.; Ganz, T. Erythroferrone contributes to recovery from anemia of inflammation. *Blood* **2014**, *124*, 2569–2574. [CrossRef]
19. Lawen, A. Is erythroferrone finally the long sought-after systemic erythroid regulator of iron? *World J. Biol. Chem.* **2015**, *6*, 78–82. [CrossRef]
20. Muckenthaler, M.U.; Rivella, S.; Hentze, M.W.; Galy, B. A Red Carpet for Iron Metabolism. *Cell* **2017**, *168*, 344–361. [CrossRef]
21. Jelkmann, W. Erythropoietin. *Front. Horm. Res.* **2016**, *47*, 115–127.
22. Lamon, S.; Russell, A.P. The role and regulation of erythropoietin (EPO) and its receptor in skeletal muscle: How much do we really know? *Front. Physiol.* **2013**, *4*, 176. [CrossRef] [PubMed]

23. Kimáková, P.; Solár, P.; Solárová, Z.; Komel, R.; Debeljak, N. Erythropoietin and Its Angiogenic Activity. *Int. J. Mol. Sci.* **2017**, *18*, 1519. [CrossRef]
24. Kasperska, A.; Zembron-Lacny, A. The effect of intermittent hypoxic exposure on erythropoietic response and hematological variables in elite athletes. *Physiol. Res.* **2020**, *69*, 283–290. [CrossRef] [PubMed]
25. Rundqvist, H.; Rullman, E.; Sundberg, C.J.; Fischer, H.; Eisleitner, K.; Ståhlberg, M.; Sundblad, P.; Jansson, E.; Gustafsson, T. Activation of the erythropoietin receptor in human skeletal muscle. *Eur. J. Endocrinol.* **2009**, *161*, 427–434. [CrossRef]
26. Edgett, B.A.; Fraquharson, L.; Romanova, N.; Brunt, K.R.; Gurd, B.J.; Simpson, J.A. Regulation of erythropoietin expression in response to exercise in mouse and human skeletal muscle. *Fed. Am. Soc. Exp. Biol. J.* **2017**, *17*, lb748.
27. Nijholt, K.T.; Meems, L.M.G.; Ruifrok, W.P.T.; Maass, A.H.; Yurista, S.R.; Pavez-Giani, M.G.; Mahmoud, B.; Wolters, A.H.G.; van Veldhuisen, D.J.; van Gilst, W.H.; et al. The erythropoietin receptor expressed in skeletal muscle is essential for mitochondrial biogenesis and physiological exercise. *Pflugers Arch. Eur. J. Physiol.* **2021**, *473*, 1301–1313. [CrossRef] [PubMed]
28. Luccia, T.P.B.; Natali, J.E.S.; Moreira, A.; Chaui-Berlinck, J.G.; Bicudo, J.E.P.W. Bouts of exercise elicit discordant testosterone:cortisol ratios in runners and non-runners. *Arch. Endocrinol. Metab.* **2018**, *62*, 325–331.
29. Hayes, L.D.; Grace, F.M.; Baker, J.S.; Sculthorpe, N. Exercise-induced responses in salivary testosterone, cortisol, and their ratios in men: A meta-analysis. *Sports Med.* **2015**, *45*, 713–726. [CrossRef]
30. Hayes, L.D.; Herbert, P.; Sculthorpe, N.F.; Grace, F.M. Exercise training improves free testosterone in lifelong sedentary aging men. *Endocr. Connect.* **2017**, *6*, 306–310. [CrossRef]
31. Molinari, P.F. Erythropoietic mechanism of androgens: A critical review and clinical implications. *Haematologica* **1982**, *67*, 442–460. [PubMed]
32. Kawabata, H. Transferrin and transferrin receptors update. *Free Radic. Biol. Med.* **2018**, *133*, 46–54. [CrossRef] [PubMed]
33. Peeling, P.; Sim, M.; Badenhorst, C.E.; Dawson, B.; Govus, A.D.; Abbiss, C.R.; Swinkels, D.W.; Trinder, D. Iron status and the acute post-exercise hepcidin response in athletes. *PLoS ONE* **2014**, *9*, e93002.
34. Kanter, M. High-Quality Carbohydrates and Physical Performance: Expert Panel Report. *Nutr. Today* **2018**, *53*, 35–39. [CrossRef] [PubMed]

biology

MDPI

Article

Effect of 12 Weeks Core Training on Core Muscle Performance in Rhythmic Gymnastics

Paula Esteban-García [1,*], José Fernando Jiménez-Díaz [1], Javier Abián-Vicén [1], Alfredo Bravo-Sánchez [1] and Jacobo Á. Rubio-Arias [2]

1 Performance and Sport Rehabilitation Laboratory, PerlaSport Group, Faculty of Physical Activity and Sport Science, University of Castilla la Mancha, 45071 Toledo, Spain; grupo.perlabsport@uclm.es (J.F.J.-D.); javier.abian@uclm.es (J.A.-V.); alfredo.bravo@uclm.es (A.B.-S.)

2 Department of Education, University of Almeria, 04120 Almería, Spain; jararias@ual.es

* Correspondence: paula.esteban@uclm.es

Simple Summary: The aim of this study was to analyze the effect of 12 weeks of core muscle training on core muscle performance in rhythmic gymnasts. Core strength training leads to improvements in body composition, as well as improvements in trunk strength and increases in muscle electromyographic activity. These improvements could therefore improve performance during competitive rhythmic gymnastics exercises.

Abstract: Background: Rhythmic gymnastics performance is characterized by technical elements involving flexibility, aerobic capacity and strength. Increased core strength in rhythmic gymnastics could lead to improved sporting performance. Objective: The aim of this study was to analyze the effect of 12 weeks of core muscle training on core muscle performance in rhythmic gymnasts. Methods: A randomized controlled study involving 24 rhythmic gymnastics was conducted. Participants were randomly assigned to a control group (CG; n = 12; age 13.50 ± 3.17 years) or a training group (TG; n = 12; age 14.41 ± 2.35 years). Body composition, isometric strength of trunk, core endurance and core muscle electromyographic activity were measured (EMG) after 12 weeks of core training. Independent sample t-tests were carried out to compare baseline values between groups. A two-way repeated-measures analysis of variance (ANOVA) (time × group) was applied. Results: The TG improved body composition, trunk lean mass (mean differences MD = −0.31; p = 0.040), lean mass (MD = 0.43; p = 0.037) and bone mass (MD = −0.06; p < 0.001) after training. Core training increased isometric strength of trunk, flexion test (MD = −21.53; p = 0.019) and extension test (MD = 22.7; p = 0.049), as well as the prone bridge core endurance test (MD = −11.27; p = 0.040). The EMG values also increased in the TG in prone bridge for front trunk (MD = −58.58; p = 0.026). Conclusions: Core strength training leads to improvements in body composition, as well as improvements in trunk strength and increases in muscle electromyographic activity. These improvements could therefore improve performance during competitive rhythmic gymnastics exercises.

Keywords: strength; muscular activity; electromyography; core endurance test; muscular performance

Citation: Esteban-García, P.; Jiménez-Díaz, J.F.; Abián-Vicén, J.; Bravo-Sánchez, A.; Rubio-Arias, J.Á. Effect of 12 Weeks Core Training on Core Muscle Performance in Rhythmic Gymnastics. *Biology* **2021**, *10*, 1210. https://doi.org/10.3390/biology10111210

Academic Editors: Filipe Manuel Clemente, Georgian Badicu, Eugenia Murawska-Ciałowicz and Johannes Vogel

Received: 3 November 2021
Accepted: 18 November 2021
Published: 19 November 2021

1. Introduction

Rhythmic gymnastics started as a sport in the 1940s and debuted as an Olympic sport at the 1984 Olympic Games [1]. Aesthetic movements, flexibility, artistic and competitive components are distinct characteristics of rhythmic gymnasts [2]. Bobo-Arce and Méndez-Rial (2013) suggested that rhythmic gymnastics is a sport with a particular training process, very young athletes, earlier specialization, a large volume of training, lots of repetition and high levels of physical and psychological stress in competition. Elements of physical fitness such as flexibility, strength and aerobic capacity have been shown to be determinants of performance in rhythmic gymnastics [3,4]. Thus, physical, technical and psychological skills, and motor control and harmony of movement are key factors in the performance of gymnasts [2].

For appropriate control and harmony of movements, gymnasts need adequate strength development, which allows them to maintain technical elements of great amplitude. In gymnastic disciplines, to perform a maximum number of strength elements in a competition routine, a high level of specific strength endurance is required [5]. Relative strength is considered to be a more important determinant of gymnastics performance than absolute strength [6], which is why many training systems use the gymnasts' own body weight to prepare them [7]. In this respect, an example of strength training with body weight is the training of the central trunk muscles (core). It is suggested that having a strong core allows for the complete transfer of forces developed with the lower extremities through the trunk to the upper extremities [7]. Many gymnastic movements are generated in the lower body, with the flexion-extension of the legs giving rise to positions held by the whole body for a few seconds, which require isometric and stabilizing strength of the central musculature, mainly. Therefore, an adequate development of the core in rhythmic gymnasts could evoke an increase in sporting performance [1], helping the execution and maintenance of technical movements. Furthermore, a link has been established between trunk stability and lower limb injuries or low back pain [8], so that specific trunk training could reduce this risk [9].

In order to be able to assess the force generated by gymnasts or athletes, there are quantitative measurements of maximal voluntary strength that can be performed with isometric testing on isokinetic dynamometers [9]. In these tests, maximum voluntary contraction (MVC) can be performed in both flexion and extension to quantify trunk strength [8]. On the other hand, for the measurement of endurance strength in athletes, trunk tests such as the McGill test are often used to assess endurance capacity and core stability [10]. Muscle activation assessment tests, such as surface electromyography (sEMG), can also be considered useful tools for assessing muscle activation [8]. In sport, the positive relationship between muscle activation and performance can be established [11].

On the other side, the study of anthropometric variables associated with sports performance is interesting because some studies associate variables such as weight, height, body mass index and lean mass with strength [12,13]. In gymnasts, a negative relationship has been established between fat mass values and improvements in strength and performance [14]; this makes it interesting to assess the gymnasts' body composition and its possible relation to training.

In some sports, improving trunk strength and endurance can increase the ability to generate and maintain strength [15]. Demand for athletic performance responses by the muscles of the whole body and core acts as a bridge between the upper and lower extremities and provides a stable base to transfer force to the extremities [16]. Strength and endurance training of the core musculature could increase trunk stability in gymnasts, facilitating the transmission of forces generated between the upper and lower limbs [16,17]. Furthermore, it has been shown that the improvement in trunk strength is positively related to the extensor strength of this musculature, allowing gymnasts to achieve greater technical performance in all their back trunk extension movements [18].

Several studies suggest that athletes should perform trunk strength training to improve their athletic performance [10,12], demonstrating the effect of trunk training on athletes' performance. However, there are few studies that analyze the effect of specific trunk training in rhythmic gymnasts on trunk muscle performance. Considering that specific trunk training, in addition to rhythmic gymnastics training, could improve trunk strength and stability and thus indirectly improve performance, the aim of this study was to analyze the effect of 12 weeks of core training in gymnasts who were still training in rhythmic gymnastics on body composition, isometric trunk strength, trunk endurance and electromyographic activity of trunk muscles.

2. Materials and Methods

2.1. Study Design

This study used a randomized, controlled single-blind design. A quasi-experimental intra- and inter-subject design with pre- and post-test, and with a control group, was used to identify the effects of 12 weeks of core training on the performance of the core muscles. Subjects were randomized into two groups: a control group (CG) or a training group (TG).

2.2. Participants

A total of 24 national women rhythmic gymnasts (n = 24; age 13.95 ± 2.77 years; height 151.39 ± 12.34 cm; weight 43.00 ± 12.82 Kg) were randomly divided into two groups: CG (n = 12; age 13.50 ± 3.17 years; height 147.87 ± 11.63 cm; weight 38.76 ± 11.91 Kg) and TG (n = 12; age 14.41 ± 2.35 years; height 154.91 ± 12.50 cm; weight 47.25 ± 12.74 Kg). The gymnasts of both groups continued their rhythmic gymnastics training on a regular basis, and core training was only applied to the gymnasts of the TG group. All participating gymnasts followed the same training, both gymnastic and core specific. The training protocols (gymnastics and core) were designed by the study researchers and subsequently applied by the trainers, previous familiarization and an informative session. In order to ensure the process, the study's principal investigator monitored the training sessions. The inclusion criteria were that they had training experience of 2 years, competed in the national category and trained ≥9 h per week. All the gymnasts and their parents received written and verbal information regarding the nature of this investigation and provided written informed consent before the beginning of the study. Ethical approval was obtained from the Clinical Research Ethics Committee of the Toledo Healthcare Area (number 112/2015). This study complied with the ethical principles of the Declaration of Helsinki.

2.3. Procedures

The week before the start of the measurements, the gymnasts performed a 90 s warm-up and then were familiarized with the isometric and core endurance tests at moderate intensity, and in addition, signed the informed consent documents. On the day of data collection all the measurements were taken by the authors and the instruments were calibrated prior to use. First of all, stature and body mass were measured on a portable scale with a stadiometer (model 700, Seca, Hamburg, Germany) and body composition and densitometry were recorded. Then the rhythmic gymnasts completed a 10 min warm-up on a bicycle ergometer, using self-chosen resistance at 40–60 rpm (20–30 watts), followed by 5 min of stretching exercises for the trunk and lower extremities, the isometric test, and McGill's core endurance test. Surface electromyography (sEMG) of the core was recorded during the isometric and McGill's core endurance tests (Table 1).

Table 1. Study protocol.

Pre-Training	Core Training 12 Weeks	Post Training
Body composition and densitometry analysis Isometric test with EMG Core endurance test EMG		Body composition and densitometry analysis Isometric test Core endurance test

Body composition and densitometry measurements were taken following the standardized techniques of the International Society for the Advancement of Kinanthropometry (ISAK), fat mass (FM, in Kg) (ICC: 0.99–0.98; CV: 2.6%), total lean mass (LM, in Kg) (ICC: 0.99–0.99; CV: 0.8%), bone mass (BM, in Kg) (ICC: 0.99–0.99; CV: 0.6%) fat tissue percentage (FT%) (ICC: 0.99–0.99; CV: 2.7%) and trunk lean mass (TLM, in Kg) (ICC: 0.99–0.98; CV: 1.6%) were assessed using dual-energy X-ray absorptiometry (DXA) (Lunar iDXA, General Electric Healthcare, Fairfield, CT, USA) [19].

The isometric tests for maximum strength of trunk were performed with a Biodex isokinetic dynamometer (Biodex System 3; Biodex Medical Systems, Inc., Shirley, NY, USA). Maximum voluntary contraction (MVC) exerted in isometric contraction for trunk flexion and extension was evaluated in terms of peak torque (PT, in N·m) (ICC: 0.87–0.92; CV: 10.5%). Isometric strength measurements were made following the protocols described by Waldhelm and Li (2012) [20] (Figure 1). Trunk flexion and extension were performed while standing, with trunk straight, looking straight ahead, pelvis stabilized, and without upper extremity support. The average of three peak torque with 2 min rest in between was taken for later analysis. The gymnasts held each contraction for 5 s with 30 s rest between trials [19].

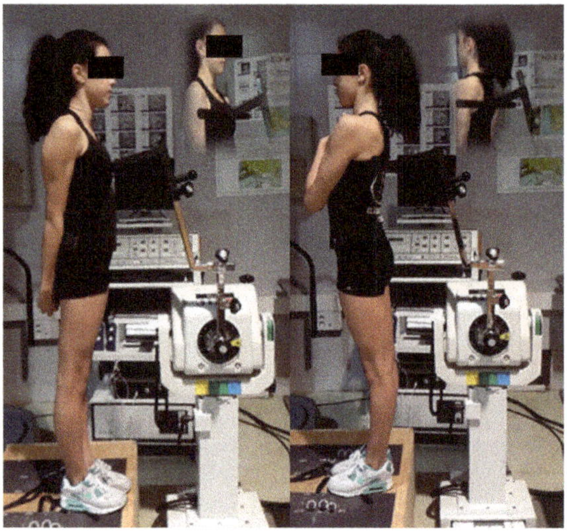

Figure 1. Isometric strength measurements.

Core endurance was measured for the same person with the McGill test [10]. The core endurance tests were the extensor endurance test or Biering-Sorensen test (Sorensen) and the prone bridge test (prone bridge). Gymnasts maintained these positions as long as possible, and the time was measured in each test in s. Both tests were considered failures when the gymnast lost the horizontal with respect to the floor. The Sorensen test began with lying prone, with the lower body manually fixed, hips extended over the edge of the test surface, and hands on the opposite shoulders. The prone bridge test was performed on the ground. The gymnasts had to maintain the prone position supporting themselves on their feet and forearms with shoulders and elbows in 90° flexion. Forearms needed to remain pronated.

sEMG was measured during McGill's core endurance and isometric tests. An 8-channel sEMG ME 6000TE (Mega Electronics, Kuopio, Finland) was used for data collection. sEMG signals from the flexor muscles of the front trunk were analyzed as a group, as were the extensor muscles of the back trunk. The average value of muscle activation (EMG root mean square (rms), EMG_{rms} in μV) (ICC: 0.87–0.94; CV: 12.8%) was measured during the middle 3 s of the 5 s of contraction. Each gymnast's skin was prepared for sEMG evaluation according to guidelines of the SENIAM organisation [21], including scrubbing and cleaning with alcohol. Electrodes were placed bilaterally on the front trunk muscles (rectus abdominis, external oblique abdominis) and back trunk muscles (erector spinae). Two 10 mm diameter Ag-AgCl surface electrodes were used on each muscle for data collection. The sampling rate was set at 1000 Hz per channel. The signals were filtered at 500 Hz, and further filtered. The raw data were stored and subsequently processed. The sEMG data were fully rectified and smoothed and the rms was normalized to the signal recorded with peak maximum value [22].

2.4. Intervention

Core muscular training was performed in two alternative sessions per week for 12 weeks, supplementary to gymnastic training, included three progressions of difficulty, periods 1, 2, and 3 (Table 2) and each period lasted for 4 weeks. The core program was based on core training by McGill, (2010), increasing the number of series and not the maintenance time of the isometry, due to the commitment to the level of tissue oxygenation in this type of prolonged contraction [23]. The core exercises were performed at the end of the rhythmic gymnastics' session. The core program comprised eight exercises, that is, hollowing (A), bracing (A), dissociation of shoulder girdle and pelvic girdle (B), Cat-Camel (C), quadrupedal stance (D), front bridge (E), side bridge (both sides) (F) and supine bridge (G) (Figure 2).

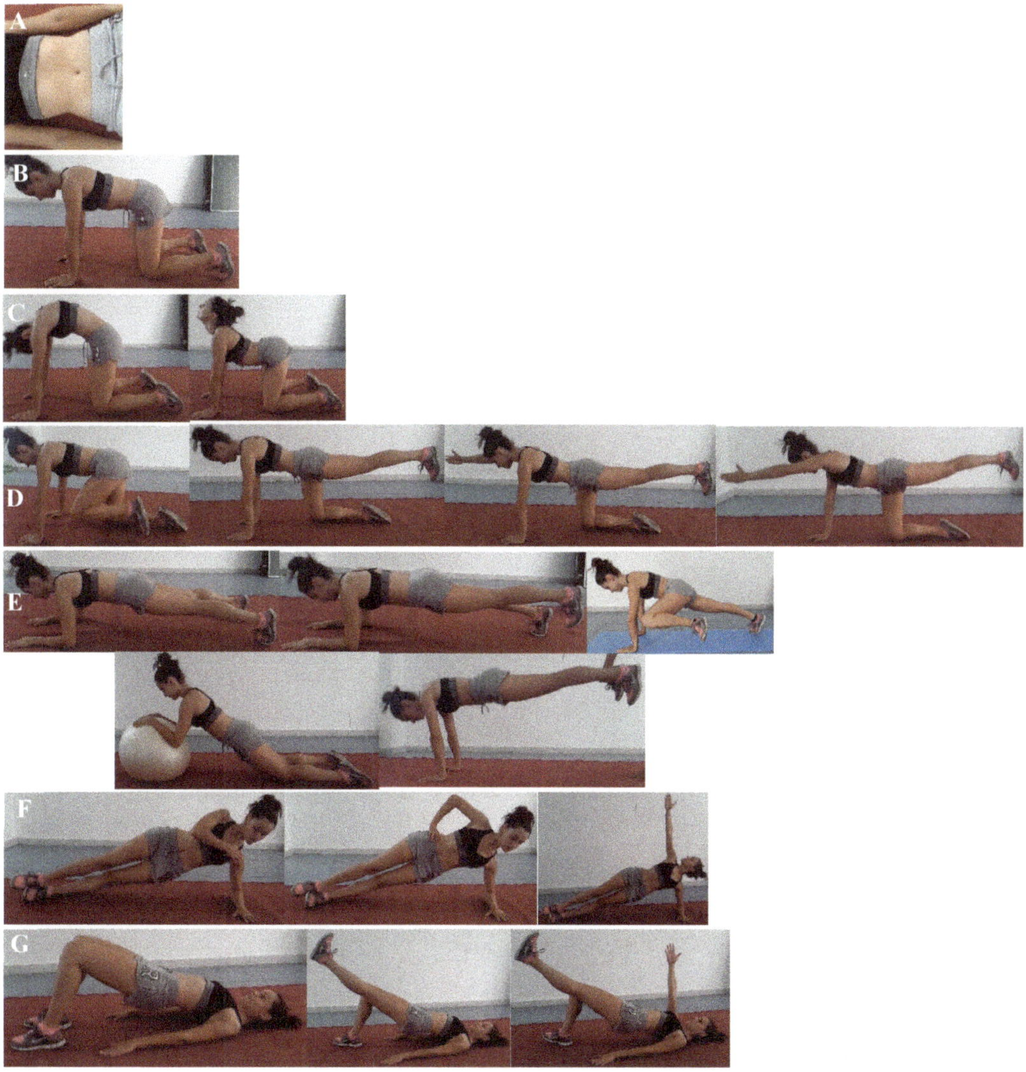

Figure 2. Core muscular training exercise. (**A**) Hollowing; (**B**) Bracing; (**C**) Dissociation; (**D**) Cat-Camel; (**D**) Quadrupedal; (**E**) Front Bridge; (**F**) Side Bridge; (**G**) Supine Bridge.

Table 2. The Core program.

Exercises	Period 1	Period 2	Period 3	
	Volume	Volume	Progress	Volume
Hollowing	10 sets	10 sets		
Bracing	10 sets	10 sets		
Dissociation	5 sets	5 sets		
Cat-Camel	10 sets	10 sets	Supine Bridge	2 × 5 sets × 20 s (15 s rest) (both legs)
Quadrupedal	8 sets of 20 s (15 s rest)	2 × 7 sets of 20 s (15 s rest)	Quadrupedal Birddog exercise	2 × 5 sets
Front Bridge			Front Bridge	2 × 5 sets (both sides)
			Front Bridge destabilisation	2 × 5 sets × 20 s) (15 s rest)
			Front Bridge on swiss ball	2 × 5 sets × 20 s) (15 s rest)
Side Bridge			Side Bridge	2 × 5 sets × 20 s (15 s rest) (both sides)
Supine Bridge			Supine Bridge	2 × 5 sets × 20 s (15 s rest) (both legs)

2.5. *Data Analysis*

Statistical analysis of data was performed with the Statistical Package for the Social Sciences (IBM Corp. IBM SPSS Statistics for Windows, Version 24.0. Armonk, NY, USA: IBM Corp.). Descriptive statistics were calculated using the mean and standard deviation and the mean difference using confidence intervals. The Shapiro–Wilk test was used to analyze data distribution, getting a normal distribution. Subsequently, independent sample *t*-tests were carried out to compare baseline values between groups. In addition, a two-way repeated-measures analysis of variance (ANOVA) (time × group) was applied to analyze the effect of the intervention on outcomes. Eta squared (η^2) effect sizes for the time × group interaction effects were calculated. An effect of $\eta^2 \geq 0.01$ indicates a small, $\geq$0.059 a medium, and $\geq$0.138 a large effect. For those variables that showed significant main effects, post-hoc tests (Bonferroni) were performed. The effect size (d) was calculated following the guidelines of Cohen [24]. The d was considered large (>0.80), moderate (0.5) and small (<0.2). An effect was considered statistically significant when $p \leq 0.05$.

3. Results

All participants completed the intervention and were included in the data analysis. No difference was observed between groups at baseline. Maximum growth velocity (MGA) was measured as a widely used indicator to assess biological maturation [25]. The age and height of the subjects were used to determine their biological maturation [26]. No significant differences in biological maturation were found between pre- and post-training in CG (p = 0.349), TG (p = 0.339) and between CG and TG in pre-training (p = 0.351).

3.1. *Body Composition and Densitometry*

Results for body composition are presented in Table 3. We observed no differences between the two groups for either of the two time-line measurements (p > 0.05). Within-group analysis showed an increase in the TG between pre- and post-core training in TLM (p = 0.040, d = −0.7; 95% confidence interval [CI] of the mean differences [MD] of the score = 0.03 Kg, 1.29 Kg), in LM (p = 0.037, d = −0.7; 95% CI of MD = 0.04 Kg, 1.30 Kg), and in BM (p < 0.001, d = −1.3; 95% CI of MD = 0.52 Kg, 2.09 Kg), and the CG showed an increase in BM ($p = 0.003$, d = −1.1; 95% CI of MD = 0.35 Kg, 1.79 Kg) and a decrease in the FT% ($p = 0.044$, d = 0.5; 95% CI of MD = −1.12%, −0.09 Kg).

Table 3. Body composition and densitometry results (Mean ± SD).

	Control Group (n = 12)				Training Group (n = 12)				
	Pre-Training	**Post-Training**	**Mean Differences**	p	**Pre-Training**	**Post-Training**	**Mean Differences**	p	**Interaction Time × Group (p)**
FM (kg)	8.74 ± 3.47	8.54 ± 3.51	0.20	*0.165*	10.41 ± 3.66	10.57 ± 3.63	−0.16	*0.365*	*0.04 (0.85)*
LM (Kg)	28.84 ± 8.58	28.89 ± 8.00	−0.43	*0.793*	34.71 ± 7.94	35.14 ± 7.89	−0.43	*0.037*	*1.83 (0.19)*
BM (Kg)	1.65 ± 0.53	1.69 ± 0.54	−0.04	*0.003*	2.04 ± 0.60	2.09 ± 0.59	−0.06	*<0.001*	*0.72 (0.41)*
%FT (%)	23.10 ± 4.69	22.47 ± 4.64	0.63	*0.044*	22.67 ± 2.73	22.71 ± 2.79	−0.04	*0.856*	*2.60 (0.12)*
TLM (Kg)	13.68 ± 4.39	13.73 ± 4.18	−0.05	*0.669*	16.79 ± 3.92	17.10 ± 4.10	−0.31	*0.040*	*2.29 (0.12)*

FM: fat mass; LM: lean mass; BM: bone mass; %FT: average fat tissue; TLM: trunk lean mass; SD: standard deviation; $p \leq 0.005$.

3.2. Isometric Tests in Isokinetic Dynamometer and Electromyography Analysis

Results for the PT and EMG_{rms} in the isometric tests are presented in Table 4. We observed no differences between the two groups for either of the two measurements ($p > 0.05$). Within-group analysis of the TG showed increases ($p < 0.05$) between pre- and post-core training in PT in the flexion isometric test ($p = 0.019$, d = 0.6; 95% CI of MD = 0.03 N·m, 1.20 N·m) and the extension isometric test ($p = 0.049$, d = 0.5; 95% CI of MD = 0.07 N·m, 1.15 N·m). In addition, the CG showed decreases of EMG_{rms} in front trunk in the flexion isometric test ($p = 0.03$, d = 0.6; 95% CI of MD = −1.19 μV, −0.04 μV) and the TG decreases of EMG_{rms} in the back trunk in the extension isometric test ($p = 0.04$, d = 0.7; 95% CI of MD = −1.326 μV, −0.054 μV).

Table 4. Performance in isometric test and electromyography values.

		Control Group (n = 12)				Training Group (n = 12)				
		Pre-Training	**Post-Training**	**Mean Differences**	p	**Pre-Training**	**Post-Training**	**Mean Differences**	p	**Interaction Time × Group (p)**
Flexion test	PT (N·m)	26.52 ± 11.26	41.78 ± 27.05	−15.26	0.086	31.56 ± 12.39	53.09 ± 41.36	−21.53	0.019	0.27 (0.61)
	EMG_{rms} Front (μV)	390.92 ± 254.19	256.58 ± 135.61	134.33	0.03	386.33 ± 205.00	345.92 ± 217.18	40.42	0.494	1.30 (0.27)
	EMG_{rms} Back (μV)	45.08 ± 34.27	57.17 ± 43.17	−12.08	0.492	66.92 ± 35.42	86.42 ± 61.21	−17.29	0.272	0.09 (0.77)
Extension test	PT (N·m)	31.75 ± 17.28	39.44 ± 34.00	−7.69	0.444	40.89 ± 19.28	63.66 ± 53.36	−22.77	0.049	0.95 (0.34)
	EMG_{rms} Front (μV)	129.58 ± 73.85	178.00 ± 155.46	−48.42	0.199	207.25 ± 123.68	232.42 ± 165.36	−25.17	0.498	0.20 (0.66)
	EMG_{rms} Back (μV)	128.83 ± 94.92	104.42 ± 41.28	24.42	0.421	163.67 ± 106.92	98.67 ± 36.61	65.00	0.04	0.93 (0.35)

PT: peak torque; EMG_{rms}: average electromyography activity; SD: standard deviation; $p \leq 0.005$.

3.3. Endurance Test and Electromyography Analysis

Results for the core endurance test are presented in Table 5. We observed no differences between the two groups for either of the two endurance core tests ($p > 0.05$). Within-group analysis of the TG showed an increase between pre- and post-training in prone bridge ($p = 0.044$, d = −0.5; 95% CI of MD = 0.083 s, 1.131 s). For EMG in the endurance test, we observed no differences between the two groups for either of the two tests ($p > 0.05$). However, within-group analysis of the TG showed an increase between pre- and post-core training in EMG_{rms} front trunk in prone bridge ($p = 0.030$, d = −0.5; 95% CI of MD = 0.035 μV, 1.197 μV) (Figure 3).

Table 5. Performance in McGill test.

	Control Group (n = 12)				Training Group (n = 12)				
	Pre-Training	**Post-Training**	**Mean Differences**	p	**Pre-Training**	**Post-Training**	**Mean Differences**	p	**Interaction Time × Group (p)**
Sorensen	32.57 ± 11.53	34.02 ± 22.32	−1.44	0.797	37.94 ± 19.86	51.00 ± 22.51	−13.06	0.172	1.23 (0.28)
Prone bridge	31.06 ± 16.57	24.74 ± 15.36	6.32	0.133	27.99 ± 13.86	39.26 ± 23.38	−11.27	0.04	5.92 (0.49)

SD: standard deviation; $p \leq 0.005$.

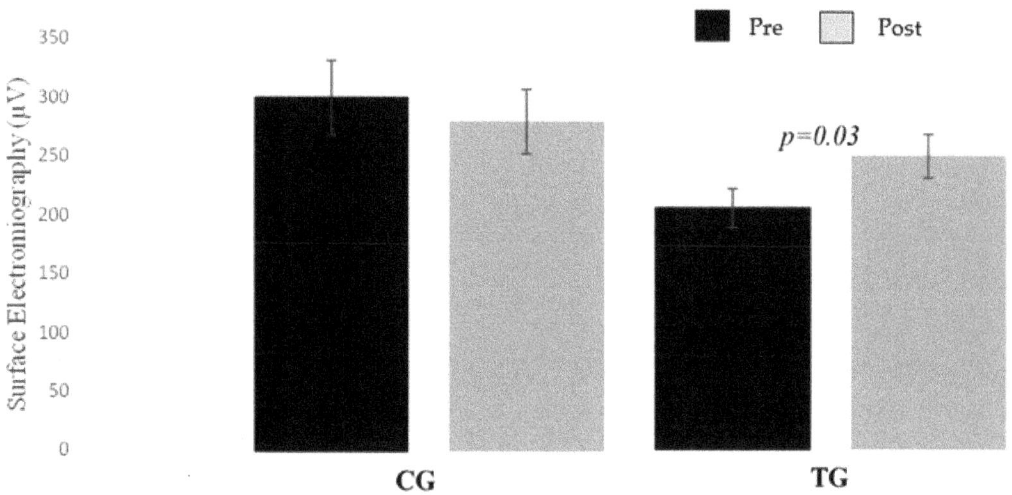

Figure 3. Significant difference EMG_{rms} front in prone bridge endurance test/TG.

4. Discussion

The aim of this study was to analyze the effect of 12 weeks of core training in gymnasts who were still training in rhythmic gymnastics on body composition, isometric and endurance strength core and core muscle electromyographic activity. The main findings were that the core training evoked an increase in trunk lean mass, lean mass and bone mass, and moreover the values of isometric strength and endurance strength and EMG in the core during the endurance test improved.

Regarding body composition, the TG showed higher values of TLM, LM and BM after core muscular training and the CG in the BM and lower values in the FT%. To our knowledge, there are no studies on the effect of core training on the body composition of gymnasts. However, it is possible to find similarities with our results in the study by Skrypnik et al. [27], where different types of interventions, resistance training and endurance strength training were compared on body composition. Only the resistance strength training group obtained a significant increase in total lean body mass (<0.001) and total fat-free body mass (<0.001). In this respect, therefore, the gains in lean mass with resistance training, used in core muscle training, would be justified. Similarly, Piacentini et al. [28], evaluated the effects of two different strength training protocols on resting metabolic rate, body composition, running economy and strength parameters, in young elite endurance athletes. Both training protocols included core muscle strength, and both also showed a decrease in body fat percentage and fat mass that reflected a significant increase in fat-free mass in the young athletes. On the evidence of these results, it can be said that the changes in body composition produced by core training in gymnasts may be due to the influence that strength training has on these parameters. In addition, the CG showed lower fat mass values after the intervention period, which can be explained by higher initial fat mass values from this group and by the CG continued with their usual gymnastic training, the effect of rhythmic gymnastics training cannot be ruled out. The effect that gymnastic training has on athletes in increasing bone mass has been demonstrated in comparison to other sports or control subjects [29,30]. This is related to the fact that both training groups in our research showed significant increases in BM. This is because the subjects were 13.95 $\pm$ 2.77 years old and in puberty, when bone mass mineral accrual increases substantially during the growing years [31]. Puberty is an opportune time for bone strengthening [32], when the mechanical loading of athletic training is a positive factor for skeletal strength, for maximizing bone mineral gain and reducing the risk of osteoporosis in later life [33,34]. Gruodyte-Raciene et al. [35], and Gruodytė et al. [30], consider that gymnastic training is especially osteogenic for bone development in children and adolescents. Therefore, although gymnastic training may already have a positive effect on the body composition of gymnasts, added core training could have greater benefits for the body composition of female athletes. A relationship is established between gymnasts' body composition and performance, with low values of fat mass being a determinant of performance [14,36].

In relation to isometric strength in the isometric test on the dynamometer, significant effects were found between pre- and post-core training in the TG. There are no studies on rhythmic gymnastics or other sports about the effect of core training on isometric trunk strength. Improvements in isometric trunk strength, both in flexion and extension, of gymnasts after training benefit these athletes, because they need upper body endurance strength and trunk muscle function to be successful in competition. Improving trunk strength and endurance would allow gymnasts to increase their ability to generate and maintain force throughout their routine. Core stability might contribute to the gymnast's performance as it would facilitate the transmission of forces generated by the lower to the upper body during technical elements and it would enhance balance control [15]. The positive data on the gymnasts' isometric strength after core training could reflect the positive effect of core training as a complementary training to gymnastic training. On the other hand, the results obtained in muscle activation, during the isometric test on the dynamometer, reflect a decrease in both study groups. This may be due to other types of neural adaptations that are not evaluated with the amplitude of the sEMG signal, such as inhibition of the antagonistic muscles, greater activation of the synergistic muscles or better inter-muscular coordination [37].

Similarly, the results obtained in the McGill endurance test and muscle activation in these tests, reflected significant effects between pre- and post-core training. The TG rhythmic gymnasts increased the maintenance time in prone bridge, as well as the muscle activation in the front trunk. In accordance with these results, previous studies have demonstrated that core training increases the maintenance time in the endurance test, and so increases trunk strength and stability strength in women collegiate gymnasts [38], dance students [39] or competitive collegiate dancers [40]. In this sense, the added and positive effect that core training could have on the gymnasts is again reflected.

Several considerations and limitations should be acknowledged. The evaluation of performance in rhythmic gymnastics was not carried out, so it cannot be confirmed that improvements in the training group had a direct influence on performance in competition. There was no control of the external activities that the participants of the sample did outside of the training. The sample size of the study can be considered small. However, the study has the strength to be considered the first to evaluate the effect of core training on national level rhythmic gymnasts. This core training program considered that the improvements found in the gymnasts are due more to core training combined with gymnastic training than to rhythmic gymnastics training alone because some improvements only occurred in the group that performed core training. Therefore, possible lines of research could analyze the effect of this type of core training on gymnastic performance, on the execution of technical gestures or on the judges' evaluation.

5. Conclusions

Our results suggest that combining a traditional rhythmic gymnastics program with a core training program could lead to increased strength and improved body composition. Additionally, core strength training produces improvements in trunk strength values in gymnasts, in addition to increasing muscle activation values.

6. Practical Applications

The proposed training is considered a useful tool for the training of gymnasts by their coaches. The improvements observed in the group that carried out a core program in addition to their traditional training presented improvements in strength and muscle activation capacity and this could have a positive transfer to competition. However, more studies analyzing the transference effect towards competition are needed.

The gains in strength and stability achieved will help coaches improve the physical preparation of gymnasts, and thus increase the technical level.

In addition, core muscle strength training may be of interest to another type of population, such as older adults, since ageing is associated with a variety of biological changes that can contribute to the decline of skeletal muscle mass, strength, and function [41].

Author Contributions: Conception: P.E.-G. and J.F.J.-D.; Performance of work: P.E.-G. and A.B.-S.; Interpretation or analysis of data: P.E.-G. and J.Á.R.-A.; Preparation of the manuscript: P.E.-G. and A.B.-S.; Revision for important intellectual content: J.F.J.-D., J.A.-V. and J.Á.R.-A.; Supervision: J.F.J.-D., J.A.-V. and J.Á.R.-A. All authors have read and agreed to the published version of the manuscript.

Funding: This research received no external funding.

Institutional Review Board Statement: Ethical approval was obtained from the Clinical Research Ethics Committee of the Toledo Healthcare Area (number 112/2015). This study complied with the ethical principles of the Declaration of Helsinki.

Informed Consent Statement: Informed consent was obtained from all subjects involved in the study. Written informed consent has been obtained from the patient(s) to publish this paper.

Data Availability Statement: Not applicable.

Conflicts of Interest: The authors declare no conflict of interest.

References

1. Gram, M.C.D.; Kari, B. High level rhythmic gymnasts and urinary incontinence: Prevalence, risk factors, and influence on performance. *Scand. J. Med. Sci. Sports* **2020**, *30*, 159–165. [CrossRef]
2. Frutuoso, A.S.; Diefenthaeler, F.; Vaz, M.A.; de la Rocha Freitas, C. Lower Limb Asymmetries in Rhythmic Gymnastics Athletes. *Int. J. Sports Phys. Ther.* **2016**, *11*, 34–43.
3. Bobo-Arce, M.; Méndez-Rial, B. Determinants of competitive performance in rhythmic gymnastics. A review. *J. Hum. Sport Exerc.* **2013**, *8*, 711–727. [CrossRef]
4. Rutkauskaitė, R.; Skarbalius, A. Interaction of Training and Performance of 13–14-Year-Old Athletes in Rhythmic Gymnastics. *Balt. J. Sport Health Sci.* **2011**, *3*, 29–36. [CrossRef]
5. Schärer, C.; Tacchelli, L.; Göpfert, B.; Gross, M.; Lüthy, F.; Taube, W.; Hübner, K. Specific eccentric–isokinetic cluster training improves static strength elements on rings for elite gymnasts. *Int. J. Environ. Res. Public Health* **2019**, *16*, 4571. [CrossRef] [PubMed]
6. Sands, W.; Mikesky, A.; Edwards, J. Physical abilities field tests US Gymnastics FederationWomen's National Teams. In *USGF Sport Science Congress Proceedings*; 1991; pp. 39–47. Available online: https://www.researchgate.net/profile/William-Sands/publication/252931619_Physical_abilities_field_tests_US_Gymnastics_Federation_Women\T1\textquoterights_National_Teams/links/54fb91580cf2c3f524204a00/Physical-abilities-field-tests-US-Gymnastics-Federation-Womens-National-Teams.pdf (accessed on 13 May 2021).
7. Sands, W.; McNeal, J.; Jemnic, M.; Delonga, T. Should Female Gymnasts Lift Weights? *Sportscience* **2000**, *4*, 1–6.
8. Ghezelbash, F.; El Ouaaid, Z.; Shirazi-Adl, A.; Plamondon, A.; Arjmand, N. Trunk musculoskeletal response in maximum voluntary exertions: A combined measurement-modeling investigation. *J. Biomech.* **2018**, *70*, 124–133. [CrossRef]
9. Bdo Amaral Benfica, P.; Aguiar, L.T.; de Brito, S.A.F.; Bernardino, L.H.N.; Teixeira-Salmela, L.F.; de Morais Faria, C.D.C. Reference values for muscle strength: A systematic review with a descriptive meta-analysis. *Braz. J. Phys. Ther.* **2018**, *22*, 355–369. [CrossRef]
10. McGill, S.M.; Childs, A.; Liebenson, C. Endurance times for low back stabilization exercises: Clinical targets for testing and training from a normal database. *Arch. Phys. Med. Rehabil.* **1999**, *80*, 941–944. [CrossRef]

11. Lynn, S.K.; Watkins, C.M.; Wong, M.A.; Balfany, K.; Feeney, D.F. Validity and reliability of surface electromyography measurements from a wearable athlete performance system. *J. Sport Sci. Med.* **2018**, *17*, 205–215.
12. Deering, R.E.; Senefeld, J.W.; Pashibin, T.; Neumann, D.A.; Hunter, S.K. Muscle function and fatigability of trunk flexors in males and females. *Biol. Sex Differ.* **2017**, *8*, 12. [CrossRef]
13. Zapolska, J.; Witczak, K.; Mańczuk, A.; Ostrowska, L. Assessment of nutrition, supplementation and body composition parameters on the example of professional volleyball players. *Rocz. Państwowego Zakładu Hig.* **2014**, *65*, 235–242.
14. Courteix, D.; Rieth, N.; Thomas, T.; Van Praagh, E.; Benhamou, C.L.; Collomp, K.; Lespessailles, E.; Jaffré, C. Preserved bone health in adolescent elite rhythmic gymnasts despite hypoleptinemia. *Horm. Res.* **2007**, *68*, 20–27. [CrossRef] [PubMed]
15. Barbado, D.; Lopez-Valenciano, A.; Juan-Recio, C.; Montero-Carretero, C.; Van Dieën, J.H.; Vera-Garcia, F.J. Trunk stability, trunk strength and sport performance level in judo. *PLoS ONE* **2016**, *11*, e0156267. [CrossRef]
16. Ahmed, S.; Saraswat, A.; Esht, V. Correlation of core stability with balance, agility and upper limb power in badminton players: A cross-sectional study. *Sport Sci. Health* **2021**. [CrossRef]
17. Kibler, W.; Press, J.; Sciascia, A. The role of core stability in athletic function. *Sports Med.* **2006**, *36*, 189–198. [CrossRef]
18. Granacher, U.; Gollhofer, A.; Hortobágyi, T.; Kressig, R.W.; Muehlbauer, T. The importance of trunk muscle strength for balance, functional performance, and fall prevention in seniors: A systematic review. *Sports Med.* **2013**, *43*, 627–641. [CrossRef]
19. Carrasco-Poyatos, M.; Ramos-Campo, D.J.; Rubio-Arias, J.A. Pilates versus resistance training on trunk strength and balance adaptations in older women: A randomized controlled trial. *PeerJ* **2019**, *2019*, e7948. [CrossRef] [PubMed]
20. Waldhelm, A.; Li, L. Endurance tests are the most reliable core stability related measurements. *J. Sport Health Sci.* **2012**, *1*, 121–128. [CrossRef]
21. Hermens, H.J.; Freriks, B.; Disselhorst-Klug, C.; Rau, G. Development of recommendations for SEMG sensors and sensor placement procedures. *J. Electromyogr. Kinesiol.* **2000**, *10*, 361–374. [CrossRef]
22. Rubio-Arias, J.A.; Ramos-Campo, D.J.; Esteban, P.; Martínez, F.; Jiménez, J.F. Effect of 6-weeks WBVT on the behaviour of the lower limb muscle fibres during vertical jumping. *J. Sports Sci.* **2018**, *36*, 398–406. [CrossRef]
23. Mcgill, S. Core training: Evidence translating to better performance and injury prevention. Strength and Conditioning Journal. *Strength Cond. J.* **2010**, *32*, 33–46. [CrossRef]
24. Cohen, J. *Statistical Power Analysis for the Behavioral Science*, 2nd ed.; Lawrence Erlbaum Associates: Hillsdale, NJ, USA, 1988.
25. Gomez Campos, R.; Hespanhol, J.E.; Portella, D.; Vargas Vitoria, R.; De Arruda, M.; Cossio-Bolanos, M.A. Predicción de la maduración somática a partir de variables antropométricas: Validación y propuesta de ecuaciones para escolares de Brasil. *Nutr. Clin. Diet. Hosp.* **2012**, *32*, 7–17.
26. Moore, S.A.; McKay, H.A.; Macdonald, H.; Nettlefold, L.; Baxter-Jones, A.D.G.; Cameron, N.; Brasher, P.M.A. Enhancing a somatic maturity prediction model. *Med. Sci. Sports Exerc.* **2015**, *47*, 1755–1764. [CrossRef] [PubMed]
27. Skrypnik, D.; Bogdański, P.; Mądry, E.; Karolkiewicz, J.; Ratajczak, M.; Kryściak, J.; Pupek-Musialik, D.; Walkowiak, J. Effects of Endurance and Endurance Strength Training on Body Composition and Physical Capacity in Women with Abdominal Obesity. *Obes. Facts* **2015**, *8*, 175–187. [CrossRef]
28. Piacentini, M.F.; Ioannon, G.; Comotto, S.; Spedicato, A.; Vernillo, G.; La Torre, A. Concurrent strenth and endurance training effects on running economy in master endurance runners. *J. Strength Cond. Res.* **2013**, *27*, 2295–2303. [CrossRef]
29. Võsoberg, K.; Tillmann, V.; Tamm, A.L.; Maasalu, K.; Jürimäe, J. Bone mineralization in rhythmic gymnasts entering puberty: Associations with jumping performance and body composition variables. *J. Sports Sci. Med.* **2017**, *16*, 99–104.
30. Gruodytė, R.; Jurimae, J.; Saar, M.; Maasalu, K.; Jurimae, T. Relationships between areal bone mineral density and jumping height in pubertal girls with different physical activity patterns. *J. Sports Med. Phys. Fit.* **2009**, *49*, 474.
31. Jackowski, S.A.; Erlandson, M.C.; Mirwald, R.L.; Faulkner, R.A.; Bailey, D.A.; Kontulainen, S.A.; Cooper, D.M.L.; Baxter-jones, A.D.G. Effect of maturational timing on bone mineral content accrual from childhood to adulthood: Evidence from 15 years of longitudinal data. *Bone* **2011**, *48*, 1178–1185. [CrossRef]
32. Jürimäe, J.; Gruodyte-Raciene, R.; Baxter-Jones, A.D.G. Effects of gymnastics activities on bone accrual during growth: A systematic review. *J. Sports Sci. Med.* **2018**, *17*, 245–258.
33. Weaver, C.M.; Gordon, C.M.; Janz, K.F.; Kalkwarf, H.J.; Lappe, J.M. The National Osteoporosis Foundation's position statement on peak bone mass development and lifestyle factors: A systematic review and implementation recommendations. *Osteoporos. Int.* **2016**, *27*, 1281–1386. [CrossRef]
34. Erlandson, M.; Kontulainen, S.A.; Baxter-Jones, A.D.G. Precompetitive and recreational gymnasts have greater bone density, mass, and estimated strength at the distal radius in young childhood. *Osteoporos. Int.* **2011**, *22*, 75–84. [CrossRef] [PubMed]
35. Gruodyte-Raciene, R.; Erlandson, M.C.; Jackowski, S.A.; Baxter-Jones, A.D.G. Structural strength development at the proximal femur in 4- to 10-year-old precompetitive gymnasts: A 4-year longitudinal hip structural analysis study. *J. Bone Miner. Res.* **2013**, *28*, 2592–2600. [CrossRef] [PubMed]
36. Purenović-Ivanović, T.; Popović, R.; Bubanj, S.; Stanković, R. Body composition in high-level female rhythmic gymnasts of different age categories. *Sci. Sports* **2019**, *34*, 141–148. [CrossRef]
37. Swinarew, A.S.; Stachura, A.; Gupta, S. Efectos de Ejercicios Core de 6 Semanas Sobre el Rendimiento de Natación de Nadadores de Nivel Nacional. *PLoS ONE* **2020**, *1*, 1–9.
38. Durall, C.J.; Udermann, B.E.; Johansen, D.R.; Gibson, B.; Reineke, D.M.; Reuteman, P. The effects of preseason trunk muscle training on low-back pain ocurrence in women collegiate gymnasts. *J. Strength Cond. Res.* **2009**, *23*, 86–92. [CrossRef] [PubMed]

39. Amorim, T.P.; Sousa, F.M.; Dos Santos, J.A.R. Influence of Pilates training on muscular strength and flexibility in dancers. *Mot. Rev. Educ. Fis.* **2011**, *17*, 660–666. [CrossRef]
40. Watson, T.; Mcpherson, S.; Edwards, J.; Melcher, I.; Burgess, T. Dance, Balance and Core Muscle Performance Measures Are Improved Following a 9-Week. *Int. J. Sports Phys. Ther.* **2017**, *12*, 25–41.
41. Fragala, M.S.; Cadore, E.L.; Dorgo, S.; Izquierdo, M.; Kraemer, W.J.; Peterson, M.D.; Ryan, E.D. Resistance Training for Older Adults: Position Statement From the National Strength and Conditioning Association. *J. Strength Cond. Res.* **2019**, *33*, 2019–2052. [CrossRef]

biology

MDPI

Article

Performance Profile among Age Categories in Young Cyclists

Cristian Marín-Pagán [1,*], Stéphane Dufour [2,3], Tomás T. Freitas [1,4,5] and Pedro E. Alcaraz [1]

1. UCAM Research Center for High Performance Sport, Catholic University of Murcia (UCAM), 30107 Murcia, Spain; tfreitas@ucam.edu (T.T.F.); palcaraz@ucam.edu (P.E.A.)
2. Faculty of Medicine, Translational Medicine Federation (FMTS), University of Strasbourg, EA 3072 Strasbourg, France; sdufour@unistra.fr
3. Faculty of Sport Sciences, University of Strasbourg, 67000 Strasbourg, France
4. NAR-Nucleus of High Performance in Sport, São Paulo 04753-060, Brazil
5. Faculty of Sport Sciences, Catholic University of Murcia (UCAM), 30107 Murcia, Spain

* Correspondence: cmarin@ucam.edu

Simple Summary: Overall, adolescence brings upon many bodily changes that modify physical capacities. To better understand these physiological changes and the characteristics of each stage of adolescent development in youth cycling, it is necessary to describe and compare cyclists that pertain to lower categories. Parameters such as maximum oxygen uptake, fat oxidation capacity, functional power threshold, and ventilatory thresholds are decisive predictors of performance in future stages. The aim of this study was to evaluate and compare the physiological profile of different road cyclist age categories (Youth, Junior, and Under-23) to obtain the performance requirements. The results suggest major differences, with the Youth group showing clear changes in all metabolic zones except in fat oxidation. The Youth group physiological profile is clearly different from the other age categories. The present results suggest that the Juniors' qualities are closer to adult performance, however, little is known about sports performance indicators in adolescent cyclists.

Abstract: Endurance profile assessment is of major interest to evaluate the cyclist's performance potential. In this regard, maximal oxygen uptake and functional threshold power are useful functional parameters to determine metabolic training zones (ventilatory threshold). The aim of this study was to evaluate and compare the physiological profile of different road cyclist age categories (Youth, Junior, and Under-23) to obtain the performance requirements. Sixty-one competitive road cyclists (15–22 years) performed a maximal incremental test on a bike in order to determine functional parameters (maximal fat oxidation zone, ventilatory thresholds, maximal oxygen uptake, and functional threshold power) and metabolic training zones. The results suggest major differences, with the Youth group showing clear changes in all metabolic zones except in fat oxidation. The main differences between Under-23 vs. Junior groups were observed in maximal relative power output (Under-23: 6.70 $W \cdot Kg^{-1}$; Junior: 6.17 $W \cdot Kg^{-1}$) and relative functional threshold power (Under-23: 4.91 $W \cdot Kg^{-1}$; Junior: 4.48 $W \cdot Kg^{-1}$). The Youth group physiological profile is clearly different to the other age categories. Some parameters normalized to body weight (maximal oxygen consumption, load and functional threshold power) could be interesting to predict a sporting career during the Junior and Under-23 stages.

Keywords: cycling; endurance; oxygen uptake; FTP; threshold; power

Citation: Marín-Pagán, C.; Dufour, S.; Freitas, T.T.; Alcaraz, P.E. Performance Profile among Age Categories in Young Cyclists. *Biology* **2021**, *10*, 1196. https://doi.org/10.3390/biology10111196

Academic Editors: Filipe Manuel Clemente, Georgian Badicu and Eugenia Murawska-Ciałowicz

Received: 11 October 2021
Accepted: 16 November 2021
Published: 17 November 2021

1. Introduction

Cycling is considered one of the most stressful and physically demanding sports, with the road stage races being its most popular modality. In professional road cyclists, values for maximal oxygen uptake (VO_2max) higher than 70–80 $mL \cdot Kg^{-1} \cdot min^{-1}$ have repeatedly been observed [1,2]. Although very high, such VO_2max values appear more as a prerequisite to achieve professional level rather than good performance predictor [3]. As such, maximal power output during an incremental test might be a better predictor than VO_2max for short efforts in flat stages, with elite cyclists achieving values between 400 and 500 W (6.0–7.5 W·Kg^{-1}), finding slight differences depending on the test characteristics [4,5]. Additionally, lactate threshold position seems to be more predictive than VO_2max for endurance cycle performance, especially in professional climber cyclists [5] where lactate thresholds (LT2) at ~90% of VO_2max have been found.

In recent years, the evaluation of cycling performance and the monitoring of cycling training load through the so-called functional threshold power (FTP) has also been of increasing scientific interest [6–8]. FTP consists of the maximum power output developed during a 1 h trial, and can be evaluated by contemplating the power developed during 20 min with the application of a correction factor of 0.95 [9]. FTP values are estimated to be around 5.0–6.0 W·Kg^{-1} and 3–5 W·Kg^{-1} for professional and trained amateur cyclists, respectively [9]. The FTP has become a supplementary parameter to the assessment of performance profile due to its applicability to the field in a non-invasive way and without the need for sophisticated equipment.

Achieving professional and world-class level in road cycling is a long-term process taking several years of regular, high volume, and high intensity training, most of the time from Youth, through to the Junior and Under-23 (U-23) categories. Therefore, important differences in physiological profile exist between competition levels and age categories in cycling [10]. For example, professional road cyclists complete approximately 30,000 to 35,000 km per season [4] while amateur competitive cyclists complete around 13,500 km [11]. Important changes in total volume progression have also been observed when comparing consecutive seasons from the Junior stages to World class level [12]. Similarly, higher VO_2max values have been reported in elite cyclists (~74 mL·Kg^{-1}·min^{-1}) when compared to amateurs (~65 mL·Kg^{-1}·min^{-1}) [13,14]. Another critical factor that has been suggested to differentiate cyclists of superior performance levels [5,13,15] is the ability to develop power, both as peak values in incremental tests or during critical power [16] tests as a FTP [9]. Due to mentioned discrepancies between cyclists, De Pauw et al. [13] proposed a five level cycling classification according to physiological demands and training loads. Nevertheless, it is worth noting that amateur cyclists are progressively showing higher performance levels, to the extent that similarities with professionals can be found, particularly in cycling economy and efficiency [17].

Regarding age categories, previous studies have reported differences in anthropometric parameters, with athletes displaying greater left–right leg length asymmetries as they progress through to older categories [18]. This unbalance could be related to an increased training duration in more experienced cyclists to meet the demands of the competition. The characteristics of Youth and Junior races are different and very stressful for the metabolic system [19] and some countries have limited the maximal number of competitions per season during the Youth categories. As the distances are usually shorter in these categories, the average race intensity is higher. For this reason, due to the progression in training volume [12] and competition characteristics [19] the recovery time necessary after an endurance exercise increases with age [20].

On a related topic, during prolonged training and competitive efforts (>4 h), an increased fatty acid contribution to total energy turnover is observed and fat oxidation capacity could be considered as a desirable adaptation for road cycling performance [21,22]. Accordingly, assessing and training to improve this capacity are important aspects of the training process in cycling [22]. Different authors have reported that the maximal fat oxidation zone (Fatmax) is achieved at approximately 45–60% of VO_2max [23,24] and that this concept is closely related to cycling economy and efficiency. The issue with the former variables is that they are somehow easy to improve in amateurs, but very difficult when it comes to professional cyclists. Thus, in highly trained cyclists, it is common to observe that endurance training is not sufficient to improve cycling economy and efficiency, which makes it necessary to rely to alternative training strategies such as resistance training to achieve this objective [25,26], with heavy strength training being recommended to achieve improvements in aerobic performance [26,27].

From the above-mentioned, it appears that endurance performance parameters are likely different among the Youth categories, but the extent of these differences remain presently unclear. Dissimilarities in some cardiorespiratory and metabolic parameters between professionals and amateur cyclists [13,14] as well as between age categories [14] can be found in the literature. However, the differences in physiological and performance profile of three different age categories (Youth, Junior, and U-23) in the pre-season phase as well as the value of FTP in these three age categories have never been documented to date. Such knowledge might have important practical applications not only for better performance evaluation (i.e., talent identification) but also to optimize training prescription and training load management in young cyclists. Currently, there are no studies that compare the physiological profile in these categories, in order to establish differences or similarities that could help determine future performance. It is also unknown which are, in these lower categories, the most important physiological characteristics to be studied and controlled. Therefore, the aim of this study was to assess the physiological parameters and performance profile that can influence cycling performance and to compare them amongst three different age categories, from Youth to U-23.

2. Materials and Methods

2.1. Participants

Sixty-one young male amateur cyclists from three distinct age categories, but with similar competitive level, participated in the study (Table 1): Youth (15–16 years); Junior (17–18 years), and U-23 (19–22 years), according to the classification of the *Union Cycliste Internacionale* (UCI). All cyclists were members of an official team, did not present any injury in the three months before the investigation and performed regular training of more than 6 h per week. Participants had at least three years of cycling training experience, were enrolled in the cycling team since "school" categories, and had previous experience in laboratory testing. Prior to study enrollment, all cyclists or the parents (of those under 18 years old) signed the consent to participate in the study (approved by the University Ethical Committee; CE022105) and obtained medical approval to participate in this study. Just before testing, weight and height were measured using a SECA 780 device (Seca, Hamburg, Germany). All tests were completed between 10 h and 14 h 2 h after breakfast intake (bread, milk or yogurt and juice). Participants did not train in the 24 h prior to testing to avoid fatigue and the test was separated from high intensity training or pre-season competitions by at least 72 h.

Table 1. General characteristics of the participants.

UCI Category	Age Range (Years)	Weight (Kg)		Height (cm)		BMI (Kg/m^2)	
	Range	Mean	SD	Mean	SD	Mean	SD
Youth (n = 24)	15–16	61.2	7.4	173.2	6.4	20.4	2.3
Junior (n = 22)	17–18	66.5	7.4	178.2	5.9	20.9	2.3
U-23 (n = 15)	19–22	64.1	4.3	176.7	5.7	20.5	0.9

UCI = Union Cycliste Internationale; SD = standard deviation (±); BMI = body mass index.

2.2. Assessments

All tests were carried out in the laboratory during pre-season (December–February). For the cardiorespiratory evaluation, a metabolic cart (Cortex Metalyzer, Leipzig, Germany) and the Cyclus2 ergometer (Cyclus, Leipzig, Germany) were used. The cyclists utilized their own bikes in all assessments. The protocol used consisted of a combined test with an initial step phase followed by final ramp. The test started at 35 W with increments of 35 W every 2 min. Then, when the respiratory exchange ratio (RER) was ≥1.05, the final ramp of 35 W per minute (~1 W each 0.583 s) was initiated. This combined protocol was applied to determine the ventilatory thresholds (VT1 = aerobic; VT2 = anaerobic) during steady states (step phase) and continued until exhaustion to assess VO_2max and maximal load (Pmax, final ramp) [28–30]. The recommended pedaling cadence was 85 to 95 rpm and the test was stopped when the participants were unable to sustain a cadence greater than 60 rpm, with permanent chainset (52–53/12 teeth). To determine blood lactate concentration, blood samples were collected from the finger at 1.5 min after exhaustion. The first blood drop was dismissed and the second was analyzed with a Lactate Pro2 (Arkray, Tokyo, Japan).

Ventilatory thresholds (VT1 and VT2) were calculated with the ventilatory equivalent method described by Wasserman [31] and using the data averaged every 20 s. The VO_2max was assumed as the maximum value of the last four data of 20 s averages. To guarantee that the VO_2max was achieved, at least three of the following criteria had to be obtained: (I) plateau in the final VO_2 values (increase $\leq$2.0 $mL \cdot kg^{-1} \cdot min^{-1}$ in the two last loads); (II) maximal theoretical HR (220–age) × 0.95) for a cycling test suggested by Millet et al. [32]; (III) RER ≥1.15; and (IV) a lactate value ≥8.0 $mmol \cdot l^{-1}$ [33]. Pmax was calculated as the maximal power achieved during the final ramp in the incremental test. Maximal oxygen uptake and load were expressed in absolute units or normalized to body weight (VO_2R and Load/BW, respectively). To determine the percentage of VO_2max at which Fatmax was achieved, the values of VO_2 corresponding to maximal fat oxidation (MFO) and normalized to VO_2max were selected.

Functional threshold power (FTP) was estimated using the equation described by Denham [34] using the maximal power output during VO_2max test.

2.3. *Statistical Analysis*

All descriptive statistics were presented as mean ± standard deviation (SD) and the statistical analysis was performed using the Statistical Package for Social Sciences (SPSS 27.0, IBM, Chicago, IL, USA). A Shapiro–Wilk test was performed to assess the normality of the variables. The between-group differences were investigated using independent *t*-tests and the statistical significance was set for a $p < 0.05$. The U-23 group was established as the "reference group" for the comparative analysis, given that it was the highest competitive level. Effect sizes (ES) were calculated utilizing Cohen's equations [35]. Threshold values for ES statistics were: >0.2 small, >0.6 moderate, and >1.2 large, >2.0, very large; and >4.0, nearly perfect [36].

3. Results

Ventilatory Threshold 1. Significant differences were obtained between the Youth and Junior groups for VO_2 ($p = 0.001$; ES = 0.98), Load ($p < 0.001$; ES = 1.21) and Load/BW ($p = 0.037$; ES = 0.62). For Youth vs. U-23 group, significant differences were obtained in HR ($p = 0.018$; ES = 0.80), %VO_2max ($p = 0.005$; ES = 0.97), Load ($p < 0.001$; ES = 1.32), and Load/BW ($p = 0.002$; ES = 1.06). For Junior vs. U-23, differences were only found in HR ($p = 0.027$; ES = 0.77). In this metabolic zone, the U-23 group showed the lowest percentage with respect to VO_2max, and the Youth group displayed the best results (Table 2).

Table 2. Performance assessments data.

Zone	Variable	Youth		Junior		U-23	
		Mean	SD	Mean	SD	Mean	SD
VT1	**HR** (bpm)	149.5	12.2	149.7	13.2	139.7 *,†	11.8
	HRmax (%)	76.1	5.6	74.9	5.9	72.6	5.4
	VO_2 ($L \cdot min^{-1}$)	2.1	0.3	2.4 *	0.3	2.3	0.2

Table 2. *Cont.*

Zone	Variable	Youth		Junior		U-23	
		Mean	SD	Mean	SD	Mean	SD
	VO_2R ($mL \cdot Kg^{-1} \cdot min^{-1}$)	35.4	3.2	37.0	5.1	35.7	4.1
	%VO_2max (%)	56.8	4.3	54.9	5.4	52.2 *	5.2
	Load (W)	152.7	20.8	179.2 *	22.3	181.3 *	22.0
	Load/BW ($W \cdot Kg^{-1}$)	2.51	0.26	2.72 *	0.40	2.84 *	0.37
VT2	**HR** (bpm)	183.8	10.2	185.1	9.1	179.2 †	6.1
	HRmax (%)	93.5	2.6	93.0	2.9	93.2	2.0
	VO_2 ($L \cdot min^{-1}$)	3.3	0.4	3.9 *	0.4	3.8 *	0.4
	VO_2R ($mL \cdot Kg^{-1} \cdot min^{-1}$)	54.2	4.3	58.6 *	7.7	59.3 *	6.9
	%VO_2max (%)	87.1	5.0	86.6	4.6	86.7	5.6
	Load (W)	252.8	35.2	303.4 *	30.3	317.9 *	36.0
	Load/BW ($W \cdot Kg^{-1}$)	4.16	0.55	4.60 *	0.57	4.98 *	0.61

SD = standard deviation (±); VT1 = ventilatory threshold 1; VT2 = ventilatory threshold 2; HR = heart rate; VO_2 = oxygen uptake; VO_2R = oxygen uptake normalized to body weight; VO_2max = maximal oxygen uptake; BW = body weight; * = Significant differences with the Youth group; † = Significant differences with the Junior group.

Ventilatory Threshold 2. For the Youth vs. Junior group comparison, significant differences were found in VO_2 ($p < 0.001$; ES = 1.47), VO_2R ($p = 0.020$; ES = 0.70), Load ($p < 0.001$; ES = 1.51) and Load/BW ($p = 0.011$; ES = 0.77). There were also significant differences between the Youth and U-23 groups in VO_2 ($p = 0.001$; ES = 1.22), VO_2R ($p = 0.007$; ES = 0.92), Load ($p < 0.001$; ES = 1.80), and Load/BW ($p < 0.001$; ES = 1.40). As for VT1, significant differences were found only in HR ($p = 0.035$; ES = 0.86) when comparing the Junior and U-23 groups (Table 2).

Maximal Zone. Significant differences were obtained between the Youth and Junior groups (Table 3 and Figure 1) for VO_2 ($p < 0.001$; ES = 1.42), VO_2R ($p = 0.003$; ES = 0.92), Load ($p < 0.001$; ES = 1.51), Load/BW ($p = 0.003$; ES = 0.89), time to exhaustion ($p < 0.001$; ES = 1.67), and blood lactate concentration ($p = 0.007$; ES = 0.86). Similar results were obtained for Youth vs. U-23 for VO_2 ($p < 0.001$; ES = 1.18), VO_2R ($p = 0.002$; ES = 1.08), Load ($p < 0.001$; ES = 1.95), Load/BW ($p < 0.001$; ES = 1.88), and time to exhaustion ($p < 0.001$; ES = 1.70) but not for blood lactate ($p = 0.059$; ES = 0.77), in which only a trend toward statistical significance was found. Finally, for Junior vs. U-23, significant differences were found for HR ($p = 0.013$; ES = 1.24) and Load/BW ($p = 0.015$; ES = 0.86).

Table 3. Maximal values in the VO_2max test.

Variable	Youth		Junior		U-23	
	Mean	SD	Mean	SD	Mean	SD
HR (bpm)	196.5	8.8	199.0	6.6	192.5 †	8.5
VO_2 ($L \cdot min^{-1}$)	3.8	0.5	4.4 *	0.3	4.4 *	0.5

Table 3. *Cont.*

Variable	Youth		Junior		U-23	
	Mean	SD	Mean	SD	Mean	SD
VO_2R ($mL \cdot Kg^{-1} \cdot min^{-1}$)	63.3	4.5	68.5 *	6.5	69.7 *	7.5
RER	1.14	0.04	1.17	0.05	1.15	0.04
Load (W)	343.2	45.6	407.5 *	37.6	428.3 *	37.7
Load/BW ($W \cdot Kg^{-1}$)	5.63	0.55	6.17 *	0.64	6.70 *,†	0.57
Time to exhaustion (second)	1107.2	124.7	1311.8 *	115.4	1323.7 *	125.5
Lactate ($mmol \cdot L^{-1}$)	12.6	2.8	15.6 *	4.0	14.6	2.8

SD = standard deviation (±); MAX = maximal value; HR = heart rate; VO_2 = oxygen uptake; VO_2R = oxygen uptake normalized to body weight; VO_2max = maximal oxygen uptake; BW = body weight; RER = respiratory exchange ratio; * = Significant differences with Youth group; † = Significant differences with Junior group.

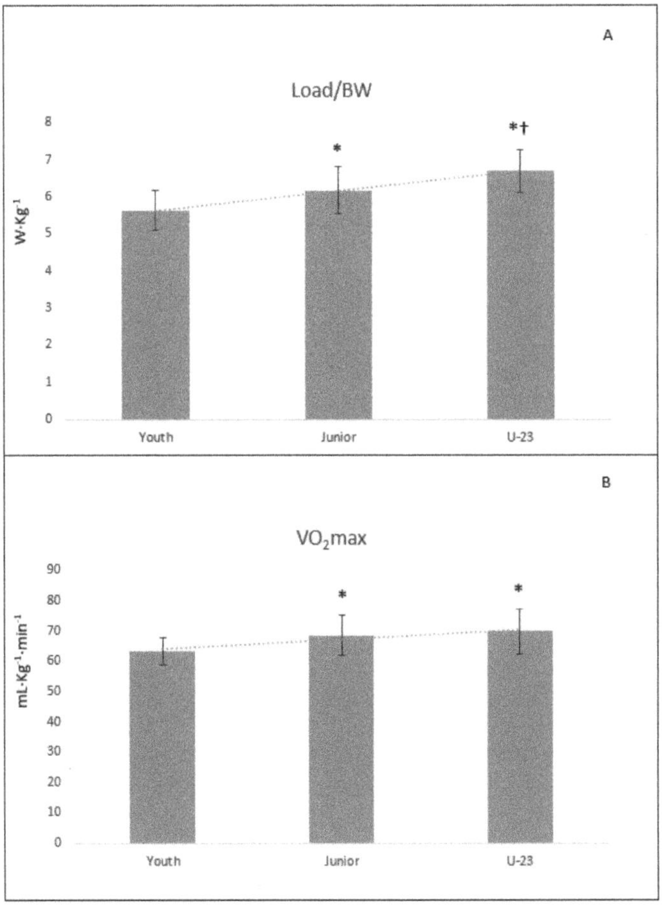

Figure 1. Maximal oxygen uptake (**A**) and Load/BW (**B**) values. Load/BW (**A**) = work load normalized to body weight; VO_2max (**B**) = maximal oxygen uptake; * = Significant differences with Youth group; † = Significant differences with Junior group.

Fatmax zone. Significant between-group differences were found in this metabolic zone (Table 4) only in VO_2 ($p = 0.007$; ES = 0.86) and Load ($p = 0.011$; ES = 1.00) for Youth vs. U-23.

Table 4. Values in the maximal fat oxidation zone.

Variable	Youth		Junior		U-23	
	Mean	SD	Mean	SD	Mean	SD
HR (bpm)	137.9	16.1	131.8	19.2	136.6	11.0
HRmax (%)	70.4	6.6	66.6	9.5	70.9	4.3
VO_2 ($L{\cdot}min^{-1}$)	1.9	0.3	2.1	0.4	2.2 *	0.4
VO_2R ($mL{\cdot}Kg^{-1}{\cdot}min^{-1}$)	31.7	4.6	32.2	5.8	35.1	5.4
%VO_2max (%)	51.2	9.0	48.9	8.5	51.2	5.4
Load (W)	147.7	30.1	163.9	35.1	181.6 *	37.6
Load/BW ($W{\cdot}Kg^{-1}$)	2.47	0.55	2.51	0.51	2.84	0.58
RER	0.88	0.03	0.87	0.03	0.89	0.03
MFO ($g{\cdot}h^{-1}$)	19.7	6.7	23.3	9.3	22.6	7.9

SD = standard deviation (±); HR = heart rate; VO_2 = oxygen uptake; VO_2R = oxygen uptake normalized to body weight; VO_2max = maximal oxygen uptake; BW = body weight; RER = respiratory exchange ratio; MFO = maximal fat oxidation; * = Significant differences with the Youth group.

Estimated functional threshold power. For the estimated FTP, significant differences were found (Table 5) between the Youth and the other two groups (Junior and U-23, $p < 0.001$; ES = 1.50 and 2.66, respectively). Additionally, for Junior vs. U-23, a significant difference was obtained in FTP/BW ($p = 0.014$; ES = 0.85).

Table 5. Estimated functional threshold power.

Variable	Youth		Junior		U-23	
	Mean	SD	Mean	SD	Mean	SD
FTP (W)	240.4	39.5	296.0 *	32.5	314.0 *	32.6
FTP (% Pmax)	69.7	2.4	72.5 *	1.3	73.2 *	1.2
FTP/BW ($W{\cdot}Kg^{-1}$)	3.93	0.46	4.48 *	0.51	4.91 *,†	0.47

SD = standard deviation (±); FTP = functional threshold power; BW = body weight; Pmax = Maximal power output; * = Significant differences with Youth group; † = Significant differences with the Junior group.

The FTP normalized to BW is a key factor in cycling, which is related to other performance parameters such as VO_2max and power output in VO_2max. Finally, in FTP/BW, a linear increase was found in relation to the age category ($R^2 = 0.995$; Figure 2). Additionally, the percentage of FTP with respect to the maximal power output (Pmax) showed a significant difference with the Youth group (Junior and U-23, $p < 0.001$; ES = 1.41 and 1.69, respectively), finding higher values in both groups (4% in Junior and 5% in U-23), but no differences were observed between the Junior and U-23 groups.

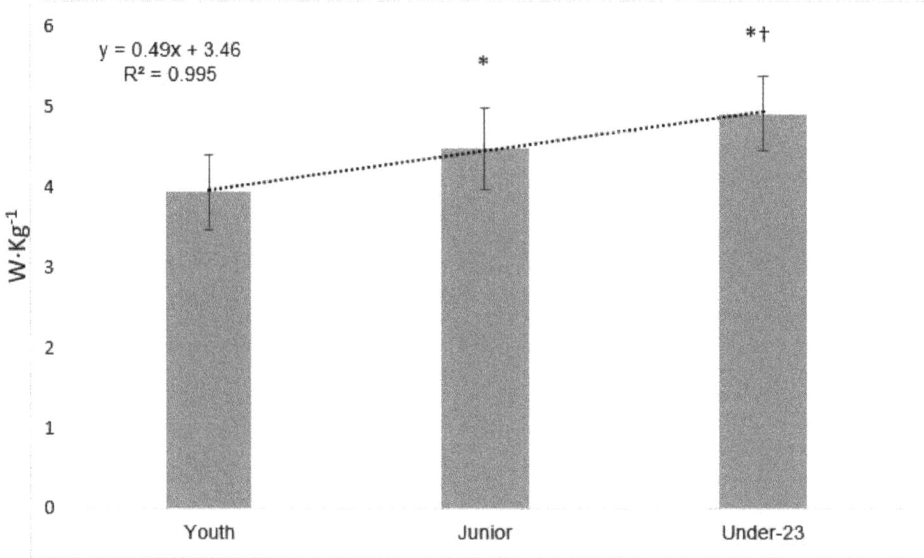

Figure 2. Estimated functional threshold power normalized to body weight. FTP = functional threshold power; BW = body weight; * = Significant differences with the Youth group; † = Significant differences with the Junior group.

4. Discussion

The aim of the present study was to assess and compare the physiological profile of different age categories. Despite cardiorespiratory testing being the most frequent procedure to assess performance in cyclists regardless of the level of competition [37], this study is the first to directly compare the Youth, Junior and U-23 categories. The main findings indicated that, during a maximal test, Junior, and U-23 group obtained values of VO_2max were slightly lower than those reported in elite and professional cyclists [1,2,37], but significantly greater than the Youth group. Similar differences were found for all performance variables analyzed in the maximal effort zone with important results in the ES analysis. These results were somehow expected due to the changes in physiological parameters with age and maturation, but could also be influenced by the athlete's training background.

Maximal values of VO_2max and Pmax showed important differences with the Youth (VO_2max = 8.2–10.2%; Pmax = 13.8–24.8% for Junior and U-23, respectively) and only for Pmax/BW were found differences between the Junior and U-23 (8.6% greater in the U-23 group). The VO_2max data obtained by the Junior and U-23 groups were similar to the values reported in professional cyclists [1,2]. However, in recent years, relative power production has proven to be a more sensitive indicator, since in professional categories, this parameter allows for better differentiating performance levels when compared to the VO_2max [4,5]. Along these lines, the present results indicated that the U-23 group outperformed the Youth (+19.0%) and the Junior (+8.6%) categories. For this reason, maximum Load/BW could be an important indicator of the competitive level in the U-23 category.

The FTP was at ~70% of Pmax with differences with the Youth group (4% in Junior and 5% in U-23). Interestingly, great differences between groups were found for Load/BW and FTP/BW, with a linear increase from Youth (19% and 14%, respectively, for Junior) to the U-23 category (25% in both for U-23). These parameters (Load/BW and FTP/BW) have been proposed to be very important for cycling performance [6–8,38,39]. Due to the duration of the stages, time under muscle tension is usually large, potentially explaining why power output normalized to BW is crucial for road cyclists. From an applied perspective, FTP/BW could be used at the beginning of the season to determine performance levels and compare them with the reference values of each category. Of note, the FTP assessments were calculated indirectly in our study using the equation proposed by Denham et al. [34], where there were similar values were obtained with direct assessments by the U-23 group in comparison with previously reported values for professional cyclists [3], which supports the notion that the athletes in the studied sample were of high competitive level. It is likely that the FTP could be the most differentiating variable, showing a large and very large ES favoring the U-23 vs. the Junior and Youth, respectively.

Regarding VT1, similarities were found between the Junior and U-23 cyclists and both groups presented greater values than the Youth (especially for Load variables with ~13% in Junior and ~16% in U-23). According to Lucia et al. [1], it is important to achieve a good "cruising speed" in this metabolic zone, because it is the predominant intensity during flat stages. The differences reported herein could be conceivably explained by the competition characteristics in the Youth categories, which are usually comprised of shorter stages. For this reason, the lower intensity profile could be optimized in Junior and U-23 and, hence, closer to the values found in professional and elite cyclists [1,2].

In the work load at VT2, similar differences to those obtained for VT1 between age categories were found, but the values reported for U-23 were clearly lower than for professional cyclists [3]. Although there were no differences between age categories in the VT2 position with respect to VO_2max, the workload developed in this metabolic zone is crucial to determine the aerobic capacity, which characterizes the professional cyclists [3,39]. Therefore, from an applied perspective, developing the aerobic capacity in younger categories should be an important objective.

Finally, when analyzing the Fatmax zone, similar results were displayed by all age groups with only small differences found between the U-23 and Youth groups for VO_2 and Load. Notably, these differences were not found for the same parameters when the values were normalized to BW (VO_2R and Load/BW), which is in line with previous findings [23,24] and had not been reported for either differences in economy and efficiency when amateur and professional cyclist were compared [17]. Based on the present results, coaches should be aware that Fatmax does not seem to be a key factor discriminating the performance level in cyclists, although it could be important in long-term modalities as demonstrated in Ironman triathletes [22].

The workload was found to be the main difference in both thresholds (VT1 and VT2) with respect to the Youth group, as displayed by the large ES. Moreover, workload seems to be the main performance determinant in both maximum and submaximal zones with respect to the Youth group. These results are supported by the large ES obtained. However, when comparing the Junior and U-23, these differences were less pronounced and are only manifested in FTP/BW and maximal Load/BW with moderate ES. Practitioners should be aware of these findings when managing workload and monitoring training through power output zones.

The main limitation of the study was the reduced number of participants. Moreover, the fact that no previous investigations have compared, among the same age categories, the physiological parameters analyzed herein limited the discussion of the findings. Future studies with longitudinal research designs comparing the evolution of the same cyclists during their career would be interesting to conduct.

5. Conclusions

The main findings in this study showed enough differences with the Youth group and minor changes in the Junior vs. U-23 group. These results suggest that the main ladder is from the Youth to Junior age category. Due to the minor differences obtained between the Junior and U-23 categories, it could be intuited that the physiological profile in the Junior stage could be predictors of performance in absolute categories. Additionally, FTP/BW showed clear differences between each age category and testing it could be a good method to determine the cycling potential for cyclists.

Author Contributions: Conceptualization, C.M.-P. and S.D.; Methodology, C.M.-P.; Formal analysis, C.M.-P., T.T.F. and P.E.A.; Investigation, C.M.-P.; Resources, C.M.-P. and P.E.A.; Data curation, C.M.-P., S.D., T.T.F. and P.E.A.; Writing—original draft preparation, C.M.-P.; Writing—review and editing, T.T.F., S.D. and P.E.A.; Visualization, C.M.-P.; Supervision, P.E.A.; Project administration, C.M.-P.; Funding acquisition, P.E.A. All authors have read and agreed to the published version of the manuscript.

Funding: This research received no external funding.

Institutional Review Board Statement: The study was conducted according to the guidelines of the Declaration of Helsinki, and approved by the Ethics Committee of Catholic University of Murcia (protocol code CE022105, 26 February 2021).

Informed Consent Statement: Informed consent was obtained from all subjects involved in the study.

Data Availability Statement: The original data report is available to reviewers by contacting the corresponding author.

Acknowledgments: We wish to thank the cyclists for the availability to carry out this study.

Conflicts of Interest: The authors declare no conflict of interest.

References

1. Luciá, A.; Hoyos, J.; Carvajal, A.; Chicharro, J.L. Heart rate response to professional road cycling: The Tour de France. *Int. J. Sports Med.* **1999**, *20*, 167–172. [CrossRef] [PubMed]
2. Lucía, A.; Joyos, H.; Chicharro, J.L. Physiological response to professional road cycling: Climbers vs. time trialists. *Int. J. Sports Med.* **2000**, *21*, 505–512. [CrossRef] [PubMed]
3. Sallet, P.; Mathieu, R.; Fenech, G.; Baverel, G. Physiological differences of elite and professional road cyclists related to competition level and rider specialization. *J. Sports Med. Phys. Fit.* **2006**, *46*, 361–365.
4. Lucia, A.; Hoyos, J.; Chicharro, J.L. Physiology of professional road cycling. *Sports Med.* **2001**, *31*, 325–337. [CrossRef] [PubMed]
5. Padilla, S.; Mujika, I.; Cuesta, G.; Goiriena, J.J. Level ground and uphill cycling ability in professional road cycling. *Med. Sci. Sports Exerc.* **1999**, *31*, 878–885. [CrossRef]
6. Jeffries, O.; Simmons, R.; Patterson, S.D.; Waldron, M. Functional threshold power is not equivalent to lactate parameters in trained cyclists. *J. Strength Cond. Res.* **2019**, *35*, 2790–2794. [CrossRef] [PubMed]
7. Morgan, P.T.; Black, M.I. Road cycle TT performance: Relationship to the power-duration model and association with FTP. *J. Sports Sci.* **2019**, *37*, 902–910. [CrossRef] [PubMed]
8. Sørensen, A.; Aune, T.K.; Rangul, V.; Dalen, T. The validity of functional threshold power and maximal oxygen uptake for cycling performance in moderately trained cyclists. *Sports* **2019**, *7*, 217. [CrossRef]
9. Allen, H.; Coggan, A. *Training and Racing with Power Meter*, 2nd ed.; Boulder: Velopress, CO, USA, 2010.
10. Spindler, D.J.; Allen, M.S.; Vella, S.A.; Swann, C. The psychology of elite cycling: A systematic review. *J. Sports Sci.* **2018**, *36*, 1943–1954. [CrossRef] [PubMed]
11. Priego Quesada, J.I.; Kerr, Z.Y.; Bertucci, W.M. The categorization of amateur cyclists as research participants: Findings from an observational study. *J. Sports Sci.* **2018**, *36*, 2018–2024. [CrossRef]
12. Pinot, J.; Grappe, F. A six-year monitoring case study of a top-10 cycling Grand Tour finisher. *J. Sports Sci.* **2015**, *33*, 907–914. [CrossRef]
13. De Pauw, K.; Roelands, B.; Cheung, S.S.; de Geus, B.; Rietjens, G.; Meeusen, R. Guidelines to classify subject groups in sport-science research. *Int. J. Sports Physiol. Perform.* **2013**, *8*, 111–122. [CrossRef] [PubMed]
14. Pérez-Landaluce, J.; Fernández-García, B.; Rodríguez-Alonso, M.; García-Herrero, F.; García-Zapico, P.; Patterson, A.M.; García-Herrero, F. Physiological differences and rating of perceived exertion (RPE) in professional, amateur and young cyclists. *J. Sports Med. Phys. Fit.* **2002**, *42*, 389–395.
15. Padilla, S.; Mujika, I.; Orbañanos, J.; Angulo, F. Exercise intensity during competition time trials in professional road cycling. *Med. Sci. Sports Exerc.* **2000**, *32*, 850–856. [CrossRef] [PubMed]
16. Hopker, J.; Jobson, S. *Performance Cycling—The Science of Succes*; Boomsbury: London, UK, 2012.
17. Moseley, L.; Achten, J.; Martin, J.C.; Jeukendrup, A.E. No differences in cycling efficiency between world-class and recreational cyclists. *Int. J. Sports Med.* **2004**, *25*, 374–379. [CrossRef] [PubMed]
18. Rauter, S.; Vodicar, J.; Simenko, J. Body asymmetries in young male road cyclists. *Int. J. Morphol.* **2017**, *35*, 907–912. [CrossRef]
19. Rodríguez-Marroyo, J.A.; Pernía, R.; Cejuela, R.; García-López, J.; Llopis, J.; Villa, J.G. Exercise intensity and load during different races in youth and junior cyclists. *J. Strength Cond. Res.* **2011**, *25*, 511–519. [CrossRef] [PubMed]
20. Fell, J.; Haseler, L.; Gaffney, P.; Reaburn, P.; Harrison, G. Performance during consecutive days of laboratory time-trials in young and veteran cyclists. *J. Sports Med. Phys. Fit.* **2006**, *46*, 395–402.
21. Cuddy, J.S.; Slivka, D.R.; Hailes, W.S.; Dumke, C.L.; Ruby, B.C. Metabolic profile of the Ironman World Championships: A case study. *Int. J. Sports Physiol. Perform.* **2010**, *5*, 570–576. [CrossRef] [PubMed]

22. Frandsen, J.; Vest, S.D.; Larsen, S.; Dela, F.; Helge, J.W. Maximal fat oxidation is related to performance in an ironman triathlon. *Int. J. Sports Med.* **2017**, *38*, 975–982. [CrossRef]
23. Achten, J.; Venables, M.C.; Jeukendrup, A.E. Fat oxidation rates are higher during running compared with cycling over a wide range of intensities. *Metab. Clin. Exp.* **2003**, *52*, 747–752. [CrossRef]
24. Chenevière, X.; Malatesta, D.; Gojanovic, B.; Borrani, F. Differences in whole-body fat oxidation kinetics between cycling and running. *Eur. J. Appl. Physiol.* **2010**, *109*, 1037–1045. [CrossRef] [PubMed]
25. Montero, D.; Lundby, C. The effect of exercise training on the energetic cost of cycling. *Sports Med.* **2015**, *45*, 1603–1618. [CrossRef] [PubMed]
26. Rønnestad, B.R.; Hansen, J.; Hollan, I.; Ellefsen, S. Strength training improves performance and pedaling characteristics in elite cyclists. *Scand. J. Med. Sci. Sports* **2015**, *25*, e89–e98. [CrossRef]
27. Aagaard, P.; Andersen, J.L.; Bennekou, M.; Larsson, B.; Olesen, J.L.; Crameri, R.; Magnusson, S.P.; Kjaer, M. Effects of resistance training on endurance capacity and muscle fiber composition in young top-level cyclists. *Scand. J. Med. Sci. Sports* **2011**, *21*, 298–307. [CrossRef]
28. Binder, R.K.; Wonisch, M.; Corra, U.; Cohen-Solal, A.; Vanhees, L.; Saner, H.; Schmid, J.P. Methodological approach to the first and second lactate threshold in incremental cardiopulmonary exercise testing. *Eur. J. Prev. Cardiol.* **2008**, *15*, 726–734. [CrossRef]
29. Larson, R.; Cantrell, G.; Ade, C.; Farrell Iii, J.; Lantis, D.; Barton, M.; Laron, D. Physiologic responses to two distinct maximal cardiorespiratory exercise protocols. *Int. J. Sports Exerc. Med.* **2015**, *1*, 3. [CrossRef]
30. Zuniga, J.M.; Housh, T.J.; Camic, C.L.; Bergstrom, H.C.; Traylor, D.A.; Schmidt, R.J.; Johnson, G.O. Metabolic parameters for ramp versus step incremental cycle ergometer tests. *Appl. Physiol. Nutr. Metab.* **2012**, *37*, 1110–1117. [CrossRef] [PubMed]
31. Wasserman, K.; Beaver, W.L.; Whipp, B.J. Gas exchange theory and the lactic acidosis (anaerobic) threshold. *Circulation* **1990**, *81* (Suppl. S1), 14–30.
32. Millet, G. Physiological differences between cycling and running. *Rev. Med. Suisse* **2009**, *5*, 1564–1567. [CrossRef] [PubMed]
33. Edvardsen, E.; Hem, E.; Anderssen, S.A. End criteria for reaching maximal oxygen uptake must be strict and adjusted to sex and age: A cross-sectional study. *PLoS ONE* **2014**, *9*, e85276. [CrossRef] [PubMed]
34. Denham, J.; Scott-Hamilton, J.; Hagstrom, A.D.; Gray, A.J. Cycling power outputs predict functional threshold power and maximum oxygen uptake. *J. Strength Cond. Res.* **2017**, *34*, 3489–3497. [CrossRef]
35. Cohen, J. *Statistical Power Analysis for the Behavioral Sciences*; Routledge: London, UK, 1988.
36. Hopkins, W.; Marshall, S.; Batterham, A.; Hanin, J. Progressive statistics for studies in sports medicine and exercise science. *Med. Sci. Sports Exerc.* **2009**, *41*, 3. [CrossRef] [PubMed]
37. Faria, I.E.; Faria, E.W.; Roberts, S.; Yoshimura, D. Comparison of physical and physiological characteristics in elite young and mature cyclists. *Res. Q. Exerc. Sport* **1989**, *60*, 388–395. [CrossRef] [PubMed]
38. Klitzke Borszcz, F.; Tramontin, A.F.; Costa, V.P. Reliability of the functional threshold power in competitive cyclists. *Int. J. Sports Med.* **2020**, *41*, 175–181. [CrossRef] [PubMed]
39. Palmer, G.S.; Hawley, J.A.; Dennis, S.C.; Noakes, T.D. Heart rate responses during a 4-d cycle stage race. *Med. Sci. Sports Exerc.* **1994**, *26*, 1278–1283. [CrossRef]

biology

MDPI

Article

The Effects of Exercise Order on the Psychophysiological Responses, Physical and Technical Performances of Young Soccer Players: Combined Small-Sided Games and High-Intensity Interval Training

Ersan Arslan [1], Bulent Kilit [2], Filipe Manuel Clemente [3], Yusuf Soylu [1], Mustafa Sögüt [4], Georgian Badicu [5], Firat Akca [6], Mine Gokkaya [6] and Eugenia Murawska-Ciałowicz [7,*]

1 Faculty of Sport Sciences, Tokat Gaziosmanpasa University, Tokat 60250, Turkey; ersanarslan1980@hotmail.com (E.A.); yusuf.soylu@gop.edu.tr (Y.S.)
2 School of Physical Education and Sports, Tekirdag Namik Kemal University, Tekirdag 59030, Turkey; bkilit@nku.edu.tr
3 Escola Superior Desporto e Lazer, Instituto Politécnico de Viana do Castelo, Rua Escola Industrial e Comercial de Nun'Álvares, 4900-347 Viana do Castelo, Portugal; filipeclemente@esdl.ipvc.pt
4 Department of Physical Education and Sports, Middle East Technical University, Ankara 06800, Turkey; msogut@metu.edu.tr
5 Department of Physical Education and Special Motricity, University Transilvania of Brasov, 500068 Brasov, Romania; georgian.badicu@unitbv.ro
6 Faculty of Sport Sciences, Ankara University, Ankara 06560, Turkey; fakca@ankara.edu.tr (F.A.); minegokkaya@ankara.edu.tr (M.G.)
7 Physiology and Biochemistry Department, University School of Physical Education, 51-612 Wroclaw, Poland
* Correspondence: eugenia.murawska-cialowicz@awf.wroc.pl

Citation: Arslan, E.; Kilit, B.; Clemente, F.M.; Soylu, Y.; Sögüt, M.; Badicu, G.; Akca, F.; Gokkaya, M.; Murawska-Ciałowicz, E. The Effects of Exercise Order on the Psychophysiological Responses, Physical and Technical Performances of Young Soccer Players: Combined Small-Sided Games and High-Intensity Interval Training. *Biology* **2021**, *10*, 1180. https://doi.org/10.3390/biology10111180

Academic Editor: Tetsuya Shiuchi

Received: 13 September 2021
Accepted: 11 November 2021
Published: 15 November 2021

Simple Summary: Small-sided games are very popular training methods, among other commonly used strategies, for enhancing the functional and sport-specific skills of young soccer players. In addition, high-intensity interval training has the potential to increase the aerobic capacity of youths. No study has compared the order effects of combined small-sided games and high-intensity interval training on the physical performances, psychophysiological responses, and technical skills of young soccer players. The results of this research show practical information that can help to design training programmes for youth soccer players.

Abstract: This study aimed to compare the order effects of combined small-sided games (SSGs) and high-intensity interval training (HIIT) on the psychophysiological responses and physical and technical performances of young soccer players. Twenty-four soccer players (aged 14.63 $\pm$ 0.71 years) were randomly divided into SSGs + HIIT (n = 12) and HIIT + SSGs (n = 12) for 6 weeks. The SSGs consisted of two 4–16 min rounds of 2, 3, and four-a-side games with 2 min of passive resting, whereas the HIIT consisted of 6–10 min of high-intensity runs at varying intensities (from 90 to 100%). Pre-test and post-test elements included a 5–30 m sprint test, countermovement jump test, zigzag agility test with the ball and without the ball, repeated sprint ability test, speed dribbling ability test, three-corner run test, and Yo-Yo Intermittent Recovery Test level 1. Both combined training interventions produced similar improvements in physical performance and technical responses ($p \geq 0.05$, d values ranging from 0.40 to 1.10). However, the combined HIIT + SSGs training produced meaningfully lower perceived exertion ($p = 0.00$, $d = 2.98$) and greater physical enjoyment ($p = 0.00$, $d = 4.28$) compared with the SSGs + HIIT intervention. Furthermore, the SSGs + HIIT group showed a higher training load than those from the HIIT + SSGs group for all weeks ($p \leq 0.05$, d values ranging from 1.36 to 2.05). The present study's results might be used by coaches and practitioners to design training programmes for youth soccer players.

Keywords: soccer; high-intensity; small-sided games; psychophysiological responses; combined training

1. Introduction

High-level performances in soccer are combined with physical performance, psychophysiological responses, and technical abilities during small-sided games [1–3]. Therefore, several alternative training methods to traditional ones have been proposed to enhance the physical and technical capabilities of young soccer players. High-intensity interval training (HIIT), one of the increasingly popular training modalities, is defined as intense and intermittent exercises interspersed with recovery periods [4]. It requires a reduced amount of time and thus allows young athletes greater time to train their sport skills [5]. Earlier studies have documented the positive influences of HIIT on various physical fitness parameters [6] and soccer-specific performance characteristics in young soccer players [7].

Small-sided games (SSGs)—training strategies that are more enjoyable, effective, and time-efficient—are another commonly used method for training the functional and sport-specific skills of young soccer players [1,8]. SSGs, which are derived from street soccer and are played with fewer players, smaller pitch areas, and modified rules [9,10], simultaneously involve actual game dynamics, technical and tactical skills, and physical demands under changeable game conditions [8,11]. Consequently, some studies have shown the contribution of SSGs to aerobic fitness, repeated sprint ability, linear sprinting, agility, change of direction, and jumping performance in young players [1,12].

A recent systematic review demonstrated the effectiveness of combined HIIT and SSGs for soccer players [13]. As a result of this study, it was discovered that combining SSGs and running-based training methods induced higher external and internal load values and greater improvements in overall fitness capacity compared to the intervention using only SSGs. On the other hand, the researchers found a larger improvement in aerobic fitness for professional players who only participated in SSGs when compared to players who participated in combined training [14]. The inconsistency among these aforementioned studies shows that more research is needed to understand the efficiency of combined training.

Several studies recently compared the effects of combined game-based and HIIT programmes in team sports [13,15]. While some coaches routinely use the combined SSGs and HIIT approach (starting with SSGs and then performing HIIT or the opposite) to optimise sport-specific technical and tactical learning without any physiological or psychological fatigue effect on performance, others may prefer the combined HIIT and SSGs approach to have players undertake game performance, including technical and tactical tasks, under fatigue conditions [16,17]. The mechanisms related to running-based and game-based training are naturally different, although both tax aerobic and anaerobic metabolisms. Running-based HIIT seems to elicit a greater proportion of anaerobic metabolism. Blood lactate concentrations vary between 4 and 9 mmol/L in short/long HIIT [6], while SSGs vary between 0.5 and 4 mmol/L [18]. Moreover, neuromuscular effects are also different. Short HIIT or nonmaximal efforts produce peripheral fatigue (e.g., alterations to muscle excitability and excitation–contraction coupling), while SSGs can produce more variability based on the type of stimulus occurring in a match [19]. Therefore, it can be expected that starting with one type of HIIT over another might constrain the physiological responses, which would interfere with the next method. In a pioneering study, researchers examined the influences of combined training with different exercise orders on semi-professional soccer players [20]. Their results indicated that changing exercise orders yielded a similar enhancement in intermittent fitness performance. However, to the best of the authors' knowledge, there are no additional data on the impacts of combination order on multiple performance parameters in soccer players. Therefore, the aim of the present study was to compare the order effects of combined SSGs and HIIT on the psychophysiological responses and physical and technical performances of young soccer players.

2. Materials and Methods

2.1. Study Design

A two-group, matched, experimental design was used in the present study. The study was completed over a total of 9 weeks, consisting of 1.5 weeks of pre-testing, 6 weeks of combined training interventions (SSGs + HIIT or HIIT + SSGs), and 1.5 weeks of post-testing. The players completed a 30–15 intermittent fitness test (30–15 IFT), speed dribbling ability (SDA) test, 5–30 m sprint test, countermovement jump (CMJ) test, repeated sprint ability (RSA) test, zigzag agility test with the ball (ZAWB) and without the ball (ZAWOB), three-corner run test (TCRT), and Yo-Yo Intermittent Recovery Test level 1 (YYIRT-1) before and after the 6-week combined intervention period. Both training interventions were performed twice a week and each daily training session was separated by a minimum of 2 days to avoid fatigue-induced adverse effects. During the present study, the players performed the same type of daily training, and combined training interventions were added to their training sessions. After 15 min of standardised warmup, which consisted of jogging and dynamic stretching at each training session, players performed combined training, including SSGs + HIIT or HIIT + SSGs. All tests and training sessions with the same order were performed on a natural grass soccer pitch.

2.2. Subjects

Twenty-four young male soccer players participated in the present study. The players were separated into two combined groups: the SSGs + HIIT group (n = 12, age: 14.67 ± 0.65 years) and the HIIT + SSG group (n = 12, age: 14.58 ± 0.79 years). All players were also members of the U-16 regional amateur league teams. They were accustomed to a training workload of $\geq$3 training units per week, consisting of core strength, plyometric and technical drills, and had been involved in soccer training and competitive soccer matches for at least 2 years. Before the study, all players and their parents were fully informed about the procedures to be used and completed voluntary written consent forms. The study was performed in accordance with the Declaration of Helsinki and the Research Ethics Committee of the local university.

2.3. Procedures

Testing Procedure. On the first day, to calculate body fat percentage, the skinfold thickness technique was used with a Holtain Tanner–Whitehouse skinfold calliper (Holtain, UK) before breakfast. Skinfold thickness was measured twice at each site and the mean of two measurements was used to calculate body fat percentage. Body fat percentage was calculated using the equation that has been validated for males aged 15 to 24 years in young Turkish athletes [21]. After anthropometric measurements, determination of individual players' high-intensity intermittent running performance with changes in direction was assessed using the 30–15 IFT. The test, which consists of 30 s of running and 15 s of passive recovery, is a reliable progressive field test according to the procedures performed by Buchheit [22]. On the third day, the SDA test was used for the evaluation of soccer-specific technical skills and according to procedures described by Rosch et al. [23]. Briefly, the test, which is available in the F-MARC test battery designed by FIFA, allows for the assessment of coordinated speed dribbling under time pressure. After the technical test, each player performed three straight 30 m sprint test (5 m, 10 m, and 20 m splits) performances with 2 min of passive resting.

On the fifth day, each player was tested on their vertical jump height using the CMJ test according to the procedures performed by Arslan et al. [1]. A portable force plate (Newtest, Finland) was used to assess the CMJ test performances. Following the CMJ test, each player performed 6 repetitions of a 30 m maximal sprint with a 180° change of direction (15 m + 15 m). Twenty seconds of recovery were allowed between shuttle sprints [24]. The ZAWB and ZAWOB tests were performed to evaluate the agility performances of the players on the seventh day. The test, which included soccer-specific movement patterns [25], consisted of four 5 m sections with each change of trajectory angled at 100° as reported by Mirkov et al. [26]. The TCRT was performed to assess the speed endurance and anaerobic endurance of the players on a natural grass pitch [23]. The running times in these tests were measured using a timing gate photocell system. We found high test–retest reliability (ICC = >0.86) for tests such as sprinting, jumping, agility, and technical skill. On the ninth day, to evaluate maximum oxygen consumption (VO_{2max}), the YYIRT-1, which is an acoustically progressive field test [27], was performed according to procedures explained by Bangsbo et al. [28]. After the test, the estimated VO_{2max} was calculated using the following formula:

$$VO_{2max} = 36.4 + (0.0084 \times \text{covered distance in YYIRT-level 1})$$

Training Interventions. The training procedure is summarised in Table 1. During the 6-week training period, young players performed 2 combined training sessions (SSGs + HIIT or HIIT + SSGs) a week in addition to their 3 days of soccer-specific training. Their weekly training routine consisted of 5 60–75 min practice sessions and 1 soccer match. During the study, their coach generally focused on developing core strength and technical and tactical skills, except for the 2 combined training sessions. After 15 min of standardised warmup, which consisted of jogging and dynamic stretching at each training session, players performed combined training, including SSGs + HIIT or HIIT + SSGs. A gradual progress plan was designed to reach maximal final performance in combined training programmes. Players performed 2, 3, and 4-a-side formats of SSGs, including free game, possession, and small goal for two 4–16 min games per training session according to the procedures detailed by Sanchez-Sanchez et al. [29]. Verbal encouragements were given by coaches throughout the SSGs. Players performed HIIT sessions, which consisted of 15 s of intermittent running at 90–100% of players' velocity at IFT (VIFT), followed by 15 s of resting (Table 1).

Table 1. Description of the 6 weeks of combined training programs.

Week	Sessions	Game Formats	Pitch Dimension	SSGs+HIIT	HIIT+SSGs
				Pre-Intervention Testing	
1	1	2 v 2	15 × 27	2 × (2 × 2 min FG), 2 min rest 2 × (5 min of 15″-15″ at 90% of V_{IFT})	2 × (5 min of 15″-15″ at 90% of V_{IFT}) 2 × (2 × 2 min FG), 2 min rest
	2	3 v 3	20 × 30	2 × (3 × 3 min FG), 2 min rest 2 × (4 min of 15″-15″ at 90% of V_{IFT})	2 × (4 min of 15″-15″ at 90% of V_{IFT}) 2 × (3 × 3 min FG), 2 min rest
2	3	4 v 4	25 × 32	2 × (4 × 4 min FG), 2 min rest 2 × (3 min of 15″-15″ at 90% of V_{IFT})	2 × (3 min of 15″-15″ at 90% of V_{IFT}) 2 × (4 × 4 min FG), 2 min rest
	4	2 v 2	15 × 27	2 × (2 × 2 min POS), 2 min rest 2 × (5 min of 15″-15″ at 90% of V_{IFT})	2 × (5 min of 15″-15″ at 90% of V_{IFT}) 2 × (2 × 2 min POS), 2 min rest
3	5	3 v 3	20 × 30	2 × (3 × 3 min POS), 2 min rest 2 × (4 min of 15″-15″ at 90% of V_{IFT})	2 × (4 min of 15″-15″ at 90% of V_{IFT}) 2 × (3 × 3 min POS), 2 min rest
	6	4 v 4	25 × 32	2 × (4 × 4 min POS), 2 min rest 2 × (3 min of 15″-15″ at 90% of V_{IFT})	2 × (3 min of 15″-15″ at 90% of V_{IFT}2 (4 × 4 min POS), 2 min rest
4	7	2 v 2	15 × 27	2 × (2 × 2 min SG), 2 min rest 2 × (5 min of 15″-15″ at 95% of V_{IFT})	2 × (5 min of 15″-15″ at 95% of V_{IFT}) 2 × (2 × 2 min SG), 2 min rest
	8	3 v 3	20 × 30	2 × (3 × 3 min SG), 2 min rest 2 × (4 min of 15″-15″ at 95% of V_{IFT})	2 × (4 min of 15″-15″ at 95% of V_{IFT}) 2 × (3 × 3 min SG), 2 min rest
5	9	4 v 4	25 × 32	2 × (4 × 4 min SG), 2 min rest 2 × (3 min of 15″-15″ at 95% of V_{IFT})	2 × (3 min of 15″-15″ at 95% of V_{IFT}) 2 × (4 × 4 min SG), 2 min rest
	10	2 v 2	15 × 27	2 × (2 × 2 min FG), 2 min rest 2 × (5 min of 15″-15″ at 100% of V_{IFT})	2 × (5 min of 15″-15″ at 100% of V_{IFT}) 2 × (2 × 2 min FG), 2 min rest
6	11	3 v 3	20 × 30	2 × (3 × 3 min POS), 2 min rest 2 × (4 min of 15″-15″ at 100% of V_{IFT})	2 × (4 min of 15″-15″ at 100% of V_{IFT}) 2 × (3 × 3 min POS), 2 min rest
	12	4 v 4	25 × 32	2 × (4 × 4 min SG), 2 min rest 2 × (3 min of 15″-15″ at 100% of V_{IFT})	2 × (3 min of 15″-15″ at 100% of V_{IFT}) 2 × (4 × 4 min SG), 2 min rest
				Post-intervention testing	

FG: free game; POS: possession; SG: small goal; V_{IFT}: Maximum speed reached in the last stage of the 30-15 Intermittent Fitness Test.

The rating of perceived exertion (RPE) was obtained using the category ratio scale (6–20) to calculate the internal training load (ITL) immediately after the completion of each session [30]. The scale was introduced at the beginning in order to familiarise the players. All players also completed a short form of the physical activity enjoyment scale (PACES). This scale includes 5 items scored on a 1–7 Likert scale and has been validated [31] as a marker of enjoyment level for physical activity by Turkish youth [1].

2.4. Statistical Analyses

Data were expressed as mean ± standard deviation (SD). Group differences in psychophysiological responses, in terms of RPE, PACES, and ITL (overall) results between SSGs + HIIT and HIIT + SSGs, were assessed using the independent sample *t*-test. A mixed ANOVA was used to test for interactions and main effects for time (pre- vs. post-test) and group (SSGs + HIIT vs. HIIT + SSGs) on the physical and technical performances. Effect sizes (Cohen's *d*) were also calculated for each dependent variable. Cohen's *d* values were considered trivial (<0.20), small (0.20–0.59), moderate (0.6–1.19), large (1.2–1.99), and very large (≥2.0) [32]. All statistical analyses were computed using SPSS version 24.0 (SPSS, Version 24.0 for Windows; SPSS Inc., Chicago, IL, USA). Statistical significance was set at the level of $p \leq 0.05$.

3. Results

Pre-test values and the effect of combined training on the body composition, physical performance responses, and technical skills of the players are summarised in Table 2.

Table 2. Effect of both training methods on physical and technical performances of the participants.

	SSGs + HIIT (*n* = 12)			HIIT + SSGs (*n* = 12)			Training Comparison		
	Pre-Test	Post-Test	Change	Pre-Test	Post-Test	Change	$F_{(1, 22)}$	p	η^2
Body fat (%)	9.78 ± 2.53	8.83 ± 2.32 *	−0.95	9.48 ± 1.57	8.58 ± 1.49 *	−0.90	0.112	0.741	0.005
5-m (s)	0.95 ± 0.05	0.90 ± 0.06 *	−0.05	0.93 ± 0.06	0.90 ± 0.06 *	−0.03	0.157	0.696	0.007
10-m (s)	1.68 ± 0.05	1.58 ± 0.06 *	−0.10	1.62 ± 0.05	1.54 ± 0.08 *	−0.08	3.737	0.066	0.145
20-m (s)	3.06 ± 0.11	2.95 ± 0.10 *	−0.11	3.05 ± 0.20	2.94 ± 0.21 *	−0.11	0.048	0.829	0.002
30-m (s)	4.42 ± 0.09	4.19 ± 0.08 *	−0.23	4.34 ± 0.26	4.14 ± 0.23 *	−0.20	0.861	0.364	0.038
CMJ (cm)	31.72 ± 2.70	33.98 ± 2.44 *	2.26	32.01 ± 2.10	33.90 ± 1.89 *	1.89	0.013	0.911	0.001
ZAWB (s)	8.44 ± 0.32	8.28 ± 0.31 *	−0.16	8.60 ± 0.23	8.40 ± 0.31 *	−0.20	1.479	0.237	0.063
ZAWOB (s)	6.92 ± 0.23	6.79 ± 0.23 *	−0.13	6.79 ± 0.36	6.66 ± 0.37 *	−0.13	1.083	0.309	0.047
RSA_{total} (s)	39.08 ± 1.01	37.55 ± 0.93 *	−1.53	38.68 ± 1.08	37.09 ± 0.73 *	−1.59	1.297	0.267	0.056
SDA (s)	25.90 ± 1.47	24.94 ± 1.49 *	−0.96	25.00 ± 1.53	24.09 ± 1.39 *	−0.91	2.113	0.160	0.088
TCRT (s)	28.63 ± 0.47	25.01 ± 0.95 *	−3.62	28.32 ± 1.10	25.14 ± 0.98 *	−3.16	0.066	0.800	0.003
YYIRTL-1 (m)	1248.3 ± 107.7	1393.0 ± 107.1 *	144. 7	1213.3 ± 95.5	1363.3 ± 87.7 *	150.0	0.816	0.376	0.036
VO_{2max} ($mL.min^{-1}.kg^{-1}$)	46.89 ± 0.90	48.10 ± 0.90 *	1.21	46.59 ± 0.80	47.85 ± 0.74 *	1.26	0.816	0.376	0.036

* $p \leq 0.05$ for within-group changes.

Both combined training interventions (SSGs + HIIT and HIIT + SSGs) showed similar improvements in body composition, physical performance responses, and technical skills ($p \geq 0.05$, *d* values ranging from 0.40 to 1.10) (Table 2) (Figure 1).

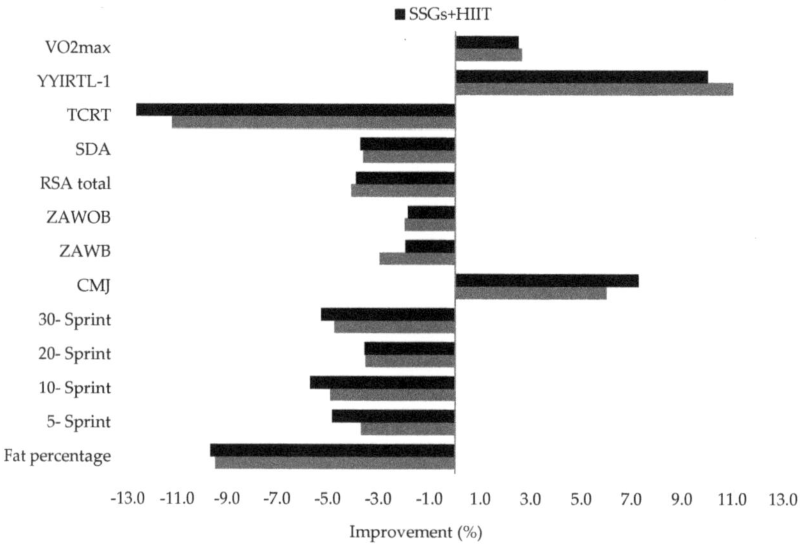

Figure 1. Improvement in body composition, physical and technical performance responses following the combined training interventions.

Overall RPE responses to HIIT + SSGs training were meaningfully lower than those from the SSGs + HIIT group (16.2 ± 0.5 vs. 17.6 ± 0.5; $p = 0.00$, $d = 2.98$). Moreover, overall PACES scores from the HIIT + SSGs training were meaningfully greater than those from the SSGs + HIIT group (30.7 ± 1.1 vs. 26.3 ± 0.9; $p = 0.00$, $d = 4.28$). Conversely, the SSGs + HIIT group demonstrated a higher training load than those from the HIIT + SSGs group for all weeks ($p \leq 0.05$, d = ranging from 1.36 to 2.05) (Figure 2).

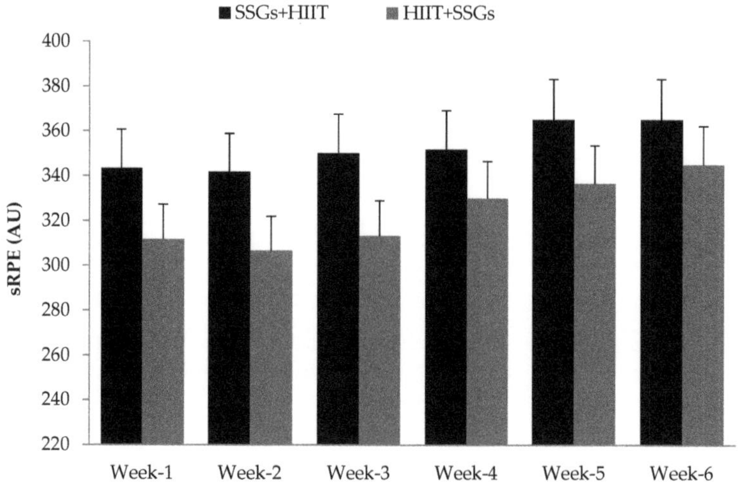

Figure 2. Weekly internal training loads during the 6 weeks combined training interventions.

4. Discussion

The aim of this study was to analyse the effects of exercise order in a combined training programme including SSGs and HIIT. The results of this parallel study revealed no significant differences between groups (SSGs + HIIT vs. HIIT + SSGs) in the fitness measures collected after the 6-week intervention. However, both combined programmes revealed significant pre–post improvements in linear sprinting, agility, vertical jump, aerobic capacity, and repeated-sprint ability.

A combination of SSGs and running-based HIIT was recently tested, aiming to provide the advantageous effect of running-based HIIT to the training programmes based on SSGs [12]. The first reported combination of SSGs and HIIT revealed the beneficial effect of the combination compared to a group using just SSGs for the improvement of VO_{2max} and 30–15 VIFT [15]. However, they did not consider how to implement the combination.

Exercise order is of paramount importance. In the first study, testing the effects of exercise order within a training session [20], it was revealed that no significant differences were found between those who completed SSGs + HIIT and those who completed HIIT + SSGs in the 30–15 VIFT [20], which revealed that the internal load imposed was similar between groups. In the present study, the measures of aerobic capacity (i.e., YYIRTL-1 and VO_{2max}) were both meaningfully improved by 6-week training interventions, with no significant difference considering the exercise order. Those findings are not surprising, since both SSGs and running-based HIIT have been repeatedly confirmed as effective in improving aerobic capacity [33,34]. The capacity to sustain high efforts while using both SSGs and running-based HIIT ensures that cardiorespiratory and aerobic systems are taxed by the training stimulus, thus promoting beneficial adaptations [35]. The nonexistence of differences between exercise order is in line with the previous work [20] and suggests that exercise intensity can be independent of the order of implementation.

Considering the effects of combined training intervention on linear sprinting, it was surprising to observe meaningful improvements independent of the exercise order, considering previous reports of combined SSGs + HIIT on such physical quality [14,15]. In fact, the results of the present study showed significant improvements of both groups in the 5, 10, 20, and 30 m sprint, thus suggesting the effectiveness of implementing SSGs and HIIT to improve linear sprinting. This fact was not observed in a previous study that combined SSGs and endurance and speed training or in the study that combined SSGs and short-interval (15′–15′) HIIT [15]. In fact, recent systematic reviews with meta-analysis revealed inconsistences and the ineffectiveness of SSGs [12] and HIIT [34] in improving linear sprinting in soccer players. One possible reason for observing improvements in the current research is due to the age effect and the capacity for improvement in this sensitive period [36].

Change of direction (COD) and agility with the ball were both capacities elicited equally by the combination of SSGs + HIIT with no difference considering the exercise order. The use of SSGs and HIIT independently has been suggested as a good way to improve COD [37], while SSGs seem to be better at improving agility with the ball compared to HIIT [38,39]. The combination of both in the present study contributed to meaningfully improving COD and agility with the ball, independent of the exercise order. Again, it seems that exercise order does not affect the capacity of both programmes and training methods to promote beneficial effects in these skills. However, it seems important to highlight the beneficial effect of combining HIIT with regular SSGs, considering a recent meta-analysis that suggested a significant favourable effect of HIIT in comparison to SSGs in improving linear sprint and COD in a within-group analysis [12]. This may be caused by the limited capacity of performing high-intensity linear or curvilinear running activities in small spaces as in the case of SSGs. On the other hand, those small spaces in SSGs can be helpful for promoting greater stimulus in COD and agility with the ball [40].

RSA was also improved after both combined interventions in which exercise order had no significant effect. Recent meta-analysis on the effects of HIIT in soccer revealed a significant favourable effect on RSA [34] in which no significant differences occurred between using HIIT or SSGs [41]. Therefore, due to the specific intermittence and energetic systems associated with both SSGs and HIIT, meaningful improvements in this capacity would be expected [42,43]. Improvements in lower-limb power and sprinting may also be factors benefiting the improvements in RSA [44]. In fact, in the current study, CMJ was also significantly improved in both combined interventions, thus suggesting possible positive effects, despite not being in line with recent meta-analysis about the use of HIIT and SSGs in soccer [12,34] and also compared with a study that combined SSGs + HIIT [45]. It is possible that the age effect and window of improvement may have caused the improvements observed in the current study.

Regarding the consequences of different exercise orders on psychophysiological responses and training load, it was found that HIIT + SSGs had meaningfully lower values of RPE, training load, and enjoyment than the group of SSGs + HIIT. The results regarding the training load are not in line with a previous study that tested the same issue (SSGs vs. HIIT and vice versa), in which no meaningful differences were found [20]. In this case, the psychological effect of more enjoyment while playing SSGs may have affected the perception of load. In fact, consistent results revealed that SSGs induce greater enjoyment than running-based HIIT [1,8,46]. Thus, for the groups ending the session with HIIT, the combination of fatigue and the least enjoyable activity may play an important role in the perceived effort and the enjoyment reported.

This study had some limitations. No control group was implemented; thus, it is not possible to compare the evolution of players without a training intervention or with different training interventions. Additionally, age may be a constraint for the possible generalisation of the findings since the study was conducted in a critical period of evolution. Future studies should compare combined interventions with single training or alternative training methods. Additionally, extending the research to more age groups and normalising the maturation status would be interesting.

5. Conclusions

The present study showed the order effects of combined SSGs and HIIT on the psychophysiological responses and physical and technical performances of young soccer players. After 6 weeks of combined training interventions, both combined training groups demonstrated similar improvements in physical performance and technical responses. However, the effects of exercise order demonstrated meaningful differences in psychophysiological responses and training load. In terms of practical implications, this study suggests that the combination of SSGs + HIIT is effective in improving the fitness status of adolescent soccer players. However, exercise order does not seem to have a determinant effect on the consequences of the changes in fitness. Therefore, coaches may organise the order based on the most appropriate plans. Future applications should consider implementing strength training in addition to the combination of SSGs + HIIT.

Author Contributions: Conceptualisation, E.A., B.K. and Y.S.; methodology, E.A., B.K. and Y.S.; formal analysis, E.A., B.K., F.M.C., Y.S. and M.S.; investigation, E.A., B.K. and M.G.; resources, E.A., B.K., F.M.C., Y.S. and G.B.; data curation, E.A., B.K., F.M.C., Y.S. and M.S.; writing—original draft preparation, E.A., B.K., F.A., F.M.C., Y.S. and M.S.; writing—review and editing, E.A., F.A., F.M.C. and E.M.-C.; visualisation, E.A., B.K. and Y.S.; supervision, E.A., B.K. and Y.S.; project administration, E.A. All authors have read and agreed to the published version of the manuscript.

Funding: Filipe Manuel Clemente: This work is funded by Fundação para a Ciência e Tecnologia/Ministério da Ciência, Tecnologia e Ensino Superior through national funds and when applicable co-funded EU funds under the project UIDB/EEA/50008/2020.

Institutional Review Board Statement: The study was conducted according to the guidelines ofthe Declaration of Helsinki and approved by the Institutional Ethics Committee of the University of Osmaniye Korkut Ata, Osmaniye, Turkey (Protocol code: 59754796-050.99-E.670, approved on 2 May 2020).

Informed Consent Statement: Informed consent was obtained from all subjects involved in the study.

Data Availability Statement: Not applicable.

Acknowledgments: The authors would like to thank all the players for their efforts.

Conflicts of Interest: The authors declare no conflict of interest.

References

1. Arslan, E.; Orer, G.; Clemente, F. Running-based high-intensity interval training vs. small-sided game training programs: Effects on the physical performance, psychophysiological responses and technical skills in young soccer players. *Biol. Sport* **2020**, *37*, 165–173. [CrossRef]
2. Rampinini, E.; Impellizzeri, F.M.; Castagna, C.; Coutts, A.J.; Wisløff, U. Technical performance during soccer matches of the Italian Serie A league: Effect of fatigue and competitive level. *J. Sci. Med. Sport* **2009**, *12*, 227–233. [CrossRef]
3. Stølen, T.; Chamari, K.; Castagna, C.; Wisløff, U. Physiology of soccer: An update. *Sports Med.* **2005**, *35*, 501–536. [CrossRef]
4. Kunz, P.; Engel, F.A.; Holmberg, H.-C.; Sperlich, B. A Meta-Comparison of the Effects of High-Intensity Interval Training to Those of Small-Sided Games and Other Training Protocols on Parameters Related to the Physiology and Performance of Youth Soccer Players. *Sports Med. Open* **2019**, *5*, 7. [CrossRef]
5. Engel, F.A.; Ackermann, A.; Chtourou, H.; Sperlich, B. High-Intensity Interval Training Performed by Young Athletes: A Systematic Review and Meta-Analysis. *Front. Physiol.* **2018**, *9*, 1012. [CrossRef] [PubMed]
6. Sperlich, B.; De Marées, M.; Koehler, K.; Linville, J.; Holmberg, H.-C.; Mester, J. Effects of 5 Weeks of High-Intensity Interval Training vs. Volume Training in 14-Year-Old Soccer Players. *J. Strength Cond. Res.* **2011**, *25*, 1271–1278. [CrossRef] [PubMed]
7. Impellizzeri, F.M.; Marcora, S.M.; Castagna, C.; Reilly, T.; Sassi, A.; Iaia, F.M.; Rampinini, E. Physiological and performance effects of generic versus specific aerobic training in soccer players. *Int. J. Sports Med.* **2006**, *27*, 483–492. [CrossRef] [PubMed]
8. Los Arcos, A.; Vázquez, J.S.; Martín, J.; Lerga, J.; Sánchez, F.; Villagra, F.; Zulueta, J.J. Effects of small-sided games vs. interval training in aerobic fitness and physical enjoyment in young elite soccer players. *PLoS ONE* **2015**, *10*, e0137224. [CrossRef] [PubMed]
9. Hill-Haas, S.V.; Dawson, B.; Impellizzeri, F.M.; Coutts, A.J. Physiology of small-sided games training in football: A systematic review. *Sports Med.* **2011**, *41*, 199–220. [CrossRef] [PubMed]
10. Hill-Haas, S.V.; Coutts, A.J.; Dawson, B.T.; Rowsell, G.J. Time-Motion Characteristics and Physiological Responses of Small-Sided Games in Elite Youth Players: The Influence of Player Number and Rule Changes. *J. Strength Cond. Res.* **2010**, *24*, 2149–2156. [CrossRef] [PubMed]
11. Owen, A.L.; Wong, D.P.; Paul, D.; Dellal, A. Effects of a Periodized Small-Sided Game Training Intervention on Physical Performance in Elite Professional Soccer. *J. Strength Cond. Res.* **2012**, *26*, 2748–2754. [CrossRef]
12. Clemente, F.M.; Ramirez-Campillo, R.; Afonso, J.; Sarmento, H. Effects of Small-Sided Games vs. Running-Based High-Intensity Interval Training on Physical Performance in Soccer Players: A Meta-Analytical Comparison. *Front. Physiol.* **2021**, *12*. [CrossRef] [PubMed]
13. Clemente, F.M.; Sarmento, H. Combining small-sided soccer games and running-based methods: A systematic review. *Biol. Sport* **2021**, *38*, 617–627. [CrossRef]
14. Castillo, D.; Raya-González, J.; Sarmento, H.; Clemente, F.; Yanci, J. Effects of including endurance and speed sessions within small-sided soccer games periodization on physical fitness. *Biol. Sport* **2021**, *38*, 291–299. [CrossRef] [PubMed]
15. Harrison, C.; Kinugasa, T.; Gill, N.; Kilding, A. Aerobic Fitness for Young Athletes: Combining Game-based and High-intensity Interval Training. *Int. J. Sports Med.* **2015**, *36*, 929–934. [CrossRef]
16. Sanchez-Sanchez, J.; Ramirez-Campillo, R.; Carretero, M.; Martín, V.; Hernández, D.; Nakamura, F.Y. Soccer Small-Sided Games Activities Vary According to the Interval Regime and their Order of Presentation within the Session. *J. Hum. Kinet.* **2018**, *62*, 167–175. [CrossRef]
17. Van Cutsem, J.; Marcora, S.; De Pauw, K.; Bailey, S.; Meeusen, R.; Roelands, B. The Effects of Mental Fatigue on Physical Performance: A Systematic Review. *Sports Med.* **2017**, *47*, 1569–1588. [CrossRef]
18. Dellal, A.; Chamari, K.; Lee Owen, A.; Wong, D.P.; Lago-Penas, C.; Hill-Haas, S. Influence of technical instructions on the physiological and physical demands of small-sided soccer games. *Eur. J. Sport Sci.* **2011**, *11*, 341–346. [CrossRef]
19. Perrey, S.; Racinais, S.; Saimouaa, K.; Girard, O. Neural and muscular adjustments following repeated running sprints. *Eur. J. Appl. Physiol.* **2010**, *109*, 1027–1036. [CrossRef]
20. Rabbani, A.; Clemente, F.M.; Kargarfard, M.; Jahangiri, S. Combined Small-Sided Game and High-Intensity Interval Training in Soccer Players: The Effect of Exercise Order. *J. Hum. Kinet.* **2019**, *69*, 249–257. [CrossRef]
21. Acikada, C.; Ergen, E.; Alpar, R.; Sarpyener, K. Study of body composition parameters in athletes. *Hacettepe J. Sports Sci. Technol.* **1991**, *2*, 1–25.
22. Buchheit, M. The 30-15 Intermittent Fitness Test: Accuracy for Individualizing Interval Training of Young Intermittent Sport Players. *J. Strength Cond. Res.* **2008**, *22*, 365–374. [CrossRef] [PubMed]
23. Rosch, D.; Hodgson, R.; Peterson, L.; Graf-Baumann, T.; Junge, A.; Chomiak, J.; Dvorak, J. Assessment and Evaluation of Football Performance. *Am. J. Sports Med.* **2000**, *28*, 29–39. [CrossRef] [PubMed]
24. Buchheit, M.; Mendez-Villanueva, A.; Delhomel, G.; Brughelli, M.; Ahmaidi, S. Improving Repeated Sprint Ability in Young Elite Soccer Players: Repeated Shuttle Sprints Vs. Explosive Strength Training. *J. Strength Cond. Res.* **2010**, *24*, 2715–2722. [CrossRef] [PubMed]

25. Little, T.; Williams, A.G. Specificity of Acceleration, Maximum Speed, and Agility in Professional Soccer Players. *J. Strength Cond. Res.* **2005**, *19*, 76.
26. Mirkov, D.; Nedeljkovic, A.; Kukolj, M.; Ugarkovic, D.; Jaric, S. Evaluation of the reliability of soccer-specific field tests. *J. Strength Cond. Res.* **2008**, *22*, 1046–1050. [CrossRef]
27. Krustrup, P.; Mohr, M.; Amstrup, T.; Rysgaard, T.; Johansen, J.; Steensberg, A.; Pedersen, P.K.; Bangsbo, J. The Yo-Yo Intermittent Recovery Test: Physiological Response, Reliability, and Validity. *Med. Sci. Sports Exerc.* **2003**, *35*, 697–705. [CrossRef]
28. Bangsbo, J.; Iaia, F.M.; Krustrup, P. The Yo-Yo Intermittent Recovery Test. *Sports Med.* **2008**, *38*, 37–51. [CrossRef]
29. Sanchez-Sanchez, J.; García, M.S.; Asián-Clemente, J.A.; Nakamura, F.Y.; Ramírez-Campillo, R. Effects of the directionality and the order of presentation within the session on the physical demands of small-sided games in youth soccer. *Asian J. Sports Med.* **2019**, *10*, 1–8. [CrossRef]
30. Foster, C.; Boullosa, D.; Mcguigan, M.; Fusco, A.; Cortis, C.; Arney, B.E.; Orton, B.; Dodge, C.; Jaime, S.; Radtke, K.; et al. 25 Years of Session Rating of Perceived Exertion: Historical Perspective and Development. *Int. J. Sports Physiol. Perform.* **2021**, *16*, 612–621. [CrossRef]
31. Mirzeoğlu, A.D. A validity and reliability study of physical activity enjoyment scale short form for Turkish children and youth. *Int. J. Hum. Sci.* **2014**, *11*, 672–687.
32. Hopkins, W.G.; Marshall, S.W.; Batterham, A.M.; Hanin, J. Progressive Statistics for Studies in Sports Medicine and Exercise Science. *Med. Sci. Sports Exerc.* **2009**, *41*, 3–13. [CrossRef]
33. Moran, J.; Blagrove, R.C.; Drury, B.; Fernandes, J.F.T.; Paxton, K.; Chaabene, H.; Ramirez-Campillo, R. Effects of Small-Sided Games vs. Conventional Endurance Training on Endurance Performance in Male Youth Soccer Players: A Meta-Analytical Comparison. *Sports Med.* **2019**, *49*, 731–742. [CrossRef] [PubMed]
34. Manuel Clemente, F.; Ramirez-Campillo, R.; Nakamura, F.Y.; Sarmento, H. Effects of high-intensity interval training in men soccer player's physical fitness: A systematic review with meta-analysis of randomized-controlled and non-controlled trials. *J. Sports Sci.* **2021**, *39*, 1202–1222. [CrossRef] [PubMed]
35. Buchheit, M.; Laursen, P.B. High-Intensity Interval Training, Solutions to the Programming Puzzle: Part I: Cardiopulmonary Emphasis. *Sports Med.* **2013**, *43*, 313–338. [CrossRef]
36. Oliver, J.L.; Lloyd, R.S.; Rumpf, M.C. Developing Speed Throughout Childhood and Adolescence. *Strength Cond. J.* **2013**, *35*, 42–48. [CrossRef]
37. Nygaard Falch, H.; Guldteig Rædergård, H.; van den Tillaar, R. Effect of Different Physical Training Forms on Change of Direction Ability: A Systematic Review and Meta-analysis. *Sports Med.Open* **2019**, *5*, 1–37. [CrossRef]
38. Chaouachi, A.; Chtara, M.; Hammami, R.; Chtara, H.; Turki, O.; Castagna, C. Multidirectional Sprints and Small-Sided Games Training Effect on Agility and Change of Direction Abilities in Youth Soccer. *J. Strength Cond. Res.* **2014**, *28*, 3121–3127. [CrossRef] [PubMed]
39. Radziminski, L.; Rompa, P.; Barnat, W.; Dargiewicz, R.; Jastrzebski, Z. A comparison of the physiological and technical effects of high-intensity running and small-sided games in young soccer players. *Int. J. Sports Sci. Coach.* **2013**, *8*, 455–465. [CrossRef]
40. Young, W.; Rogers, N. Effects of small-sided game and change-of- direction training on reactive agility and change-of-direction speed. *J. Sports Sci.* **2014**, *32*, 307–314. [CrossRef]
41. Hill-Haas, S.V.; Coutts, A.J.; Rowsell, G.J.; Dawson, B.T. Generic versus small-sided game training in soccer. *Int. J. Sports Med.* **2009**, *30*, 636–642. [CrossRef] [PubMed]
42. Girard, O.; Mendez-Villanueva, A.; Bishop, D. Repeated-Sprint Ability—Part I: Factors contributing to fatigue. *Sports Med.* **2011**, *41*, 673–694. [CrossRef] [PubMed]
43. Buchheit, M.; Laursen, P.B. High-intensity interval training, solutions to the programming puzzle: Part II: Anaerobic energy, neuromuscular load and practical applications. *Sports Med.* **2013**, *43*, 927–954. [CrossRef]
44. Bishop, D.; Girard, O.; Mendez-Villanueva, A. Repeated-Sprint Ability—Part II. *Sports Med.* **2011**, *41*, 741–756. [CrossRef] [PubMed]
45. Paul, D.J.; Marques, J.B.; Nassis, G.P. The effect of a concentrated period of soccer-specific fitness training with small-sided games on physical fitness in youth players. *J. Sports Med. Phys. Fitness* **2019**, *59*, 962–996. [CrossRef]
46. Selmi, O.; Ouergui, I.; Levitt, D.E.; Nikolaidis, P.T.; Knechtle, B.; Bouassida, A. Small-Sided Games are More Enjoyable Than High-Intensity Interval Training of Similar Exercise Intensity in Soccer. *Open Access J. Sports Med.* **2020**, *11*, 77. [CrossRef]

biology

MDPI

Article

Effects of the COVID-19 Lockdown on Body Composition and Bioelectrical Phase Angle in Serie A Soccer Players: A Comparison of Two Consecutive Seasons

Francesco Campa [1,†], Tindaro Bongiovanni [2,3,†], Athos Trecroci [2], Alessio Rossi [4], Gianpiero Greco [5,*], Giulio Pasta [6] and Giuseppe Coratella [2]

1 Department for Life Quality Studies, University of Bologna, 47921 Rimini, Italy; francesco.campa3@unibo.it
2 Department of Biomedical Sciences for Health, University of Study of Milano, 20133 Milano, Italy; tindaro.bongiovanni@unimi.it (T.B.); athos.trecroci@unimi.it (A.T.); giuseppe.coratella@unimi.it (G.C.)
3 Department of Health, Nutrition and Exercise Physiology, Parma Calcio 1913, 43100 Parma, Italy
4 Department of Computer Science, University of Pisa, 56126 Pisa, Italy; alessio.rossi@di.unipi.it
5 Department of Basic Medical Sciences, Neuroscience and Sense Organs, University of Study of Bari, 70121 Bari, Italy
6 Medical Department, Parma Calcio 1913, 43100 Parma, Italy; ghitopasta@hotmail.com
* Correspondence: gianpierogreco.phd@yahoo.com
† These authors have contributed equally.

Simple Summary: In 2020, the first Italian soccer league (Serie A) was canceled due to the COVID-19 pandemic. Consequently, a detraining process was triggered in soccer players, leading coaches and sports scientists to implement alternative training strategies to prevent a remodeling in body composition. This study tested the hypothesis that male elite soccer players, when confined to their home during the coronavirus disease 2019 pandemic, will display unfavorable trends in bioelectrical and body composition parameters. The results of the present study showed that reduction in phase angle and muscle mass occurred in soccer players during the coronavirus disease 2019 pandemic lockdown. Recognizing these adverse effects of a detraining period is critical in avoiding adverse effects on body composition in soccer players. In addition, the bioelectrical phase angle has been identified as a valid predictor of muscle mass changes during the competitive soccer season. Considerably, the phase angle represents a parameter that can be measured directly through bioelectrical impedance analysis, and it is independent of predictive equations such as those that quantify muscle mass.

Abstract: The present study compared changes in body composition during the COVID-19-associated lockdown with the same period of the following season in elite soccer players. Fifteen elite male soccer players (30.5 ± 3.6 years.) underwent a bioelectrical impedance analysis (BIA) before (end of February) and after (end of May) the lockdown, which occurred during the 2019/2020 season, and at the same period during the following competitive season in 2020/2021, when restrictions were lifted. Fat and muscle mass were estimated using predictive equations, while phase angle (PhA) and bioelectrical impedance vector analysis (BIVA) patterns were directly measured. After lockdown, fat mass remained unchanged ($p > 0.05$), while muscle mass (95%CI = −1.12/−0.64; ES = −2.04) and PhA (95%CI = 0.51/−0.24, ES = −1.56) decreased. A rightward displacement of the BIVA vector was also found ($p < 0.001$, ES = 1.50). After the same period during the regular season, FM% and muscle mass did not change ($p > 0.05$), while the PhA increased (95%CI = 0.01/0.22; ES = 0.63). A leftward vector displacement ($p < 0.001$, ES = 1.05) was also observed. The changes in muscle mass correlated with changes in PhA ("lockdown" season 2019/2020: ß = −1.128, $p = 0.011$; "regular" season 2020/21: ß = 1.963, $p = 0.011$). In conclusion, coaches and strength conditioners should monitor muscle mass in soccer players during detraining periods as this parameter appears to be mainly affected by changes in training plans.

Keywords: BIA; BIVA; coronavirus disease; detraining; fat mass; football; muscle mass; team sports

Citation: Campa, F.; Bongiovanni, T.; Trecroci, A.; Rossi, A.; Greco, G.; Pasta, G.; Coratella, G. Effects of the COVID-19 Lockdown on Body Composition and Bioelectrical Phase Angle in Serie A Soccer Players: A Comparison of Two Consecutive Seasons. *Biology* **2021**, *10*, 1175. https://doi.org/10.3390/biology10111175

Academic Editors: Filipe Manuel Clemente, Georgian Badicu and Eugenia Murawska-Ciałowicz

Received: 26 October 2021
Accepted: 11 November 2021
Published: 13 November 2021

1. Introduction

The coronavirus disease 2019 (COVID-19) pandemic has had far-reaching social and health implications affecting the worldwide population, including athletes [1]. Soccer players experienced an initial lockdown, during which traditional training activities were suspended and players had to train individually at home. During this lockdown, drastic change in training plans led to a reduction in training activities, mimicking a detraining condition [2], consequently resulting in a reduction in performance parameters [3,4].

Previous studies with professional soccer players have shown that the detraining that occurs during the off-season break may impair body composition, increasing fat mass (FM) and reducing fat-free mass [5,6]. As for detraining context, the lockdown was also shown to affect body composition, increasing FM [4]. However, no further information on the effects of lockdown on body composition is available in soccer players. This may be of interest, since changes in fat-free mass, which includes muscle mass and body fluids, could negatively affect tolerance to high training exposure, possibly increasing the risk of injury [7]. In addition, muscle mass seems related to anaerobic performance in soccer players, given that its decrements correlated to decrements in sprinting and jumping capacity [8]. Therefore, monitoring these components of fat-free mass may also be relevant.

Among the methods for assessing body composition, bioelectrical impedance analysis (BIA) has recently gained relevance, especially in soccer [9–15], where field and user-friendly methods are warranted [16]. Through the measurement and use of bioelectrical resistance (R) and reactance (Xc) in predictive equations, quantification of a wide range of body composition parameters is possible [17,18].

The combined evaluation of R and Xc as a vector within a graph results in a qualitative evaluation of body composition, defined as bioelectrical impedance vector analysis (BIVA) [19]. In BIVA, the position of the vector is interpreted in relation to its lateral and vertical displacements, which are reflected by phase angle (PhA) and vector length, respectively [15,20–22]. Several studies on soccer players have shown that PhA is related to sprint performance [11,12] and that BIVA patterns allow for the evaluation of nutritional status and physical condition during the competitive season [13,21,23,24].

Considering that a decline in body composition features impairs health and sports performance, exploring the effect of the COVID-19 lockdown could help practitioners better understand which parameters of body composition are most affected during a detraining period in soccer. Therefore, the present study aimed to examine whether or not BIA-derived FM and muscle mass and PhA were affected during the pandemic-associated lockdown in Series A soccer players. We compared the lockdown period with the same in-season period during the following 2020/21 regular season. We hypothesized that the lockdown affected body composition parameters as detected by BIA and BIVA.

2. Materials and Methods

2.1. Participants

A total of 15 male soccer players (age 30.5 ± 3.6 years.; body mass 79.6 ± 7.6 kg; height 1.82 ± 0.1 m), currently competing in the Italian First division team (Serie A), voluntarily participated in the study. The exclusion criteria were age <18 years and muscle injury in the previous 6 months. After a detailed explanation of the procedures, participants signed informed consent. All research procedures were reviewed and approved by the Bioethics Committee of the University of Milan (approval number: 32/16) and conformed to the Declaration of Helsinki (1964 and further updates) concerning studies involving human subjects.

2.2. Study Design and Procedures

The present investigation was designed as an observational, two-condition, two-time, one-group study. The exclusion criteria were age <18 years and muscle injury in the previous 6 months. The initial assessments were performed at the end of February 2020 as scheduled during the regular season, immediately before the COVID-19 lockdown. The post-lockdown assessments were performed at the end of May 2020, when the Italian Government lifted the restrictions. During the following 2020/21 season, we repeated the same assessment in the same period, i.e., end of February 2021 and end of May 2021. After the 2020 season, a total of 15 participants underwent all assessments and were included in analysis.

Table 1 provides an overview of training contents within a typical weekly routine during both lockdown and competitive periods. During lockdown, each player followed an individualized nutritional and supplementation plan developed by the team's nutritionist and adjusted every week.

Table 1. Overview of training contents during the COVID-19 lockdown and competitive period.

Period	No of Training Weeks	No of Training Sessions/Matches	Weekly Training Contents *
Competitive	14	84	*Monday:* rest day. *Tuesday and Thursday:* sessions based on TT and aerobic training. *Wednesday:* session based on a combination of strength-related stimuli with a special emphasis on lower limbs. *Friday and Saturday:* sessions based on a combination of low-intensity TT, attacking and defending maneuvers, and SAQ training. *Sunday:* match day.
Lockdown	14	84	*Monday, Wednesday, and Saturday:* sessions based on a combination of aerobic drills suitable for the home-confinement condition. *Tuesday and Thursday:* sessions based on a combination of aerobic (mainly running/cycling-based MIIT) training and strength-related stimuli suitable for the home-confinement condition. *Friday:* sessions based on a combination of aerobic drills (mainly running/cycling-based LIT) suitable for the home-confinement condition. *Sunday:* rest day.

Note: TT = technical and tactical, HIIT = high-intensity interval training, MIIT = moderate-intensity interval training, LIT = low-intensity interval training, SAQ = speed, agility, and quickness, and COD = change of direction. * Additional individual training, injury prevention program, and warm-up sessions are not included.

All anthropometric and BIA measurements were made in a resting and fasted state at least 24 h after the last exercise session and were profiled by an accredited and trained anthropometrist (T.B.). Height was recorded to the nearest 0.1 cm with a standing stadiometer (Seca 217, Basel, Switzerland), and body mass was measured to the nearest 0.1 kg with a high-precision mechanical scale (Seca 877, Basel, Switzerland). Body mass index (BMI) was calculated as the ratio of body mass to height squared (kg/m^2). The impedance measurements were performed with a previously validated [19,20] bioimpedance analyzer (BIA 101 Anniversary, Akern, Florence, Italy) at a frequency of 50 kHz. Before each testing session, the analyzer was checked with a calibration circuit of known impedance (resistance = 500.0 Ω; reactance = 0.1 Ω; 0.9% error). The participants were assessed in the supine position with legs (45° compared to the median line of the body) and arms (30° from the trunk) abducted. After cleansing the skin with alcohol, two electrodes were placed on the right hand and two on the right foot. Bioimpedance values were analyzed according to the BIVA methods [16,19] and analyzed in relation to the distribution of the reference population (tolerance ellipses of Serie A soccer players) [9]. PhA was calculated as the arctangent of $Xc/R*180°/\pi$. Quantitative body composition was evaluated according to a three-compartment tissue model where body mass is given by the sum of FM, muscle mass, and residual components [16]. Body composition parameters were estimated using bioimpedance-derived equations [18,25] as follows:

%FM = [(Body mass − fat-free mass)/body mass] × 100, with fat-free mass = $-2.261 + 0.327 \times H^2/R + 0.525 \times \text{body mass} + 5.462 \times 1$

Muscle mass (kg) = $-4.211 + (0.267 \times H^2/R) + (0.095 \times \text{body mass}) + (1.909 \times s1) + (-0.012 \times \text{age}) + (0.058 \times Xc)$

2.3. Statistical Analysis

Data were analyzed using SPSS v. 27.0 (SPSS, IBM Corp., Armonk, NY, USA). The Kolmogorov–Smirnov test was used to ensure normal distribution of data. A one-way analysis of variance (ANOVA) was used to assess whether participants differed in the investigated parameters at baseline. A condition x time repeated-measures ANOVA was performed to determine changes in dependent parameters over time (two levels: PRE and POST) and condition (two levels: lockdown and regular seasons). Multiple comparisons were performed using Bonferroni's correction. The paired, one-sample Hotelling's T^2 test, a multivariate extension of the Student's *t*-test for paired data, was performed to determine if the changes in BIVA vectors were different from zero (null vector). Partial eta squared ($\eta_p{}^2$) was calculated to estimate the degree of variance of the dependent factor due to independent factors and interpreted as follows: <0.059: small; 0.06 to 0.12: medium; >0.13: large [26]. Cohen's *d* effect size (ES) with 95% confidence interval (CI) was reported for significant pairwise comparisons, and Mahalanobis distance (D^2), which represents a multivariate measure of effect and a multivariate measure of distance, was calculated to determine the magnitude of changes in the mean group vectors. Cohen's ES was interpreted according to the following Hopkins' recommendations: 0–0.19: trivial; 0.20–0.59: small; 0.60–1.19: moderate; 1.20–1.99: large; ≥2.00: very large [27]. Mahalanobis D^2 was interpreted according to the following Stevens's [28] guidelines: 0.25–0.49: small; 0.5–0.99; ≥1: large. Single and multiple regression analyses were performed to determine the correlation between the changes in FM% and muscle mass, considering the age of participants as an independent variable in the multiple regression. Data were reported as mean ± standard deviation, and significance was set at $p < 0.05$.

3. Results

No between-group difference ($p > 0.05$) in body mass or BMI was found at baseline. Table 2 shows change over time for each condition and statistical analysis for all dependent parameters. A time effect ($p < 0.05$) was found for body mass and BMI that decreased from PRE to POST during the "lockdown" season 2019/2020 (Table 2).

Table 2. Baseline (PRE) and post values (mean ± standard deviation) of dependent parameters are shown. Differences over time are reported as mean with 95% confidence interval (CI). Effect size is also reported.

Variable		"Lockdown" 2019/2020 Season	"Regular" 2020/2021 Season	Time Effect	Condition x Time Interaction
Body mass (kg)	PRE	79.6 ± 7.6	79.7 ± 7.9	F = 4.40, $p = 0.045$, $\eta_p{}^2 = 0.136$	F = 8.26, $p = 0.008$, $\eta_p{}^2 = 0.228$
	POST	78.6 ± 7.8 *	79.8 ± 7.9		
	95% CI	−1.66/−0.22	−0.21/0.51		
	Effect size	−0.72	0.22		
Body mass index (kg/m^2)	PRE	23.7 ± 1.0	23.7 ± 1.1	F = 1.43, $p = 0.242$, $\eta_p{}^2 = 0.049$	F = 9.66, $p = 0.004$, $\eta_p{}^2 = 0.257$
	POST	23.4 ± 1.1 *	23.8 ± 1.1		
	95% CI	−0.06/−0.49	−0.05/0.30		
	Effect size	−0.73	0.38		
R/H (Ohm/m)	PRE	258.9 ± 22.4	253.7 ± 19.7	F = 1.09, $p = 0.305$, $\eta_p{}^2 = 0.038$	F = 13.11, $p = 0.001$, $\eta_p{}^2 = 0.319$
	POST	263.9 ± 23.5 *	250.9 ± 19.3 *		
	95% CI	0.83/9.15	−4.69/−0.81		
	Effect size	0.66	−0.78		
Xc/H (Ohm/m)	PRE	35.6 ± 3.6	35.4 ± 3.1	F = 6.53, $p = 0.016$, $\eta_p{}^2 = 0.189$	F = 10.88, $p = 0.003$, $\eta_p{}^2 = 0.028$
	POST	34.6 ± 3.5 *	35.6 ± 3.1		
	95% CI	−1.75/−0.37	−0.22/0.49		
	Effect size	−0.85	0.20		
PhA (degree)	PRE	8.0 ± 0.5	7.9 ± 0.5	F = 10.74, $p = 0.003$, $\eta_p{}^2 = 0.277$	F = 39.24, $p < 0.001$, $\eta_p{}^2 = 0.584$
	POST	7.5 ± 0.5 *	8.1 ± 0.5 *		
	95% CI	−0.51/−0.24	0.01/0.22		
	Effect size	−1.56	0.63		

Table 2. *Cont.*

Variable		"Lockdown" 2019/2020 Season	"Regular" 2020/2021 Season	Time Effect	Condition x Time Interaction
Fat mass (%)	PRE	14.1 ± 1.7	13.6 ± 1.5	F = 0.51, p = 0.823, η_p^2 = 0.002	F = 1.31, p = 0.261, η_p^2 = 0.045
	POST	14.3 ± 1.9	13.4 ± 1.3		
	95% CI	0.69/−0.41	−0.51/0.10		
	Effect size	0.13	−0.35		
Muscle mass (kg)	PRE	28.2 ± 2.3	28.1 ± 2.3	F = 29.54, p < 0.001, η_p^2 = 0.513	F = 16.29, p < 0.001, η_p^2 = 0.368
	POST	27.4 ± 2.7 *	27.9 ± 2.7		
	95% CI	−1.12/−0.64	−0.45/0.19		
	Effect size	−2.04	−0.22		

Note: * = $p < 0.05$ vs. PRE. R/H: resistance adjusted for stature, Xc/H: reactance adjusted for stature, PhA: phase angle.

3.1. *Quantitative Analysis*

No between-group difference ($p > 0.05$) in FM% and muscle mass was found at baseline. Table 2 shows change over time for each condition and statistical analysis for all dependent parameters. A time effect ($p < 0.01$) was observed for muscle mass that decreased from PRE to POST during the "lockdown" season 2019/2020 (Table 2).

3.2. *Qualitative Analysis*

No between-group difference ($p > 0.05$) in R/H, Xc/H, or PhA was found at baseline. Table 2 shows change over time for each condition and statistical analysis for all dependent parameters. A time effect ($p < 0.05$) was observed for R/H, Xc/H, PhA, and muscle mass (Table 2). Figure 1 shows individual data for PhA, FM%, and muscle mass during the "lockdown" and the "regular" seasons.

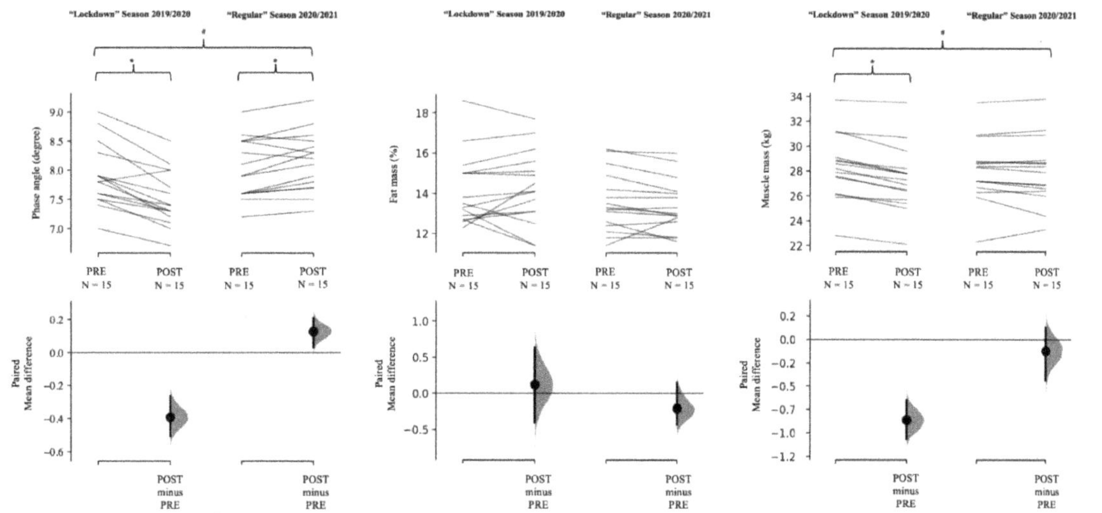

Figure 1. Paired mean differences for the comparisons are shown in the above Cumming estimation plots. Raw data are plotted on the upper axes; each paired set of observations is connected by a line. On the lower axes, each paired mean difference is plotted as a bootstrap sampling distribution. Mean differences are depicted as dots; 95% confidence intervals are indicated by the ends of the vertical error bars. * = significant ($p < 0.05$) time effect; # = significant ($p < 0.05$) condition by time interaction.

Multivariate analysis of the combined change in R/H and Xc/H showed differences from PRE to POST in both seasons, as shown in Figure 2. During the "lockdown" season 2019/2020, there was a rightward displacement of the BIVA vector from PRE to POST

($T^2 = 33.7$, $F = 15.7$, $p < 0.001$, $D^2 = 1.50$), due to a simultaneous reduction in R/H and Xc/H (Figure 2). A leftward BIVA vector displacement was observed during the "regular" season 2020/2021 from PRE to POST ($T^2 = 16.6$, $F = 7.7$, $p < 0.001$, $D^2 = 1.05$) due to an increase in Xc/H and a decrease in R/H (Figure 2).

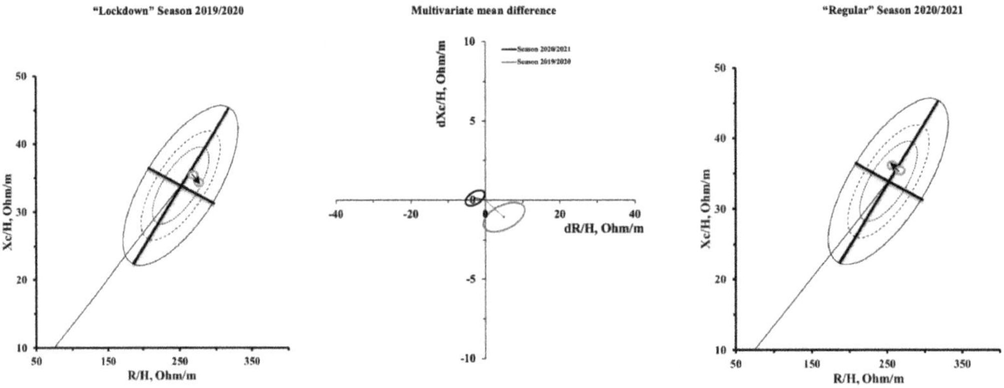

Figure 2. R-Xc and paired graphs for multivariate changes in specific resistance and reactance are shown on the left and right sides. In the R-Xc graphs, bioimpedance data are plotted on the tolerance ellipses of the reference population [9]. In the middle of the figure, mean vector displacements with 95% confidence ellipses are shown.

3.3. Correlations between Qualitative and Quantitative Data

During both seasons, changes in PhA were correlated with changes in muscle mass ("lockdown" season 2019/2020: $R^2 = 0.403$, $ß = -1.128$, $p = 0.011$; "regular" season 2020/21: $R^2 = 0.404$, $ß = 1.963$, $p = 0.011$) (Figure 3). PhA remains a significant predictor even when adjusted for age ("lockdown" season 2019/2020: $R^2 = 0.404$, $ß = -1.134$, $p = 0.019$; "regular" season 2020/21: $R^2 = 0.404$, $ß = 1.964$, $p = 0.015$). No significant ($p > 0.05$) correlation was found for PhA and FM%.

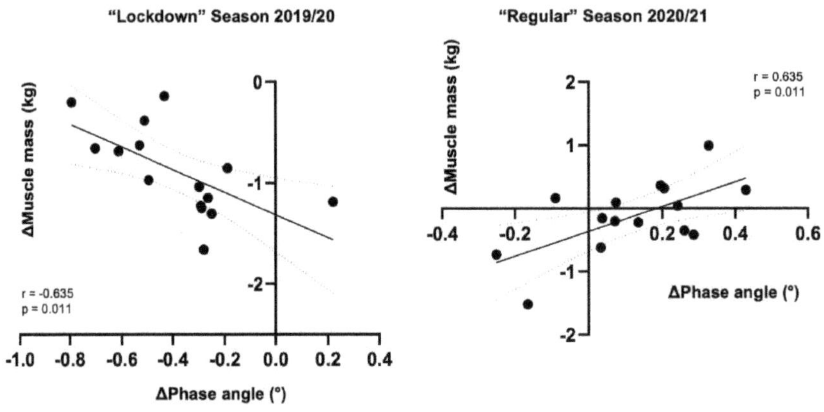

Figure 3. Scatterplots showing the associations between change in phase angle and change in muscle mass.

4. Discussion

The aim of the present study was to examine whether or not bioelectrical FM%, muscle mass, and PhA were affected during the COVID-19 lockdown in Serie A soccer players. The main findings indicated that (i) FM% was not affected during home confinement but decreases in muscle mass were observed during the quarantine period; (ii) a decrease in

PhA was found after the quarantine period, while an increase occurred when the same participants were tested during the successive regular season. Additionally, the change in PhA was correlated with change in muscle mass in both "lockdown" 2019/2020 and "regular" 2020/2021 seasons. To the best of our knowledge, this is the first study that examined the effect of the COVID-19 lockdown on a wide range of body composition parameters in a group of elite soccer players.

The COVID-19 lockdown did not lead to any change in FM%, as also happened for the same period during the regular season. The exceptionality of the pandemic context and its impact on sports has been extensively studied recently, albeit only one study assessed body composition in soccer players [4]. Contrarily to what was observed here, the authors observed an increase in FM% after the lockdown [4]. However, first, FM% was estimated using skinfolds, so different procedures may somehow explain this inconsistency [29].

Second, the authors included lower-level soccer players, so it is possible that the nutritional and training strategies during lockdown were less controlled by the team staff. While FM% remained unchanged, lockdown led to a very large reduction in muscle mass. Such a reduction was probably the result of inadequate training stimulus sustained during forced home confinement. In accordance, previous studies have shown a reduction in fat-free mass in soccer players following the off-season period [5,30]. It should be noted that fat-free mass is not equivalent to muscle mass, but includes a wide range of elements (e.g., bone) not related to any physical capacity [16]. Having assessed muscle mass here allows the determination of a body composition parameter strictly related to sports performance in soccer players [31,32]. Indeed, increasing and preserving muscle mass is warranted across seasons [33], and the data collected here during the successive regular in-season period seems to confirm that. In this regard, the detraining-like lockdown period was already associated with deleterious effects on a wide range of neuromuscular and performance parameters in soccer players [3,34,35]. As such, the lockdown brought unfavorable changes in body composition in soccer players. Interestingly, muscle mass returned to baseline values at the end of the 2019/2020 season, showing how resumption of training and change in lifestyle can be decisive for the maintenance of muscle mass.

The PhA is calculated as the arctangent between R and Xc and is graphically reflected as the distance of the vector from the X-axis in BIVA [16]. Its interpretation allows qualitative estimation of the intracellular/extracellular water ratio, a biomarker of cellular integrity [20]. In the present study, the PhA decreased during the "lockdown" period, represented graphically by a rightward vector displacement, while it did not change during the successive regular season. In this regard, it has previously been shown that PhA remained stable during the in-season period [13]. The decrease in PhA can be ascribed to compromised cellular health, an accumulation of extracellular fluids, or a loss of body cell mass [13,24,36,37]. This may, in turn, depend on the loss of muscle mass observed during the detraining-like period. Indeed, we found a correlation between the simultaneous decrease in PhA and muscle mass during the lockdown. In contrast, previous studies have shown increases in PhA following the pre-season period in soccer players [9,13], as well as in other sports [21,38], possibly representing an increase in muscle mass, although the authors did not calculate the correlation. This evidence can be useful to practitioners who try to interpret changes in PhA and, therefore, in vector position in BIVA during the competitive period. As such, BIVA may be assessed regularly to evaluate body composition in sports practice. In this regard, phase angle and BIVA patterns represent direct bioelectrical measures, and this allows for the avoidance of predictive equations such as those used for quantifying fat and muscle mass.

The present study has some limitations. First, we acknowledge that the sample size is limited; however, it refers to elite soccer players, who are not easily recruited in practice. Second, the participants may have trained differently during home confinement, as the coach provided them with individualized training. Furthermore, different home confinement restrictions imposed by other countries may have resulted in different outcomes. Lastly, BIA is not identified as the gold standard for measuring fat and muscle mass.

5. Conclusions

During the lockdown period experienced in the 2019/2020 season, elite soccer players showed no change in FM%, while muscle mass decreased along with PhA. When returning to regular training in the same period of the successive season, muscle mass remained stable, while PhA increased. The change in bioelectric properties resulted in a rightward shift of the BIVA vector during the lockdown, and a leftward displacement during the regular season. In soccer practice, BIA and BIVA could be regularly used to assess body composition during the season. In particular, BIVA represents a qualitative analysis of body composition, which, through bioelectrical vector and PhA assessment, allows the evaluation of directly obtained measurements. The results of this study could lead coaches and strength conditioners to monitor muscle mass during detraining periods, since this parameter appears to be mainly affected in elite soccer players.

Author Contributions: Conceptualization, F.C., G.C., T.B. and A.R.; data curation, F.C. and A.R.; formal analysis, T.B., G.P. and A.T.; investigation, F.C. and G.C.; methodology, T.B. and G.G.; project administration; supervision, G.C.; writing-original draft, F.C., A.T. and G.C.; writing-review and editing, G.P., G.G. and G.C. All authors have read and agreed to the published version of the manuscript.

Funding: This research received no external funding.

Institutional Review Board Statement: The study was conducted according to the guidelines of the Declaration of Helsinki and approved by the Ethics Committee of the University of Milan (approval number: 32/16).

Informed Consent Statement: Informed consent was obtained from all subjects involved in the study.

Data Availability Statement: The data presented in this study are available on request from the corresponding author. The data are not publicly available due to privacy.

Acknowledgments: The authors express their gratitude to all the participants involved in this study.

Conflicts of Interest: The authors declare no conflict of interest.

References

1. World Health Organization. *WHO Novel Coronavirus (2019-nCoV): Situation Report*, 3; World Health Organization, 2020. Available online: https://apps.who.int/iris/handle/10665/330762 (accessed on 10 September 2021).
2. Mujika, I.; Padilla, S. Detraining: Loss of Training-Induced Physiological and Performance Adaptations. Part I. *Sports Med.* **2000**, *30*, 79–87. [CrossRef]
3. Rampinini, E.; Donghi, F.; Martin, M.; Bosio, A.; Riggio, M.; Maffiuletti, N.A. Impact of COVID-19 Lockdown on Serie A Soccer Players' Physical Qualities. *Int. J. Sports Med.* **2021**, *42*, 917–923. [CrossRef]
4. Grazioli, R.; Loturco, I.; Baroni, B.M.; Oliveira, G.S.; Saciura, V.; Vanoni, E.; Dias, R.; Veeck, F.; Pinto, R.S.; Cadore, E.L. Coronavirus Disease-19 Quarantine Is More Detrimental Than Traditional Off-Season on Physical Conditioning of Professional Soccer Players. *J. Strength Cond. Res.* **2020**, *34*, 3316–3320. [CrossRef] [PubMed]
5. Suarez-Arrones, L.; Lara-Lopez, P.; Maldonado, R.; Torreno, N.; De Hoyo, M.; Nakamura, F.Y.; Di Salvo, V.; Mendez-Villanueva, A. The effects of detraining and retraining periods on fat-mass and fat-free mass in elite male soccer players. *PeerJ* **2019**, *7*, e7466. [CrossRef]
6. Koundourakis, N.; Androulakis, N.; Malliaraki, N.; Tsatsanis, C.; Venihaki, M.; Margioris, A.N. Discrepancy between Exercise Performance, Body Composition, and Sex Steroid Response after a Six-Week Detraining Period in Professional Soccer Players. *PLoS ONE* **2014**, *9*, e87803. [CrossRef]
7. Silva, J.R.; Brito, J.; Akenhead, R.; Nassis, G.P. The Transition Period in Soccer: A Window of Opportunity. *Sports Med.* **2015**, *46*, 305–313. [CrossRef]
8. Bongiovanni, T.; Rossi, A.; Iaia, F.M.; Alberti, G.; Pasta, G.; Trecroci, A. Association of phase angle and appendicular upper and lower body lean soft tissue with physical performance in young elite soccer players: A pilot study. *J. Sports Med. Phys. Fit.* **2021**. Online first. [CrossRef] [PubMed]
9. Bongiovanni, T.; Mascherini, G.; Genovesi, F.; Pasta, G.; Iaia, F.M.; Trecroci, A.; Ventimiglia, M.; Alberti, G.; Campa, F. Bioimpedance Vector References Need to Be Period-Specific for Assessing Body Composition and Cellular Health in Elite Soccer Players: A Brief Report. *J. Funct. Morphol. Kinesiol.* **2020**, *5*, 73. [CrossRef] [PubMed]
10. Micheli, M.L.; Pagani, L.; Marella, M.; Gulisano, M.; Piccoli, A.; Angelini, F.; Burtscher, M.; Gatterer, H. Bioimpedance and Impedance Vector Patterns as Predictors of League Level in Male Soccer Players. *Int. J. Sports Physiol. Perform.* **2014**, *9*, 532–539. [CrossRef] [PubMed]

11. Nabuco, H.C.G.; Silva, A.M.; Sardinha, L.; Rodrigues, F.B.; Tomeleri, C.M.; Ravagnani, F.C.P.; Cyrino, E.; Ravagnani, C.F.C. Phase Angle is Moderately Associated with Short-term Maximal Intensity Efforts in Soccer Players. *Int. J. Sports Med.* **2019**, *40*, 739–743. [CrossRef]
12. Martins, P.; Teixeira, A.; Guglielmo, L.; Francisco, J.; Silva, D.; Nakamura, F.; Lima, L. Phase Angle Is Related to 10 m and 30 m Sprint Time and Repeated-Sprint Ability in Young Male Soccer Players. *Int. J. Environ. Res. Public Health* **2021**, *18*, 4405. [CrossRef]
13. Mascherini, G.; Gatterer, H.; Lukaski, H.; Burtscher, M.; Galanti, G. Changes in hydration, body-cell mass and endurance performance of professional soccer players through a competitive season. *J. Sports Med. Phys. Fit.* **2014**, *55*, 749–755.
14. Campa, F.; Micheli, M.L.; Pompignoli, M.; Cannataro, R.; Gulisano, M.; Toselli, S.; Greco, G.; Coratella, G. The Influence of Menstrual Cycle on Bioimpedance Vector Patterns, Performance, and Flexibility in Elite Soccer Players. *Int. J. Sports Physiol. Perform.* **2021**, *1*, 1–9. [CrossRef] [PubMed]
15. Campa, F.; Matias, C.N.; Marini, E.; Heymsfield, S.B.; Toselli, S.; Sardinha, L.; Silva, A. Identifying Athlete Body Fluid Changes During a Competitive Season With Bioelectrical Impedance Vector Analysis. *Int. J. Sports Physiol. Perform.* **2020**, *15*, 361–367. [CrossRef]
16. Campa, F.; Toselli, S.; Mazzilli, M.; Gobbo, L.; Coratella, G. Assessment of Body Composition in Athletes: A Narrative Review of Available Methods with Special Reference to Quantitative and Qualitative Bioimpedance Analysis. *Nutrients* **2021**, *13*, 1620. [CrossRef] [PubMed]
17. Coratella, G.; Campa, F.; Matias, C.N.; Toselli, S.; Koury, J.C.; Andreoli, A.; Sardinha, L.B.; Silva, A.M. Generalized bioelectric impedance-based equations underestimate body fluids in athletes. *Scand. J. Med. Sci. Sports* **2021**, *31*, 2123–2132. [CrossRef]
18. Matias, C.N.; Campa, F.; Santos, D.A.; Lukaski, H.; Sardinha, L.B.; Silva, A.M. Fat-free Mass Bioelectrical Impedance Analysis Predictive Equation for Athletes using a 4-Compartment Model. *Int. J. Sports Med.* **2021**, *42*, 27–32. [CrossRef] [PubMed]
19. Piccoli, A.; Rossi, B.; Pillon, L.; Bucciante, G. A new method for monitoring body fluid variation by bioimpedance analysis: The RXc graph. *Kidney Int.* **1994**, *46*, 534–539. [CrossRef]
20. Marini, E.; Campa, F.; Buffa, R.; Stagi, S.; Matias, C.N.; Toselli, S.; Sardinha, L.; Silva, A. Phase angle and bioelectrical impedance vector analysis in the evaluation of body composition in athletes. *Clin. Nutr.* **2020**, *39*, 447–454. [CrossRef]
21. Reis, J.F.; Matias, C.N.; Campa, F.; Morgado, J.P.; Franco, P.; Quaresma, P.; Almeida, N.; Curto, D.; Toselli, S.; Monteiro, C.P. Bioimpedance Vector Patterns Changes in Response to Swimming Training: An Ecological Approach. *Int. J. Environ. Res. Public Health* **2020**, *17*, 4851. [CrossRef]
22. Campa, F.; Matias, C.; Nunes, C.; Monteiro, C.; Francisco, R.; Jesus, F.; Marini, E.; Sardinha, L.; Martins, P.; Minderico, C.; et al. Specific Bioelectrical Impedance Vector Analysis Identifies Body Fat Reduction after a Lifestyle Intervention in Former Elite Athletes. *Biology* **2021**, *10*, 524. [CrossRef] [PubMed]
23. Nescolarde, L.; Yanguas, J.; Lukaski, H.; AlOmar, X.; Rosell-Ferrer, J.; Rodas, G. Effects of muscle injury severity on localized bioimpedance measurements. *Physiol. Meas.* **2014**, *36*, 27–42. [CrossRef] [PubMed]
24. Nescolarde, L.; Yanguas, J.; Terricabras, J.; Lukaski, H.; AlOmar, X.; Rosell-Ferrer, J.; Rodas, G. Detection of muscle gap by L-BIA in muscle injuries: Clinical prognosis. *Physiol. Meas.* **2017**, *38*, L1–L9. [CrossRef] [PubMed]
25. Kyle, U.; Genton, L.; Hans, D.; Pichard, C. Validation of a bioelectrical impedance analysis equation to predict appendicular skeletal muscle mass (ASMM). *Clin. Nutr.* **2003**, *22*, 537–543. [CrossRef]
26. Cohen, J. *Statistical Power Analysis for the Behavioral Sciences*; Routledge: New York, NY, USA, 2013.
27. Hopkins, W.G.; Marshall, S.W.; Batterham, A.M.; Hanin, J. Progressive Statistics for Studies in Sports Medicine and Exercise Science. *Med. Sci. Sports Exerc.* **2009**, *41*, 3–12. [CrossRef] [PubMed]
28. Stevens, J. *Applied Multivariate Statistics for the Social Sciences*; Lawrence E.: Mahwah, NJ, USA, 2002.
29. Kasper, A.; Langan-Evans, C.; Hudson, J.; Brownlee, T.; Harper, L.; Naughton, R.; Morton, J.; Close, G. Come Back Skinfolds, All Is Forgiven: A Narrative Review of the Efficacy of Common Body Composition Methods in Applied Sports Practice. *Nutrients* **2021**, *13*, 1075. [CrossRef] [PubMed]
30. Owen, A.L.; Lago-Peñas, C.; Dunlop, G.; Mehdi, R.; Chtara, M.; Dellal, A. Seasonal Body Composition Variation Amongst Elite European Professional Soccer Players: An Approach of Talent Identification. *J. Hum. Kinet.* **2018**, *62*, 177–184. [CrossRef]
31. Brocherie, F.; Girard, O.; Forchino, F.; Al Haddad, H.; Dos Santos, G.A.; Millet, G.P. Relationships between anthropometric measures and athletic performance, with special reference to repeated-sprint ability, in the Qatar national soccer team. *J. Sports Sci.* **2014**, *32*, 1243–1254. [CrossRef]
32. Campa, F.; Piras, A.; Raffi, M.; Toselli, S. Functional Movement Patterns and Body Composition of High-Level Volleyball, Soccer, and Rugby Players. *J. Sport Rehabil.* **2019**, *28*, 740–745. [CrossRef]
33. Collins, J.; Maughan, R.J.; Gleeson, M.; Bilsborough, J.; Jeukendrup, A.; Morton, J.P.; Phillips, S.M.; Armstrong, L.; Burke, L.M.; Close, G.L.; et al. UEFA expert group statement on nutrition in elite football. Current evidence to inform practical recommendations and guide future research. *Br. J. Sports Med.* **2021**, *55*, 416. [CrossRef]
34. Moreno-Pérez, V.; Del Coso, J.; Romero-Rodríguez, D.; Marcé-Hernández, L.; Peñaranda, M.; Madruga-Parera, M. Effects of home confinement due to COVID-19 pandemic on eccentric hamstring muscle strength in football players. *Scand. J. Med. Sci. Sports* **2020**, *30*, 2010–2012. [CrossRef] [PubMed]
35. Villaseca-Vicuña, R.; Pérez-Contreras, J.; Merino-Muñoz, P.; González-Jurado, J.; Aedo-Muñoz, E. Effects of COVID-19 confinement measures on training loads and the level of well-being in players from Chile women's national soccer team. *Rev. Fac. Med.* **2021**, *69*, e88480. [CrossRef]

36. Matias, C.N.; Monteiro, C.P.; Santos, D.A.; Martins, F.; Silva, A.M.; Laires, M.J.; Sardinha, L.B. Magnesium and phase angle: A prognostic tool for monitoring cellular integrity in judo athletes. *Magnes. Res.* **2015**, *28*, 92–98. [CrossRef]
37. Nescolarde, L.; Yanguas, J.; Lukaski, H.; AlOmar, X.; Rosell-Ferrer, J.; Rodas, G. Localized bioimpedance to assess muscle injury. *Physiol. Meas.* **2013**, *34*, 237–245. [CrossRef] [PubMed]
38. Lukaski, H.; Raymond-Pope, C.J. New Frontiers of Body Composition in Sport. *Int. J. Sports Med.* **2021**, *42*, 588–601. [CrossRef] [PubMed]

biology

MDPI

Article

Effect of Physical Exercise Program Based on Active Breaks on Physical Fitness and Vigilance Performance

Francisco Tomás González-Fernández [1,2], Sixto González-Víllora [3], Salvador Baena-Morales [4], Juan Carlos Pastor-Vicedo [3], Filipe Manuel Clemente [5,6], Georgian Badicu [7] and Eugenia Murawska-Ciałowicz [8,*]

1 Department of Physical Activity and Sports Sciences, Pontifical University of Comillas (Centro de Estudios Superiores Alberta Giménez), 07013 Palma de Mallorca, Spain; francis.gonzalez.fernandez@gmail.com
2 SER Research Group, Pontifical University of Comillas (Centro de Estudios Superiores Alberta Giménez), 07013 Palma, Spain
3 Department of Physical Education, Arts and Music, University of Castilla-La Mancha, 02006 Albacete, Spain; sixto.gonzalez@uclm.es (S.G.-V.); juancarlos.pastor@uclm.es (J.C.P.-V.)
4 EDUCAFD Research Group, Department of General Didactic and Specific Didactics, Faculty of Education, University of Alicante, 03690 Alicante, Spain; salvador.baena@ua.es
5 Escola Superior Desporto e Lazer, Instituto Politécnico de Viana do Castelo, Rua Escola Industrial e Comercial de Nun'Álvares, 4900-347 Viana do Castelo, Portugal; filipe.clemente5@gmail.com
6 Instituto de Telecomunicações, Delegação da Covilhã, 1049-001 Lisboa, Portugal
7 Department of Physical Education and Special Motricity, University Transilvania of Brasov, 500068 Brasov, Romania; georgian.badicu@unitbv.ro
8 Department of Physiology and Biochemistry, University School of Physical Education, 51-612 Wrocław, Poland
* Correspondence: eugenia.murawska-cialowicz@awf.wroc.pl

Citation: González-Fernández, F.T.; González-Víllora, S.; Baena-Morales, S.; Pastor-Vicedo, J.C.; Clemente, F.M.; Badicu, G.; Murawska-Ciałowicz, E. Effect of Physical Exercise Program Based on Active Breaks on Physical Fitness and Vigilance Performance. *Biology* **2021**, *10*, 1151. https://doi.org/10.3390/biology10111151

Academic Editor: Paulo Gentil

Received: 7 September 2021
Accepted: 5 November 2021
Published: 8 November 2021

Simple Summary: Our study aimed to analyze the effects of 8 weeks physical training on vigilance performance in high school students. Forty-two healthy students were assigned for convenience and matched into two groups, a Control Group (CG) and an Active-Break Group (ABG). The participants were assessed before the training program using the Alpha-Fitness test battery and Psychomotor Vigilance Task (PVT) to observe their physical fitness and vigilance performance. Compared with the pre-test, significant different were observed in the post-test PVT. Results showed a main effect of ABG responding faster than students in the CG group. This demonstrated that 8 weeks physical training have an effect on vigilance performance and improve the efficiency of vigilance in high school students.

Abstract: The scientific literature has shown the beneficial effects of chronic Physical Exercise (PE) on a wide range of tasks that involve high-order functioning. For this reason, the present study aimed to investigate the effects of active breaks on physical fitness and vigilance performance in high school students through eight weeks of physical training. A total of 42 healthy students (age = 16.50 ± 0.59 years; height = 171.08 ± 8.07 cm; weight = 67.10 ± 13.76 kg) from one Andalusian high school (Spain) were assigned for convenience and matched into two groups, a Control Group (CG) and an Active-Break Group (ABG). The ABG performed two active breaks (based on strength and self-loading exercises) during the school day, first at 10:00 a.m. and second at 12:30 p.m. The participants were assessed before and after the training program using the Alpha-Fitness test battery and the Psychomotor Vigilance Task (PVT). Significant differences were observed in the post-training PVT results, compared with the pretraining PVT, showing ABG responding faster than CG. Thus, the presents study demonstrated that eight weeks of physical training affects vigilance performance (compared to CG) and improves the efficiency of vigilance in high school students, contributing to enhancement of quality of education.

Keywords: physical activity; executive functions; cognitive performance; youth; physical education

1. Introduction

In the last decade, there has been a rapidly growing interest in the scientific knowledge that links chronic physical exercise (PE) and cognitive performance [1–5]. A comprehensive review of the scientific literature has shown the beneficial effects of chronic PE on a wide range of tasks involving high-order functioning, such as attention, cognitive control, memory, and perception, among others [6]. The vast majority of the studies in this field have focused on the impact of chronic PE on executive functions [5,7,8], and to a lesser extent, on tasks that involve short-term memory [9,10], attention [11], and language processing [12]. However, current research has shown that regular PE produces different constant changes, such as those at the structural level involving angiogenesis or neurogenesis in different areas of the brain, especially in the hippocampus [13,14]. There is also an increase in blood vessels in the hippocampus, cortex, and cerebellum, which increase the supply of nutrients and energy in these neural areas [15]. It has been widely demonstrated that performing regular exercise at moderate aerobic intensities (40% to 80% of maximum oxygen consumption ($VO_{2máx}$)) acts positively on cognitive tasks such as processing speed, selective attention, and short-term memory [3,5]. Finally, there is an increase in brain structures due to neuronal plasticity, increased vascularization, and neurogenesis (brain plasticity). The evidence suggests that these adaptations produce a better cognitive response in various tasks, including memory, attention, processing speed, cognitive flexibility, and inhibition.

Vigilance refers to the cognitive (attentional) function that determines the capacity to respond appropriately (quickly and accurately) to relevant stimuli [16]. In the laboratory, vigilance is typically investigated using tasks involving the monotonous presentation of stimuli for a relatively long period of time, requiring participants to detect rare events [17] or to simply respond to unpredictable target onsets [18]. Low levels of vigilance result in slow reaction time (RT), response anticipation, or even failure to detect the target. Consistent findings in sustained attention research show a decline in performance with time-on-task, the so-called vigilance decrement. Researchers have suggested that this performance decrement over time reflects a decrease in attentional resources [19–21]. A cursory look at the literature reveals studies investigating vigilance primarily in the context of various everyday activities [22,23]. However, scientific research on the relationship between regular exercise (based on ABs) and vigilance in the high school setting is lacking. In this respect, ABs have been applied in classrooms using different motor games and including varied coordination abilities, locomotor skills (e.g., running, jumping, or sliding), and stability skills (e.g., balance, bending, or turning). Moreover, the results of previous research obtained after physical interventions have found a positive relationship between physical fitness acquired on intervention and cognitive performance [24–26]. Therefore, we suggest that vigilance could be exponentially affected during the school day, and more specifically, that ABs could contribute positively to the vigilance performance, both immediately after ABs as well as chronically, i.e., with a long-lasting effect caused by the chronic implementation of ABs. This may be significant, especially in the academic context, given the importance of vigilance for maintaining optimal performance during the course of the school day [27].

The scientific literature reported to date points to the positive influence of physical fitness on vigilance [28–30]. A small number of studies carried out have been conducted at early ages (i.e., in children from 3 to 12 years old) and show a positive relationship between the level of physical fitness and vigilance. For example, Pontifex et al.'s [31] study examined the performance of vigilance as a function of time-on-task, using an Eriksen flanker task in two age groups (preadolescents with a low level of physical fitness and preadolescents with a high level of physical fitness). Their results showed an increase in the rate of omission error and the number of sequential omissions as a function of time-on-task in preadolescents with low physical fitness. Another study investigated the time course of behavioral performance and brain functioning in preadolescents with high and low levels of physical fitness [31]. The study found a decrease in performance reflected in the incongruent trials in the low physical fitness group, evidenced by an increase in bilateral activation of frontal and parietal brain regions. In contrast, participants with high physical fitness showed a decrease in physical activity as a function of time-on-task, although in the initial time block, they showed better activity compared with participants with low physical fitness. Bunce [28] carried out research with young and older adults that attempted to analyze the performance of vigilance according to a degree of complexity of the task and a level of physical fitness. Results indicated a smaller decrease in vigilance in the group of older adults with a high level of physical fitness than in the low-level physical activity group. These results are reflected in tasks with high demands on attention resources, e.g., when information or complexity is high.

In summary, the existing literature confirms the important role that PE and the level of physical fitness [10,32,33] play in cognitive performance in areas involving vigilance. However, research on this is still scarce, and a number of questions still need to be answered. The main purpose of this research was to examine the chronic effects of eight weeks of physical training on cognition, specifically on vigilance performance, in high school students. It should be noted that, based on the evidence found in the literature on regular PE and vigilance, most studies are limited insofar as they show experimental designs between groups and can only look for relationships between vigilance and physical condition indirectly. The central hypothesis of the current study is that an 8-week physical training based on AB would enhance physical fitness and vigilance performance, and in doing so, draw a direct link between cognition and PE.

2. Materials and Methods

2.1. Study Design

The study was conducted between January and February of 2020. At the time of these observations, the students had completed three months of training and familiarization with training protocols, tasks, and rating of perceived Exertion (RPE) during Physical Education class. A quasi-experimental pre–post design control group (CG) and an active break group (ABG) were used in the present research. The high school students were selected for convenience, assigned and matched into two groups, an ABG and a CG, based on the class to which they belonged. To investigate the effects of an 8-week AB-based Physical Exercise Program on physical fitness and vigilance performance, those in the CG were asked to maintain their ordinary routines and training practices, while those in the ABG modified their training sessions by introducing ten minutes of AB before 10:00 a.m. and another AB before 12:30 p.m.

2.2. Participants

A total of 42 healthy students from a high school in the region of Andalusia, Spain, participated in this study: 25 girls (age = 16.42 ± 0.50 years; height = 167.66 ± 6.30 cm; weight = 64.13 ± 11.78 kg) and 17 boys (age = 16.62 ± 0.71 years; height = 176.53 ± 7.73 cm; weight = 72.06 ± 15.58 kg) as shown in Table 1. Concerning the sample size, the next equation was used: $Sample\ Size = Z2 \times (p) \times (1 - p)/C2$, where Z = confidence level (95%); p = 0.05 and C = margin of error 0.05. Participants were recruited from the city of Granada, which has a population in the range from 100,000 to 1,000,000 inhabitants according to the National Institute of Statistics from the Spanish Government (http://www.ine.es/ accessed on 1 March 2021).

Inclusion criteria for the participants in this study were: (i) reporting normal vision and no history of any neuropsychological impairments that could affect the results of the experiment, (ii) not presenting any injuries during the previous two months, (iii) giving consent, and (iv) participating in 85% of the AB during the study period.

Table 1. Participants' characteristics (mean ± SD) in the present study.

	Participants		
	CG n = 21	**ABG n = 21**	**Total n = 42**
Age (years)	16.52 ± 0.60	16.48 ± 0.60	16.50 ± 0.59
Height (cm)	170.32 ± 7.94	171.80 ± 8.34	171.08 ± 8.07
Weight (kg)	66.79 ± 9.59	67.40 ± 17.07	67.10 ± 13.76
BMI (kg·m^{-2})	22.90 ± 2.44	22.41 ± 3.74	22.64 ± 3.16
Agility test (s)	10.98 ± 1.23	10.55 ± 2.21	10.77 ± 1.78
Standing broad jump (cm)	177.00 ± 0.44	173.00 ±0.36	175.00 ± 0.40
20-m shuttle run test [VO2max (mL/kg/min)]	43.08 ± 7.25	43.91 ± 6.75	43.50 ± 6.94
IPAQ-SF (Mets)	712.85 ± 435.40	686.85 ± 402.99	699.85 ± 419.28

Note. CG: Control Group and ABG: Active Break Group. $VO2_{max}$ was estimated with the equation by Legger et al., 1982 [34]: $VO2_{max}$ = 5.857 × velocity (km/h) − 19.45.

In addition, all participants completed a healthy lifestyle questionnaire and the short version of the international physical activity questionnaire (IPAQ-SF) [35]. In the first document, they were asked about current sports habits, addictions, and diseases that could impede the practice of physical exercise, while, in the second, we recorded their level of physical activity (PA).

The participants were informed about the main goals of the investigation and signed informed consent forms. Families were informed that they could revoke the participation agreement at any time. The students were treated according to the American Psychological Association (APA) guidelines, which ensured the anonymity of participants' responses. In addition, the study was conducted following the ethical principles of the 1964 Helsinki declaration for human research and was approved by the Research Ethics Committee of the University of Castilla-La Mancha (Hospital Universitario de Albacete, Record April 2020, and internal project n° 2020/05/052).

2.3. *Procedure*

2.3.1. Preintervention

First, the school management team was informed about the study objectives and ensured that the students' parents or guardians had signed informed consent forms detailing the possible benefits and risks. Subsequently, the study and the plan for the structure of every class day were designed together with the school's teachers and advisors. In order for the students not to miss any classes, we use the ABs to improve performance in the classroom. Every task or a test was monitored by one lead researcher assisted by a physical education teacher responsible for the groups, specially trained for accurate and reliable data recording, especially in Physical Fitness assessment.

In the first session, participants filled out the questionnaires (≈10 min to assess the entire group), had the anthropometry assessment (≈15 min to assess the entire group), and tests from the ALPHA-Fitness test battery in the following order: 4 × 10 m speed-agility (≈10 min to assess the entire group) and standing broad jump (≈15 min to assess the entire group). Participants were given adequate time to fully recover due to the neuromuscular characterization of the different tests applied in this session. In addition, the application order that the students followed during the research was always the same. Therefore, the measures taken ensured the recovery. In the second session, participants performed the 20 m shuttle run test (≈30 min to assess the entire group) see Section 2.4. Finally, in the third session, students performed the PVT to determine their basal level of vigilance.

All measures were recorded at the same time of day, between 10:30 a.m. and 14:30 p.m., in the same space and time with the same humidity conditions (30–40%). The students were familiarized with the use of the subjective rating of perceived exertion scale (RPE) during their physical education classes. In addition, the ABG group was instructed in an 8-week PE program based on ABs in the physical education class for three months. Every task or a test was monitored by one lead researcher assisted by a physical education teacher responsible for the groups specially trained for accurate and reliable data recording, especially in physical fitness assessment.

2.3.2. Intervention

The students completed eight weeks of training based on functional training. In the case of ABG, they performed two ABs (based on strength and self-loading exercises) during the school day, AB1 at 10:00 a.m. and AB2 at 12:30 p.m. (Figure 1). The training program was supervised by the lead researcher and teachers responsible for the ABG and was designed individually for each participant, considering their individual physical characteristics (each exercise was designed with an adaptation in difficulty and intensity). We followed the guidelines of the American College of Sports Medicine [36] to ensure the safety of the participants.

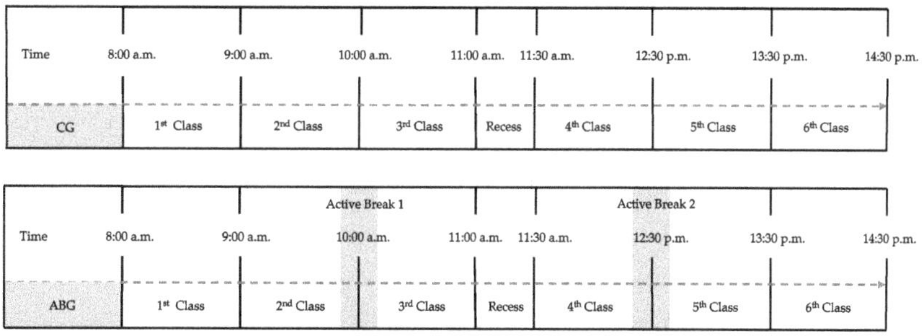

Figure 1. Schematic representation of the school day procedure of one session of the experimental group.

A total of 80 ABs was performed by each participant. Training sessions in both ABs were divided into three phases: (i) warm-up 2 min (dynamic stretching and joint mobility exercises); (ii) training activity ≈8 min (more information in Figures 2 and 3); and (iii) cooldown 2 min (stretching exercises). According to the RPE scale, the average intensity of the sessions was between hard and very hard (16.10 ± 1.12). To control the intensity of the training program, we used a google questionnaire that recorded RPE immediately after each AB. Crucially, the CG were instructed to maintain their classroom routines while the ABG performed the AB. The very light average intensity of sessions (6.29 ± 0.42) was also recorded.

In the training activity, students worked in a circuit. Thus, equipment was distributed around the classroom, and students performed the AB exercises in sequence until finished. The exercises of AB1 (4 sets ×12 s a recovery time between exercises of 10 s and between sets of 20 s) and AB2 (4 sets × 10 s and a recovery time between exercises of 10 s and between sets of 20 s) were the same during the whole intervention. However, the progress of all sessions varied by weeks due to the heterogeneity of the sample and the individuals' varied physical conditions. All students were familiarized with the exercises and the exercise intensity (between hard and very hard).

Active Break 1 consisted of the following exercises: Push-ups progresssion: (i) performed against a wall, (ii) performed on the knees, and (iii) push-ups complete; Dynamic plank: (i) reach and touch plank; (ii) clock plank, and (iii) side plank (left: 2 × 12 + right: 2 × 12); Unilateral jump: (i) lateral jump (left-right), (ii) box step-up performed in chairs of 46 cm (left-right), and (iii) single-leg squat (left: 2 × 12 + right: 2 × 12); Squats: (i) wall squat, (ii) counterbalance air squat, and (iii) bodyweight prisoner squat; Lunges with MB: (i) stationary lunge (without MB and left-right), (ii) stationary lunge with medicine ball of 4, 6 or 8 kg (left-right), and (iii) lateral lunge with MB (left-right).

Figure 2. Description of Active Break 1 for high school students. (Arrows represent the movement direction, and S was used for static and D for dynamic movement).

Figure 3. Description of Active Break 2 on high school students. (Arrows represent the movement direction, and S was used for static and D for dynamic movement).

Active Break 2 consisted of Unilateral jump: (i) lateral jump (left-right), (ii) Box step-up performed in chairs of 46 cm (left-right), and (iii) single-leg squat (left: 2 × 12 + right: 2 × 12); Shoulder bridge: (i) Bridge stationary, (ii) bridge with single leg static hold (left-Right), and (iii) bridge with marching (left-right). Lunges: (i) side-to-side samurai lunges (Left-right), (ii) Knee thrust lunges (left-right), and (iii) jump switch lunges (left-right); Nordic hamstring curl: (i) band-assisted nordic hamstring curl, (ii) bodyweight and support hands, and (iii) inclined curls positioning; TRX biceps bilateral curl: (i) inclined to 75%, (ii) inclined to 60%, and (iii) inclined to 45%; Balance: (i) balance without movement (left-right), (ii) standing crunch with under-the-leg clap (left-right), and (iii) T-Stand with hinge and side bend (left-right).

See Figures 2 and 3 to see the variations provided to students to adjust the level of difficulty to their ability in order to execute the exercise.

2.3.3. Post-Intervention

After eight weeks, both groups were evaluated at the same time of day as in the pre-intervention session (between 10:30 a.m. and 14:30 p.m.), in the same space, with the same humidity conditions. Everything was the same as in the preintervention session except t that questionnaires and the ALPHA-Fitness test battery were not completed. Notably, after the termination of the study, students from the CG were also given the opportunity to perform the same program as the experimental group.

2.4. *Measures*

2.4.1. Anthropometry

Height and body weight were measured before the start of the intervention. A bioelectrical impedance analysis (BIA) device (Tanita BC-730) was used to calculate body weight (kg) to the nearest 0.1 kg. Both measures were assessed by one main researcher (Pre and Post). A stadiometer (Type SECA 225, Hamburg, Germany) was utilized to measure the height (cm) to the nearest 0.1 cm. Participants were asked to remove their shoes and other accessories that could influence the assessment. They also had to be vertical and immobile, with arms extended along the body and looking straight ahead in an upright position. For each parameter, only one measurement was collected.

2.4.2. Physical Fitness Assessment

The level of physical fitness was assessed using the ALPHA-Fitness test battery [37]. We followed the protocol established for this test battery and the guidelines of the ACSM [36] to ensure the safety of the participants. In addition, to ensure successful performance in the ALPHA-Fitness test battery, all the students were informed about the protocol to ensure an adequate data collecting process in both groups.

20-m Shuttle Run Test

Cardiorespiratory fitness (CRF) was assessed with the 20 m shuttle run test [34]. Students were required to run between two lines separated by 20 m while keeping pace with audio signals emitted from a USB Player with the test protocol. Students started the test with an initial speed of 8.5 km/h^{-1}, which was increased by 0.5 km/h^{-1} min^{-1} (stage duration = 1 min). We recorded the last one-half stage completed as an indicator of CRF. In addition, $VO2_{max}$ was estimated with the equation $VO2_{max}$ = 5.857 × velocity (km/h) − 19.45.

4 × 10 m Speed-Agility Test

Coordination, agility, and speed were evaluated with this test. The aim of the test was to run four repetitions of 10 m distance. Students had to run at a maximum speed, and they had two attempts. We recorded the best of the two attempts, and results were measured in seconds with a Casio handheld stopwatch (HS-3V-1).

Standing Broad Jump

This test has been successfully used for measuring lower limb explosive strength. Students jumped horizontally to achieve maximum distance (in centimeters). Participants performed the standing broad jump three times, with 20 s of recovery between attempts to minimize the effect of fatigue. The best jump was considered as the final outcome. The test was performed in the school gym to avoid falls caused by slipping [38].

2.4.3. Rating of Perceived Exertion (RPE)

The RPE was measured with the Borg scale [39] immediately after the exercise in ABG and CG. The RPE scale ranged from 6 (no exertion) to 20 (maximal exertion).

2.4.4. Cognitive Measurement: Psychomotor Vigilance Task

iPhones 5s (iOS 12.4.5) were used to present the stimuli of the PVT. Performance in the PVT has been shown to be valid to control vigilance [18,40] and was linked to the control of cardiovascular fitness [41]. The devices were previously blocked to any other type of notification. The center of the mobile screen was placed about 50–80 cm from the participants' heads at eye level (aiming to help everyone feel as comfortable as possible during the duration of the task). The PVT presents a grey screen with a chronometer at the center, which begins the countdown at the speed of a real stopwatch and could be presented on the screen after a random time interval ranging between 2000 and 10,000 ms. Verbal and written instructions were given to the participant prior to the start of the PVT in every session, stressing that they had to fixate on the center of the screen, try not to move their eyes, and respond as quickly as possible (while avoiding anticipation errors) as soon as the chronometer starts. The task included a single block lasting 10 min. The exact number of trials of each participant depended on the latency of the individual's response.

The task duration in both preintervention and postintervention was a 10-min test [42]. Students completed the first PVT, and five trials were excluded from the analysis. In addition, these trials were considered as practice in the preintervention for both groups (See Figure 4, for more information).

Figure 4. Experimental Set. Student performing the PVT (see text for full description).

2.5. Data Analysis

For data processing and mean and standard deviation were used. Descriptive statistics were calculated for each variable. For the comparison of samples and to observe statistically significant differences between groups, ABG vs. CG, was used as the between-subjects factor, and time of measurement, baseline vs. eight weeks, as a within-subject factor. We performed a paired-sample *t*-test in body composition characteristics (body weight, BMI) and physiological parameters (RPE). Effect size is indicated with Cohen's d for *t*-tests [0.2 (small); 0.5 (medium) and >0.8 (large)] and partial eta squared for Fs. In addition, confidence intervals (95%) were calculated.

Analyses of variance (ANOVA) were used to analyze the RTs. Trials with RTs below 100 ms in the experimental group (preintervention = 11.38% and postintervention = 7.79%) were assumed to represent anticipation errors and were discarded from the analysis [18].

Statistically significant effects were further analyzed with paired-sample *t*-tests corrected by Holm-Bonferroni for multiple comparisons. The Greenhouse–Geisser correction was applied when sphericity was violated. Data were analyzed using Statistical software (version 10.0; Statsoft, Inc., Tulsa, OK, USA). For all analyses, significance was accepted at $p < 0.05$.

3. Results

3.1. Anthropometrical Characteristics

A paired sample *t*-test with body weight between CG (66.79 ± 9.59; CI 95%: 4.68) and ABG (67.40 ± 17.07; CI 95%: 8.38) was not significant [t(21) = 0.05, $p > 0.05$, d = 0.04]. Another *t*-test with BMI between CG (22.90 ± 3.74; CI 95%: 1.25) and ABG (22.41 ± 3.74; CI 95%: 1.83) also was not significant [t(21) = 0.40, $p > 0.05$, d = −0.13]. These results confirmed that there was no statistically significant difference between the groups, therefore, both groups were equal.

3.2. Physical Fitness Assessment

The level of physical fitness was assessed by means of the ALPHA-Fitness test battery. A paired sample *t*-test with *Standing broad jump* between ABG (172.71 ± 35.84; CI 95%: 14.28) and CG (177.38 ± 43.96; CI 95%: 21.09) was not significant [t(21) = 0.05, $p > 0.05$, d= −0.09]. Another *t*-test with 4 × 10m speed-agility test between ABG (10.55 ± 2.2; CI 95%: 0.58) and CG (10.98 ± 1.23; CI 95%: 1.05) also was not significant [t(21) = 0.05, $p > 0.05$, d= −0.24]. Finally, a *t*-test with 20-m shuttle run test between ABG (43.91 ± 6.75; CI 95%: 3.41) and CG (43.08 ± 7.25; CI 95%: 2.99) also was not significant [t(21) = 0.05, $p > 0.05$, d = 0.12]. As was the case in anthropometrical characteristics, results confirmed both groups were equal at the start.

3.3. Rating of Perceived Exertion (RPE)

A paired sample *t*-test with RPE scale showed higher values in the ABG (16.10 ± 1.21; CI 95%: 0.52) than in CG (6.29 ± 0.42; CI 95%: 0.18) [t(21) = 35, 35, $p < 0.001$, d = −10.83]. Previous results confirmed that effort (CG vs. ABG) was different in terms of physical demands.

3.4. Psychomotor Vigilance Task

A different analysis of variance of repeated measures (ANOVA) was performed with the average of the participants' RTs with the groups (CG vs. ABG) and time-on-task (10 min). First, an ANOVA with participants' mean RT [Pre-CG (380.08 ± 59.41 ms; CI 95%: 17.27) and Pre-ABG (375.97 ± 57.09 ms; CI 95%: 14.56)] and time-on-task, was not significant in any effects or interactions [$F < 1$ in all cases]. Second, an ANOVA with participants' mean RT [Pre-CG (380.08 ± 59.41 ms; CI 95%: 17.27) and Post-CG (382.05 ± 53.21 ms; CI 95%: 20.33)] also was not significant in any effects or interactions [$F < 1$ in all cases]. Finally, a new ANOVA with participants' mean RT [(Post-CG (382.05 ± 53.21 ms; CI 95%: 20.33) and Post-ABG (359.76 ± 62.89 ms; CI 95%: 22.91)] revealed a significant main effect of group condition [F = 4.89, p = 0.03, $\eta 2$ = 0.19]. Participants responded faster in the ABG than in the CG. The effect of time-on-task and interaction between the control condition and time-on-task was insignificant ($F < 1$). More information is in Figure 5.

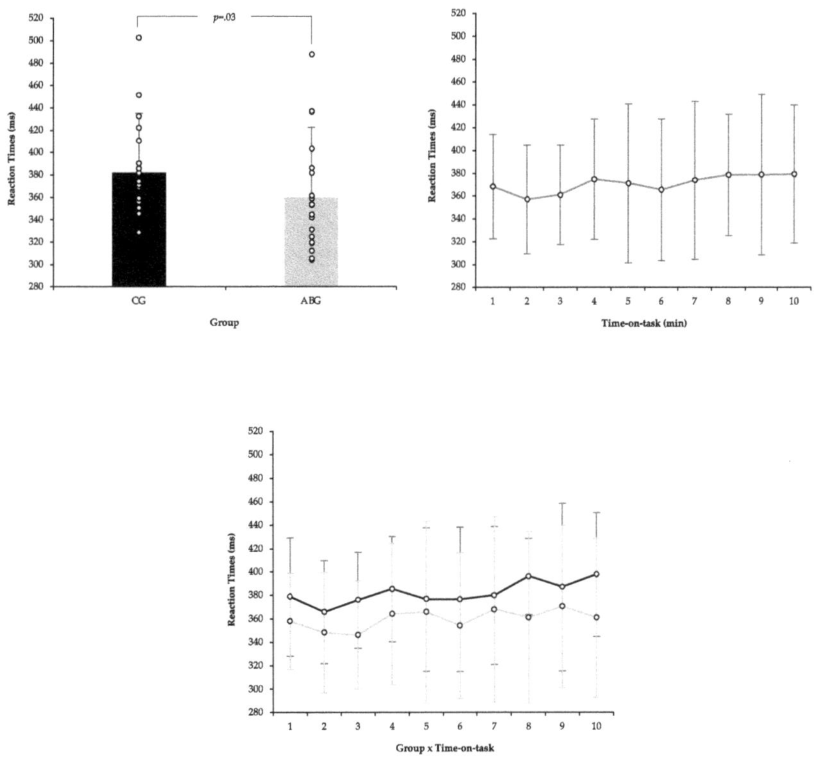

Figure 5. Mean RT (±SE) as a function of Group Condition, time-on-task and Group x time-on-task.

4. Discussion

The present study investigated the chronic effects of an eight-week training program on vigilance performance in high school students. The results revealed faster RTs in the experimental group than in the CG. However, the effect of time-on-task and interaction between the control condition and time-on-task was not significant ($F < 1$). Crucially, our results showed a significant main effect of the group with faster RTs in the ABG than in the CG. This result suggests a facilitation effect on vigilance in the ABG and provides support to previous research that showed moderate aerobic exercise had a selective impact on cognitive processing [43,44]. Therefore, the inclusion of uncertainty regarding the appearance of the target in the PVT makes it different from simple RT tasks and provides a reliable instrument to measure vigilance. Thus, in our study, the PE not only improved the nonspecific response speed but rather improved participants' vigilance.

As previously noted, practicing regular PE has been shown to produce changes to structural and functional levels of the brain [3,44,45]. Chronic PE produces lasting physiological adaptations [46]. Therefore, the body will naturally adjust, finally producing different anthropometric and physiological changes, thus causing an increase in the individual functional level (improved capacity and effectiveness in exercise). Considering the above, the conclusion from the literature is that physical fitness is one of the moderators between the effect of PE and cognitive function [40]. In this respect, we can explain the changes produced by chronic PE in the present experiment, based on the "cardiovascular hypothesis". Significantly, based on our results, the benefits found for cognitive functions usually associated with the regular practice of PE are moderated by the improvement of physical fitness [3,47,48]. In addition, physiological adaptations at the cardiovascular level, which we suggest occurred due to improvement in vigilance values, are associated with regular PE and have also been associated with adaptation at the brain level, which have been related to improvements in cognitive performance [47,48]. This could be considered a potential limitation of our study since the healthy lifestyle questionnaire and the level of physical fitness of the ALPHA-Fitness test battery were only applied in the preintervention. Consequently, we can only suggest the facilitation effect on vigilance in the ABG.

In the same context, we found interesting studies suggesting that regular aerobic PE is a good stimulus for triggering structural changes at the neural level [3,49] and therefore appears to positively impact cognitive performance [50,51]. Within this specific framework, the new research performed with magnetic resonance techniques [9,14,44,45,47,48] has been linked to adaptations at the brain level, which seem to have a positive impact on cognitive performance. In this respect, the literature revealed that chronic exercise leads to maintenance and neuronal proliferation in different brain areas (especially the hippocampus) and causes the growth of new blood capillaries through the action of brain-derived neurotrophic factor (BDNF) and insulin-like growth type 1 or somatomedin (IGF-1) in the hippocampus, cortex, and cerebellum, which has consequently been shown to have repercussions at the level of cognitive function [52]. Both proteins have shown a permanent increase in their production with the lasting intervention of regular physical exercise [15,53] and could be decisive preventive factors for brain degeneration, long-term enhancers and the development and protein for new neurons.

Finally, regarding the relationship between the chronic practice of PE or the level of physical fitness and general cognitive functioning, it should be noted that practically all of the literature explains the association between these variables based on the premise of the cardiovascular hypothesis, and mainly shows studies in children and older adults. According to this hypothesis, the cognitive function benefits associated with regular exercise are mediated by improving physical fitness. In addition, physiological adaptations attributed to chronic PE have also been linked to adaptations at the brain level, which seem to have a positive impact on cognitive performance [47,48].

Regarding the absence of fitness improvements, such a fact can be determined by the limited volume and intensity of practice [54–56]. Some fitness tests are also strength and power-dependent, such as sprinting, jumping and change-of-direction [57,58]. The program provided was based on strength endurance; however, intensity and intention were not controlled, which may cause a bias in the results as intensity may be critical for improvements [59]. Additionally, extra activities performed outside were not controlled, which may constrain the effects of parallel stimulus on the final outcomes.

This study has some limitations. One of the limitations is the absence of a counterbalanced intervention aiming to test different AB effects for the same target group. An additional study limitation is not controlling the extra activities and the effects of baseline levels of students. Baseline levels may play an important role in the progression since being a good or bad responder can be constrained by the starting point and trainability. Despite these limitations, this study provides an important and innovative approach to a micro-dose strategy for improving the quality of life and health of populations. This is one of the few studies dedicated to active break effects in a programmed approach that may help better understand the minimal effective dose that can be applied in students. Future research may compare different micro-doses and intensities while extending the approach to working, elderly and other populations).

5. Conclusions

The outcome of the present study suggests that an eight-week PE program based on AB of 16.10 ± 1.21 of the RPE scale improves vigilance performance. The importance of these findings is partly due to the sample of adolescent participants since most previous research has been done on children and adults. In addition, our study highlights a potential finding that locates the basis of dose-response on AB studies. Taken together, the current dataset extends this topic of research and contributes to demonstrating the evidence of the effect of chronic exercise on cognition. It is suggested, however, that future research should systematize greater monitoring of training, not just pre and postintervention. Consequently, another important factor is to analyze the characteristic of ABs (physical exercise, technical exercise, mindfulness, integration in the classroom contents, etc.), as they must be understood in order to assess the best impact on vigilance during the class. In addition, it is recommended that training interventions be carried out for more extended periods of time so that it will be possible to investigate the behavior of vigilance capacity as training time increases. It also contributes to the extant research on cognitive performance during the PA performed in the classroom and opens up exciting avenues for future research.

Author Contributions: Conceptualization, F.T.G.-F. and S.G.-V.; methodology, F.T.G.-F. and J.C.P.-V.; formal analysis, S.B.-M. and F.T.G.-F.; investigation, F.T.G.-F. and S.G.-V.; resources, F.T.G.-F.; writing—original draft preparation, S.B.-M., F.M.C. and G.B.; writing—review and editing, E.M.-C.; project administration, J.C.P.-V. and S.G.-V.; funding acquisition, E.M.-C. All authors have read and agreed to the published version of the manuscript.

Funding: The research was supported by the Consejería de Educación, Cultura y Deportes de Castilla-La Mancha, co-financed with European Union ERDF funds, under Grant number: SB-PLY/19/1805001/000147). This work is part of a two-year project entitled "Active Breaks Influence on Attentional Primary School Students".

Institutional Review Board Statement: The study was conducted in accordance with the ethical principles of the 1964 Helsinki declaration for human research and was approved by the Research Ethics Committee of the University of Castilla-La Mancha (Hospital Universitario de Albacete, Record 04/2020, and internal project n° 2020/05/052).

Informed Consent Statement: The participants' parents were informed about the objectives of the investigation and signed consent forms detailing their possible benefits and risks. Families were informed that they could revoke the participation agreement at any time. All participants were verbally informed and asked to provide consent prior to the intervention. The participants were fully debriefed about the purpose of the study at the end of the experiments.

Data Availability Statement: Not applicable.

Acknowledgments: We thank the schoolchildren, teachers, schools, and families for their collaboration in the study.

Conflicts of Interest: The authors declare no conflict of interest.

References

1. Etnier, J.L.; Chang, Y.K. Exercise, cognitive function, and the brain: Advancing our understanding of complex relationships. *J. Sport Health Sci.* **2019**, *8*, 299–300. [CrossRef] [PubMed]
2. Guiney, H.; Machado, L. Benefits of regular aerobic exercise for executive functioning in healthy populations. *Psychon. Bull.* **2013**, *20*, 73–86. [CrossRef] [PubMed]
3. Hillman, C.H.; Erickson, K.I.; Kramer, A.F. Be smart, exercise your heart: Exercise effects on brain and cognition. *Nat. Rev. Neurosci.* **2008**, *9*, 58–65. [CrossRef]
4. Mandolesi, L.; Polverino, A.; Montuori, S.; Foti, F.; Ferraioli, G.; Sorrentino, P.; Sorrentino, G. Effects of Physical Exercise on Cognitive Functioning and Wellbeing: Biological and Psychological Benefits. *Front. Psychol.* **2018**, *9*, 509. [CrossRef]
5. Tomporowski, P.D.; Lambourne, K.; Okumura, M.S. Physical activity interventions and children's mental function: An introduction and overview. *Prev. Med.* **2011**, *52*, S3–S9. [CrossRef] [PubMed]
6. Smith, P.J.; Blumenthal, J.A.; Hoffman, B.M.; Cooper, H.; Strauman, T.A.; Welsh-Bohmer, K.; Browndyke, J.N.; Sherwood, A. Aerobic exercise and neurocognitive performance: A meta-analytic review of randomized controlled trials. *Psychosom Med.* **2010**, *72*, 239–252. [CrossRef] [PubMed]
7. Chaddock, L.; Erickson, K.I.; Prakash, R.S.; Voss, M.W.; VanPatter, M.; Pontifex, M.B.; Hillman, C.H.; Kramer, A.F. A functional MRI investigation of the association between childhood aerobic fitness and neurocognitive control. *Biol. Psychol.* **2012**, *89*, 260–268. [CrossRef]

8. Best, J.R. Effects of Physical Activity on Children's Executive Function: Contributions of Experimental Research on Aerobic Exercise. *Dev. Rev.* **2010**, *30*, 331–551. [CrossRef]
9. Chaddock, L.; Erickson, K.I.; Prakash, R.S.; Kim, J.S.; Voss, M.W.; Vanpatter, M.; Pontifex, M.B.; Raine, L.B.; Konkel, A.; Hillman, C.H.; et al. A neuroimaging investigation of the association between aerobic fitness, hippocampal volume, and memory performance in preadolescent children. *Brain Res.* **2010**, *1358*, 172–183. [CrossRef] [PubMed]
10. Chaddock, L.; Hillman, C.H.; Buck, S.M.; Cohen, N.J. Aerobic fitness and executive control of relational memory in preadolescent children. *Med. Sci. Sports Exec.* **2011**, *43*, 344–349. [CrossRef] [PubMed]
11. Trudeau, F.; Shephard, R.J. Physical education, school physical activity, school sports and academic performance. *Int. J. Beh. Nut. Phys. Act.* **2008**, *5*, 10. [CrossRef] [PubMed]
12. Scudder, M.R.; Federmeier, K.D.; Raine, L.B.; Direito, A.; Boyd, J.K.; Hillman, C.H. The association between aerobic fitness and language processing in children: Implications for academic achievement. *Brain Cogn.* **2014**, *87*, 140–152. [CrossRef]
13. Erickson, K.I.; Gildengers, A.G.; Butters, M.A. Physical activity and brain plasticity in late adulthood. *Dialogues Clin. Neurosci.* **2013**, *15*, 99–108. [CrossRef] [PubMed]
14. Pereira, A.C.; Huddleston, D.E.; Brickman, A.M.; Sosunov, A.A.; Hen, R.; McKhann, G.M.; Sloan, R.; Gage, F.H.; Brown, T.R.; Small, S.A. An in vivo correlate of exercise-induced neurogenesis in the adult dentate gyrus. *Proc. Natl. Acad. Sci. USA* **2007**, *104*, 5638–5643. [CrossRef] [PubMed]
15. Cotman, C.W.; Berchtold, N.C.; Christie, L.A. Exercise builds brain health: Key roles of growth factor cascades and inflammation. *Trends Neurosci.* **2007**, *30*, 464–472. [CrossRef]
16. Sarter, M.; Givens, B.; Bruno, J.P. The cognitive neuroscience of sustained attention: Where top-down meets bottom-up. *Brain Res. Rev.* **2001**, *35*, 146–160. [CrossRef]
17. Parasuraman, R. Human-Computer Monitoring. *Hum. Factors* **1987**, *29*, 695–706. [CrossRef]
18. Basner, M.; Dinges, D.F. Maximizing sensitivity of the psychomotor vigilance test (PVT) to sleep loss. *Sleep* **2011**, *34*, 581–591. [CrossRef]
19. Davies, D.R.; Parasuraman, R. *The Psychology of Vigilance*; Academic Press: Cambridge, MA, USA, 1982.
20. Helton, W.S.; Warm, J.S. Signal salience and the mindlessness theory of vigilance. *Acta Psychol.* **2008**, *129*, 18–25. [CrossRef]
21. Warm, J.S.; Parasuraman, R.; Matthews, G. Vigilance requires hard mental work and is stressful. *Hum. Factors* **2008**, *50*, 433–441. [CrossRef]
22. Formentin, C.; De Rui, M.; Zoncapè, M.; Ceccato, S.; Zarantonello, L.; Senzolo, M.; Burra, P.; Angeli, P.; Amodio, P.; Montagnese, S. The psychomotor vigilance task: Role in the diagnosis of hepatic encephalopathy and relationship with driving ability. *J. Hepatol.* **2019**, *70*, 648–657. [CrossRef]
23. Weinberg, R.S.; Gould, D. *Foundations of Sport and Exercise Psychology*; Human Kinetics: Melbourne, Australia, 1995.
24. Hillman, C.H.; Belopolsky, A.V.; Snook, E.M.; Kramer, A.F.; McAuley, E. Physical activity and executive control: Implications for increased cognitive health during older adulthood. *RQES* **2004**, *75*, 176–185. [CrossRef]
25. Kamijo, K.; Pontifex, M.B.; O'Leary, K.C.; Scudder, M.R.; Wu, C.T.; Castelli, D.M.; Hillman, C.H. The effects of an afterschool physical activity program on working memory in preadolescent children. *Dev. Sci.* **2011**, *14*, 1046–1058. [CrossRef]
26. Predovan, D.; Fraser, S.A.; Renaud, M.; Bherer, L. The effect of three months of aerobic training on stroop performance in older adults. *J. Aging Res.* **2012**, *2012*, 269815. [CrossRef] [PubMed]
27. Steinmayr, R.; Ziegler, M.; Träuble, B. Do intelligence and sustained attention interact in predicting academic achievement? *Learn. Individ. Differ.* **2010**, *20*, 14–18. [CrossRef]
28. Bunce, D. Age differences in vigilance as a function of health-related physical fitness and task demands. *Neuropsychology* **2001**, *39*, 787–797. [CrossRef]
29. Hillman, C.H.; Castelli, D.M.; Buck, S.M. Aerobic fitness and neurocognitive function in healthy preadolescent children. *Med. Sci. Sports Exerc.* **2005**, *37*, 1967–1974. [CrossRef]
30. Pontifex, M.B.; Scudder, M.R.; Brown, M.L.; O'Leary, K.C.; Wu, C.T.; Themanson, J.R.; Hillman, C.H. On the number of trials necessary for stabilization of error-related brain activity across the life span. *Psychophysiology* **2010**, *47*, 767–773. [CrossRef]
31. Pontifex, M.B.; Scudder, M.R.; Drollette, E.S.; Hillman, C.H. Fit and vigilant: The relationship between poorer aerobic fitness and failures in sustained attention during preadolescence. *Neuropsychology* **2012**, *26*, 407–413. [CrossRef] [PubMed]
32. Sibley, B.A.; Etnier, J.L.; Le Masurier, G.C. Effects of an Acute Bout of Exercise on Cognitive Aspects of Stroop Performance. *J. Sport Exerc. Psychol.* **2006**, *28*, 285–299. [CrossRef]
33. Verburgh, L.; Königs, M.; Scherder, E.J.; Oosterlaan, J. Physical exercise and executive functions in preadolescent children, adolescents and young adults: A meta-analysis. *Brit. J. Sports Med.* **2014**, *48*, 973–979. [CrossRef] [PubMed]
34. Léger, L.A.; Lambert, J.A. maximal multistage 20-m shuttle run test to predict VO2 max. *Eur J. Appl. Physiol. Occup. Physiol.* **1982**, *49*, 1–12. [CrossRef] [PubMed]
35. Lee, P.H.; Macfarlane, D.J.; Lam, T.H.; Stewart, S.M. Validity of the International Physical Activity Questionnaire Short Form (IPAQ-SF): A systematic review. *Int. J. Behav. Nut. Phys. Act.* **2011**, *8*, 115. [CrossRef]
36. American College of Sports Medicine. *ACSM's Guidelines for Exercise Testing and Prescription*, 9th ed.; Lippincott Williams: Philadelphia, PA, USA, 2010.
37. Ruiz, J.R.; España Romero, V.; Castro Piñero, J.; Artero, E.G.; Ortega, F.B.; Cuenca García, M.; Jiménez Pavón, D.; Chillón, P.; Girela Rejón, M.J.; Mora, J.; et al. Batería ALPHA-Fitness: Test de campo para la evaluación de la condición física relacionada con

la salud en niños y adolescentes [ALPHA-fitness test battery: Health-related field-based fitness tests assessment in children and adolescents]. *Nutr. Hosp.* **2011**, *26*, 1210–1214. [CrossRef]
38. Ortega, F.B.; Ruiz, J.R.; Castillo, M.J.; Sjöström, M. Physical fitness in childhood and adolescence: A powerful marker of health. *Int. J. Obes. Lond.* **2008**, *32*, 1–11. [CrossRef]
39. Borg, G. *Borg's Perceived Exertion and Pain Scales*; Human Kinetics: Melburne, Australia, 1998.
40. Drummond, S.P.; Bischoff-Grethe, A.; Dinges, D.; Ayalon, L.; Mednick, S.C.; Meloy, M.J. The Neural Basis of the Psychomotor Vigilance Task. *Sleep* **2005**, *28*, 1059–1068.
41. Luque-Casado, A.; Perakakis, P.; Hillman, C.H.; Kao, S.C.; Llorens, F.; Guerra, P.; Sanabria, D. Differences in Sustained Attention Capacity as a Function of Aerobic Fitness. *Med. Sci. Sports Exerc.* **2015**, *48*, 887–895. [CrossRef]
42. Loh, S.; Lamond, N.; Dorrian, J.; Roach, G.; Dawson, D. The validity of psychomotor vigilance tasks of less than 10-minute duration. *Behav. Res. Methods* **2004**, *36*, 339–346. [CrossRef] [PubMed]
43. Lambourne, K.; Tomporowski, P. The effect of exercise-induced arousal on cognitive task performance: A meta-regression analysis. *Brain Res.* **2010**, *1341*, 12–24. [CrossRef] [PubMed]
44. Colcombe, S.; Kramer, A.F. Fitness effects on the cognitive function of older adults: A meta-analytic study. *Psychol. Sci.* **2003**, *14*, 125–130. [CrossRef] [PubMed]
45. Erickson, K.I.; Voss, M.W.; Prakash, R.S.; Basak, C.; Szabo, A.; Chaddock, L.; Kim, J.S.; Heo, S.; Alves, H.; White, S.M.; et al. Exercise training increases size of hippocampus and improves memory. *Proc. Natl. Acad. Sci. USA* **2011**, *108*, 3017–3022. [CrossRef] [PubMed]
46. Kenney, W.L.; Willmore, J.H.; Costill, D.L. *Physiology of Sport and Exercise*, 5th ed.; Human Kinetics: Melbourne, Australia, 2013.
47. Colcombe, S.J.; Erickson, K.I.; Scalf, P.E.; Kim, J.S.; Prakash, R.; McAuley, E.; Elavsky, S.; Marquez, D.X.; Hu, L.; Kramer, A.F. Aerobic exercise training increases brain volume in aging humans. *J. Ger. Ser. A Biol. Sci. Med. Sci.* **2006**, *61*, 1166–1170. [CrossRef]
48. Colcombe, S.J.; Kramer, A.F.; Erickson, K.I.; Scalf, P.; McAuley, E.; Cohen, N.J.; Webb, A.; Jerome, G.J.; Marquez, D.X.; Elavsky, S. Cardiovascular fitness, cortical plasticity, and aging. *Proc. Natl. Acad. Sci. USA* **2004**, *101*, 3316–3321. [CrossRef] [PubMed]
49. Thomas, A.G.; Dennis, A.; Bandettini, P.A.; Johansen-Berg, H. The effects of aerobic activity on brain structure. *Front. Psychol.* **2012**, *3*, 86. [CrossRef]
50. Acevedo, E.O.; Ekkekakis, P. *Psychobiology of Physical Activity*; Human Kinetics: Melbourne, Australia, 2006.
51. McMorris, T.; Tomporowski, P.; Audiffren, M. *Exercise and Cognitive Function*; Wiley-Blackwell: Hoboken, NJ, USA, 2009.
52. Voss, M.W.; Nagamatsu, L.S.; Liu-Ambrose, T.; Kramer, A.F. Exercise, brain, and cognition across the life span. *J. Appl. Phys.* **2011**, *111*, 1505–1513. [CrossRef] [PubMed]
53. Ferris, L.T.; Williams, J.S.; Shen, C.L. The effect of acute exercise on serum brain-derived neurotrophic factor levels and cognitive function. *Med. Sci. Sport Exerc.* **2007**, *39*, 728–734. [CrossRef]
54. Gillen, J.B.; Gibala, M.J. Is high-intensity interval training a time-efficient exercise strategy to improve health and fitness? *Appl. Phys. Nutr. Metab.* **2014**, *39*, 409–412. [CrossRef] [PubMed]
55. Lee, I. Dose-Response Relation Between Physical Activity and Fitness: Even a Little Is Good; More Is Better. *JAMA* **2007**, *297*, 2137–2139. [CrossRef]
56. de Geus, B.; Joncheere, J.; Meeusen, R. Commuter cycling: Effect on physical performance in untrained men and women in Flanders: Minimum dose to improve indexes of fitness. *Scand. J. Med. Sci. Sports* **2009**, *19*, 179–187. [CrossRef]
57. Loturco, I.; Contreras, B.; Kobal, R.; Fernandes, V.; Moura, N.; Siqueira, F.; Winckler, C.; Suchomel, T.; Pereira, L.A. Vertically and horizontally directed muscle power exercises: Relationships with top-level sprint performance. *PLoS ONE* **2018**, *13*, e0201475. [CrossRef]
58. Feeney, D.; Stanhope, S.J.; Kaminski, T.W.; Machi, A.; Jaric, S. Loaded Vertical Jumping: Force-Velocity Relationship, Work, and Power. *J. Appl. Biomech.* **2016**, *32*, 120–127. [CrossRef] [PubMed]
59. Behringer, M.; Vom Heede, A.; Matthews, M.; Mester, J. Effects of strength training on motor performance skills in children and adolescents: A meta-analysis. *Pediatr. Exerc. Sci.* **2011**, *23*, 186–206. [CrossRef] [PubMed]

biology

MDPI

Article

Performance Prediction Equation for 2000 m Youth Indoor Rowing Using a 100 m Maximal Test

Luiz Felipe da Silva [1,*], Paulo Francisco de Almeida-Neto [1], Dihogo Gama de Matos [2], Steven E. Riechman [3], Victor de Queiros [1], Joseane Barbosa de Jesus [4,5], Victor Machado Reis [6], Filipe Manuel Clemente [7,8], Bianca Miarka [9], Felipe J. Aidar [4,5,10,11], Paulo Moreira Silva Dantas [1] and Breno Guilherme de Araújo Tinoco Cabral [1]

1 Health Sciences Center, Department of Physical Education, Federal University of Rio Grande do Norte, Natal 59078-970, Brazil; paulo220911@hotmail.com (P.F.d.A.-N.); victorsabino97@ufrn.edu.br (V.d.Q.); pgdantas@icloud.com (P.M.S.D.); brenotcabral@gmail.com (B.G.d.A.T.C.)
2 Cardiorespiratory & Physiology of Exercise Research Laboratory, Faculty of Kinesiology and Recreation Management, University of Manitoba, Winnipeg, MB R3T 2N2, Canada; dihogogmc@hotmail.com
3 Department of Health and Kinesiology, Texas A&M University, College Station, TX 77843, USA; sriechman@tamu.edu
4 Group of Studies and Research of Performance, Sport, Health and Paralympic Sports GEPEPS, The Federal University of Sergipe, UFS, São Cristovão 49100-000, Brazil; josybj@hotmail.com (J.B.d.J.); fjaidar@academico.ufs.br (F.J.A.)
5 Graduate Program in of Physical Education, Federal University of Sergipe—UFS, São Cristovão 49100-000, Brazil
6 Research Center in Sports Sciences, Health Sciences, and Human Development (CIDESD), Trás os Montes and Alto Douro University, 5001-801 Vila Real, Portugal; victormachadoreis@gmail.com
7 Sports and Leisure, Polytechnic Institute of Viana do Castelo, Rua Industrial and Commercial School of Nun'Álvares, 4900-347 Viana do Castelo, Portugal; filipe.clemente5@gmail.com
8 Telecommunications Institute, Delegation of Covilhã, 1049-001 Lisbon, Portugal
9 Laboratory of Psychophysiology and Performance in Sports & Combats, Postgraduate Program in Physical Education, School of Physical Education and Sport, Federal University of Rio de Janeiro, Rio de Janeiro 21941-901, Brazil; miarkasport@hotmail.com
10 Department of Physical Education, Federal University of Sergipe—UFS, São Cristovão 49100-000, Brazil
11 Program of Physiological Science, Federal University of Sergipe—UFS, São Cristovão 49100-000, Brazil
* Correspondence: feliperowing@ufrn.edu.br; Tel.: +55-8499408-8453

Citation: Silva, L.F.d.; de Almeida-Neto, P.F.; de Matos, D.G.; Riechman, S.E.; de Queiros, V.; de Jesus, J.B.; Reis, V.M.; Clemente, F.M.; Miarka, B.; Aidar, F.J.; et al. Performance Prediction Equation for 2000 m Youth Indoor Rowing Using a 100 m Maximal Test. *Biology* **2021**, *10*, 1082. https://doi.org/10.3390/biology10111082

Academic Editor: Tetsuya Shiuchi

Received: 4 October 2021
Accepted: 14 October 2021
Published: 22 October 2021

Simple Summary: The 2000 m tests, usually applied in indoor rowing, during weeks of evaluation and selection of young rowing athletes, often discourage participation or are performed by athletes without a previously established strategy (i.e., execution strategy, according to an estimated performance expectation) which may underestimate the performance of young athletes. Thus, the mathematical model developed in this research can contribute to the selection of athletes in Olympic rowing by providing a low-cost tool with a significant level of reliability and performance prediction of 2000 m. Furthermore, the mathematical model could help to propose highly reliable assessment strategies following coaches. This model could be used as an alternative to traditional ways of evaluating training progression up to 2000 m, thus contributing to the strategic planning of the tests applied and the development of athletes.

Abstract: Background: The exhaustive series of tests undergone by young athletes of Olympic rowing prior to important competitions imply loads of physical stress that can ultimately impact on mood and motivation, with negative consequences for their training and performance. Thus, it is necessary to develop a tool that uses only the performance of short distances but is highly predictive, offering a time expectancy with high reliability. Such a test must use variables that are easy to collect with high practical applicability in the daily routine of coaches. Objective: The objective of the present study was to develop a mathematical model capable of predicting 2000 m rowing performance from a maximum effort 100 m indoor rowing ergometer (IRE) test in young rowers. Methods: The sample consisted of 12 male rowing athletes in the junior category (15.9 ± 1.0 years). A 100 m time trial was performed on the IRE, followed by a 2000 m time trial 24-h later. Results: The 2000 m mathematical model to predict performance in minutes based on the maximum 100 m test demonstrated a high correlation (r = 0.734; p = 0.006), strong reliability index (ICC: 0.978; IC95%: [0.960; 0.980]; p = 0.001) and was within usable agreement limits (Bland -Altman Agreement: −0.60 to 0.60; 95% CI [−0.65; 0.67]). Conclusion: The mathematical model developed to predict 2000 m performance is effective and has a statistically significant reliability index while being easy to implement with low cost.

Keywords: athletic performance; rowing; sport; young athlete; mathematical model

1. Introduction

Olympic Rowing is characterized by a high physical demand in which a high aerobic and anaerobic capacity are required for optimal performance [1,2]. Depending on the athletes' gender and age and the type of the boat, a 2000 m Olympic rowing race can last from 5 to 8 min [3,4]. During a rowing race, the metabolic source of energy is predominantly aerobic [3,4]. Yet, both in the first ~100-m and at the last ~200-m race, athletes tend to perform maximal output sprints that can be decisive at the finish line [3,4]. Those sprints require a rapid and high load of metabolic energy, with the anaerobic system taking over. Therefore, a significant contribution of the anaerobic capability and efficiency of this metabolic system can also predict rowing performance [3,4]. For instance, the modified Wingate performance for rowing in national-level adolescent rowers could be used to predict rowing performance [3]. Accordingly, the anaerobic stimuli, as an indicator of the anaerobic capacity, indirectly reflects rowers' performance at 2000 m [3–7]. Previous studies have suggested that parameters such as race time and short tests [i.e., 50-m, 100-m, or 500-m anaerobic stimuli performed on a rowing ergometer] should be taken into consideration during the process of selection and orientation of young athletes [5–8].

Several studies have reported a positive correlation between 2000 m indoor rowing performance and power produced during 20 s, 30 s, and 60 s rowing tests [5–8]. Cerasola et al. [8] developed a mathematical model for predicting 2000 m performance in indoor rowing using the combination of anthropometric variables, VO2max, and 60 s maximum sprint. In a study conducted by Maciejewski [9], the researchers found a strong correlation of the Wingate test adapted for rowing, with 1500 m indoor rowing performance in competitive adolescent rowers, and the results point out that rowing coaches can use the modified Wingate test to potentially identify talented young rowers. In the same study, the authors found a strong correlation of the 60 s test with 2000 m performance and point out that 60 s performance can be considered a valuable tool to predict 2000 m performance of elite young rowers, not requiring expensive and long duration [7].

Anaerobic energy pathways can be entered into regression models to predict ergometer performance at 2000 m and identify rowing talent [3]. Regression analysis models for performance prediction have already been suggested for the rowing modality [3,5–7]. Nonetheless, although results of studies have contributed to the scientific knowledge of rowing, few coaches use these methods due to the requirement of sophisticated equipment for evaluation, which is not always available to them. Therefore, they choose to use traditional evaluation methods in the selection of young athletes [3,5–7]. In addition, reliable mathematical models that use only short distance performance to predict 2000 m performance in indoor rowing are not yet available in the scientific literature [5–10]. Therefore, relating practice tests to success in rowing may be beneficial in adapting and constructing training plans to optimize athlete development [9,10]. Thus, developing specific tests for rowers with good reproducibility could help predict performance, and training progression in the training environment becomes necessary.

A mathematical model that could predict the performance and progression of training would help coaches in the initial planning of training objectives, also being useful to assess the anaerobic capacity of rowers indirectly, in addition to providing accurate information for estimation of performance concerning increasing the distance of the tests applied in 2000 m with low cost, high practical applicability, and easy collection. [7,9,10].

Sports scientists have suggested complete indoor rowing tests of 20 s and 30 s [3,5,6]. However, this assessment does not reflect the initial phases and final sprints that last about 60 s, with an estimated energy expenditure of 500–700 Watt [11–13]. Therefore, a maximum 100 m ergometer rowing test might be better suited to monitor rowers' ability to sustain energy expenditure during the actual start and finish phases of a 2000 m rowing.

The present study aimed to develop a mathematical model capable of predicting 2000 m performance for young rowers from a 100 m maximal effort test. The present study raised the hypothesis that the performance of 2000 m in rowing could be predicted with a mathematical model using parameters easily collected by the coaches and reproducible in their daily environment.

2. Materials and Methods

2.1. Participants

The research has a cross-sectional design with a sample selected from February to March 2020 at the "Sports Club de Natal" (Brazil). The sample was composed of twelve young male rowing athletes (15.9 ± 1.0 years; body weight (kg): 66.5 ± 13.1; height: (cm) (170.7 ± 7.0, wingspan (cm): 161.5 ± 43.2, body mass index (kg/m^2): 22.6 ± 3.4, time of practical experience in rowing: (1.3 ± 2.0 years), who were ranked on the national level were selected for the study. Based on criteria from Matsudo, Rivet, and Pereira [14], the sample is classified as national-level athletes. These athletes were among the top 20 in the positions between 7th and 12th place (final B) and between 13th and 19th place (final C) national ranking during 2020.

Table 1 reports the characteristics of the subjects about body composition and power during 100-m and 2000-m tests.

Table 1. Subject Characteristics.

Variables	Mean ± SD
Fat mass (kg)	16.5 ± 6.7
Lean mass (kg)	47.4 ± 8.1
Mean Power in 100 m (watts)	376.9 ± 62.7
Mean Power in 2000 m (watts)	235.9 ± 29.0

kg = kilograms. kg/m^2 = kilograms per square meter. m = Meters. SD = Standard deviation.

Inclusion criteria were: (i) Being affiliated to a state federation and national Olympic rowing confederation; (ii) at least one year of training experience; (iii) age 14 to 16 years, and; (iv) a minimum training frequency of six times a week (≥60 min per training session)

Exclusion criteria were: (i) Presence of osteoarticular lesions (i.e., lesions in bone tissues and joints) or muscle injuries in the last six months before the research. (ii) could not complete all the tests proposed by this study. (iii) use of ergogenic substances (i.e., supplements) that could enhance physical performance (i.e., caffeine, taurine, creatine, and others).

The Ethics approved this study, and Research Committee with Human Beings of the Federal University of Rio Grande do Norte (technical advice: 3.552.010), respecting the national and international ethical principles in the declaration Helsinki and ethical standards in sport and exercise science research [15]. In addition, the present study complies with items on the STROBE checklist for observational studies [16]. All participants and their guardians received information about the research objectives and methodological procedures adopted. Informed consent terms (TALE) and a free and informed consent term (ICF) were signed by the volunteers and their respective legal guardians, according to Resolution 466/12.

2.2. Procedures

On the first day, all procedures for the experiment were explained to participants and their respective guardians. On the second day, anthropometric and body composition tests were performed for sample characterization purposes. On the third day, the athletes performed the 100 m sprint. Finally, on day 4, a 2000 m time trial was performed (See Figure 1). All rowing testing was conducted on an indoor rowing ergometer (Concept® model-D equipped with PM5 digital monitor; Morrisville, CT USA) [17] in a temperature-controlled environment (26 °C). The equipment was calibrated with a resistance factor of 125 (N s^2/m^2) (i.e., Air System specific to Indoor Rowing) according to the specifications of the international rowing federation. All tests were performed on consecutive days starting at 8:00 am.

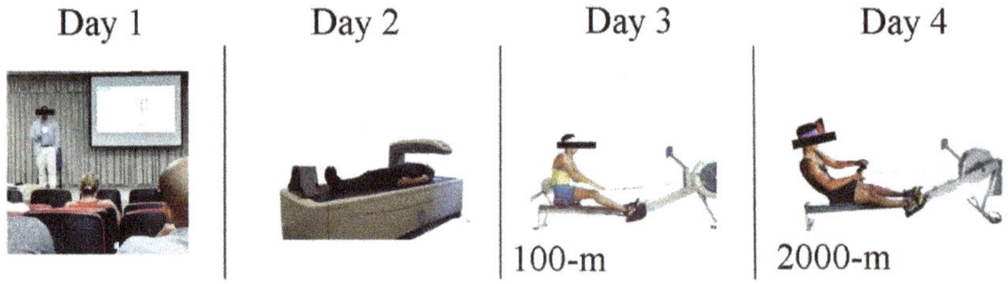

Figure 1. Study design.

2.3. Analysis of Body Morphology

For sample characterization, body composition was determined with dual-energy X-ray absorptiometry (DEXA) (LUNAR®/GE PRODIGY—LNR 41,990, Software enCORE version 18; United States–Washington DC) using specific algorithms for the pediatric population [18].

2.4. Development of the Mathematical Model

A mathematical model was developed to predict 2000 m performance based on the performance of the maximum effort 100 m Indoor Rowing Test with consideration of a priori theoretical model [19–21]. Thus, based on traditional physics, we developed a mathematical model of approaching the distance from the sprint time of 100 m (in seconds) [19–21]. In this sense, we consider the time in seconds of the sprint of 100 m multiplied by constant 22 (number determined by algorithms for an approximation with the time of 2000 m). Subsequently, the correction factor +18 was identified by algorithms to equalize the results of the predictive equation with those of indoor rowing. Thus, to convert the results into minutes, we divide the final product by 60 [19–21]. Subsequently, regression analyses were performed, in sequence and the theoretical model was tested using confirmatory factor analyses and the reproducibility index [19].

The following is the mathematical model developed to predict 2000 m performance from the performance in the 100-m test:

$$\text{Time in Minutes 2000 m} = [(\text{Time in Seconds 100 m} \times 22) + 18]/60$$

2.5. Statistics

To determine a priori the minimum sample size to develop the mathematical model, we used the effect size of 0.915, referring to the linear regression result (mathematical model for peak power X peak power in indoor rowing) found by Almeida-Neto et al. [22]. We used the G*Power ® software (Version 3.0; Berlin, Germany) in the configuration "T family statistics for regressions" and an $\alpha = 0.05$ and a $\beta = 0.80$ considering a single variable to perform the prediction. A minimum sample size of 10 subjects (t (2.0) = 2.91) was determined to be an acceptable sample size with power estimated to be 0.90. The normality of the data was analyzed by the Shapiro-Wilk tests and z-score for asymmetry and kurtosis (−1.96 to 1.96).

Pearson's linear correlation test determined the data correlations. The correlational magnitude thresholds used were those proposed by: Insignificant: $r < 0.10$; Weak: $r = 0.10–0.39$; Moderate: $r = 0.40–0.69$; Strong: $r = 0.70–0.89$; Very strong: $r = 0.90–1.00$ [23]. The Breush-Pegan test tested the homogeneity of the models, and the assumptions of normality, variance, and independence of the data were confirmed. The Durbin–Watson test was used to test the multicollinearity of the regression models. By White's test, we checked the heteroscedasticity of the regression models.

To measure the reproducibility and reliability of the mathematical model, an analysis of the intraclass correlation coefficient was performed, with magnitude of absence: ICC ≤ 0; poor: ICC = 0–0.19; weak: ICC = 0.20–0.39; moderate: ICC = 0.30–0.59; substantial: ICC = 0.60–0.79; and almost complete: ICC ≥ 0.80 [24]. The Bland-Altman method was used to verify the degree of agreement between the models. By proportion bias analysis, we checked for heteroscedasticity of the Bland-Altman concordance.

For the comparative analysis, the Student t-test was used. The size of the effect of the differences was calculated by the Cohen test (d). The magnitude of the Effect Size followed the classification described by Espírito Santo and Daniel [25]: insignificant <0.19; 0.20–0.49 small; mean 0.50–0.79; large 0.80–1.29; very large <1.30). For the technical error of anthropometric measurements, the following magnitude was used: Acceptable for skin folds $\leq$ 5.0% and other anthropometric measurements $\leq$1.0% [26]. All analyses were performed using open-source software R (version 4.0.1; R Foundation for Statistical Computing®, Vienna, Austria) with a significance threshold of $p < 0.05$.

3. Results

The 100 m (seconds) performance correlated significantly with the 2000 m (minutes) performance. The linear regression model in Table 2 shows that the 100 m performance was also efficient in predicting the 2000 m performance. We emphasize that for the linear regression, no significant heteroscedasticity or significant multicollinearity was detected.

Table 2. Correlations and regressions of variables with a performance at different distances in rowing.

Variable	Rowing 2000 m					
	Correlation			Regression		
Rowing 100 m	r	r^2	*p* Value	(r^2)	β	*p* Value
	0.734 *	0.538	0.006	0.539 *	15.42	0.006

* Statistically significant; r = correlation coefficient; r^2 = squared correlation coefficient. (r^2) = regression determination coefficient; β = angular regression coefficient in relation to the dependent variable. m = Meters.

The G * Power software (Version 3.0; Berlin, Germany) was used to check the power of the post-hoc results (Table 2), and in the "T" statistic configuration for correlations and regressions, reporting the effect as r^2 (see Table 2), adopting $\alpha = 0.05$. Thus, for the correlation analysis, a sampling power of 0.950 (t (10.0) = 1.75) and regression of 0.960 (t (11.0) = 1.80) suggests that the findings of the present study are reliable.

The only variable used in the model was the 100 m sprint test time in seconds. In addition, the result predicted by the mathematical model showed a substantial reliability index and a significant agreement index (see Figure 2), with the result of the 2000 m indoor rowing performance (CCI = 0.978; IC95%: [0.960; 0.980]; $p = 0.001$); (Bland-Altman Agreement: −0.60 to 0.60; IC 95%: [−0.65; 0.67]). Figure 2 shows the limits of agreement of the performance in minutes predicted by the mathematical model with the actual performance in minutes. No significant heteroscedasticity was observed, and the difference between the methods, was values between −0.60-min and 0.60-min (the differences ranging between 7% and 11%).

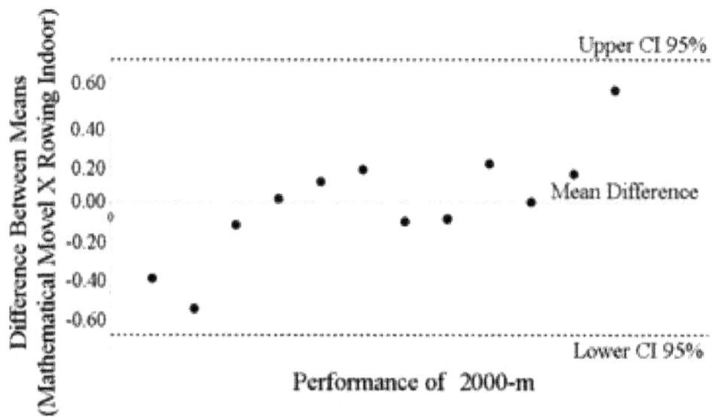

Figure 2. Bland-Altman Plot. CI 95%: Confidence Interval of 95%. M: Meters.

Figure 3 shows that no significant differences were observed when comparing the result of the 2000 m test performed on an indoor ergometer with the result of the mathematical model developed presently (Real Performance (minutes): 7.49 ± 0.39, Coefficient of variation: 5.2%, standard error: 0.11; Predicted performance (minutes): 7.50 ± 0.41 Coefficient of variation: 5.4%, standard error: 0.12; Effect Size = 0.01; IC 95%: [−0.83; 0.85]; $p = 0.9$). It should be noted that the mathematical model showed no significant bias in relation to the performance of 2000 m (difference between the methods = 0.00 ± 0.30; $r^2 = 0.094$; $\beta = -1.72$, CI 95% β: [−5.02; 0.09]; $p = 0.33$), which suggests that there is no systematic bias in the model developed by this study. However, based on the individual data, the mathematical model points out the limitation of underestimating or overestimating the 2000 m time by ~12-s in 7 out of 12 athletes (58%).

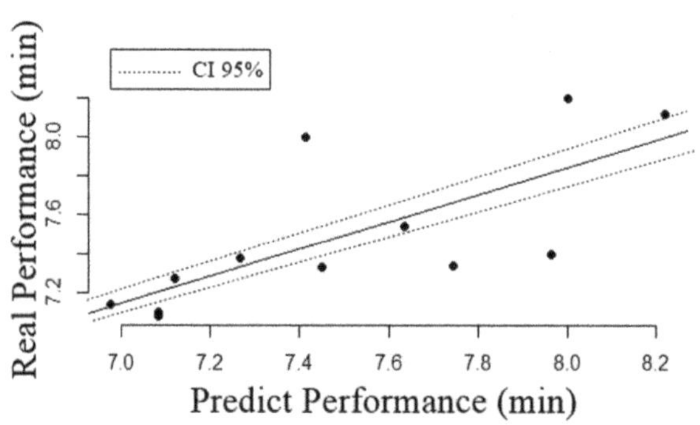

Figure 3. Comparison of the result predicted by the mathematical model with the result of the 2000 m test in indoor rowing. CI 95%: Confidence Interval of 95%. (min): Minutes.

4. Discussion

The present study aimed to develop a mathematical model to predict rowing performance at a distance of 2000 m in an indoor maximum rowing test at a distance of 100 m. This study pioneered the use of a 100 m top speed test to predict 2000 m rowing performance in young males. The main finding of this study demonstrates that the mathematical model based on a maximum effort of 100 m was moderately significant to predict the performance of 2000 m in indoor rowing. However, the mathematical model of the present study underestimated or overestimated the 2000-m time of 58% of athletes in ~12 s.

Previously, in a study conducted by Šmída et al. [27], a significant relationship was observed between the peak anaerobic power and 2000 m indoor rowing performance. This result suggests that a test with a predominant anaerobic energy requirement could be a good predictor of aerobic performance in the 2000 m rowing test [27]. It is known that in rowers, the energy demands of a standard test of 2000 m are predominantly aerobic. However, one-third and a quarter of the total energy demand comes from anaerobic sources [28,29]. Due to the high resistance and physical strength required by this modality, both energetic (i.e., aerobic and anaerobic) pathways end up being stressed submaximally or maximally [30].

In a similar study, Cataldo et al. [5] developed a 2000 m rowing performance prediction model in 20 young male athletes (average age 15.2 ± 1.3) and concluded that the 2000 m rowing performance could be estimated from a 20 s indoor rowing sprint test. However, this mathematical model requires the use of sophisticated and costly testing, such as fat-free mass index and maximum VO2 (VO2max) levels consumed during the 20-s sprint. The mathematical model created by Riechman et al. [6] proved to effectively predict 2000 m performance in highly trained young women from a maximum 30 s sprint performance (Rowing Wingate). Riechman et al. [6] reported statistically significant correlations in the range of $r = 0.84$ to $r = 0.89$ between the results of the Wingate anaerobic rowing test and the performance of the 2000 m ergometer rowing test. The significant relationship between the Wingate test results and the performance of the 2000-m rowing ergometer is likely to be explained, in part, by a substantial aerobic contribution to a 30 s test [6]. However, this study used the mean power variables of 30 s maximum, VO2max, and fatigue percentage of the Wingate test. Therefore, the equations developed by Cataldo et al. [5] and Riechman et al. [6] require a sophisticated laboratory evaluation before using the 2000 m performance prediction equation.

Given this assumption, it is suggested that the time in seconds of a short performance in rowers is a reliable predictor of competition distance performances [28–30], which eliminates the need to perform sophisticated analyses such as those mentioned above. Thus, the mathematical model presented in this study used only the 100 m sprint time to estimate the 2000 m performance, with significant reliability. The use of the 100-m test becomes practical for the rower training environment, being a significant anaerobic stimulus that can help in predicting the 2000-m challenge. Billat et al., [31] point out that anaerobic capacity is predictive of rowing performance, attributing this fact to sprints often performed during rowing race. Thus, the results of the present study indicate moderately significant relative reliability. The coefficient of variation values was low (<6%), as well as the standard error values (<0.15) and the difference in relation to the base method (indoor rowing) were less than 1%, demonstrating that the mathematical model based in one sprint of 100-m, can be helpful for sports. According to Atkinson and Nevill [32], the estimate of performance in sports needs to have high reliability and an agreement above 95% in relation to the basic method used. Thus, the present study showed a significant reliability index (ICC: 0.978; IC95%: [0.960; 0.980]; $p = 0.001$) and thus can contribute to the selection of athletes in Olympic rowing by providing a low-cost tool with a significant level of applicability and prediction of 2000 m performance.

Seeking to propose a highly reliable assessment strategy for rowing coaches, the strength of this study was to present an equation developed as an effective tool to predict the athlete's performance at a distance of 2000 m. Therefore, this research can contribute to the monitoring and evaluation of young rowers, providing a tool with a significant level of reliability to predict 2000 m performance. In addition, our equation may be an alternative to traditional ways of evaluating training progression for the 2000 m, thus contributing to planning and development parameters for athletes and coaches. Therefore, this research can contribute to the monitoring and evaluation of young rowers, providing a tool with a significant level of reliability to predict 2000 m performance.

The strengths of the present research were: (i) the study design was adequate to an-swer the research question by presenting an assessment model with high reliability; (ii) gender was not a divergent factor in the current sample as we used only male athletes; (iii) Significant practical applicability for rowing coaches as it was easy to use and repeated frequently due to its short duration.

Despite the relevance of the results, this study is limited by the fact that it was per-formed only with a small number of young male athletes, and athletes from other catego-ries, ages, and sex may present different results. In addition, we highlight that although the mathematical model has practical applicability, its predicting power is limited and the absence of comparisons with more sophisticated mathematical models that use variables such as VO2max [5,33,34] is also a limitation to accurately determine the quality of the mathematical model developed by the present study. In addition, the mathematical model points to an estimation error of ~12-s in 58% of the sample (underestimating or overestimating). This suggests that a single stimulus of 100-m should not be the only criteria used to evaluate rowers. To improve the model, it may be necessary to use more physiological variables (i.e., heart rate, cardiorespiratory capacity, lactate threshold, etc.) or non-physiological variables (i.e., body composition, rowing power, etc.) that correlate with the performance of 2000-m.

This, results suggest that future studies seek to improve the mathematical model of the present study, increasing physical fitness tests that are easy to use in the coaches' work environment. Therefore future studies that seek to demonstrate the reliability of the mathematical model of the present study with inclusion of metabolic variables from indirect and direct methods are encouraged.

5. Conclusions

The results of this study show that the predictive equation proposed for performance of 2000 m is moderately reliable and predicts performance within 5% of actual performance. Thus, the equation model presented is low cost, and favors time savings and lower physical wear for athletes. It is necessary that future studies improve the mathematical model in order to provide a tool with lower estimation errors, thus providing an option to evaluate the performance prediction and monitor training using a maximum 100-m indoor rowing test.

Author Contributions: Conceptualization, L.F.d.S., P.F.d.A.-N., P.M.S.D. and B.G.d.A.T.C.; methodology, L.F.d.S.; software, P.F.d.A.-N.; validation, D.G.d.M., V.M.R., F.M.C. and F.J.A.; formal analysis, J.B.d.J.; investigation, L.F.d.S., and P.F.d.A.-N.; resources, D.G.d.M.; data curation, P.F.d.A.-N., and S.E.R.; writing—original draft preparation, L.F.d.S., V.d.Q., and F.J.A., writing—review and editing, B.M., D.G.d.M., P.F.d.A.-N., S.E.R., V.M.R., F.M.C., P.M.S.D., F.J.A. and B.G.d.A.T.C., visualization, B.M., V.d.Q. and B.G.d.A.T.C., supervision, B.M., B.G.d.A.T.C.; project administration, B.M., P.M.S.D. All authors have read and agreed to the published version of the manuscript.

Funding: Victor Machado Reis was funded by FCT Fundação para a Ciência e Tecnologia (UID04045/2020).

Institutional Review Board Statement: This study was approved by the Ethics and Research Committee with Human Beings of the Federal University of Rio Grande do Norte (technical advice: 3.552.010), respecting the national and international ethical principles contained in the declaration of Helsinki.

Informed Consent Statement: Informed consent was obtained from all subjects involved in the study.

Data Availability Statement: The data that support the findings of this study are available from the corresponding author upon reasonable request.

Acknowledgments: For your support and encouragement for the development of this academic article, we thank the Federal University of Rio Grande do Norte (UFRN), the Physical Activity and Health (AFISA) research base, the Child and Adolescent Maturation Research Group (GEPMAC). The National Council for Scientific Development (CNPQ), the Higher Education Personnel Improvement Coordination (CAPES), the Olympic rowing school, Sport Club de Natal and Scientist Gilmara Gomes de Assis (ORCID: 0000-0002-7198-3601).

Conflicts of Interest: The authors declare no conflict of interest.

References

1. Bourdin, M.; Lacour, J.-R.; Imbert, C.; Messonnier, L.A. Factors of Rowing Ergometer Performance in High-Level Female Rowers. *Int. J. Sports Med.* **2017**, *38*, 1023–1028. [CrossRef]
2. Mikulic, P.; Bralic, N. Elite status maintained: A 12-year physiological and performance follow-up of two Olympic champion rowers. *J. Sports Sci.* **2017**, *36*, 660–665. [CrossRef]
3. Akça, F. Prediction of Rowing Ergometer Performance from Functional Anaerobic Power, Strength and Anthropometric Components. *J. Hum. Kinet.* **2014**, *41*, 133–142. [CrossRef]
4. Gee, T.I.; Caplan, N.; Gibbon, K.C.; Howatson, G.; Thompson, K.G. Investigating the Effects of Typical Rowing Strength Training Practices on Strength and Power Development and 2000 m Rowing Performance. *J. Hum. Kinet.* **2016**, *50*, 167–177.
5. Cataldo, A.; Cerasola, D.; Russo, G.; Zangla, D.; Traina, M. Mean power during 20 sec all-out test to predict 2000 m rowing ergometer performance in national level young rowers. *J. Sports Med. Phys. Fit.* **2015**, *55*, 872–877.
6. Riechman, S.E.; Zoeller, R.F.; Balasekaran, G.; Goss, F.L.; Robertson, R.J. Prediction of 2000 m indoor rowing performance using a 30 s sprint and maximal oxygen uptake. *J. Sports Sci.* **2002**, *20*, 681–687. [CrossRef]
7. Cerasola, D.; Bellafiore, M.; Cataldo, A.; Zangla, D.; Bianco, A.; Proia, P.; Traina, M.; Palma, A.; Capranica, L. Predicting the 2000-m Rowing Ergometer Performance from Anthropometric, Maximal Oxygen Uptake and 60-s Mean Power Variables in National Level Young Rowers. *J. Hum. Kinet.* **2020**, *75*, 77–83. [CrossRef] [PubMed]
8. Jensen, K.; Frydkjær, M.; Jensen, N.M.; Bannerholt, L.M.; Gam, S. A Maximal Rowing Ergometer Protocol to Predict Maximal Oxygen Uptake. *Int. J. Sports Physiol. Perform.* **2021**, *16*, 382–386. [CrossRef]
9. Maciejewski, H.; Rahmani, A.; Chorin, F.; Lardy, J.; Giroux, C.; Ratel, S. The 1,500-m Rowing Performance is Highly Dependent on Modified Wingate Anaerobic Test Performance in National-Level Adolescent Rowers. *Pediatr. Exerc. Sci.* **2016**, *28*, 572–579. [CrossRef] [PubMed]
10. Rahmani, A.; Giroux, C.; Ben Abdessamie, A.; Chorin, F.; Lardy, J.; Maciejewski, H. Anaerobic physical evaluation of young national rowers. *Comput. Methods Biomech. Biomed. Eng.* **2015**, *18*, 2034–2035. [CrossRef] [PubMed]
11. Hartmann, U.; Mader, A.; Wasser, K.; Klauer, I. Peak Force, Velocity, and Power During Five and Ten Maximal Rowing Ergometer Strokes by World Class Female and Male Rowers. *Int. J. Sports Med.* **1993**, *14*, S42–S45. [CrossRef]
12. Lawton, T.W.; Cronin, J.B.; McGuigan, M.R. Strength, power, and muscular endurance exercise and elite rowing ergometer per-formance. *J. Strength Cond. Res.* **2013**, *27*, 1928–1935. [CrossRef]
13. Steinacker, J.M. Physiological aspect of training in rowing. *Int. J. Sports Med.* **1993**, *1*, s3–s10.
14. Matsudo, V.K.; Rivet, R.E.; Pereira, M.H. Standard score assessment on physique and performance of Brazilian athletes in a six tiered competitive sports model. *J. Sports Sci.* **1987**, *5*, 49–53. [CrossRef] [PubMed]
15. Harriss, D.; MacSween, A.; Atkinson, G. Ethical Standards in Sport and Exercise Science Research: 2020 Update. *Int. J. Sports Med.* **2019**, *40*, 813–817. [CrossRef] [PubMed]
16. Strobe Foundation. Checklist Strengthening the Reporting of Observational Studies in Epidemiology. (1 September 2014). Retrieved 20 June 2020, from STROBE Statement. Available online: https://www.strobe-statement.org/index.php?id=strobe-home (accessed on 20 January 2021).
17. Ingham, S.A.; Whyte, G.P.; Jones, K.; Nevill, A.M. Determinants of 2000 m rowing ergometer performance in elite rowers. *Eur. J. Appl. Physiol.* **2002**, *88*, 243–246. [PubMed]
18. Wasserman, H.; O'Donnell, J.M.; Gordon, C.M. Use of dual energy X-ray absorptiometry in pediatric patients. *Bone* **2017**, *104*, 84–90. [CrossRef]
19. Segel, L.A. *Mathematical Models in Molecular Cellular Biology*; Cambridge University Press Archive: Cambridge, UK, 1980.
20. Neimark, J.I. *Mathematical Models in Natural Science and Engineering; Gabler*; Springer Science & Business Media: Berlin, Germany, 2012.
21. Paragios, N.; Chen, Y.; Faugeras, O.D. *Handbook of Mathematical Models in Computer Vision*; Springer Science & Business Media: Berlin, Germany, 2006.
22. de Almeida-Neto, P.F.; da Silva, L.F.; de Matos, D.G.; Jeffreys, I.; de Cesário, T.M.; Neto, R.B.; de Almeida Barbosa, W.; Aidar, F.J.; Silva Dantas, P.M. Equation for analyzing the peak power in aquatic environment: An alternative for olympic rowing athletes. *PLoS ONE* **2020**, *15*, e0243157.
23. Schober, P.; Boer, C.; Schwarte, L.A. Correlation Coefficients: Appropriate Use and Interpretation. *Anesth. Analg.* **2018**, *126*, 1763–1768. [CrossRef] [PubMed]
24. Miot, H.A. Agreement analysis in clinical and experimental studies. *J. Vasc. Bras.* **2016**, *15*, 89–92. [CrossRef]
25. Espírito Santo, H.; Daniel, F. Calculating and Reporting Effect Sizes on Scientific Papers (1): P < 0.05 Limitations in the Analysis of Mean Differences of Two Groups. *Port. J. Behav. Soc. Res.* **2017**, *1*, 3–16.
26. Perini, T.A.; Oliveira, G.L.; Ornellas, J.S.; Oliveira, F.P. Calculation of the technical error of examination in anthropometry. *Braz. J. Sports Med.* **2005**, *11*, 81–85.
27. Šmída, M.; Clementis, M.; Hamar, D.; Macejková, Y. Relation between Maximal Anaerobic Power Output and Tests on Rowing Ergometer. *Acta Fac. Educ. Phys. Univ. Comen.* **2017**, *57*, 68–75. [CrossRef]
28. Maestu, J.; Jiirimae, J.; Jiirimae, T. Monitoring of performance and training in rowing. *Sports Med.* **2005**, *35*, 597–617. [CrossRef]
29. Secher, N.H. Rowing. In *Endurance in Sport*; Shephard, R.J., Astrand, P.O., Eds.; Blackwell Science: Oxford, UK, 2000; pp. 836–843.

30. Steinacker, J.M.; Lormes, W.; Lehmann, M.; Altenburg, D. Training of rowers before world championships. *Med. Sci. Sports Exerc.* **1998**, *30*, 1158–1163. [CrossRef]
31. Penichet-Tomás, A.; Pueo, B.; Jiménez-Olmedo, J.M. Physical performance indicators in traditional rowing championships. *J. Sports Med. Phys. Fit.* **2019**, *59*, 767–773. [CrossRef] [PubMed]
32. Billat, V.; Hamard, L.; Koralsztein, J.P.; Morton, R.H. Differential modeling of anaerobic and aerobic metabolism in the 800-m and 1500 m run. *J. Appl. Physiol.* **2009**, *107*, 478–487. [CrossRef] [PubMed]
33. Atkinson, G.; Nevill, A. Statistical Methods for Assessing Measurement Error (Reliability) in Variables Relevant to Sports Medicine. *Sports Med.* **1998**, *26*, 217–238. [CrossRef]
34. Russell, A.P.; Le Rossignol, P.F.; Sparrow, W.A. Prediction of elite schoolboy 2000-m rowing ergometer performance from metabolic, anthropometric and strength variables. *J. Sports Sci.* **1998**, *16*, 749–754. [CrossRef] [PubMed]

biology

MDPI

Review

The Role of Satellite Cells in Skeletal Muscle Regeneration—The Effect of Exercise and Age

Agnieszka Kaczmarek [1,*], Mateusz Kaczmarek [2], Maria Ciałowicz [3], Filipe Manuel Clemente [4], Paweł Wolański [1,5], Georgian Badicu [6] and Eugenia Murawska-Ciałowicz [1,*]

1 Department of Physiology and Biochemistry, University School of Physical Education, 51-612 Wroclaw, Poland; pawel.wolanski@onet.pl
2 Gynecology and Obstetrics Department, St. Hedwig's of Silesia Hospital, 55-100 Trzebnica, Poland; mateusz.kaczmarek@yahoo.com
3 Physiotherapy Faculty, University School of Physical Education, 51-612 Wroclaw, Poland; marysia.cialowicz@gmail.com
4 Escola Superior Desporto e Lazer, Instituto Politécnico de Viana do Castelo, Rua Escola Industrial e Comercial de Nun'Álvares, 4900-347 Viana do Castelo, Portugal; Filipe.clemente5@gmail.com
5 Department of Physiology, Gdansk University of Physical Education and Sport, 80-336 Gdansk, Poland
6 Department of Physical Education and Special Motricity, Transilvania University of Brasov, 500086 Brasov, Romania; georgian.badicu@unitbv.ro
* Correspondence: agnieszka.kaczmarek@awf.wroc.pl (A.K.); eugenia.murawska-cialowicz@awf.wroc.pl (E.M.-C.)

Simple Summary: Studies describing the effects of various forms of exercise and age on muscle regeneration were reviewed. Satellite cells are a heterogeneous group of cells that includes stem cells and skeletal muscle progenitor cells. Each skeletal muscle fiber has its own pool of satellite cells that remain inactive until the muscle is damaged. Minor damage within the cell membrane of muscle fibers is patched by fusing intracellular vesicles with the damaged sarcolemma. More severe muscle damage initiates a multistep regeneration process in which satellite cells play an essential role. The condition that initiates the cascade of reactions is the formation of inflammation at the structural discontinuity site, resulting in satellite cell activation. The multitude of reactions and pathways occurring during this process means that many different substances are involved in it and control it. Not all of them are well-understood yet. In parallel, the body's own population of satellite cells is being rebuilt so that more fibers can be regenerated in the future. Athletes and the elderly are primarily at risk for muscle damage, and pathologies in muscle fiber regeneration cause serious diseases.

Abstract: The population of satellite cells (mSCs) is highly diversified. The cells comprising it differ in their ability to regenerate their own population and differentiate, as well as in the properties they exhibit. The heterogeneity of this group of cells is evidenced by multiple differentiating markers that enable their recognition, classification, labeling, and characterization. One of the main tasks of satellite cells is skeletal muscle regeneration. Myofibers are often damaged during vigorous exercise in people who participate in sports activities. The number of satellite cells and the speed of the regeneration processes that depend on them affect the time structure of an athlete's training. This process depends on inflammatory cells. The multitude of reactions and pathways that occur during the regeneration process results in the participation and control of many factors that are activated and secreted during muscle fiber damage and at different stages of its regeneration. However, not all of them are well understood yet. This paper presents the current state of knowledge on satellite cell-dependent skeletal muscle regeneration. Studies describing the effects of various forms of exercise and age on this process were reviewed.

Keywords: satellite cells; muscle regeneration; myogenic regulatory factors; inflammation; exercise; age

Citation: Kaczmarek, A.; Kaczmarek, M.; Ciałowicz, M.; Clemente, F.M.; Wolański, P.; Badicu, G.; Murawska-Ciałowicz, E. The Role of Satellite Cells in Skeletal Muscle Regeneration—The Effect of Exercise and Age. *Biology* **2021**, *10*, 1056. https://doi.org/10.3390/biology10101056

Academic Editor: Brian McDonagh

Received: 27 September 2021
Accepted: 12 October 2021
Published: 18 October 2021

1. History of Studies on Satellite Cells

The invention of the electron microscope in the 1930s sparked a revolutionary advance in research. The same was true for studies on mammalian muscle tissues. Specifically, satellite cells were first reported in the second half of the 20th century. In 1961, Alexander Mauro, a Rockefeller Institute scientist who studied frogs' tibial muscles under an electron microscope, published a paper demonstrating the existence of mononuclear cells located peripherally between the sarcolemma and the basement membrane. He called them satellite cells [1]. That same year, in London, Bernard Katz also observed similar-looking cells during his studies on afferent nerve fiber endings in frog muscles [2]. Subsequent scientists tried to find out more about the structure, genesis, and role of these newly discovered cells. In 1966, Ishikawa made an unsuccessful attempt to describe the mSC structure [3]. Three years later, Kelly and Zacks showed that Ishikawa had confused satellite cells with connective tissue cells by calling them "fibroblast-like." In the same paper, they described the histogenesis of the muscle tissues of rats' intercostal muscles as an example [4].

In the following years, by labeling the tibialis anterior muscle cells of 14- to 17-day-old rats with a radioactive nucleoside (3H-thymidine), Moss and Leblond (1971) observed that the nuclei of multinucleated myofibers were derived from mSCs that were capable of mitotic divisions [5]. Observations of the mouse upper limb lumbrical cells made seven and 30 days after birth led to the conclusion that fewer and fewer satellite cells undergo fusion with muscle fibers as mice age. The rate of division of these cells did not decrease until the mice were three weeks old. In contrast, after the third week, mSC proliferation ceased, and they entered a quiescent state [6]. As early as during the discovery of satellite cells, Mauro postulated that they could participate in skeletal muscle regeneration [1]. This was eventually confirmed in 1975 by studies using phase-contrast microscopy [7].

In 1986, it was discovered that muscle injury causes satellite cells to exit the G0 phase and resume cell division [8]. Following these studies, researchers searched for factors that affect differentiation and the activation and deactivation of mitotic divisions of muscle satellite cells. One of the earliest agents studied was transforming growth factor-β (TGFβ). Studies have investigated its inhibitory effect on mSC differentiation [9]. The course of myogenesis in mammals has been established by observing mouse embryogenesis. During these observations, it was found that the first mSCs appeared under the basement membrane of the muscle on day 17 of fetal life [4].

Throughout the 1990s, many published studies described the role of various proteins in myogenesis, including myogenic regulatory factors (MRFs) and Pax7 in mSCs [10,11]. A 1990 paper by Bischoff addressed how the position of mSCs in the muscle affects their sensitivity to myogenic factors. It was shown that the niche in which satellite cells reside has an important influence on their performance [12].

In 2005, single myofibers with adherent mSCs were transplanted into irradiated mice that could not regenerate muscle. In this study, satellite cells were shown to have stem cell characteristics in that they were capable of regenerating their own population. Subsequent experiments helped understand the mechanism of mSC populations' self-renewal process (asymmetric division) [13].

2. Pax7 Characterizes Inactive mSCs

Pax7 transcription factor belongs to a group of nine proteins encoded by *Pax* genes whose name comes from the DNA-binding domain called *"paired."* These proteins have important functions during organogenesis, cell division, and differentiation. Over the centuries, the structures of these proteins have changed little, and their homologs are found in the proteomes of numerous animals, from simple organisms, such as nematodes, to insects, amphibians, fish, birds, or mammals [14]. The Pax7 protein is found in the cell nuclei of inactive satellite cells and in all human mSCs. This protein is responsible for regulating the division and differentiation of these cells [15]. Its effects include influencing myogenic determination protein 1 (MyoD1) and myogenic factor 5 (Myf5), which belong to MRFs [16]. The expression of the gene encoding this protein decreases in activated satellite cells [15].

In a study in mice that lacked the operational Pax7 gene (Pax7$^{-/-}$), this factor was shown to maintain satellite cell populations after birth [16]. Although no significant differences in muscle thickness or appearance were observed in Pax7$^{-/-}$ and wild-type mice, only a small percentage of the Pax7$^{-/-}$-type mice reached adulthood. Survivors showed impaired growth and significant muscle tissue loss caused by a lack of functional mSCs [17].

3. Inflammatory Processes Involved in Muscle Regeneration

Mature skeletal muscle cells are characterized by high stability and are not subject to mitotic divisions [18]. Minor damage within the cell membranes of muscle fibers is patched by the fusion of intracellular vesicles with damaged sarcolemma. Caveolin 3, dysferlin, and one of two calpains (m or μ) are involved in this process. [19,20]. Severe muscle damage initiates a multistep regeneration process in which satellite cells play an essential role [21] (Figure 1).

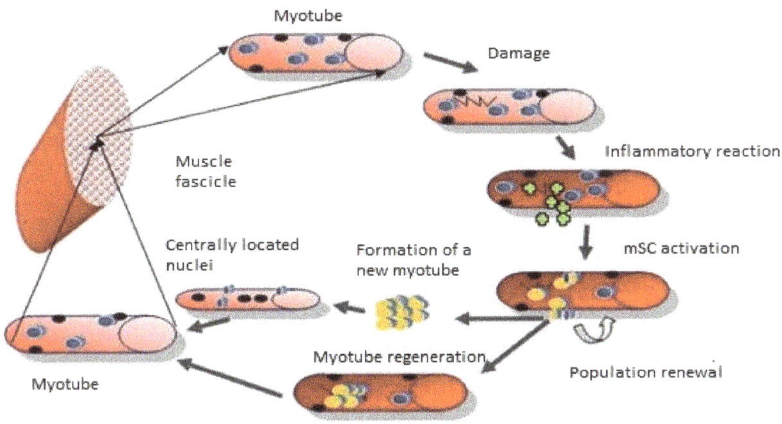

Figure 1. Schematic representation of the muscle fiber regeneration stages involving satellite cells. —quiescent mSC; —nuclei; —active mSCs; —myofiber damage; —inflammatory cells.

Skeletal muscle fibers can be damaged by denervation, ischemia, or mechanical damage caused by (for example) the effect of high temperature. Intensive training, exposure to toxins, and genetic mutations that cause degenerative muscle diseases (including muscular dystrophies) may also cause this kind of damage [22]. Due to the disruption of the cell membrane of the muscle fiber, the concentration of Ca^{2+} ions increases rapidly inside the muscle fiber. This, in turn, activates proteolytic enzymes (calpain-calcium-dependent, non-lysosomal cysteine proteases) that digest the structural protein molecules of the damaged fibers. From damaged cells, the molecules of compounds normally present in the sarcoplasm enter the bloodstream. These compounds include creatine kinase, myosin heavy chain, lactate dehydrogenase, troponins, myoglobin, and beta-galactosidase [23].

Necrosis occurring in damaged muscle stimulates inflammatory processes—for example, the cascade activation of the complement system and an increase in the synthesis of proinflammatory cytokines [24]. Destroyed cells are phagocytosed by neutrophils and macrophages that migrate to the damage site [22,23]. The entire process begins with the stimulation of selectin family protein synthesis by tumor necrosis factor-α (TNF-α) and interleukin-1β (IL-1β). These substances promote the influx of neutrophils to the injury site, which takes 1–6 h [25]. The chemokines interleukin-6 (IL-6) and interleukin-8 (IL-8) also contribute.

Furthermore, neutrophil granulocytes secrete chemoattractants, resulting in the accumulation of two macrophage subpopulations at the injured muscle site [26]. The first of these is characterized by proinflammatory properties due to the secreted cytokines (TNF-α, IL-1β). The number of proinflammatory macrophages peaks approximately 24 h after fiber injury.

A defining feature of proinflammatory macrophages is the expression of the CD68 membrane protein, and their function is to remove the debris remaining after cell necrosis. The second subpopulation of macrophages is the anti-inflammatory group, which reaches its maximum concentration within two to four days after injury. Their properties are due to the secretion of cytokines that inhibit inflammatory processes—for example, interleukin-10 (IL-10). Unlike proinflammatory macrophages, they synthesize the CD163 protein (which is their marker) instead of the CD68 protein. The function of these macrophages is to protect newly created structures from proteolytic agents. In addition, macrophages stimulate satellite cell division [27]. Importantly, when the immune response is not triggered, satellite cells do not appear to be involved in regeneration [28].

4. Presence of MyoD1 and Myf5 Is Characteristic of Proliferating Satellite Cells

Satellite cells start dividing once anti-inflammatory scavenger cells appear in the muscle (i.e., on the second day after the injury occurs). Activation includes the satellite cells in the immediate vicinity of the resulting damage, as well as all cells adjacent to the damaged fiber. Activated mSCs migrate to the damage site, where the regeneration process continues [29]. The membranes of quiescent satellite cells contain the CD34 glycoprotein, which belongs to a family of adhesion proteins, the sialomucins. Immediately after satellite cell activation, there is a sharp decrease in CD34 production, allowing the molecules to adhere to each other less strongly and move more easily [30]. Researchers have observed that satellite cells can move not only within the same fiber but also between fibers and even between muscles. Such movement allows these cells to overcome basement membrane and connective tissue barriers [31].

Satellite cells undergoing division—called myogenic precursor cells (MPCs)—produce MRFs [21]. MRFs are characteristic of mSCs that have awakened from their quiescent state, and the expression of these factors occurs in an ordered sequence. In one study, MyoD1 and Myf5 were the first MRFs observed during the initial phase of mSC activation and division. In-depth analyses of single cells on the first day after injury show that some satellite cells begin to synthesize MyoD1 first, while others synthesize Myf5 first. On the other hand, the simultaneous expression of both factors occurs on the following day [10]. The appearance of both MRFs is necessary for satellite cells to exit the quiescent state. Cells in which no further MRFs are produced—and in which MyoD1 and Myf5 levels are decreased—revert to the pool of quiescent cells [32]. It is believed that MyoD1 expression determines MPC differentiation [33].

Megeney et al. [34] showed that cell division is not impaired in MyoD1$^{-/-}$ mutant mice, though an accumulation of myoblasts was observed at the muscle injury site. However, these myoblasts do not differentiate into mature muscle cells. After examining them under an electron microscope, the authors did not report any abnormalities in their appearance [34]. However, Sabourin, et al. [35] demonstrated a different appearance of in vitro cultured mouse MyoD1$^{-/-}$ cells. Compared to MyoD1$^{+/+}$ cells, which were rounded, the mutant cells were flat and star-shaped [35]. In another study, another team of researchers showed that MyoD1$^{-/-}$ muscles have branched muscle fibers, indicating an abnormal regeneration process within these myofibers [36]. These abnormalities may result from an imbalance between the satellite cells that stop proliferating, returning to the G0 phase of the cell cycle, and those that rapidly divide [33–35].

Asakura et al. [37] made an interesting discovery after injecting satellite cells into the tibialis anterior muscles of mice from the SCID/beige model that had been damaged by cardiotoxin (CTX). A significantly higher number of transplanted MyoD1$^{-/-}$ cells survived in the regenerating muscle compared with the wild type. The difference was threefold after

24 h. Not only were there more of them, but they easily formed multinucleated myofibers and replenished the pool of quiescent satellite cells. The results of this experiment suggest that MyoD1 may regulate apoptosis, as a vast number of transplanted wild-type myofibers underwent programmed death during differentiation. The researchers tested this by subjecting the cells of both lines to UV radiation, which damages DNA. More apoptotic cells were shown in the wild-type MyoD1$^{+/+}$ line than in fibers lacking this factor [37].

The characteristic phenotype of mice lacking Myf5 is muscle fiber hypertrophy, and proliferation is impaired in Myf5$^{-/-}$ myoblasts. In contrast, Myf5 deletion does not affect the initial abundance of satellite cells. It is likely that the presence of Myf5 in the mSC determines which cell type it will differentiate into. In vivo studies have revealed that mSCs lacking Myf5 are more likely to differentiate into fibroblasts or adipocytes than into myofibers [38]. Additionally, it is postulated that Myf5 promotes the restoration of its own cell population by inhibiting the expression of MyoD1, thereby inhibiting further differentiation [31].

Previous studies suggest a distinct role of MyoD1 and Myf5 in muscle regeneration. MyoD1 has an overarching function in the initiation of cell differentiation, whereas Myf5 is involved in myoblast proliferation. The functions of these factors in adult myoblasts overlap with their roles during embryogenesis [39,40]. It can be speculated that the determination of muscle precursor cell development depends on whether MyoD1 or Myf5 expression is predominant. As an example of the dominance of MyoD1 factor expression, Myf5$^{-/-}$ myoblasts showed early cell differentiation [21]. In contrast, the behavior of MyoD1$^{-/-}$ myoblasts—in which increased proliferation and differentiation occurred with a significant delay—may serve as an example of a program path that is presumably followed by cells with Myf5 overexpression [32,33].

The expression of a protein that is crucial for cell differentiation—and, therefore, the regeneration of the MyoD1 proteins—is strictly controlled by the serum response factor (SRF). It binds to the DNA sequence recognized by SRF (SRE), which is located in the promoter of the myoD1 gene [41,42]. In proliferating myoblasts, low levels of MyoD1 are maintained by specific cyclin-induced reactions between cyclin-dependent kinase-4 and MyoD1 [43]. These compounds lead to the phosphorylation and subsequent degradation of the MyoD1 factor [44]. A decrease in the expression of genes, inducing cell division, results in the expression of myogenesis-enhancing factor-2 (MEF-2). As it competes with SRF for a binding site to the SRE sequence, it promotes MyoD1 expression and makes the cells enter the differentiation pathway [42]. By binding to gene promoters, MyoD1 facilitates the transcription of proteins that are characteristic of striated muscle tissue, such as M-cadherin. It also causes changes in the spatial arrangement of chromatin, thus allowing transcription proteins to subsequently bind to it [45].

5. Myogenin and Myf6 have Essential Roles in mSCs Differentiation

Early in cellular differentiation, MyoD1 initiates the formation of another MRF: myogenin. This occurs through the interaction of MyoD1 with enzymes that acetylate and deacetylate the gene promoter of this protein [46,47], and this process is controlled by Myf5 and MEF2 [31,47]. Myogenin expression is responsible for Pax7 inactivation in differentiating myoblasts [48]. This factor was also shown to enhance the formation of muscle cell-specific proteins whose expression was initiated by MyoD1 [49].

Myogenin is the only one among the four MRFs whose absence in fetal life leads to severe abnormalities in the development of muscle tissue, resulting in death soon after birth. The conclusion is that, just as the other factors compensate for each other's effects, myogenin cannot be replaced by any of them. This situation changes immediately after birth. When gene encoding becomes inactive, this protein does not significantly affect the function of either myofibers or satellite cells. As a result, its role is taken over by other MRFs [50].

The last of the MRFs synthesized during differentiation is myogenic factor-6 (Myf6), the name of which is used interchangeably with muscle regulatory factor-4 (MRF4). Similar to myogenin, this MRF promotes cell division. Myf6 expression is regulated by MyoD1 and Myf5, and it is characteristic of terminal differentiation stages [36,46].

6. Fusion Is the Final Stage of Muscle Fiber Regeneration

MPC fusion is the final stage of muscle fiber regeneration following muscle damage. These cells either integrate with damaged fibers or fuse to form new syncytia (Figure 1). The latter process occurs in two stages. First, several myoblasts fuse to form a small *de novo* filament. Then, during maturation, more cells are recruited to the newly formed fiber, increasing the diameter of the spindle; this process is accompanied by increased contractile protein expression [21]. This process is complex and requires the participation of numerous molecules capable of reorganizing the structure of the intracellular cytoskeleton.

Over the years, it has been shown that the myoblast fusion process cannot take place without the involvement of transmembrane proteins responsible for cell-cadherin adhesion. Their action, in turn, is dependent on Ca^{2+} ions. Among the members of this family, M-cadherin plays the most important role in the formation of multinucleated myotubes during both embryonic myogenesis and regeneration [51,52]. This has been confirmed by numerous in vivo and in vitro studies.

This protein is found in mice and humans in most myofibers and satellite cells at different development stages. Muscle damage increases M-cadherin expression [52,53]. A significant reduction in M-cadherin expression was observed in injury-activated MyoD1$^{-/-}$ satellite cells that could not integrate into fibers [33,34]. However, this protein is not irreplaceable, as the authors of the study in mice with disabled M-cadherin expression demonstrated. No abnormalities in muscle recovery after CTX injection were observed in these animals. The role of cadherin has been most likely compensated by other proteins from this family [54].

M-catherin is not the only adhesive protein responsible for the interaction of myoblasts. The most important adhesion proteins involved in the muscle fiber regeneration (without which the fusion of newly formed cells with existing fibers or with each other would not take place) are CD9, CD81, alpha-3, alpha-7, alpha-10, and beta-1 integrins, as well as vascular cell adhesion molekule-1 (VCAM-1) [55]. It is also likely that M-calpain, a Ca^{2+} ion-dependent proteinase, contributes to MPC fusion by reorganizing the cytoskeleton [56]. However, the mechanism of this process is still unknown.

One of the substrates for calpains is desmin, and mice with the desmin$^{-/-}$ phenotype have been shown to have impaired regeneration and delayed cell fusion [57]. It has also been shown that satellite cells associated with the fiber differentiate into several myoblasts (equal to the number of nuclei of each satellite cell) within four to five days of fiber injury [58]. In mice, the first regenerated or newly formed muscle fibers are observed approximately five to seven days after injury [59]. These fibers are typically thinner than others, and their cell nuclei are centrally located. This changes during myofiber maturation as they move to the peripheral parts of the myotome [60].

In addition to muscle and satellite cells, connective tissue, an integral part of the muscle, is involved in mSC-dependent skeletal muscle regeneration. More specifically, these are mesenchymal stem cell-derived fibroblasts (collagen synthesis, organization of extracellular structures), adipocytes (which replace myofibers in some muscle diseases and during aging), and adipocyte and fibroblast progenitor cells [61].

7. Satellite Cell Self-Renewal

Satellite cells can divide in three different ways. First, symmetric division may result in two undifferentiated cells with stem cell characteristics. Second, two differentiating cells may be formed. Third, asymmetric division results in the formation of one undifferentiated and one differentiating cell [62]. The ability to reconstitute their own populations via undifferentiated progenitor cells is a distinctive feature of satellite cells. Presumably, maternal DNA strands are responsible for determining the fate of newly formed cells. One of them provides immortality [63].

To revise this hypothesis, Conboy et al. [64] damaged mouse muscles and injected them with a thymidine analog compound two days later, which the cells used during replication. Before the next cell division, they repeated this activity using a different thymidine analog. By analyzing mSCs extracted from muscle prepared with this method, the researchers found that selective inheritance of maternal DNA strands occurred in some of them. In addition, they observed that in most cases, the parental strands were found in cells that reproduced the mSC population [64].

The mechanism that controls the proliferation of mSCs is the Notch protein signaling pathway (Figure 2). It is activated by Delta and Jagged ligands. Their interaction causes the Notch intracellular domain (NICD) to detach from the participation of γ-secretase. NICD enters the cell nucleus, where it induces the transcription of target genes with the participation of other factors. The Numb protein, a Notch receptor antagonist, is one of the compounds unevenly distributed between progenitor cells during asymmetric division, which may determine their fate. That cell exhibits a high concentration of Numb ligand and inhibits the Notch pathway, causing it to increase Myf5 factor and desmin levels and differentiate. The second Numb-depleted cell recreates a population of inactive satellite cells.

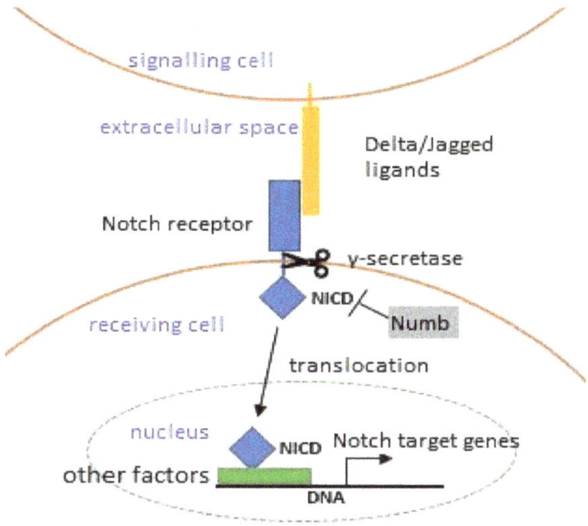

Figure 2. Notch signaling pathway (The description is in the text).

This was demonstrated in an ex vivo cell culture that contained myofibers and mSCs [65]. Studies on mice in which green fluorescence protein (GFP) synthesis depended on *Pax7* gene expression confirmed the effect of the Notch signaling pathway on its future. Notch pathway activity was high in cells that contained abundant GFP (Pax7+). GFP (Pax7-) detection decreased as Notch pathway activity decreased, whereas MyoD1 and myogenin expression increased as cells differentiated [66].

Whether a cell replenishes the mSC pool or undergoes differentiation is influenced by the division plane of the maternal cell. This has been observed in muscle cell cultures with mSCs attached to them. A progeny cell composition analysis for Myf5 factor revealed that one in ten Pax^+ cells lacked Myf5 expression ($Pax^+/Myf5^-$). The majority of the remaining cells were $Pax^+/Myf5^+$ mSCs. Cells lacking Myf5 divide symmetrically and asymmetrically, but they do not differentiate and renew the mSC population. Furthermore, they do not lose contact with the basement membrane of the myofiber (Figure 3).

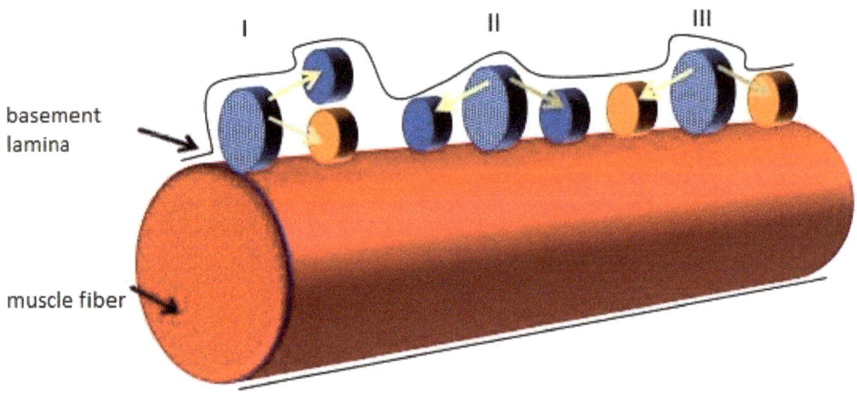

Figure 3. Renewing satellite cell populations. Blue cells—Pax7+/Myf5−; orange cells—$Pax7^+/Myf5^+$ cells; I—asymmetric division resulting in one differentiating cell ($Pax7^+/Myf5^+$) (orange cell), and another renewing a population of undifferentiated cells ($Pax7^+/Myf5^-$) (blue cell); II—symmetrical division resulting in two cells renewing the population of undifferentiated cells (2x $Pax7^+/Myf5^-$); III—symmetrical division resulting in two differentiating cells (2x $Pax7^+/Myf5^+$).

The researchers decided to use $Pax7^+/Myf5^+$ and $Pax7^+/Myf5^-$ cells in another experiment to test their biological properties. Each was separately transplanted into $Pax7^{-/-}$ mice that were mSC−deficient due to impaired proliferation. Between 20 and 60 newly created fibers were formed when $Myf5^+$ mSCs were administered, which is significantly more than in muscles into which $Myf5^-$ mSCs were transplanted. In this case, the number of new fibers did not exceed 20. Administered $Pax7^+/Myf5^+$ were far more likely to differentiate than to participate in niche cell restoration. Additionally, when comparing $Pax7^+/Myf5^-$ cells with $Pax7^+/Myf5^+$, only the former could move in the muscle [13].

The Pax7 protein is also vital to the process of self-renewal; its high level is maintained both in inactive cells and proliferating ones, after which its expression decreases under the influence of myogenin, which is characteristic of differentiating cells [67]. Among the cells synthesizing the Pax7 protein and the MyoD1 factor, some did not reach the next stage (i.e., differentiation). A closer look using a thymidine analog revealed that the $Pax7/MyoD1^-$ cells were the same cells that had normally expressed MyoD1 previously. MyoD1 activity decreased in these cells, and subsequent MRFs specific for cellular differentiation were not synthesized.

The researchers assumed that this cell lineage is responsible for self-renewal [67–69]. The same conclusions were reached by other authors, who additionally observed changes in the expression of nestin—a protein whose high level, as with Pax7, persists in inactive cells and those undergoing division, while decreasing during differentiation. Some researchers used transgenic mice in which GFP synthesis was coupled to nestin synthesis for this experiment. They found that nestin and Pax7 were present in population-renewing mSCs [40,68].

8. Satellite Cells and Physical Activity

After intense exercise—specifically, exercise that causes eccentric stretching of the muscle fibers—people may experience pain, increased tension, and limited mobility in the area of the exercised muscle. This phenomenon is referred to as delayed-onset muscle soreness syndrome (DOMS) [70]. Pain discomfort usually appears 12 to 48 h after exercise and reaches its maximum on the second or third day, after which it gradually subsides until the pain is completely gone. This lasts anywhere from a few days to a week [71]. Micro-damage occurs in the muscles due to intense exercise, constituting a signal to start the fiber regeneration process, which depends on satellite cells. The timing of pain coincides with inflammatory processes, which are the first stage of regeneration. The influx of neutrophils and (subsequently) macrophages to the injury site—along with the presence of inflammatory mediators—increases the sensitivity of neural tissue (specifically, nociceptors), which is thought to be the cause of pain [72].

The mitotic activity of mSCs can be stimulated by exercise in the form of strength (resistance) or endurance training (e.g., running). The changes in mSC activity and muscle appearance observed in subjects during a cycle of progressive running training indicate alternating fiber damage with regenerative processes [73]. This is reflected in the increased number of mSCs [74]. In an experiment in which subjects underwent regular exercise, an increase in mSCs was demonstrated as early as four days after the first exercise unit. Elevated levels of these mSCs were maintained in skeletal muscles throughout the entire training period. mSC levels started declining after completion of the exercises [75].

Human trials have demonstrated similarities and differences in the relationship between the type of exercise performed and the number of satellite cells. Therefore, clear conclusions cannot be drawn about these trials.

Resistance training is considered the primary method for increasing muscle mass. Nederveen et al. [76] analyzed the effects of a single bout of exercise and 16 weeks of resistance training on satellite cell activation. Satellite cell activity was detected when Pax7 and 4′, 6-diamidino-2-phenylindole (DAPI) (a marker used to stain cell nuclei) and/or Pax7, MyoD, and DAPI were observed together, as visualized by immunofluorescence.

An examination of the MyoD protein's presence showed that, when tested before the 16-week training regimen, the number of cells with an activated MyoD gene ($Pax7^+/MyoD^+$) increased significantly 24 h after a single strength training session. In contrast, at the end of the 16-week experiment, significant differences were obtained after 24 and 72 h had passed. The increase was greater in the second case, suggesting that the organism had adapted to chronic exercise training, and indicating improved muscle regeneration processes due to a cycle of resistance training. The same study also found a positive effect from the training cycle on muscle capillarization. This improves the availability of ingredients necessary for muscle regeneration with the participation of mSC, thus improving the overall process [76].

The problems associated with satellite cell activity's dependence on the type of exercise were also analyzed by Hyldahl et al. [77]. However, this study focused on comparing the effects of concentric and eccentric contraction on mSCs. The number of satellite cells increased in muscles subjected to an eccentric workout, but not in those subjected to a concentric workout.

A similar dependence on the type of contraction was observed in several parameters related to muscle damage, which were taken to be the expression of Xin (a skeletal muscle-specific protein whose concentration increases in proportion to the degree of muscle damage) and the disruption of extracellular matrix adhesion. The authors hypothesized that muscle injury-related changes in the extracellular matrix might regulate satellite cell activation and proliferation [77].

Differences in eccentric and concentric training were also studied by Farup et al. [78]. However, they broadened the issue to include protein supplementation. Twenty-two young men were divided into two groups of 11 participants each. Each followed the same resistance training program for 12 weeks, with each participant performing a concentric workout with one leg and an eccentric workout with the other (legs were chosen randomly to avoid the possible influence of single-leg dominance). The differences between the groups consisted of different dietary supplementation—the study group received 19.5 g of whey protein in combination with 19.5 g of carbohydrates, while the control group received a calorically equivalent placebo (39 g of carbohydrates).

Concentric training induced a significant increase in the number of satellite cells in the test and control groups and in both types of muscle fibers. Eccentric training induced an increase in mSCs in type I fibers in both groups; no significant change occurred in the number of mSCs in type II fibers, regardless of supplementation. There was also an increase in the ratio of mSCs to muscle cell nuclei in the group receiving protein supplementation, which was not the case in the control group.

The number of nuclei increased in type I fibers in all studied groups when the densities of cell nuclei were compared among different types of muscle fibers. Interestingly, concerning type II fibers, the group taking protein supplementation experienced an increase in the number of nuclei only for those who performed concentric training. Conversely, the increase occurred only in the group performing eccentric training in the placebo group.

These results indicate a clear dependence of mSC activity and cell nuclei accumulation on both types of resistance training and supplementation. This is especially true for type II fibers, as changes in type I fibers occurred regardless of the intervention type [78].

The reciprocal effect of different forms of exercise is also not insignificant. Babcock et al. [79] conducted a study in which a group of eight young men was subjected to two types of training: resistance training and mixed training (resistance training combined with aerobic exercise). Resistance training increased the number of satellite cells in the muscle fibers significantly more than mixed training. It was observed that aerobic exercise co-occurring in the training cycle with resistance exercises lessened the increase in the number of mSC in relation to the response after the resistance training itself. Therefore, the authors suggested that adding aerobic exercise to resistance training may reduce the muscle gains expected after resistance exercise [79].

Also, the ages of the people undergoing training can influence the results. Snijders et al. [80] studied the effects of a single resistance training session on satellite cell numbers and activation in relation to age. The number of satellite cells in type II fibers, relative to baseline values, in young and elderly subjects changed significantly 48 and 72 h after exercise, respectively. Myostatin content in both fiber types decreased 12, 24, and 48 h after exercise. However, at one and two days, the decrease was significantly higher among the younger subjects. The response of satellite cells to microdamage associated with completed muscle work was delayed with age. However, it was not impossible, confirming the presence of satellite cell activity in adult muscles and its potential involvement in muscle hypertrophy [80].

In an effort to compare the effects of resistance and endurance training on human skeletal muscles, Verney et al. [81] conducted a study involving elderly participants. Their upper bodies were subjected to resistance training, and their lower bodies were subjected to endurance training. The entire experiment lasted 14 weeks. Similar increases were observed in the number of mSCs in the deltoid muscle and the vastus lateralis muscle. However, as there was no repeatability in this experiment (because different muscles were tested), it cannot be assumed that the outcome was affected solely by the type of training.

A similar study in rodents also showed that changes in satellite cell numbers did not depend on the type of training, as both endurance and strength training increased the mSC pool [82]. The regenerative potential of the exercised muscle increased along with this pool. However, this does not always translate directly into muscle mass gains. Resistance training increases such gains [72,83], while endurance training can sometimes decrease them [84]. This correlates with a change in the number of muscle fiber nuclei [81,82], meaning that whether muscle mass increases or decreases does not depend entirely on the number of mSCs in the muscle.

Based on the effect of workout type on muscle mass gain and the problem of the occurrence of sarcopenia in the elderly, proper form and the reciprocal relationship between resistance and endurance training must be considered when selecting exercises for them. Inappropriately designed endurance exercise can have an effect opposite to the desired effect [85].

Shefer et al. [86] obtained different results when they studied the effects of endurance training on rats of different ages and both sexes. The number of satellite cells was higher in young animals of both sexes than in old animals (exercise and non-exercise groups, respectively). There was also an apparent sex difference, as young and elderly males had more satellite cells than the corresponding female age groups. More satellite cells were observed in animals subjected to exercise in each of the studied groups, with the increase being the greatest in young males [86]. It seems that a moderate intensity of endurance training is essential to increasing the number of mSCs.

The study by Murach et al. [87] showed that the impact of resistance training on the mSC population depended on the muscle fitness level. The number of satellite cells after resistance training in the untrained muscle increased at a much higher rate than after a cycle of endurance exercise. This demonstrates the evident influence of the cellular environment on the proliferative activity of satellite cells. It is also noteworthy that the reserve of inactive satellite cells was significantly higher after the 12-week endurance training cycle than before it. The authors hypothesized that resistance training programs presented a greater challenge for the untrained muscle at the cellular level. In a muscle accustomed to concentric endurance exercise, an increased reserve of satellite cells correlates with relatively little change in proliferative activity [87].

In addition to the type of exercise and age of the patients, the activity and number of stem cells depend on the broader environment. It appears that the type of muscle fiber may affect the number of activated mSCs that occur with it, although previous studies have not demonstrated conclusive results. Experiments in rodents comparing the satellite cell content of muscles predominantly composed of fast-contracting fibers with muscles composed mostly of slow-contracting fibers showed that type I fibers have more mSCs in untrained muscle than type II myofibers [82,86]. In contrast, no difference was found between type I and type II fibers when mSC content was tested in the vastus lateralis muscles of young, untrained humans [88–90]. However, after exercise, the number of mSCs increases significantly more in type II myofibers than in type I [82,91] owing to the observed hypertrophy of type II fibers after resistance training [85].

Cermak et al. [92] analyzed changes in the numbers of satellite cells before and after exercise in relation to individual types of muscle fibers in a group of young men practicing sports recreationally. There were no statistically significant differences in satellite cell content or activity in the type I and type II fibers in the biopsy taken before training. In the specimens obtained 24 h after training (when considering the mixed muscle type), there were no significant changes in the number of cell nuclei, the number of satellite cells, or their activity. However, when considering these values in the context of individual fiber types, the number and activity of satellite cells increased significantly in type II fibers (Delta like non-canonical Notch ligand 1 (DLK1) expression was taken as its determinant). Meanwhile, the values did not significantly change in type I fibers [92].

The number and function of mSCs also depend on other non-muscle cells. Mackey et al. [93] studied the interaction between satellite cells and fibroblasts in muscles damaged by electrostimulation-induced contractions. The numbers of fibroblasts and satellite cells increased at successive time intervals, and their mutual ratio changed. It was approximately 1.8 in favor of fibroblasts in the control group, increasing to 2.7 after 30 days. High levels of satellite cells after 30 days corresponded with a significant number of immature muscle fibers (labeled due to expression of neonatal or embryonic myosin).

The authors also cultured muscle cells in media containing different ratios of fibroblasts and satellite cells. The experiment showed that fibroblasts and mSCs interact not only on muscle fibers but also on each other. Specifically, satellite cells suppress the expression of certain collagen types, and fibroblasts regulate the maturation of satellite cells [93]. These findings are supported by a study in which the fibroblast pool was genetically reduced in mice, resulting in premature maturation of satellite cells [94]. These findings suggest the need to further investigate the interrelationship between satellite cells and fibroblasts, as well as the role they play in muscle recovery after exercise-induced damage.

Nielsen et al. [95] studied the effect of restricted blood flow in exercised muscles on mSC population. Twenty young, healthy men were divided into two groups: a study group that exercised with partial blood flow restriction (achieved by a compression cuff placed on the proximal aspect of the thigh) and a control group that exercised identically but without blood flow restriction. There was a three- to four-fold increase in $Pax7^+$ expression in the study group when compared to their pre-workout values. The number of cell nuclei per muscle fiber also increased. Meanwhile, none of these values showed significant statistical changes in the control group.

Moreover, the muscle fiber cross-sectional area increased by 30–40% from baseline values. This increase was comparable in both groups in the biopsy taken on the eighth day of training, with the study group sustaining the increase on the third and tenth days after stopping training. Conversely, there was a marked decrease in the control group.

Having observed changes in the cross-sectional area of muscle fibers and in the number of cell nuclei, the authors hypothesized that the number of nuclei is the limiting factor for skeletal muscle hypertrophy, as it determines the volume of cytoplasm at which the cell will lose its ability to transcribe mRNA. However, the mechanism of such a significant effect of hypoxia on skeletal muscle hyperoxia itself remains unexplored [95].

The influence of physical activity on satellite cell population is multifactorial. It depends on type of activity, age of subject, protein supplementation, sex, and many other variables, some of them remaining unclear. Some recent papers oppose each other in terms of results, which suggests the need for further investigation.

9. Satellite Cells and Age

Aging processes occur in all living organism tissues. The changes that can be observed involve both morphology and function. Mass and contractile force decrease, and the muscle nerves in skeletal muscles degenerate. These processes start, and are reflected at, the level of reactions and transformations occurring in a single cell.

Rosenberg introduced the term "sarcopenia" in 1989, modifying it eight years later. Currently, sarcopenia is defined as a decrease in muscle mass and an accompanying decrease in muscle strength. Sarcopenia is a physiological sign of aging [96]. Studies show that it affects about 25% of people between the ages of 50 and 70 and 40% of people above the age of 80 [97]. The first symptoms are usually noticed after the age of 60. It is believed that the degeneration of α-motoneurons, resulting in the atrophy of entire motor units, is the cause of age-related reduction in muscle mass. These reductions are accompanied by demyelinating changes within axons. Furthermore, the numbers of nerve endings, vesicles, and receptors at synapses decrease.

Degenerative changes include motor end plates, which result in impaired and slowed nerve conduction, followed by denervation, which, combined with muscle cell atrophy, leads to permanent damage within the striated muscle tissue [98]. The denervated fibers are reinnervated by undamaged neighboring neurons. This process is called innervation, or reinnervation, and is often observed in older people. Unfortunately, it is not flawless. Occasionally, a small motoneuron starts innervating a fiber that is capable of rapid conduction, resulting in impaired muscle fiber recruitment during contraction and, consequently, impaired strength [99].

Experiments in rats have shown that, in many cases, the denervated fast-contracting fiber is reinnervated by motoneurons of slow-contracting units, causing their phenotype to change [100]. Histological observations have confirmed that large type II b fibers are the first to be damaged, and their mass is reduced significantly before the age of 80 years. Degradation of the slow-contracting units is also evident later. In cases where reinnervation cannot occur (or where denervation occurs too rapidly and the body cannot keep up with reinnervating the damaged areas), muscle cells are subject to atrophy and are replaced by fibrous cartilage tissue and adipose tissue [101]. These muscle changes, among other things, are responsible for the slower and less precise motions experienced by the elderly.

Testosterone also decreases with age, and such decreases are associated with muscle mass loss, reduced function, and decreased muscle strength. Testosterone hormone replacement therapy suppresses these phenomena, at least partially. The same is true for estrogen in women [102]. After the age of 30, the concentration of growth hormones in the human body begins to decline. In the elderly, low levels of these hormones increase myostatin expression, which inhibits mSC division. The regeneration of damaged muscle is also impaired, contributing to sarcopenia [103].

IGF-I levels in satellite cells also decrease in old age. Studies in which IGF-I was administered into atrophically changed aged rat muscles confirm that this factor increases the number of satellite cells and promotes muscle regeneration [104]. Observations are inconsistent regarding whether the number of satellite cells decreases with age.

Reductions in the number of mSCs have been observed in studies conducted in mice [105]. However, in another study, no significant differences were found in the number of mSCs when comparing the muscles of young and old rats [106]. The fact that different species of animals, different muscle types, and different age groups have been tested makes it difficult to draw conclusions from these experiments.

However, researchers agree that the proliferative potential of mSCs decreases with age [104]. This is related not so much to a decrease in their number as to changes in the microenvironment of aging muscles and, thus, in the niche specific for mSCs [107]. These changes are primarily associated with decreased concentrations of various factors that disrupt the regeneration process at various stages [104,105].

On the other hand, the sensitivity of mSCs to environmental signals also decreases. A group of scientists analyzed one of these signaling agents in 2003 and described the effect of age on Notch pathway activation. The muscles of young subjects, in which Numb protein levels were artificially increased, showed impaired muscle regeneration associated with early mSC differentiation.

The same regeneration pattern occurred in the muscles of elderly individuals. Based on this, the researchers ventured a step further: using an artificial ligand for Notch receptors in aged muscle, they obtained the same regeneration level as in young individuals. Per the above discussion, the number of Notch receptors does not change with age, and the impaired regeneration in aging muscles is associated with decreases in the number of Delta ligands [108].

Heterochronic parabiosis is a treatment that seems to support the hypothesis that the mSC environment, not the processes within it, has the greatest influence on skeletal muscle regeneration activity. This procedure involves fusing the circulatory systems of two organisms. When an old mouse was paired with a young individual, the former experienced a better regeneration of the previously damaged muscle than before the circulatory fusion. Conversely, the efficiency of the restoration process decreased in the young mouse. To ensure that regeneration processes in older mice were not related to whole cell transport, the researchers labeled young individuals "GFP." Satellite cells from old mice were labeled "GFP$^-$." After the treatment, GFP detection did not demonstrate the presence of this protein in older mice. Hence, the environmental influences changed the mSCs' fate. The same effects were obtained in the 1980s by researchers who transplanted part of the muscle of an older mouse into a young one and vice versa. However, they could not exclude the influence of intracellular processes in their experiments, which was finally achieved in 2005 [107].

Changes in the Wnt pathway—more specifically, the presence of its activators in the bloodstream—are observed in older subjects. Consequently, muscle progenitor cells lose their linear specificity with age. Therefore, instead of becoming myofibers, they increasingly start differentiating into fibroblasts or adipocytes, which contributes to the decreased functionality of the human muscular system [109].

Some researchers have traced the loss of function in aging mSCs to changes at the proteome level. The system that degrades damaged proteins includes lysosomes, autolysosomes, proteasomes, and chaperones. As people age, these structures no longer function as efficiently as they used to, causing residual degraded proteins to accumulate in cells, which often have toxic effects on those cells [110]. Cells that remain in a resting phase, including mSCs, are the most vulnerable to toxins.

The body's aging process is often attributed to free radicals. Their destructive effects on mitochondrial membranes appear to increase with age as the efficiency of antioxidant mechanisms decreases. Studies have shown that type II fibers are more prone to free radical destruction than type I fibers, and damaged mitochondria are more frequently observed in them [111].

Author Contributions: Conceptualization, A.K. and E.M.-C.; Writing—original draft preparation: A.K., M.K., M.C., F.M.C., P.W., G.B. and E.M.-C.; Writing—review and editing: A.K., M.K., M.C., F.M.C., P.W., G.B. and E.M.-C.; Project administration: A.K. and E.M.-C.; Visualization: A.K., M.K. and M.C. Funding acquisition: E.M.-C. All authors have read and agreed to the published version of the manuscript.

Funding: This research received no external funding. It was funded only by an internal grant from the University School of Physical Education. Project No. 503 62/05 "Effectiveness of various therapeutic forms and their influence on nervous, muscular and vascular plasticity in patients after ischemic stroke".

Institutional Review Board Statement: Not applicable.

Informed Consent Statement: Not applicable.

Data Availability Statement: Not applicable.

Conflicts of Interest: The authors declare no conflict of interest. The funders had no role in the design of the study; the collection, analyses, or interpretation of data; the writing of the manuscript; or the decision to publish the results.

References

1. Mauro, A. Satellite cell of skeletal muscle fibers. *J. Cell Biol.* **1961**, *9*, 493–495. [CrossRef] [PubMed]
2. Katz, B. The terminations of the afferent nerve fiber in the muscle spindle of the frog. *Philos. Transac. R. Soc. Lond. B Biol. Sci.* **1961**, *243*, 221–240.
3. Ishikawa, H. Electron microscopic observations of satellite cells with special reference to the development of mammalian skeletal muscles. *Brain Struct. Funct.* **1966**, *125*, 43–63. [CrossRef] [PubMed]
4. Kelly, A.M.; Zacks, S.I. The histogenesis of rat intercostal muscle. *J. Cell Biol.* **1969**, *42*, 135–153. [CrossRef] [PubMed]
5. Moss, F.P.; Leblond, C.P. Satellite cells as the source of nuclei in muscles of growing rats. *Anat. Rec. Adv. Integr. Anat. Evol. Biol.* **1971**, *170*, 421–435. [CrossRef]
6. Shultz, E. A quantitative study of the satellite cell population in postnatal mouse lumbrical muscle. *Anat. Rec.* **1974**, *180*, 589–595. [CrossRef]
7. Konigsberg, U.R.; Lipton, B.H.; Konigsberg, I.R. The regenerative response of single mature muscle fibers isolated in vitro. *Dev. Biol.* **1975**, *45*, 260–275. [CrossRef]
8. Bischoff, R. A satellite cell mitogen from crushed adult muscle. *Dev. Biol.* **1986**, *115*, 140–147. [CrossRef]
9. Allen, R.E.; Boxhorn, L.K. Inhibition of skeletal muscle satellite cell differentiation by transforming growth factor-beta. *J. Cell. Physiol.* **1987**, *133*, 567–572. [CrossRef]
10. Cornelison, D.; Wold, B.J. Single-Cell Analysis of Regulatory Gene Expression in Quiescent and Activated Mouse Skeletal Muscle Satellite Cells. *Dev. Biol.* **1997**, *191*, 270–283. [CrossRef]
11. Maroto, M.; Reshef, R.; Munsterberg, A.; Koester, S.; Goulding, M.; Lassar, A.B. Ectopic Pax-3 Activates MyoD and Myf-5 Expression in Embryonic Mesoderm and Neural Tissue. *Cell* **1997**, *89*, 139–148. [CrossRef]
12. Bischoff, R. Interaction between satellite cells and skeletal muscle fibers. *Development* **1990**, *109*, 943–952. [CrossRef]
13. Kuang, S.; Kuroda, K.; Le Grand, F.; Rudnicki, M.A. Asymmetric Self-Renewal and Commitment of Satellite Stem Cells in Muscle. *Cell* **2007**, *129*, 999–1010. [CrossRef]
14. Przewoźniak, M.; Brzóska, E. Białka Pax w różnicowaniu komórek i organogenezie. *Post. Biol. Kom.* **2008**, *2*, 229–242.
15. Seale, P.; Sabourin, L.A.; Girgis-Gabardo, A.; Mansouri, A.; Gruss, P.; Rudnicki, M.A. Pax7 Is Required for the Specification of Myogenic Satellite Cells. *Cell* **2000**, *102*, 777–786. [CrossRef]
16. Relaix, F.; Montarras, D.; Zaffran, S.; Gayraud-Morel, B.; Rocancourt, D.; Tajbakhsh, S.; Mansouri, A.; Cumano, A.; Buck-ingham, M. Pax3 and Pax7 have distinct and overlapping functions in adult muscle progenitor cells. *J. Cell Biol.* **2006**, *17*, 91–102. [CrossRef] [PubMed]
17. Kuang, S.; Chargé, S.B.; Seale, P.; Huh, M.; Rudnicki, M.A. Distinct roles for Pax7 and Pax3 in adult regenerative myogenesis. *J. Cell Biol.* **2006**, *172*, 103–113. [CrossRef] [PubMed]
18. Schmalbruch, H.; Lewis, D. Dynamics of nuclei of muscle fibers and connective tissue cells in normal and denervated rat muscles. *Muscle Nerve* **2000**, *23*, 617–626. [CrossRef]
19. Bansal, D.; Miyake, K.; Vogel, S.; Groh, S.; Chen, C.-C.; Williamson, R.A.; McNeil, P.L.; Campbell, K. Defective membrane repair in dysferlin-deficient muscular dystrophy. *Nat. Cell Biol.* **2003**, *423*, 168–172. [CrossRef] [PubMed]

20. Mellgren, R.L.; Miyake, K.; Kramerova, I.; Spencer, M.J.; Bourg, N.; Bartoli, M.; Richard, I.; Greer, P.A.; McNeil, P.L. Cal-cium-dependent plasma membrane repair requires m- or μ-calpain, but not calpain-3, the proteasome, or caspases. *Biochim. Biophys. Acta* **2009**, *1793*, 1886–1893. [CrossRef]
21. Ciciliot, S.; Schiaffino, S. Regeneration of mammalian skeletal muscle. Basic mechanisms and clinical implications. *Curr. Pharma. Des.* **2010**, *16*, 906–914. [CrossRef]
22. Chargé, S.B.P.; Rudnicki, M. Cellular and Molecular Regulation of Muscle Regeneration. *Physiol. Rev.* **2004**, *84*, 209–238. [CrossRef]
23. Yin, H.; Price, F.; Rudnicki, M.A. Satellite Cells and the Muscle Stem Cell Niche. *Physiol. Rev.* **2013**, *93*, 23–67. [CrossRef] [PubMed]
24. Orimo, S.; Hiyamuta, E.; Arahata, K.; Sugita, H. Analysis of inflammatory cells and complement C3 in bupivacaine-induced myonecrosis. *Muscle Nerve* **1991**, *14*, 515–520. [CrossRef] [PubMed]
25. Fielding, R.A.; Manfredi, T.J.; Ding, W.; Fiatarone, M.A.; Evans, W.J.; Cannon, J.G. Acute phase response in exercise. III. Neutrophil and IL-1 beta accumulation in skeletal muscle. *Am. J. Physiol. Regul. Integr. Comp. Physiol.* **1993**, *265*, R166–R172. [CrossRef] [PubMed]
26. Ceafalan, L.C.; Popescu, B.O.; Hinescu, M.E. Cellular Players in Skeletal Muscle Regeneration. *BioMed Res. Int.* **2014**, *2014*, 1–21. [CrossRef] [PubMed]
27. Sonnet, C.; Lafuste, P.; Arnold, L.; Brigitte, M.; Poron, F.; Authier, F.J.; Chrétien, F.; Gherardi, R.K.; Chazaud, B. Human macrophages rescue myoblasts and myotubes from apoptosis through a set of adhesion molecular systems. *J. Cell Sci.* **2006**, *119*, 2497–2507. [CrossRef]
28. Lescaudron, L.; Peltékian, E.; Fontaine-Pérus, J.; Paulin, D.; Zampieri, M.; Garcia, L.; Parrish, E. Blood borne macrophages are essential for the triggering of muscle regeneration following muscle transplant. *Neuromuscul. Disord.* **1999**, *9*, 72–80. [CrossRef]
29. Schultz, E.; Jaryszak, D.L.; Valliere, C.R. Response of satellite cells to focal skeletal muscle injury. *Muscle Nerve* **1985**, *8*, 217–222. [CrossRef]
30. Alfaro, L.A.S.; Dick, S.A.; Siegel, A.L.; Anonuevo, A.S.; McNagny, K.; Megeney, L.; Cornelison, D.D.; Rossi, F.M. CD34 Promotes Satellite Cell Motility and Entry into Proliferation to Facilitate Efficient Skeletal Muscle Regeneration. *Stem Cells* **2011**, *29*, 2030–2041. [CrossRef]
31. Hughes, S.; Blau, H.M. Migration of myoblasts across basal lamina during skeletal muscle development. *Nat. Cell Biol.* **1990**, *345*, 350–353. [CrossRef] [PubMed]
32. Yoshida, N.; Koishi, K.; Masuda, K.; Nabeshima, Y. Cell heterogeneity upon myogenic differentiation: Down-regulation of MyoD and Myf-5 generates 'reserve cells'. *J. Cell Sci.* **1998**, *111*, 769–779. [CrossRef]
33. Yablonka-Reuveni, Z.; Rudnicki, M.A.; Rivera, A.J.; Primig, M.; Anderson, J.E.; Natanson, P. The transition from proliferation to differentiation is delayed in satellite cells from mice lacking MyoD. *Develop. Biol.* **1999**, *210*, 440–455. [CrossRef]
34. Megeney, L.A.; Kablar, B.; Garrett, K.; Anderson, J.E.; Rudnicki, M.A. MyoD is required for myogenic stem cell function in adult skeletal muscle. *Genes Dev.* **1996**, *10*, 1173–1183. [CrossRef]
35. Sabourin, L.A.; Girgis-Gabardo, A.; Seale, P.; Asakura, A.; Rudnicki, M.A. Reduced Differentiation Potential of Primary MyoD−/− Myogenic Cells Derived from Adult Skeletal Muscle. *J. Cell Biol.* **1999**, *144*, 631–643. [CrossRef]
36. Cornelison, D.D.; Olwin, B.B.; Rudnicki, M.A.; Wolb, B.J. MyoD(−/−) satellite cells in single- fiber culture are differentiation defective and MRF4 deficient. *Develop. Biol.* **2000**, *224*, 122–137. [CrossRef]
37. Asakura, A.; Hirai, H.; Kablar, B.; Morita, S.; Ishibashi, J.; Piras, B.A.; Christ, A.J.; Verma, M.; Vineretsky, K.A.; Rudnicki, M.A. Increased survival of muscle stem cells lacking the MyoD gene after transplantation into regenerating skeletal muscle. *Proc. Natl. Acad. Sci. USA* **2007**, *104*, 16552–16557. [CrossRef]
38. Gayraud-Morel, B.; Chrétien, F.; Flamant, P.; Gomès, D.; Zammit, P.; Tajbakhsh, S. A role for the myogenic determination gene Myf5 in adult regenerative myogenesis. *Dev. Biol.* **2007**, *312*, 13–28. [CrossRef]
39. Gensch, N.; Borchardt, T.; Schneider, A.; Riethmacher, D.; Braun, T. Different autonomous myogenic cell populations revealed by ablation of Myf5-expressing cells during mouse embryogenesis. *Development* **2008**, *135*, 1597–1604. [CrossRef] [PubMed]
40. Kablar, B.; Rudnicki, M.A. Skeletal muscle development in the mouse embryo. *Histol. Histopathol.* **2000**, *15*, 649–656. [PubMed]
41. Li, S.; Czubryt, M.; McAnally, J.; Bassel-Duby, R.; Richardson, J.A.; Wiebel, F.F.; Nordheim, A.; Olson, E.N. Requirement for serum response factor for skeletal muscle growth and maturation revealed by tissue-specific gene deletion in mice. *Proc. Natl. Acad. Sci. USA* **2005**, *102*, 1082–1087. [CrossRef] [PubMed]
42. L'Honoré, A.; Rana, V.; Arsic, N.; Franckhauser, C.; Lamb, N.J.; Fernandez, A. Identification of a New Hybrid Serum Response Factor and Myocyte Enhancer Factor 2-binding Element in MyoD Enhancer Required for MyoD Expression during Myogenesis. *Mol. Biol. Cell* **2007**, *18*, 1992–2001. [CrossRef] [PubMed]
43. Zhang, J.-M.; Wei, Q.; Zhao, X.; Paterson, B.M. Coupling of the cell cycle and myogenesis through the cyclin D1-dependent interaction of MyoD with cdk4. *EMBO J.* **1999**, *18*, 926–933. [CrossRef] [PubMed]
44. Kitzmann, M.; Fernández, A. Crosstalk between cell cycle regulators and the myogenic factor MyoD in skeletal myoblasts. *Cell. Mol. Life Sci.* **2001**, *58*, 571–579. [CrossRef] [PubMed]
45. Berkes, C.A.; Tapscott, S.J. MyoD and the transcriptional control of myogenesis. *Semin. Cell Dev. Biol.* **2005**, *16*, 585–595. [CrossRef] [PubMed]
46. Tomczak, K.K.; Marinescu, V.D.; Ramoni, M.F.; Sanoudou, D.; Montanaro, F.; Han, M.; Kunkel, L.M.; Kohane, I.S.; Beggs, A.H. Expression profiling and identification of novel genes involved in myogenic differentiation. *FASEB J.* **2004**, *18*, 1–23. [CrossRef] [PubMed]

47. Yablonka–Reuveni, Z.; Paterson, B.M. MyoD and Myogenin Expression Patterns in Cultures of Fetal and Adult Chicken Myoblasts. *J. Histochem. Cytochem.* **2001**, *49*, 455–462. [CrossRef] [PubMed]
48. Chao, D.; Gorospe, J.R.; Brenman, J.E.; Rafael, J.A.; Peters, M.F.; Froehner, S.C.; Hoffman, E.; Chamberlain, J.S.; Bredt, D.S. Selective loss of sarcolemmal nitric oxide synthase in Becker muscular dystrophy. *J. Exp. Med.* **1996**, *184*, 609–618. [CrossRef]
49. Cao, Y.; Kumar, R.M.; Penn, B.H.; Berkes, C.A.; Kooperberg, C.; Boyer, L.A.; Young, R.A.; Tapscott, S.J. Global and gene-specific analyses show distinct roles for Myod and Myog at a common set of promoters. *EMBO J.* **2006**, *25*, 502–511. [CrossRef]
50. Meadows, E.; Cho, J.-H.; Flynn, J.M.; Klein, W.H. Myogenin regulates a distinct genetic program in adult muscle stem cells. *Dev. Biol.* **2008**, *322*, 406–414. [CrossRef]
51. Kaufmann, U.; Martin, B.; Link, D.; Witt, K.; Zeitler, R.; Reinhard, S.; Starzinski-Powitz, A. M-cadherin and its sisters in de-velopment of striated muscle. *Cell Tiss.* **1999**, *296*, 191–198. [CrossRef] [PubMed]
52. Moore, R.; Walsh, F. The cell adhesion molecule M-cadherin is specifically expressed in developing and regenerating, but not denervated skeletal muscle. *Development* **1993**, *117*, 1409–1420. [CrossRef] [PubMed]
53. Jankowski, R.J.; Deasy, B.M.; Cao, B.; Gates, C.; Huard, J. The role of CD34 expression and cellular fusion in the regeneration capacity of myogenic progenitor cells. *J. Cell Sci.* **2002**, *115*, 4361–4374. [CrossRef]
54. Hollnagel, A.; Grund, C.; Franke, W.W.; Arnold, H.H. The cell adhesion molecule M-cadherin is not essential for muscle de-velopment and regeneration. *Mol. Cell. Biol.* **2002**, *22*, 4760–4770. [CrossRef]
55. Brzoska, E.; Ciemerych, M.A.; Przewozniak, M.; Zimowska, M. Regulation of muscle stem cells activation: The role of growth factors and extracellular matrix. *Vitam. Horm.* **2011**, *87*, 239–276.
56. Brustis, J.J.; Elamrani, N.; Balcerzak, D.; Safwate, A.; Soriano, M.; Poussard, S.; Cottin, P.; Ducastaing, A. Rat myoblast fusion requires exteriorized m-calpain activity. *Eur. J. Cell Biol.* **1994**, *64*, 320–327.
57. Smythe, G.; Davies, M.; Paulin, D.; Grounds, M. Absence of desmin slightly prolongs myoblast proliferation and delays fusion in vivo in regenerating grafts of skeletal muscle. *Cell Tissue Res.* **2001**, *304*, 287–294. [CrossRef]
58. Zammit, P.S.; Heslop, L.; Hudon, V.; Rosenblatt, J.D.; Tajbakhsh, S.; Buckingham, M.E.; Beauchamp, J.R.; Partridge, T.A. Kinetics of myoblast proliferation show that resident satellite cells are competent to fully regenerate skeletal muscle fibers. *Exp. Cell Res.* **2002**, *281*, 39–49. [CrossRef]
59. Czerwińska, A.M.; Streminska, W.; Ciemerych, M.A.; Grabowska, I. Mouse gastrocnemius muscle regeneration after me-chanical or cardiotoxin injury. *Folia Histochem. Cytobiol.* **2012**, *50*, 144–153. [CrossRef] [PubMed]
60. Schultz, E.; McCormick, K.M. Skeletal muscle satellite cells. *Rev. Physiol. Biochem. Pharmacol.* **1994**, *123*, 213–257. [PubMed]
61. Moyer, A.L.; Wagner, K.R. Regeneration versus fibrosis in skeletal muscle. *Curr. Opin. Rheumatol.* **2011**, *23*, 568–573. [CrossRef] [PubMed]
62. Molofsky, A.V.; Pardal, R.; Morrison, S.J. Diverse mechanisms regulate stem cell self-renewal. *Curr. Opin. Cell Biol.* **2004**, *16*, 700–707. [CrossRef]
63. Lark, K.G.; Consigli, R.A.; Minocha, H.C. Segregation of Sister Chromatids in Mammalian Cells. *Science* **1966**, *154*, 1202–1205. [CrossRef]
64. Conboy, M.J.; Karasov, A.O.; Rando, T.A. High incidence of non-random template strand regregation and asymmetric fate determination in dividing stem cells and their progeny. *PLoS Biol.* **2007**, *5*, e102.
65. Conboy, I.M.; Rando, T.A. The Regulation of Notch Signaling Controls Satellite Cell Activation and Cell Fate Determination in Postnatal Myogenesis. *Dev. Cell* **2002**, *3*, 397–409. [CrossRef]
66. Mourikis, P.; Sambasivan, R.; Castel, D.; Rocheteau, P.; Bizzarro, V.; Tajbakhsh, S. A Critical Requirement for Notch Signaling in Maintenance of the Quiescent Skeletal Muscle Stem Cell State. *Stem Cells* **2012**, *30*, 243–252. [CrossRef]
67. Zammit, P.S.; Golding, J.P.; Nagata, Y.; Hudon, V.; Partridge, T.A.; Beauchamp, J.R. Muscle satellite cells adopt divergent fates: A mechanism for self-renewal? *J. Cell Biol.* **2004**, *166*, 347–357. [CrossRef]
68. Day, K.; Shefer, G.; Richardson, J.B.; Enikolopov, G.; Yablonka-Reuveni, Z. Nestin-GFP reporter expression defines the quies-cent state of skeletal muscle satellite cells. *Dev. Biol.* **2007**, *304*, 246–259. [CrossRef]
69. Day, K.; Shefer, G.; Shearer, A.; Yablonka-Reuveni, Z. The depletion of skeletal muscle satellite cells with age is concomitant with reduced capacity of single progenitors to produce reserve progeny. *Dev. Biol.* **2010**, *340*, 330–343. [CrossRef] [PubMed]
70. Bąkowski, P.; Musielak, B.; Sip, P.; Biegański, G. Wpływ masażu na powysiłkową bolesność mięśni. *Chir. Narz. Ruchu Ortop. Pol.* **2008**, *73*, 261–265.
71. Barnett, A. Using Recovery Modalities between Training Sessions in Elite Athletes. *Sports Med.* **2006**, *36*, 781–796. [CrossRef]
72. Proske, U.; Morgan, D.L. Muscle damage from eccentric exercise: Mechanism, mechanical signs, adaptation and clinical applications. *J. Physiol.* **2001**, *537*, 333–345. [CrossRef]
73. McCormick, K.M.; Thomas, D.P. Exercise-induced satellite cell activation in senescent soleus muscle. *J. Appl. Physiol.* **1992**, *72*, 888–893. [CrossRef]
74. Kadi, F.; Thornell, L.-E. Concomitant increases in myonuclear and satellite cell content in female trapezius muscle following strength training. *Histochem. Cell Biol.* **2000**, *113*, 99–103. [CrossRef] [PubMed]
75. Kadi, F.; Charifi, N.; Denis, C.; Lexell, J.; Andersen, J.L.; Schjerling, P.; Olsen, S.; Kjaer, M. The behaviour of satellite cells in response to exercise: What have we learnd from human studies? *Pflug. Arch.* **2005**, *451*, 319–327. [CrossRef] [PubMed]

76. Nederveen, J.P.; Snijders, T.; Joanisse, S.; Wavell, C.G.; Mitchell, C.J.; Johnston, L.M.; Baker, S.K.; Phillips, S.M.; Parise, G. Altered muscle satellite cell activation following 16 wk of resistance training in young men. *Am. J. Physiol. Integr. Comp. Physiol.* **2017**, *312*, R85–R92. [CrossRef] [PubMed]
77. Hyldahl, R.D.; Olson, T.; Welling, T.; Groscost, L.; Parcell, A.C. Satellite cell activity is differentially affected by contraction mode in human muscle following a work-matched bout of exercise. *Front. Physiol.* **2014**, *5*, 485. [CrossRef] [PubMed]
78. Farup, J.; Rahbek, S.K.; Riis, S.; Vendelbo, M.H.; de Paoli, F.; Vissing, K. Influence of exercise contraction mode and protein supplementation on human skeletal muscle satellite cell content and muscle fiber growth. *J. Appl. Physiol.* **2014**, *117*, 898–909. [CrossRef] [PubMed]
79. Babcock, L.; Escano, M.; D'Lugos, A.; Todd, K.; Murach, K.; Luden, N. Concurrent aerobic exercise interferes with the satellite cell response to acute resistance exercise. *Am. J. Physiol. Integr. Comp. Physiol.* **2012**, *302*, R1458–R1465. [CrossRef]
80. Snijders, T.; Verdijk, L.; Smeets, J.; McKay, B.R.; Senden, J.M.G.; Hartgens, F.; Parise, G.; Greenhaff, P.; Van Loon, L.J.C. The skeletal muscle satellite cell response to a single bout of resistance-type exercise is delayed with aging in men. *Age* **2014**, *36*, 1–15. [CrossRef]
81. Verney, J.; Kadi, F.; Charifi, N.; Féasson, L.; Saafi, M.A.; Castells, J.; Piehl-Aulin, K.; Denis, C. Effects of combined lower body endurance and upper body resistance training on the satellite cell pool in elderly subjects. *Muscle Nerve* **2008**, *38*, 1147–1154. [CrossRef]
82. Smith, H.K.; Merry, T.L. Voluntary resistance wheel exercise during post-natal growth in rats enhances skeletal muscle satellite cell and myonuclear content at adulthood. *Acta Physiol.* **2012**, *204*, 393–402. [CrossRef]
83. Bruusgaard, J.C.; Johansen, I.B.; Egner, I.M.; Rana, Z.A.; Gundersen, K. Myonuclei acquired by overload exercise precede hypertrophy and are not lost on detraining. *Proc. Natl. Acad. Sci. USA* **2010**, *107*, 15111–15116. [CrossRef] [PubMed]
84. Hoppeler, H.; Howald, H.; Conley, K.; Lindstedt, S.L.; Claassen, H.; Vock, P.; Weibel, E.R. Endurance training in humans: Aerobic capacity and structure of skeletal muscle. *J. Appl. Physiol.* **1985**, *59*, 320–327. [CrossRef]
85. Martin, N.R.; Lewis, M.P. Satellite cell activation and number following acute and chronic exercise: A mini review. *Cell. Mol. Exerc. Physiol.* **2012**, *1*, e3. [CrossRef]
86. Shefer, G.; Rauner, G.; Yablonka-Reuveni, Z.; Benayahu, D. Reduced Satellite Cell Numbers and Myogenic Capacity in Aging Can Be Alleviated by Endurance Exercise. *PLoS ONE* **2010**, *5*, e13307. [CrossRef] [PubMed]
87. Murach, K.A.; Walton, R.G.; Fry, C.S.; Michaelis, S.L.; Groshong, J.S.; Finlin, B.S.; Kern, P.A.; Peterson, C.A. Cycle training modulates satellite cell and transcriptional responses to a bout of resistance exercise. *Physiol. Rep.* **2016**, *4*, e12973. [CrossRef]
88. Kadi, F.; Charifi, N.; Henriksson, J. The number of satellite cells in slow and fast fibres from human vastus lateralis muscle. *Histochem. Cell Biol.* **2006**, *126*, 83–87. [CrossRef] [PubMed]
89. Snijders, T.; Verdijk, L.B.; Beelen, M.; McKay, B.R.; Parise, G.; Kadi, F.; van Loon, L.J. A single bout of exercise activates skeletal muscle satellite cells during subsequent overnight recovery. *Exp. Physiol.* **2012**, *97*, 762–773. [CrossRef] [PubMed]
90. Verdijk, L.B.; Koopman, R.; Schaart, G.; Meijer, K.; Savelberg, H.H.; van Loon, L.J. Satellite cell content is specifically reduced in type II skeletal muscle fibers in the elderly. *Am. J. Physiol. Endocrinol. Metab.* **2007**, *292*, 151–157. [CrossRef]
91. Verdijk, L.B.; Gleeson, B.G.; Jonkers, R.A.M.; Meijer, K.; Savelberg, H.H.C.M.; Dendale, P.; van Loon, L.J. Skeletal Muscle Hypertrophy Following Resistance Training Is Accompanied by a Fiber Type-Specific Increase in Satellite Cell Content in Elderly Men. *J. Gerontol. Ser. A Boil. Sci. Med. Sci.* **2009**, *64*, 332–339. [CrossRef] [PubMed]
92. Cermak, N.M.; Snijders, T.; McKay, B.R.; Parise, G.; Verdijk, L.B.; Tarnopolsky, M.A.; Gibala, M.J.; Loon, L.J. Eccentric exercise increases satellite cell content in type II muscle fibers. *Med. Sci. Sports Exerc.* **2013**, *45*, 230–237. [CrossRef]
93. Mackey, A.; Magnan, M.; Chazaud, B.; Kjaer, M. Human skeletal muscle fibroblasts stimulate in vitro myogenesis and in vivo muscle regeneration. *J. Physiol.* **2017**, *595*, 5115–5127. [CrossRef] [PubMed]
94. Murphy, M.M.; Lawson, J.A.; Mathew, S.; Hutcheson, D.A.; Kardon, G. Satellite cells, connective tissue fibroblasts and their interactions are crucial for muscle regeneration. *Development* **2011**, *138*, 3625–3637. [CrossRef] [PubMed]
95. Nielsen, J.L.; Aagaard, P.; Bech, R.D.; Nygaard, T.; Hvid, L.G.; Wernbom, M.; Suetta, C.; Frandsen, U. Proliferation of myogenic stem cells in human skeletal muscle in response to low-load resistance training with blood flow restriction. *J. Physiol.* **2012**, *590*, 4351–4361. [CrossRef]
96. Fulle, S.; Protasi, F.; Di Tano, G.; Pietrangelo, T.; Beltramin, A.; Boncompagni, S.; Vecchiet, L.; Fanò, G. The contribution of reactive oxygen species to sarcopenia and muscle ageing. *Exp. Gerontol.* **2004**, *39*, 17–24. [CrossRef]
97. Faulkner, J.A.; Larkin, L.M.; Claflin, D.R.; Brooks, S.V. Age-related changes in the structure and function of skeletal muscles. *Clin. Exp. Pharmacol. Physiol.* **2007**, *34*, 1091–1096. [CrossRef]
98. Jang, Y.C.; van Remmen, H. Age-associated alterations of the neuromuscular junction. *Exp. Gerontol.* **2011**, *46*, 193–198. [CrossRef]
99. Erim, Z.; Beg, M.F.; Burke, D.T.; De Luca, C.J. Effects of Aging on Motor-Unit Control Properties. *J. Neurophysiol.* **1999**, *82*, 2081–2091. [CrossRef]
100. Pette, D.; Staron, R.S. Myosin isoforms, muscle fiber types, and transitions. *Microsc. Res. Tech.* **2000**, *50*, 500–509. [CrossRef]
101. Lexell, J. Evidence for Nervous System Degeneration with Advancing Age. *J. Nutr.* **1997**, *127*, 1011S–1013S. [CrossRef] [PubMed]
102. Araujo, A.B.; Wittert, G.A. Endocrinology of the aging male. *Best Pr. Res. Clin. Endocrinol. Metab.* **2011**, *25*, 303–319. [CrossRef] [PubMed]
103. Ryall, J.; Schertzer, J.D.; Lynch, G.S. Cellular and molecular mechanisms underlying age-related skeletal muscle wasting and weakness. *Biogerontology* **2008**, *9*, 213–228. [CrossRef] [PubMed]

104. Chakravarthy, M.V.; Davis, B.S.; Booth, F.W. IGF-I restores satellite cell proliferative potential in immobilized old skeletal muscle. *J. Appl. Physiol.* **2000**, *89*, 1365–1379. [CrossRef] [PubMed]
105. Shefer, G.; Van de Mark, D.P.; Richardson, J.B.; Yablonka-Reuveni, Z. Satellite-cell pool size does matter: Defining the myogenic potency of aging skeletal muscle. *Dev. Biol.* **2006**, *294*, 50–66. [CrossRef] [PubMed]
106. Brooks, N.E.; Schuenke, M.D.; Hikida, R.S. No change in skeletal muscle satellite cells in young and aging rat soleus muscle. *J. Physiol. Sci.* **2009**, *59*, 465–471. [CrossRef]
107. Conboy, I.M.; Conboy, M.J.; Wagers, A.J.; Girma, E.R.; Weissman, I.L.; Rando, T. Rejuvenation of aged progenitor cells by exposure to a young systemic environment. *Nat. Cell Biol.* **2005**, *433*, 760–764. [CrossRef]
108. Conboy, I.M.; Conboy, M.J.; Smythe, G.M.; Rando, T.A. Notch-Mediated Restoration of Regenerative Potential to Aged Muscle. *Science* **2003**, *302*, 1575–1577. [CrossRef]
109. Waś, H.; Czarnecka, J. Komórki macierzyste a starzenie. *Post. Biochem.* **2014**, *60*, 161–176.
110. Koga, H.; Kaushik, S.; Cuervo, A.M. Protein homeostasis and aging: The importance of exquisite quality control. *Ageing Res. Rev.* **2011**, *10*, 205–215. [CrossRef]
111. Conley, K.E.; Amara, C.E.; Jubrias, S.A.; Marcinek, D.J. Mitochondrial function, fibre types and ageing: New insights from human muscle in vivo. *Exp. Physiol.* **2007**, *92*, 333–339. [CrossRef] [PubMed]

biology

MDPI

Article

Effects of Ibuprofen Use on Lymphocyte Count and Oxidative Stress in Elite Paralympic Powerlifting

Felipe J. Aidar [1,2,3,4,*], Guacira S. Fraga [1], Márcio Getirana-Mota [1,2], Anderson Carlos Marçal [1,2], Jymmys L. Santos [1,2], Raphael Fabricio de Souza [1,2,3], Alexandre Reis Pires Ferreira [5], Eduardo Borba Neves [6], Aristela de Freitas Zanona [7], Alexandre Bulhões-Correia [8], Paulo Francisco de Almeida-Neto [8], Tulio Luiz Banja Fernandes [9], Nuno Domingos Garrido [10], Maria do Socorro Cirilo-Sousa [11,12], María Merino-Fernández [13], Alfonso López Díaz-de-Durana [14], Eugenia Murawska-Ciałowicz [15], Breno Guilherme de Araújo Tinoco Cabral [8] and Filipe Manuel Clemente [16,17]

1 Graduate Program of Physical Education, Federal University of Sergipe (UFS), São Cristovão 49100-000, Brazil; guacirafraga@yahoo.com.br (G.S.F.); marcio_getirana@hotmail.com (M.G.-M.); acmarcal@yahoo.com.br (A.C.M.); jymmys.lopes@gmail.com (J.L.S.); raphaelctba20@hotmail.com (R.F.d.S.)
2 Group of Studies and Research of Performance, Sport, Health and Paralympic Sports (GEPEPS), Federal University of Sergipe (UFS), São Cristovão 49100-000, Brazil
3 Department of Physical Education, Federal University of Sergipe (UFS), São Cristovão 49100-000, Brazil
4 Graduate Program of Physiological Science, Federal University of Sergipe (UFS), São Cristovão 49100-000, Brazil
5 College of Physical Education and Exercise Science, University of Brasília (UnB), Brasília 70910-900, Brazil; alexandreispf@gmail.com
6 Graduate Program in Biomedical Engineering, Federal Technological University of Paraná (UTFPR), Curitiba 80230-901, Brazil; eduardoneves@utfpr.edu.br
7 Department of Occupational Therapy, Federal University of Sergipe (UFS), Lagarto 49400-000, Brazil; arisz_to@yahoo.com.br
8 Department of Physical Education, Federal University of Rio Grande do Norte, Natal 59064-741, Brazil; alexandrebulhoescorreia@gmail.com (A.B.-C.); paulo220911@hotmail.com (P.F.d.A.-N.); brenotcabral@gmail.com (B.G.d.A.T.C.)
9 Institute of Physical Education and Sport, Federal University of Ceará (UFC), Fortaleza 60020-181, Brazil; banja@ufc.br
10 Research Center in Sports Sciences, Health Sciences and Human Development (CIDESD), University of Trás-os-Montes e Alto Douro, 5001-801 Vila Real, Portugal; ngarrido@utad.pt
11 Graduate Program Association of Physical Education, Federal University of Paraíba (UFPB), João Pessoa 58051-900, Brazil; helpcirilo@yahoo.com.br
12 Department of Physical Education, Regional of University (URCA), Crato 63105-010, Brazil
13 Faculty of Health Sciences, Universidad Francisco de Vitoria (UFV), 28223 Madrid, Spain; m.merino.prof@ufv.es
14 Sports Department, Physical Activity and Sports Faculty-INEF, Universidad Politécnica de Madrid, 28040 Madrid, Spain; alfonso.lopez@upm.es
15 Physiology and Biochemistry Department, University School of Physical Education, 51-612 Wroclaw, Poland; eugenia.murawska-cialowicz@awf.wroc.pl
16 Escola Superior Desporto e Lazer, Instituto Politécnico de Viana do Castelo, Rua Escola Industrial e Comercial de Nun'Álvares, 4900-347 Viana do Castelo, Portugal; filipe.clemente5@gmail.com
17 Instituto de Telecomunicações, Delegação da Covilhã, 1049-001 Lisboa, Portugal
* Correspondence: fjaidar@gmail.com; Tel.: +55-(79)-99685-7777

Citation: Aidar, F.J.; Fraga, G.S.; Getirana-Mota, M.; Marçal, A.C.; Santos, J.L.; de Souza, R.F.; Ferreira, A.R.P.; Neves, E.B.; Zanona, A.d.F.; Bulhões-Correia, A.; et al. Effects of Ibuprofen Use on Lymphocyte Count and Oxidative Stress in Elite Paralympic Powerlifting. *Biology* **2021**, *10*, 986. https://doi.org/10.3390/biology10100986

Academic Editors: Redha Taiar and Anne Delextrat

Received: 16 July 2021
Accepted: 23 September 2021
Published: 30 September 2021

Simple Summary: Paralympic Powerlifting (PP) is a strength sport and training tends to promote fatigue. Ten national-level PP athletes were evaluated concerning post-training oxidative stress using Ibuprofen and a placebo. Strength indicators were evaluated. The training consisted of five sets of five repetitions (80–90% 1-Repetition Maximum) in the bench press. The IBU had a positive effect on strength indicators, with decreased fatigue and increased lymphocyte count. There were no differences in oxidative stress. The use of IBU provided improvements in strength and fatigue reduction and did not protect against oxidative stress.

Abstract: Background: Paralympic Powerlifting (PP) training tends to promote fatigue and oxidative stress. Objective: To analyze the effects of ibuprofen use on performance and oxidative stress in post-training PP athletes. Methodology: Ten national level PP athletes (age: 27.13 ± 5.57) were analyzed for oxidative stress in post-training. The study was carried out in three weeks, (1) familiarization and (2 and 3) evaluated the recovery with the use of a placebo (PLA) and ibuprofen (IBU), 800 mg. The Peak Torque (PT), Torque Development Rate (TDR), Fatigue Index (FI), reactive substances to thiobarbituric acid (TBARS) and sulfhydryl groups (SH) were evaluated. The training consisted of five sets of five repetitions (80–90%) 1-Repetition Maximum (1-RM) in the bench press. Results: The IBU showed a higher PT (24 and 48 h, $p = 0.04$, $\eta^2 p = 0.39$), a lower FI (24 h, $p = 0.01$, $\eta^2 p = 0.74$) and an increased lymphocyte count ($p < 0.001$; $\eta^2 p = 4.36$). There was no change in oxidative stress. Conclusions: The use of IBU provided improvements in strength and did not protect against oxidative stress.

Keywords: Paralympic Powerlifting; ibuprofen; muscle strength; oxidative stress; recovery of function

1. Introduction

Paralympic Powerlifting is characterized by being a sport that demands high intensities during competitions and training, and the training demands progressive overloads to take athletes to the peak of the required physical performance [1,2]. Due to training overloads, it is necessary to perform an adequate recovery so that athletes have performance gains. When insufficient recovery occurs, tissue injury may be induced and may lead athletes to overtrain (loss of performance due to the accumulation of training shifts without adequate recovery) [1,3].

It is noteworthy that intense physical exercise is a physiological stress capable of altering immune responses and blood biomarkers [4]. Scientific studies show that intense physical exercise can modulate the leukocyte count in the bloodstream and the interaction of these leukocytes (neutrophils and monocytes/macrophages) with endothelial cells in the muscle and consequent transmigration to the damaged skeletal muscle tissue [5,6].

Physical exercise, immune system and oxidative stress indicate that volume and intensity are directly related to alterations in the redox balance, and the excessive increase in production or the reduction of antioxidant capacity, which can induce oxidative damage to lipids, proteins and nucleic acids [7,8]. Excessive loads of physical exercise can generate oxidative stress, considering that physical exercise can promote the formation of ROSs (i.e., reactive oxygen species) in the human body [9,10]. It is noteworthy that ROSs can cause tissue damage and, in high concentrations, damage cellular organelles, nucleic acids, lipids and proteins, causing harm to human health [6]. In the same direction, it has been reported that strenuous exercise, as in strength training, tends to increase stress biomarkers. Thus, oxidative stress has been associated with strength training [11–13]. Exercise-induced oxidative stress has been associated with reactive oxygen species (ROSs), especially during exercise [14], as well as post-intensive exercise muscle damage and inflammation that tend to contribute to increased oxidative stress [12,15]. In this sense, it has been suggested that elite powerlifters may benefit from blunted responses to oxidative stress after intensive weightlifting sessions, which may have implications for recovery between training sessions (Ammar et al., 2017a). Therefore, to reduce oxidative stress and protect athletes' bodies with the objective of enhancing the recovery process, many methods have been proposed and used, among which ibuprofen (IBU), which is a non-steroidal anti-inflammatory drug (NSAID), stands out for self-administrative use [1,16]. However, the use of NSAIDs can inhibit the muscle myofibers regeneration, the proliferation and the differentiation of satellite cells, and muscle hypertrophy induced by an adaptation to training overload [17–19].

For this reason, despite the aforementioned information about ROSs in response to physical activity, there is no consensus of what the best post-workout recovery would be to minimize oxidative stress in the athletes' body [6,9,10], especially because the physical exercises models and their evaluation methods have not been standardized, which makes a conclusive analysis difficult [6,9,10]. In this sense, the present study raised the hypothesis that using IBU during the recovery period of Paralympic Powerlifting athletes is beneficial for the parameters of sports performance, immunity maintenance and the reduction of oxidative stress.

In this regard, the objective of the present study was to analyze the effects of the use of ibuprofen on performance parameters, cell count and oxidative stress in national level Paralympic Powerlifting athletes in the period of resisted post-training recovery.

2. Materials and Methods

2.1. Study Design

The study design is shown in Figure 1. The study was carried out in three weeks, using the adapted bench press [20], with the first week being aimed at familiarization and the second and the third at the recovery method with the use of placebo (PLA) and Ibuprofen (IBU), the collections of the Peak Torque (PT), Rate of Torque Development (RTD), Fatigue Index (FI), Oxidative Stress Assessment through Thiobarbituric Acid Reactive Substances (TBARS) and Sulfidril Groups (SH), in addition to the blood indicators performed through the blood count and ammonia after training.

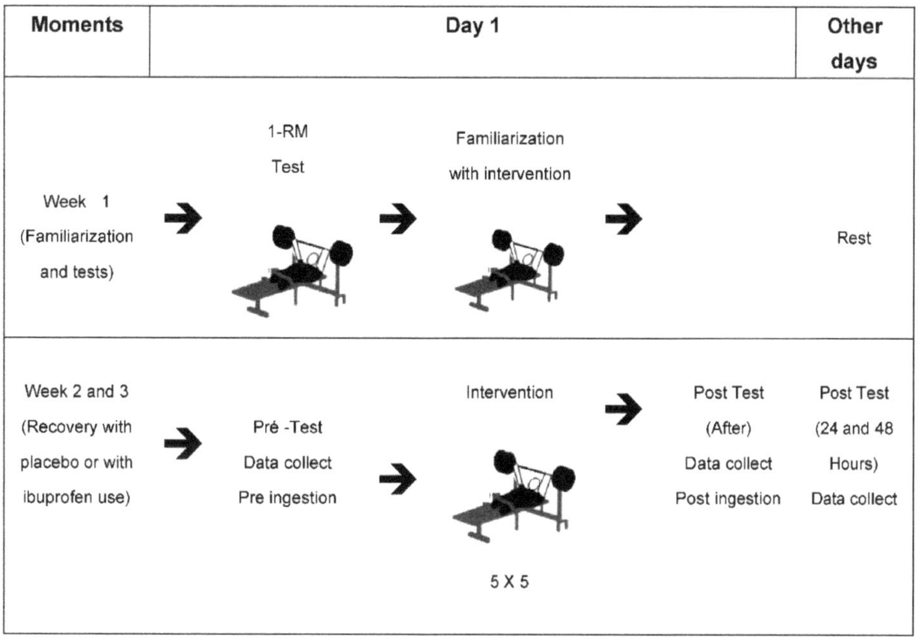

Figure 1. Experimental drawing. Weekly training schedule.

The order of the PLA or IBU conditions was determined randomly through a draw, considering 50% for each condition.

Week 1: familiarization; Week 2 and 3: recovery with the use of Ibuprofen or Placebo (week 2, 50% PLA and 50% IBU, changing in week 3). Pre-ingestion: 400 mg Ibuprofen/Placebo ingestion 15 min before training; Intervention: training length (3 h); Post-ingestion: 400 mg Ibuprofen/Placebo ingestion 5 h after training (5 h); Data collection: measures of strength in the Adapted Bench Press (FI: Fatigue Index, RTD: Rate of Torque Development, PT: Peak Torque), Oxidative Stress, TBARS: Thiobarbituric Acid Reactive Substances; SH: Sulfhydryl groups. After training, Ammonia and CBC were evaluated. 1-RM: One Maximum Repetition.

Collections were carried out between 9 a.m. and 12 p.m., according to the participants' availability. All assessments were carried out 30 min before the training started and immediately after, 24 h and 48 h after the training. The participants were evaluated for the rate of torque development, peak torque and fatigue index. The blood variables evaluated were the cell count of the immune system and markers of oxidative stress {thiobarbituric acid (TBARS) and sulfhydryl groups (SH)} were performed before, after, two hours later, 24 h and 48 h after.

Before the intervention began, the athletes performed a previous warm-up for the upper limbs, using three exercises: (1) pulley elbow extension, (2) shoulders rotations with dumbbells, (3) shoulders abduction with dumbbells. Three sets of 10 to 20 maximum repetition (1-RM) were carried out; the warm-up lasted approximately 10 min [3,21]. Then, a specific warm-up was performed on the bench press itself with 30% of 1-RM where: 10 slow repetitions (eccentric 3-s × concentric 1-s) and 10 rapid repetitions (eccentric 1-s × concentric 1-s) were performed before the intervention started. It is noteworthy that during the specific warm-up, athletes received verbal encouragement to give their maximum performance [3,21].

Subsequently, the athletes were submitted to an intervention of five sets of five maximum repetitions (5 repetitions with 80–90% of 1-RM). In the intervention, the traditional method was applied, using only fixed loads (invariable resistance). Two types of recovery were applied: one using the wheat flour placebo and the other using the IBU (400 mg) where both groups ingested the tablet 15 min before and 5 h after training.

2.2. Sample

The sample was entirely composed of male athletes [20]. Forty percent of the athletes had spinal cord injury below the eighth thoracic vertebra, 20% had sequelae due to polio, 20% had a malformation of the lower limbs and 20% had disabilities due to brain injury. The athletes were of Brazilian nationality and competed on a national level with rankings in the top 10 of their respective categories. Exclusion criteria were adopted: (1) not participating in any phase of monitoring and data collection, (2) in the 24 h prior to the collection, strenuous exercise, (3) consumption of alcohol, caffeine, non-steroidal anti-inflammatory drugs (including IBU), nutritional supplements (confirmed by interview), (4) be allergic to Ibuprofen, (5) having any muscle or joint injuries and/or reporting a change in arterial hypertension.

The sample size was determined a priori based on a previous study [1], which found an effect size of partial squared eta (η^2p) = 0.6 for the analyses of the influence of ibuprofen on neuromuscular aspects in Paralympic Powerlifting athletes (in this case the variable was creatine kinase). Thus, the open-source G* Power software (Version 3.0; Berlin, Germany) was used in the statistical configuration for family tests "F" (ANOVA two way), considering an $\alpha < 0.05$ and a $\beta = 0.80$. In addition, two groups (placebo x ibuprofen) in four distinct measures (Before × After × After 24 Hs × After 48 Hs) were considered. Thus, a minimum sample size of six subjects was indicated for the present study, with the sample power estimated at 0.80.

Table 1 shows the sample characterization.

Table 1. Sample characterization.

Variables	(Mean ± Standard Deviation)
n	10
Age (years)	27.13 ± 5.57
Body Weight (kg)	79.25 ± 25.51
Experience (years)	2.99 ± 0.51
1-RM/Bench press (kg)	137.13 ± 30.53 *
1-RM/Body Weight	1.80 ± 0.31 **

* All athletes with loads that keep them in the top 10 of their categories nationwide. ** Athletes with values above 1.4 in the Bench Press (1-RM/Body Weight) would be considered elite athletes, according to Ball & Wedman, [22].

2.3. Ethics

The athletes participated in the study voluntarily and signed a free and informed consent term, in accordance with resolution 466/2012 of the National Research Ethics Commission—CONEP, of the National Health Council, following the ethical principles expressed in the Helsinki Declaration (1964, reformulated in 1975, 1983, 1989, 1996, 2000, 2008 and 2013), by the World Medical Association. In addition, the present clinical trial was previously registered (CAEE ID: 79909917.0.0000.55.46) and approved by the Human Research Ethics Committee of the Federal University of Sergipe (UFS), under Statement Number 2637882/2018.

2.4. Body Mass Analysis

Body mass was measured while sitting on a Micheletti Electronic Wheelchair Scale (Model Mic Wheelchair) of the digital electronic platform type (Micheletti®, São Paulo, Brazil) with a maximum weight capacity of 500 kg (dimensions of 5.0 cm thickness, with a diameter of 102 × 120 cm).

2.5. Maximum Training Load Analysis

In order to determine the maximum training load, the 1-Repetition Maximum (1-RM) test was performed and, because the individuals evaluated were familiarized with the 1-RM test, they were not submitted to familiarization sessions. In the test, each subject started the attempts with a weight that could be lifted using maximum effort. Afterwards, weight increments were added until reaching the maximum load that could be lifted only once. If the practitioner was unable to perform a single repetition, 2.4 to 2.5% of the load used in the test were subtracted. The subjects rested 3–5 min between attempts [3,22].

2.6. Upper Limbs Muscle Strength

To measure muscle strength, the Fatigue Index (FI), the Peak Torque (PT), and the Rate of Torque Development (RTD) were determined by a Chronojump load cell (Chronojump®, BoscoSystem, Madrid, Spain), fixed on the Straight Bench Press, using Spider HMS Simond carabiners (Chamonix, France), with a breaking load of 21 KN, approved for climbing by the Union International des Associations d'Alpinisme (UIAA). A steel chain with a breaking load of 2300 kg was used to secure the load cell to the bench. The perpendicular distance between the load cell and the center of the joint was determined and used to calculate joint torques and fatigue index [21,23].

The isometric peak torque (PT) was measured by the maximum torque generated by the muscles of the upper limbs. PT was determined by the product of the peak isometric force, measured between the load cell cable fixation point and the adapted bench press, which was adjusted so that there was an angle close to 90° at the elbow, at a 15 cm distance from the starting point (chest to bar), verified with a device for measuring the angular amplitude, Model FL6010 (Sanny®, São Bernardo do Campo, Brazil). Participants were instructed to perform a single maximum movement until elbow extension (as fast as possible) and then relax, for PT evaluation.

As for the Fatigue Index (FI) assessment, the same exercise was performed and it was determined that the subjects maintained the maximum contraction for 5.0 s, where the index was determined by dividing the initial PT in relation to the final PT, subtracted from one. FI = {(Maximum PT − Minimum PT/Maximum Pt) × 100}. Thus, the results in Newton (Nm) were conceived by the formula Nm = (M) × (C) × (H), where M = Body mass in kg, C = 9.80665, H = Height of the bar in relation to the cell load (0.45 m), corresponding to the height that the equipment was fixed, adopting an angle of 90° between the forearm and the arm. The Rate of Torque Development (RTD) was determined using the Peak Torque to time ratio until reaching the Peak Torque (RTD = ΔPeak Torque/ΔTime), in 300 ms [21].

2.7. Blood Sample Collection

Blood samples were collected in the antecubital vein of the forearm and immediately transferred to tubes with EDTA. Blood collection (10 mL) was performed by a health professional (two nursing technicians). The samples were placed in vacuum blood collection tubes and sent to the Clinical Laboratory of the University Hospital of the Federal University of Sergipe (Aracaju, Sergipe, Brazil), where biochemical analyzes were performed by a laboratory biochemist.

2.8. Blood Cell and Leukocyte Count (HEMOGRAM)

Cell Dyn Ruby Abbott—The Cell-Dyn Ruby is an automatic hematology analyzer and multiparameter. The machine uses three basic methods. (1) The optical method is used to obtain red blood cell, platelet and leukocyte global counts. A photosensitive detector measures light scattering. The detected pulse length is proportional to the particle size (leukocyte, erythrocyte or platelet), which enables the identification of the volume of each of the formed elements of the blood. (2) In Laser Flow Cytometry, a flow of particulate matter passes through the laser beam crossing at an angle of 90 degrees, dispersing the light to a photomultiplier, which generates pulses in the histograms by determining the size and granularity of cells. This analysis is used to perform global and differential counts of leukocytes in 5 parts (neutrophils, lymphocytes, monocytes, eosinophils and basophils). (3) Colorimetry is a method that utilizes a chemical reaction color change and the final absorbance reading reaction, and is used for hemoglobin. Full blood counts were performed using a five-part differential hematology analyzer (Beckman Coulter AcT 5 diff AL Hematology Analyzer, CA, USA). The hematology analyzer uses a sequential dilution system and a dual-focused flow fluid dynamics technology, employing the Coulter Principle of impedance to count and measure the size of the cells.

2.9. Oxidative Stress

For oxidative stress, the tubes were centrifuged (3000× g for 10 min), and the plasma and serum were then aliquoted and stored at 4 °C for further analysis. To prevent the loss of volatile compounds, plasma ammonia was immediately measured using a spectrophotometric assay (Randox®, Crumlin, UK). The blood was centrifuged at 800× g for 15 min at 4 °C and the was serum stored at −80 °C.

In the serum, oxidative stress markers were evaluated. Lipoperoxidation was determined by measuring substances reactive to thiobarbituric acid (TBARS), according to the method described by Lapenna et al. [24]. For TBARS, 200 µL aliquots of the blood samples were added to a 400 µL mixture formed by equal parts of 15% trichloroacetic acid (TCA), 0.25N HCl and 0.375% TBA, plus 2.5 mM hydroxytoluene butylate (BHT) and 40 µL of 8.1% sodium dodecyl sulfate (SDS), being heated for 30 min at 95 °C in an oven. The pH of the mixture was adjusted to 0.9 with concentrated HCl. BHT was used to prevent lipid peroxidation during heating. After cooling to room temperature and adding 4 mL of butanol, the material was centrifuged at 800× g for 15 min at ±4 °C and the absorbance of the supernatant was measured at 532 nm. The molar extinction coefficient used was $1.54 \times 105\ M^{-1}\ cm^{-1}$ and the TBARS result was expressed in nmol Eq MDA/mL for the plasma samples.

The determination of sulfhydryl groups (SH) was carried out according to the methodology described by Faure and Lafond [25]. A 50 µL aliquot of the plasma was mixed in 1 mL of Tris-EDTA buffer (1 mM), and the first reading was taken at 412° (A1 reading). After this reading, 20 µL of 10 mM 5,5′-dithiobis 2-nitrobenzoic acid (DTNB) diluted in methanol were added. A new reading was taken after 15 min at room temperature (A2 reading). The Blank (B) contained only DTNB and Tris-EDTA buffer. The final unit was expressed in mM. The total sulfhydryl groups were calculated according to the molar absorption coefficient = $13{,}600\ cm^{-1}\ M^{-1}$: $(A2 - A1 - B) \times 1.57$ mM [25].

2.10. *Post-Workout Recovery Using a Placebo*

The control group received two sugar capsules, with packages identical to the IBU, 15 min before and 5 h after the resistance training. The same protocol as the IBU was followed, as described below.

2.11. *Post-Workout Recovery Using Ibuprofen*

This study followed the protocol used by De Souza et al. [16] and Fraga et al. [1], which consisted of administering IBU 15 min before and 5 h after resistance training. The experimental group received two capsules of IBU (400 mg) adding up to a total of 800 mg. Both IBU and PLA were packaged in identical capsules and the experiment was double-blind (i.e., participants and evaluators were unaware of what the capsules' substance was). Upon receiving the capsules, all volunteers were instructed on the ingestion procedures. Follow-up calls from the research team ensured compliance.

2.12. *Statistical Analysis*

The normality of the data was tested by the Shapiro Wilk test and the assumption was not denied. Descriptive statistics were used with measures of central tendency, mean (X) $\pm$ Standard Deviation (SD). Comparisons with ammonia were performed using the paired Student's t test. For the t-test, an effect size (Cohen's "*d*") was considered, adopting values of low effect ($\leq$0.20), medium effect (0.20 to 0.80), high effect (0.80 to 1.20) and very high effect (>1.20) [26–28]. For performance comparisons between time periods (Before $\times$ After $\times$ After 24 Hs $\times$ After 48 Hs), the assumptions were complicit for the use of the ANOVA test (Two Way). Point differences were verified by Bonferroni's Post Hoc. For ANOVA, the effect size was verified by the "partial squared eta" ($\eta 2p$), adopting values of low effect ($\leq$0.05), medium effect (0.05 to 0.25), high effect (0.25 at 0.50) and very high effect (>0.50) [28]. All statistical treatment was performed using the computerized package Statistical Package for the Social Science (SPSS; version 22.0) considering that the level of significance adopted was $p < 0.05$.

3. Results

It is noteworthy that, based on the effect size results of the present study, the calculation of the sampling power through the open-source software G* Power software (Version 3.0; Berlin, Germany) was performed, considering an $\alpha < 0.05$ and a $\beta = 0.80$. In this sense, the sample showed a power of >0.80 for the variables PT, FI, lymphocyte count, TBARS and SH.

Figure 2 presents the data related to the isometric strength through the Peak Torque (PT), Rate of Torque Development (RTD) and Fatigue Index (FI).

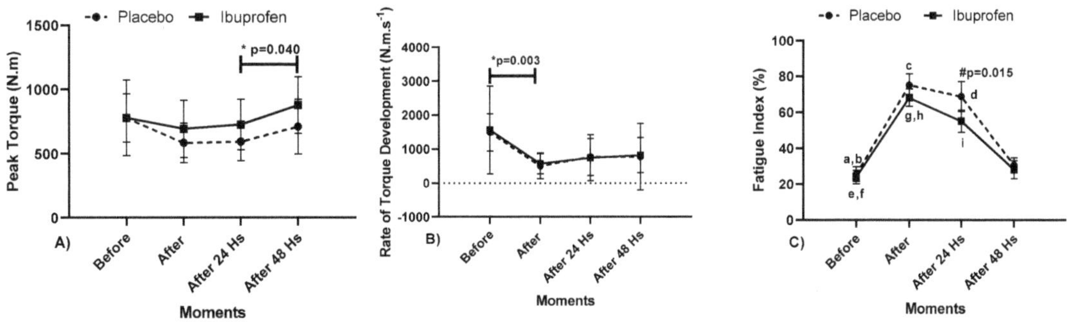

Figure 2. (**A**) Peak of Torque (PT), (**B**) Rate of Torque Development (RTD) e (**C**) Fatigue Index (FI) in diverse moments with Placebo (PLA) and Ibuprofen (IBU) used in recovery. Legend: "*": Indicates IntraClass difference in (**A**,**B**); "a–h": Indicates IntraClass differences in (**C**) and "#": Indicates InterClass difference (**C**) ($p < 0.05$).

Regarding the results, the data presented point to: Figure 2A PT—"*" Indicates a difference (IntraClass) in Ibuprofen (IBU) between the moments 24 and 48 h later (p = 0.040, η2p = 0.399, high effect), in PLA there were no differences. Figure 2B) RTD—"*" Indicates a difference (IntraClass) in Placebo (PLA) between the moments before and after (p = 0.003, η2p = 0.542, very high effect). Figure 2C FI—"#" Indicates the difference between PLA and IBU (InterClass), at the moment 24 h later, (p = 0.015, η2p = 0.745, very high effect). Regarding the supplement (IntraClass), in PLA—"a" Indicates a difference in the moments before and after; "b" in the moments before and 24 h after; "c" in the moments after and after 48 h and "d" in the moments 24 and 48 h afterwards (p <0.001). In the IBU supplement—"e" Indicates a difference between before and after; "f" in the moments before and 24 h after, "g" in the moments after and after 48 h and in the moments 24 and 48 h later (p < 0.001) and "h" in the moments after and 24 h after (p = 0.012). For PLA and IBU, η2p = 0.982 (very high effect).

Table 2 shows the results of the blood count and blood variables, concerning to the changes in the cell counts of the volunteers' immune system when comparing the post-PLA and the post-IBU.

Table 2. Blood markers in the presence and the absence of Ibuprofen.

Variables	Placebo	Ibuprofen	p	Cohen's d
CPR (mg/dL)	1.80 ±1.47	3.55 ± 2.37	0.031 *	1.80 d
Urea (mg/dL)	28.88 ± 6.66	24.38 ± 6.74	0.074	3.35 d
Uric acid (mg/dL)	5.14 ± 1.10	5.50 ± 1.21	0.140	1.41 d
Leukocytes (mm^3)	7.41 ± 1.80	6.64 ± 1.67	0.415	2.08 d
Neutrophils (%)	3.72 ± 1.22	4.88 ± 1.14	0.151	4.66 d
Lymphocytes (%)	2.43 ± 0.58	3.48 ± 0.78	0.001 *	4.36 d
Erythrocytes (million/mm^3)	5.06 ± 0.39	5.13 ± 0.46	0.221	0.64 b
Hemoglobin (g/mL)	15.08 ± 1.12	15.00 ± 1.43	0.767	0.20 a
Hematocrit (%)	42.63 ± 3.30	43.95 ± 4.00	0.019 *	1.31 d
MCV (U3)	84.29 ± 2.21	85.67 ± 3.36	0.090	1.08 c
MCH (UUG)	29.79 ± 0.52	29.23 ± 0.89	0.123	1.42 d
MCHC (%)	35.36 ± 0.94	34.12 ± 1.14	0.007 *	4.31 d
RDW (%)	10.24 ± 3.35	11.50 ± 0.54	0.305	0.45 b

* $p \leq 0.05$ (ANOVA two way and Post Hoc de Bonferroni). "a" small effect (≤0.20), "b" medium effect (0.20 a 0.80), "c" high effect (0.80 a 1.20) and "d" very high effect (>1.20). Legend: MCV: Mean Corpuscular Volume, MCHC: Mean Corpuscular Hemoglobin Concentration, RDW: Erythrocyte anisocytosis index.

The levels of C-Reactive Protein (CRP) increased (1.80 ± 1.47 to 3.55 ± 2.37 mg/dL, p = 0.031). There was no significant decrease in the total leukocyte count from 7.41 ± 1.80 to 6.64 ± 1.67 (mm^3) (p = 0.415) and a raise in the percentage of neutrophils 3.72 ± 1.22 (%) for 4.88 ± 1.14 (%) (p = 0.151) did not suffer a statistical difference, the percentage of lymphocytes from 2.43 ± 0.58 to 3.48 ± 0.78 (%) was increased (p = 0.001). All values remained within the reference values for cell counts for the adult population.

Figure 3 shows Oxidative Stress (TBARS and SH) at different times with the use of a placebo (PLA) and Ibuprofen (IBU) at different times.

Regarding Oxidative Stress, the following differences were presented: Figure 3A TBARS, "#" Difference between PLA and IBU after 48 h (p = 0.010), "a" Difference in PLA between Before and 24 h after (p = 0.023), "B" Difference in PLA between 2 and 24 h after (p < 0.001), and "c" Difference in PLA between 24 and 48 h after (p = 0.034), η2p = 0.173 (InterClass, medium effect) and η2p = 0.479 (Intra Group, high effect). Figure 3B SH, "a" Difference in PLA Before and 24 h after (p = 0.030), and "b" Difference in IBU Before and 2 h after (p = 0.001), η2p = 0.484 (IntraClass, high effect).

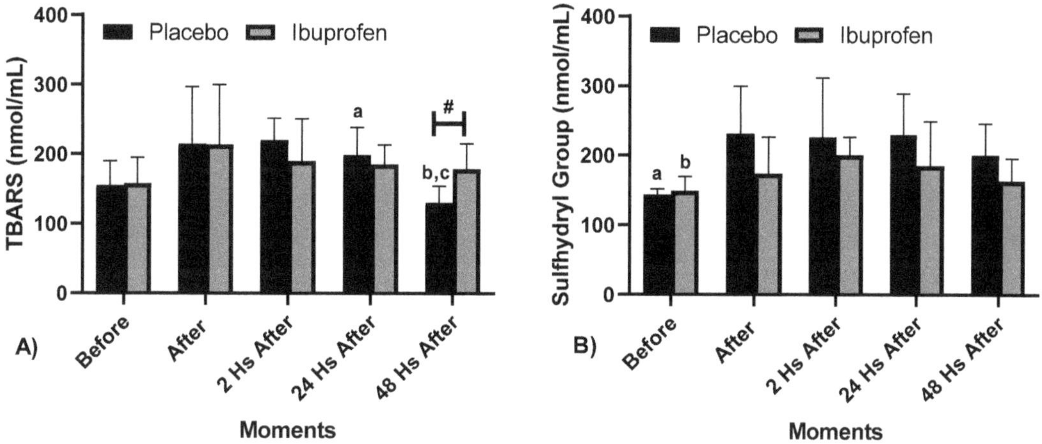

Figure 3. Oxidative Stress (**A**) Thiobarbituric Acid Reactive Substance (TBARS) e (**B**) Sulfhydrys Group (SH), at diverse moments with Placebo (PLA) and Ibuprofen (IBU) use at recovery. Legend: "a–c": Indicates IntraClass differences, and "#": Indicates InterClass difference C) ($p < 0.05$).

4. Discussion

This study aimed to analyze the effect of IBU on resisted post-workout recovery in PP athletes, by biomechanical variables and through biochemical indicators for muscle damage in the blood. The results highlighted that the Peak Torque with the use of IBU between 24 e 48 h after presented a significant difference, which resulted in better athlete performance. When evaluating the RTD, there was a decrease in the rate before and after training in the recovery method with PLA, and there were no differences in the IBU. The Fatigue Index was higher in recovery with the use of PLA after training compared to the use of IBU afterwards.

The results after the use of the IBU contributed to an improvement in the maximum isometric strength in relation to the use of the IBU 48 h after the training and the PLA 24 h after. A significant difference was also found with the use of the IBU 48 h after and PLA after the training. Therefore, it can be noticed that there was a maintenance of muscle function in the recovery with the use of IBU in the adapted bench press in Paralympic Powerlifting athletes concerning the PLA. This result can also be seen in the study by De Souza et al. [16] who demonstrated to mitigate fatigue in the gastrocnemius muscle in competing male runners who used IBU. Thus, when evaluating the FIM, the participants found better performance in the squat jump after the race than the control group.

The fatigue index showed significant differences in the results at all times of recovery. The moment that showed the highest peak was right after training with the use of PLA in relation to the use of IBU and in 48 h the values started to normalize. This result demonstrates that the recovery with the use of the IBU decreased the Fatigue Index and at the same time increased the Maximum Isometric Strength.

The muscle's ability to generate strength can be decreased due to muscle fatigue, and damage the movement's motor control [29,30]. It is noteworthy that the cause of fatigue can be central (i.e., when it affects the nervous system linked to muscle contraction) or peripheral (i.e., inhibitions in the contraction mechanisms of skeletal muscles). Therefore, fatigue when installed can disorder the movement sequence of the muscle segment's movements [31], and the recovery becomes an important factor to observe [32].

In addition, this exercise protocol also demonstrated changes in the number of lymphocytes (immunological parameter). These data indicate that these extra blood cells were mobilized from the cell matrix because there was not enough time to produce new cells in the bone marrow [6]. The specific mechanisms by which leukocyte counts increase have been intensively discussed and some studies have suggested that exercise induces an increase in circulating stress hormones (growth hormone, epinephrine and norepinephrine) and that these hormones may play a role in the mobilization of white blood cells [33,34]. Increased production and the release of these hormones at the beginning of exercise can also stimulate the initial increase in the number of circulating leukocytes [35].

Recently, in a systematic review, Gonçalves et al. [6], exposed that many studies showed that intense physical activity increases the ROSs production in the human body. The results of this study do not show that intense physical activity, (represented here by the bench press) with five sets of five maximum repetitions (80–90% of 1-RM) was not capable to increase the ROSs production. In the present work, ROSs production was evaluated by the levels of TBARS and Sulfhydrys Group (SH). It is noteworthy that no other work had investigated the ROSs production using a similar protocol. Most studies use indirect methods to evaluate an increased ROSs production, for example, by measuring malonaldehyde (MDA), which is a marker of lipid peroxidation and reacts with thiobarbituric acid reactive substances (TBARS), signaling the existence of oxidative stress [36,37].

Barili et al. [38], found that the test on the treadmill was a sufficient stimulus to increase the peroxides production in elderly subjects. Wang et al. [39] investigated how the exercise intensity impacts redox status mediated by oxidation of Low-Density Lipoprotein (LDL) in monocytes. The aforementioned authors concluded the work by stating that high-intensity physical activity (80% VO_2 max) increases ROSs production. Miyazaki et al. [40] investigated whether the high-intensity training (80% HRmax), during twelve weeks, would alter the oxidative stress induced by exercise after an event until the fatigue, verifying that exercising until the fatigue increases the ability of the neutrophils to produce ROSs and the training decreases this ability.

Studies measuring oxidative stress between different exercise models, such as aerobic exercise to fatigue and isometric exercise, and even associations between systemic oxidative stress, exercise intolerance and skeletal muscle abnormalities in patients with cardiac problems [41]. Another study comparing before and after with three different exercise protocols with trained subjects showed an increase of oxidative stress after intervention compared to pre-exercise [42]. Conversely, physical inactivity can reduce the body's antioxidant systemic defense capacity [43].

It has also been shown that the immobilization of a leg for two weeks tends to induce the production of ROSs and impaired mitochondrial breathing capacity in the immobilized muscles [44]. Studies in humans indicate that exercise tends to be beneficial in the defense and prevention of oxidative stress, dependent on an inflammatory process [45,46] since, during exercise, the inner membrane of the mitochondria interferes with ROSs, and the intensity or volume of exercise leads to an impact in the activity of free radical production that can interfere with the degrees of oxidative damage [47]. It seems that only a single session of acute exercise is able to increase the total antioxidant capacity [42]. Muscle damage tends to induce the build-up of neutrophils and cytokines, inducing oxidative stress [46]. On the other hand, researches indicate that chronic physical activities tend to increase adaptive and antioxidant defense systems [47,48].

Regarding the increase in free radicals, there is an indication that the antioxidant activity in the body tends not to decrease after intense chronic and acute exercises [46]. De Souza et al., [49] demonstrated lipid peroxidation in high intensity and long duration exercises in healthy individuals. Plasma MDA levels were measured before and after exercise until fatigue and did not undergo any significant changes. In the same direction, high intensity or exhaustive strength exercises tend to cause injuries and chronic fatigue. This would happen due to the imbalance between the production of reactive oxygen species (ROSs) and the endogenous antioxidant activity. Although ideal ROS production is important for muscle contraction, high ROSs concentrations tend to promote exercise-induced fatigue [50,51].

Skeletal musculature is reported to produce greater amounts of superoxide anion during training [52]. However, improvements have been reported concerning the oxidative stress provided by strength training [53,54]. Nevertheless, a higher training volume tends not to alter oxidative stress markers [55]. In this direction, studies indicate that in trained weightlifting athletes, high-intensity strength training tends to increase oxidative stress and decrease the antioxidant capacity of these athletes [56], which tends to lead to unfavorable effects of exercise in relation to health. In training with loads above 70% of 1 RM, the oxidative stress markers did not change. In contrast, high-intensity strength training, such as the one in the study, tends to increase the level of oxidative markers, as well as tends to decrease the production of antioxidants in powerlifting athletes [56], despite moderate to high-intensity training tends to improve oxidative stress [53,54]. Thus, it appears that strength training tends to improve oxidative stress among athletes [57]. The use of anti-inflammatory drugs, such as ibuprofen, tends to delay the anti-inflammatory response after exercise, helping the performance of powerlifting athletes [1], and this would explain the decrease in fatigue in the condition with ibuprofen use found in our study.

As is already widely discussed in the literature, high intensity or exhaustive physical exercise is recognized for increasing oxygen consumption resulting in a greater formation of reactive oxygen species (ROSs), greater susceptibility to muscle injuries and chronic fatigue [58]. In turn, non-steroidal anti-inflammatory agents (NSAIDs) became the most widely prescribed and used drugs worldwide [59,60], the use of IBU Non-steroidal anti-inflammatory drugs (NSAIDs) constitute one of the most consumed drug classes in the world. They have analgesic, antipyretic and anti-inflammatory effects that are used to treat acute pain arising from inflammation. Its effects occur through the reduction of the enzyme cyclooxygenase (COX), resulting in a decrease in precursors of prostaglandins and thromboxanes. The use of NSAIDs, when administered orally, is generally rapidly absorbed, it was found that the 400 mg tablet of IBU showed a peak concentration of 20–40 mg/mL in 1–2 h and decreasing to 5 mg/mL at the end of 6 h [61].

In this sense, the rapid absorption of IBU, which leads to rapid lowering of (MDA or TBARS) levels, occurs because it is subject to N-hydroxylation in the liver with the involvement of cytochrome P450 enzymes to form a toxic metabolite (NAPQI), which is rapidly inactivated by glutathione sulfhydryl (GSH) groups [62]. In large amounts of NAPQI, there is depletion of endogenous GSH in the liver and favors the binding of NAPQI with cellular biological macromolecules, such as proteins, nucleic acids and lipids, resulting in mitochondrial damage, endoplasmic reticulum stress and necrotic cell death. Then, in the toxicity phase, mitochondrial dysfunction increased oxidative stress occurs (damaged mitochondria lead to overproduction of reactive oxygen species (ROSs) [63,64]. As previously mentioned, the prophylactic use of IBU has a rapid absorption by the body, and as the levels of (MDA or TBARS) remain high as shown 48 h later. Finally, in studies carried out with animals that used ibuprofen, a cyclooxygenase inhibitor, the hematocrit and platelet counts were similar to those that did not receive ibuprofen [65].

As previously shown, the results of the present study agree with several other studies. However, it is necessary to emphasize that further research is still needed to recognize which model of physical activity really increases the production of ROSs without being neutralized by the antioxidant defense system, resulting in oxidative stress, which could cause damage to biomolecular structures.

Nevertheless, despite the important findings, the present study has limitations. The absence of monitoring food and quality of sleep of athletes during the recovery period made it impossible to control the intake of any food that would help to reduce oxidative stress and in the approach regarding sleep interference on oxidative stress.

5. Conclusions

It was concluded that the recovery with the use of Ibuprofen (IBU) presented a lower fatigue index and a lower decrease in strength when compared to the recovery with the placebo (PLA). There was also a reduction of muscle damage with the use of IBU compared to the recovery with PLA. This study demonstrated that, in the two forms of recovery, with

PLA and with the use of the IBU after strength training, there was no protective effect of the anti-inflammatory on oxidative stress markers.

From the point of view of practical applications of the findings, the results indicate that the use of ibuprofen can be a good strategy in the recovery of athletes aiming for a new training session, when it is necessary to recover more effectively after a workout, and even for competitions, as it allows an improved recovery in relation to the recovery without the use of supplements.

Author Contributions: Conceptualization, F.J.A., G.S.F., M.G.-M. and A.C.M., methodology, F.J.A., G.S.F., J.L.S. and R.F.d.S., software, R.F.d.S.; and A.R.P.F., validation, E.B.N., A.B.-C., T.L.B.F. and P.F.d.A.-N., formal analysis, T.L.B.F., investigation, M.G.-M., A.C.M. and J.L.S., resources, A.R.P.F., and N.D.G.; data curation, A.L.D.-d.-D.; A.d.F.Z. and M.d.S.C.-S., writing—original draft preparation, M.M.-F., E.B.N., A.d.F.Z., A.B.-C., A.L.D.-d.-D. And B.G.d.A.T.C.; writing—review and editing, P.F.d.A.-N., N.D.G., M.d.S.C.-S., B.G.d.A.T.C. and F.M.C., visualization, M.M.-F., E.M.-C., supervision, F.M.C., project administration, F.M.C.; funding acquisition, E.M.-C. All authors have read and agreed to the published version of the manuscript.

Funding: This research received no external funding. Nuno Domingos Garrido was funded by FCT—Fundação Portuguesa para a Ciência e Tecnologia (UID04045/2020).

Institutional Review Board Statement: The study was conducted according to the guidelines of the Declaration of Helsinki, and approved by the Institutional Review Board and approved by the Human Research Ethics Committee of the Federal University of Sergipe (UFS), under Statement Number 2637882/2018.

Informed Consent Statement: Informed consent was obtained from all subjects involved in the study.

Data Availability Statement: The data that support this study can be obtained from the address: https://www.ufs.br/Department%20of%20Physical%20Education, accessed on 12 July 2021.

Conflicts of Interest: The authors declare no conflict of interest.

References

1. Fraga, G.S.; Aidar, F.J.; Matos, D.G.; Marçal, A.C.; Santos, J.L.; Souza, R.F.; Carneiro, A.L.; Vasconcelos, A.B.; Da Silva-Grigoletto, M.E.; Tillaar, R.V.D.; et al. Effects of Ibuprofen Intake in Muscle Damage, Body Temperature and Muscle Power in Paralympic Powerlifting Athletes. *Int. J. Environ. Res. Public Health* **2020**, *17*, 5157. [CrossRef]
2. Hecksteden, A.; Skorski, S.; Schwindling, S.; Hammes, D.; Pfeiffer, M.; Kellmann, M.; Ferrauti, A.; Meyer, T. Blood-Borne Markers of Fatigue in Competitive Athletes—Results from Simulated Training Camps. *PLoS ONE* **2016**, *11*, e148810. [CrossRef] [PubMed]
3. Sampaio, C.R.S.F.; Aidar, F.J.; Ferreira, A.R.P.; Santos, J.L.D.; Marçal, A.C.; Matos, D.G.; Souza, R.F.; Moreira, O.C.; Guerra, I.; Fernandes Filho, J.; et al. Can Creatine Supplementation Interfere with Muscle Strength and Fatigue in Brazilian National Level Paralympic Powerlifting? *Nutrientes* **2020**, *12*, 2492. [CrossRef] [PubMed]
4. Bessa, A.L.; Oliveira, V.N.; Agostini, G.G.; Oliveira, R.J.; Oliveira, A.C.; White, G.E.; Wells, G.D.; Teixeira, D.N.; Espindola, F.S. Exercise Intensity and Recovery. *J. Strength Cond. Res.* **2016**, *30*, 311–319. [CrossRef]
5. Nunes-Silva, A.; Bernardes, P.T.T.; Rezende, B.M.; Lopes, F.; Gomes, E.C.; Marques, P.E.; Lima, P.M.A.; Coimbra, C.C.; Menezes, G.B.; Teixeira, M.M.; et al. Treadmill Exercise Induces Neutrophil Recruitment into Muscle Tissue in a Reactive Oxygen Species-Dependent Manner. An Intravital Microscopy Study. *PLoS ONE* **2014**, *9*, e96464. [CrossRef]
6. Gonçalves, C.A.M.; Dantas, P.M.S.; Dos Santos, I.K.; Dantas, M.P.; Da Silva, D.C.P.; Cabral, B.G.D.A.T.; Guerra, R.O.; Júnior, G.B.C. Effect of Acute and Chronic Aerobic Exercise on Immunological Markers: A Systematic Review. *Front. Physiol.* **2020**, *10*, 1602. [CrossRef]
7. Macedo, N.C.D.; Iessi, I.L.; Gallego, F.Q.; Netto, A.O.; Sinzato, Y.K.; Volpato, G.T.; Zambrano, E.; Damasceno, D.C. Swimming Program on Mildly Diabetic Rats in Pregnancy. *Reprod. Sci.* **2021**, *28*, 2223–2235. [CrossRef]
8. Yang, D.K.; Lee, S.-J.; Adam, G.O.; Kim, S.-J. Aralia continentalis kitagawa Extract Attenuates the Fatigue Induced by Exhaustive Exercise through Inhibition of Oxidative Stress. *Antioxidants* **2020**, *9*, 379. [CrossRef]
9. Speer, H.; D'Cunha, N.M.; Alexopoulos, N.I.; McKune, A.J.; Naumovski, N. Anthocyanins and Human Health—A Focus on Oxidative Stress, Inflammation and Disease. *Antioxidants* **2020**, *9*, 366. [CrossRef] [PubMed]
10. Varamenti, E.; Nikolovski, Z.; Elgingo, M.I.; Jamurtas, A.Z.; Cardinale, M. Training-induced variations in haematological and bio-chemical variables in adolescent athletes of arab origin throughout an entire athletic season. *J. Hum. Kinet.* **2018**, *64*, 123–135. [CrossRef]
11. Ammar, A.; Turki, M.; Hammouda, O.; Chtourou, H.; Trabelsi, K.; Bouaziz, M.; Abdelkarim, O.; Hoekelmann, A.; Ayadi, F.; Souissi, N.; et al. Effects of Pomegranate Juice Supplementation on Oxidative Stress Biomarkers Following Weightlifting Exercise. *Nutritients* **2017**, *9*, 819. [CrossRef] [PubMed]

12. Ammar, A.; Chtourou, H.; Souissi, N. Effect of Time-of-Day on Biochemical Markers in Response to Physical Exercise. *J. Strength Cond. Res.* **2017**, *31*, 272–282. [CrossRef] [PubMed]
13. Ammar, A.; Chtourou, H.; Hammouda, O.; Turki, M.; Ayedi, F.; Kallel, C.; Abdelkarim, O.; Hoekelmann, A.; Souissi, N. Relationship between biomarkers of muscle damage and redox status in response to a weightlifting training session: Effect of time-of-day. *Acta Physiol. Hung.* **2016**, *103*, 243–261. [CrossRef] [PubMed]
14. Finaud, J.; Lac, G.; Filaire, E. Oxidative stress: Relationship with exercise and training. *Sports Med.* **2006**, *36*, 327–358. [CrossRef] [PubMed]
15. Ammar, A.; Chtourou, H.; Hammouda, O.; Trabelsi, K.; Chiboub, J.; Turki, M.; Abdelkarim, O.; El Abed, K.; ben Ali, M.; Hoekelmann, A.; et al. Acute and delayed responses of C-reactive protein, malondialdehyde and antioxidant markers after resistance training session in elite weightlifters: Effect of time of day. *Chronobiol. Int.* **2015**, *32*, 1211–1222. [CrossRef]
16. De Souza, R.F.; de Matos, D.G.; Ferreira, A.R.P.; Chilibeck, P.; Barros, N.D.A.; Oliveira, A.S.; Cercato, L.M.; da Silva, D.S.; Aidar, F.J. Effect of Ibuprofen on Muscle, Hematological and Renal Function, Hydric Balance, Pain, and Performance During Intense Long-Distance Running. *J. Strength Cond. Res.* **2020**, *34*, 2076–2083. [CrossRef]
17. Mishra, D.K.; Fridén, J.; Schmitz, M.C.; Lieber, R.L. Anti-inflammatory medication after muscle injury. A treatment resulting in short-term improvement but subsequent loss of muscle function. *J. Bone Jt. Surg. Am.* **1995**, *77*, 1510–1519. [CrossRef]
18. Bondesen, B.A.; Mills, S.T.; Kegley, K.M.; Pavlath, G.K. The COX-2 pathway is essential during early stages of skeletal muscle re-generation. *Am. J. Physiol. Cell Physiol.* **2004**, *56*, C475–C483. [CrossRef]
19. Bondesen, B.A.; Mills, S.T.; Pavlath, G.K. The COX-2 pathway regulates growth of atrophied muscle via multiple mechanisms. *Am. J. Physiol. Cell Physiol.* **2006**, *290*, C1651–C1659. [CrossRef]
20. International Paralympic Committee (IPC). World Para Powerlifiting. Rules & Regulations. Available online: https://www.paralympic.org/sites/default/files/document/180215210800620_World%2BPara%2BPowerlifting%2BRules%2Band%2BRegulations_Feb%2B2018_0.pdf (accessed on 13 October 2020).
21. Resende, M.; Resende, R.V.; Reis, G.; Barros, L.; Bezerra, M.; Matos, D.; Marçal, A.; Almeida-Neto, P.; Cabral, B.; Neiva, H.; et al. The Influence of Warm-Up on Body Temperature and Strength Performance in Brazilian National-Level Paralympic Powerlifting Athletes. *Medicina* **2020**, *56*, 538. [CrossRef]
22. Ball, R.; Weidman, D. Analysis of USA Powerlifting Federation Data From 1 January 2012–11 June 2016. *J. Strength Cond. Res.* **2018**, *32*, 1843–1851. [CrossRef]
23. Souza, R.F.; Nogueira, A.C.; Ferreira, A.R.P.; Matos, D.G.; Oliveira, L.S.L.; Zanona, A.F.; Aidar, F.J. Analysis of muscle recovery time after acute stretching at peak torque of the hamstring muscles. *Med. Dello Sport* **2019**, *72*, 171–180.
24. Lapenna, D.; Ciofani, G.; Pierdomenico, S.D.; Giamberardino, M.A.; Cuccurullo, F. Reaction conditions affecting the relationship between thiobarbituric acid reactivity and lipid peroxides in human plasma. *Free Radic. Biol. Med.* **2001**, *31*, 331–335. [CrossRef]
25. Faure, P.; Lafond, J.L. Measurement of plasma sulfhydryl and carbonyl groups as a possible indicator of protein oxidation. In *Analysis of Free Radicals in Biological Systems*; Favier, A.E., Cadet, J., Kalyanaraman, B., Fontecave, M., Pierre, J.L., Eds.; Birkhäuser Verlag: Basel, Switzerland, 1995; pp. 237–248.
26. Lakens, D. Calculating and reporting effect sizes to facilitate cumulative science: A practical primer for t-tests and ANOVAs. *Front. Psychol.* **2013**, *4*, 863. [CrossRef]
27. Lakens, D. Equivalence tests: A practical primer for t tests, correlations, and meta-analyses. *Soc. Psychol. Personal. Sci.* **2017**, *8*, 355–362. [CrossRef] [PubMed]
28. Cohen, J. Statistics a power primer. *Psychol. Bull.* **1992**, *112*, 155–159. [CrossRef] [PubMed]
29. Elshafei, M.; Shihab, E. Towards Detecting Biceps Muscle Fatigue in Gym Activity Using Wearables. *Sensors* **2021**, *21*, 759. [CrossRef]
30. Mushtaq, W.; Hamdani, N.; Noohu, M.M.; Raghavan, S. Effect of Modified Constrain Induced Movement Therapy on Fatigue and Motor Performance in Sub Acute Stroke. *J. Stroke Cerebrovasc. Dis.* **2020**, *29*, 105378. [CrossRef] [PubMed]
31. Kerhervé, H.A.; Stewart, D.G.; McLellan, C.; Lovell, D. Fatigue Indices and Perceived Exertion Highlight Ergometer Specificity for Repeated Sprint Ability Testing. *Front. Sports Act. Living* **2020**, *2*, 45. [CrossRef] [PubMed]
32. Chen, C.-Y.; Chou, W.-Y.; Ko, J.-Y.; Lee, M.S.; Wu, R.-W. Early Recovery of Exercise-Related Muscular Injury by HBOT. *BioMed Res. Int.* **2019**, *2019*, 1–10. [CrossRef] [PubMed]
33. Stolk, R.F.; Van Der Pasch, E.; Naumann, F.; Schouwstra, J.; Bressers, S.; Van Herwaarden, A.E.; Gerretsen, J.; Schambergen, R.; Ruth, M.M.; Van Der Hoeven, J.G.; et al. Norepinephrine Dysregulates the Immune Response and Compromises Host Defense during Sepsis. *Am. J. Respir. Crit. Care Med.* **2020**, *202*, 830–842. [CrossRef]
34. Basile, U.; Bruno, C.; Napodano, C.; Vergani, E.; Pocino, K.; Brunetti, A.; Gulli, F.; Santini, S.A.; Mancini, A. Plasmatic free light chains as inflammatory marker in insulin resistance: Comparison of metabolic syndrome with adult growth hormone deficiency. *BioFactors* **2018**, *44*, 480–484. [CrossRef] [PubMed]
35. Moro, T.; Tinsley, G.; Longo, G.; Grigoletto, D.; Bianco, A.; Ferraris, C.; Guglielmetti, M.; Veneto, A.; Tagliabue, A.; Marcolin, G.; et al. Time-restricted eating effects on performance, immune function, and body composition in elite cyclists: A randomized controlled trial. *J. Int. Soc. Sports Nutr.* **2020**, *17*, 1–11. [CrossRef] [PubMed]
36. Barbouti, A.; Vasileiou, P.V.S.; Evangelou, K.; Vlasis, K.G.; Papoudou-Bai, A.; Gorgoulis, V.G.; Kanavaros, P. Implications of Oxidative Stress and Cellular Senescence in Age-Related Thymus Involution. *Oxid. Med. Cell. Longev.* **2020**, *2020*, 7986071. [CrossRef] [PubMed]

37. Sun, W.; Wang, Z.; Sun, M.; Huang, W.; Wang, Y.; Wang, Y. Aloin antagonizes stimulated ischemia/reperfusion-induced damage and inflammatory response in cardiomyocytes by activating the Nrf2/HO-1 defense pathway. *Cell Tissue Res.* **2021**, *384*, 735–744. [CrossRef] [PubMed]
38. Barili, A.; Corralo, V.D.S.; Cardoso, A.M.; Mânica, A.; Bonadiman, B.D.S.R.; Bagatini, M.D.; Da Silva-Grigoletto, M.E.; De Oliveira, G.G.; De Sá, C.A. Acute responses of hemodynamic and oxidative stress parameters to aerobic exercise with blood flow restriction in hypertensive elderly women. *Mol. Biol. Rep.* **2018**, *45*, 1099–1109. [CrossRef] [PubMed]
39. Wang, J.-S.; Lee, T.; Chow, S.-E. Role of exercise intensities in oxidized low-density lipoprotein-mediated redox status of monocyte in men. *J. Appl. Physiol.* **2006**, *101*, 740–744. [CrossRef] [PubMed]
40. Miyazaki, H.; Oh-Ishi, S.; Ookawara, T.; Kizaki, T.; Toshinai, K.; Ha, S.; Haga, S.; Ji, L.L.; Ohno, H. Strenuous endurance training in humans reduces oxidative stress following exhausting exercise. *Eur. J. Appl. Physiol.* **2001**, *84*, 1–6. [CrossRef]
41. Yokota, T.; Kinugawa, S.; Hirabayashi, K.; Yamato, M.; Takada, S.; Suga, T.; Nakano, I.; Fukushima, A.; Matsushima, S.; Okita, K.; et al. Systemic oxidative stress is associated with lower aerobic capacity and impaired skeletal muscle energy metabolism in heart failure patients. *Sci. Rep.* **2021**, *11*, 1–8. [CrossRef]
42. Souza, A.V.; Giolo, J.S.; Teixeira, R.R.; Vilela, D.D.; Peixoto, L.G.; Justino, A.B.; Caixeta, D.C.; Puga, G.M.; Espindola, F.S. Salivary and plas-matic antioxidant profile following continuous, resistance, and high-intensity interval exercise: Preliminary study. *Oxid. Med. Cell. Longev.* **2019**, *2019*, 5425021. [CrossRef]
43. Bouzid, M.A.; Hammouda, O.; Matran, R.; Robin, S.; Fabre, C. Influence of physical fitness on antioxidant activity and malondialdehyde level in healthy older adults. *Appl. Physiol. Nutr. Metab.* **2015**, *40*, 582–589. [CrossRef]
44. Gram, M.; Vigelsø, A.; Yokota, T.; Helge, J.W.; Dela, F.; Hey-Mogensen, M. Skeletal muscle mitochondrial H2O2 emission increases with immobilization and decreases after aerobic training in young and older men. *J. Physiol.* **2015**, *593*, 4011–4027. [CrossRef]
45. Koubaa, A.; Triki, M.; Trabelsi, H.; Baati, H.; Sahnoun, Z.; Hakim, A. The Effect of a 12-Week Moderate Intensity Interval Training Program on the Antioxidant Defense Capability and Lipid Profile in Men Smoking Cigarettes or Hookah: A Cohort Study. *Sci. World J.* **2015**, *2015*, 1–9. [CrossRef]
46. Nobari, H.; Nejad, H.; Kargarfard, M.; Mohseni, S.; Suzuki, K.; Adsuar, J.C.; Pérez-Gómez, J. The Effect of Acute Intense Exercise on Activity of Antioxidant Enzymes in Smokers and Non-Smokers. *Biomolecules* **2021**, *11*, 171. [CrossRef]
47. Pingitore, A.; Lima, G.P.P.; Mastorci, F.; Quinones, A.; Iervasi, G.; Vassalle, C. Exercise and oxidative stress: Potential effects of anti-oxidant dietary strategies in sports. *Nutrition* **2015**, *31*, 916–922. [CrossRef]
48. Bogdanis, G.; Stavrinou, P.; Fatouros, I.; Philippou, A.; Chatzinikolaou, A.; Draganidis, D.; Ermidis, G.; Maridaki, M. Short-term high-intensity interval exercise training attenuates oxidative stress responses and improves antioxidant status in healthy hu-mans. *Food Chem. Toxicol.* **2013**, *61*, 171–177. [CrossRef]
49. De Souza, R.F.; Augusto, R.L.; De Moraes, S.R.A.; De Souza, F.B.; Gonçalves, L.V.D.P.; Pereira, D.D.; Moreno, G.M.M.; De Souza, F.M.A.; Andrade-Da-Costa, B.L.D.S. Ultra-Endurance Associated with Moderate Exercise in Rats Induces Cerebellar Oxidative Stress and Impairs Reactive GFAP Isoform Profile. *Front. Mol. Neurosci.* **2020**, *13*, 157. [CrossRef]
50. De Araújo, S.S.; Aidar, F.J.; De Matos, D.G.; Santos, J.L.; Souza, L.M.V.; Silva, A.N.; Marçal, A.C.; Mourão, D.M.; Júnior, A.L.; Durães, G.M.; et al. Does Croton Argyrophyllus Extract Has an Effect on Muscle Damage and Lipid Peroxidation in Rats Submitted to High Intensity Strength Exercise? *Int. J. Environ. Res. Public Health* **2019**, *16*, 4237. [CrossRef] [PubMed]
51. Korivi, M.; Chen, C.-T.; Yu, S.-H.; Ye, W.; Cheng, I.-S.; Chang, J.-S.; Kuo, C.-H.; Hou, C.-W. Seaweed Supplementation Enhances Maximal Muscular Strength and Attenuates Resistance Exercise-Induced Oxidative Stress in Rats. *Evid. Based Complement. Altern. Med.* **2019**, *2019*, 3528932-9. [CrossRef] [PubMed]
52. Mastaloudis, A.; Yu, T.-W.; O'Donnell, R.P.; Frei, B.; Dashwood, R.H.; Traber, M.G. Endurance exercise results in DNA damage as detected by the comet assay. *Free Radic. Biol. Med.* **2004**, *36*, 966–975. [CrossRef] [PubMed]
53. Azizbeigi, K.; Azarbayjani, M.A.; Atashak, S.; Stannard, S.R. Effect of Moderate and High Resistance Training Intensity on Indices of Inflammatory and Oxidative Stress. *Res. Sports Med.* **2015**, *23*, 73–87. [CrossRef]
54. Ryrsø, C.K.; Thaning, P.; Siebenmann, C.; Lundby, C.; Lange, P.; Pedersen, B.K.; Hellsten, Y.; Iepsen, U.W. Effect of endurance versus resistance training on local muscle and systemic inflammation and oxidative stress in COPD. *Scand. J. Med. Sci. Sports* **2018**, *28*, 2339–2348. [CrossRef]
55. Koozehchian, M.S.; Daneshfar, A.; Fallah, E.; Agha-Alinejad, H.; Samadi, M.; Kaviani, M.; B., M.K.; Jung, Y.P.; Sablouei, M.H.; Moradi, N.; et al. Effects of nine weeks L-Carnitine supplementation on exercise performance, anaerobic power, and exercise-induced oxidative stress in resistance-trained males. *J. Exerc. Nutr. Biochem.* **2018**, *22*, 7–19. [CrossRef]
56. Jürgenson, J.; Serg, M.; Kampus, P.; Kals, J.; Zagura, M.; Viru, M.; Zilmer, K.; Zilmer, M.; Eha, J.; Unt, E. Oxidative Stress Parameters and Its Associations with Arterial Stiffness in Competitive Powerlifting Athletes After 12-Week Supervised Strength Training. *J. Strength Cond. Res.* **2019**, *33*, 1816–1822. [CrossRef]
57. Prasertsri, P.; Boonla, O. Upper and lower limb resistance training program versus high-intensity interval training on oxidative stress markers in university athletes. *J. Exerc. Rehabil.* **2021**, *17*, 198–206. [CrossRef] [PubMed]
58. Alves, N.F.; Porpino, S.K.; Monteiro, M.M.; Gomes, E.; Braga, V.A. Coconut oil supplementation and physical exercise improves baroreflex sensitivity and oxidative stress in hypertensive rats. *Appl. Physiol. Nutr. Metab.* **2015**, *40*, 393–400. [CrossRef] [PubMed]
59. Lima, F.D.; Stamm, D.N.; Della Pace, I.D.; Ribeiro, L.R.; Rambo, L.M.; Bresciani, G.; Ferreira, J.; Rossato, M.F.; Silva, M.A.; Pereira, M.E.; et al. Ibuprofen intake increases exercise time to exhaustion: A possible role for pre-venting exercise-induced fatigue. *Scand. J. Med. Sci. Sports* **2016**, *26*, 1160–1170. [CrossRef]

60. McAnulty, S.R.; Owens, J.T.; McAnulty, L.S.; Nieman, D.C.; Morrow, J.D.; Dumke, C.L.; Milne, G.L. Ibuprofen use during extreme exercise: Effects on oxidative stress and PGE2. *Med. Sci. Sports Exerc.* **2007**, *39*, 1075–1079. [CrossRef] [PubMed]
61. López-Durán, J.L.; Frías-Iniesta, J. Drug synergism in analgesia treated with ibuprofen and codeine. *SEMERGEN-Med. Fam.* **2014**, *38*, 24–32. [CrossRef]
62. Trappe, T.A.; White, F.; Lambert, C.P.; Cesar, D.; Hellerstein, M.; Evans, W.J. Effect of ibuprofen and acetaminophen on post exercise muscle protein synthesis. *Am. J. Physiol.* **2002**, *282*, E551–E556.
63. Trappe, T.A.; Fluckey, J.D.; White, F.; Lambert, C.P.; Evans, W.J. Skeletal Muscle PGF2α and PGE2 in Response to Eccentric Resistance Exercise: Influence of Ibuprofen and Acetaminophen. *J. Clin. Endocrinol. Metab.* **2001**, *86*, 5067–5070. [CrossRef] [PubMed]
64. Peterson, J.; Trappe, T.A.; Mylona, E.; White, F.; Lambert, C.P.; Evans, W.J.; Pizza, F.X. Ibuprofen and Acetaminophen: Effect on Muscle Inflammation after Eccentric Exercise. *Med. Sci. Sports Exerc.* **2003**, *35*, 892–896. [CrossRef] [PubMed]
65. Almqvist, P.M.; Ekström, B.; Kuenzig, M.; Haglund, U.; Schwartz, S.I. Increased survival of endotoxin-injected dogs treated with methylprednisolone, naloxone, and ibuprofen. *Circ. Shock* **1984**, *14*, 129–136. [PubMed]

biology MDPI

Article

Associations of Vitamin D Levels with Physical Fitness and Motor Performance; A Cross-Sectional Study in Youth Soccer Players from Southern Croatia

Barbara Gilic [1,2], Jelena Kosor [3], David Jimenez-Pavon [4,5,6], Josko Markic [3,7], Zeljka Karin [8], Daniela Supe Domic [9,10] and Damir Sekulic [1,*]

1 Faculty of Kinesiology, University of Split, 21000 Split, Croatia; barbaragilic@gmail.com
2 Faculty of Kinesiology, University of Zagreb, 10000 Zagreb, Croatia
3 Department of Pediatrics, University Hospital of Split, 21000 Split, Croatia; jelena.kosor1@gmail.com (J.K.); jmarkic@mefst.hr (J.M.)
4 MOVE-IT Research Group, Department of Physical Education, Faculty of Education Sciences, University of Cadiz, 11519 Cadiz, Spain; david.jimenez@uca.es
5 Biomedical Research and Innovation Institute of Cadiz (INiBICA) Research Unit, Puerta del Mar University Hospital University of Cadiz, 11009 Cadiz, Spain
6 CIBER of Frailty and Healthy Aging (CIBERFES), 28001 Madrid, Spain
7 School of Medicine, University of Split, 21000 Split, Croatia
8 Teaching Institute of Public Health of Split Dalmatian County, 21000 Split, Croatia; karinzeljka@gmail.com
9 Department of Medical Laboratory Diagnostics, University Hospital of Split, 21000 Split, Croatia; daniela.supedomic@gmail.com
10 Department of Health Studies, University of Split, 21000 Split, Croatia
* Correspondence: dado@kifst.hr

Citation: Gilic, B.; Kosor, J.; Jimenez-Pavon, D.; Markic, J.; Karin, Z.; Domic, D.S.; Sekulic, D. Associations of Vitamin D Levels with Physical Fitness and Motor Performance; A Cross-Sectional Study in Youth Soccer Players from Southern Croatia. *Biology* **2021**, *10*, 751. https://doi.org/10.3390/biology10080751

Academic Editors: Filipe Manuel Clemente, Badicu Georgian and Eugenia Murawska-Ciałowicz

Received: 14 June 2021
Accepted: 2 August 2021
Published: 5 August 2021

Simple Summary: Vitamin D is a fat-soluble prohormone crucial for bone mineralization, muscle contractility, and neurological conductivity. It is theorized that Vitamin D plays an important role in sport performances, especially in young athletes. In this study we examined the associations of levels of 25-hydroxyvitamin D (25(OH)D) with physical fitness and motor-performance achievements in youth soccer players from Southern Croatia. Participants were tested on physical fitness, motor performance and vitamin D at the end of the winter period, when levels of vitamin D are known to be lowest due to low exposure to sunlight. Results showed that deficiency of 25(OH)D was widespread among youth soccer players living in Southern Croatia. Low 25(OH)D levels were associated with lower results in fitness tests (i.e., tests of energetic capacities), but there was no correlation between 25(OH)D levels and the results in motor performance tests (i.e., skill tests). Our results support the theory of the association between vitamin D and energetic capacities of athletes, but there is no evidence on association between vitamin D and skill-based capacities.

Abstract: Vitamin D level is known to be a factor potentially influencing physical fitness, but few studies have examined this phenomenon among youth athletes. We aimed to evaluate the associations of vitamin D levels (as measured by 25-hydroxyvitamin D concentrations—25(OH)D) with various physical fitness and motor performance tests in youth football (soccer) players. This cross-sectional study included a total of 52 youth soccer players (15.98 ± 2.26 years old) from Southern Croatia. The participants were evaluated at the end of the winter period and data were collected of anthropometric measures (body mass and body height), vitamin D status (25(OH)D levels), physical fitness tests (sprints of 10 and 20 m, 20 yards test, the countermovement jump, the reactive strength index (RSI)) and motor performance tests (the soccer-specific CODS, the soccer-specific agility, and static balance). Among the studied players, 54% had 25(OH)D insufficiency/deficiency, showing a lack of 25(OH)D is widespread even in youth athletes living at a southern latitude. The 25(OH)D level was correlated with sprint 20 m, 20 yards tests, and RSI, showing a greater role of 25(OH)D in physical fitness tests where energetic capacity is essential than in sport-related motor performance tests where skills are crucial. Our results support the idea that vitamin D can play a determinant role in physical fitness tests with a clear physiological component, but is not crucial in motor performance tests related to specific sports where skills are a key component. Future studies should investigate the effects of vitamin D supplementation on the performance in physical fitness and motor performance tests among youth athletes.

Keywords: 25(OH)D; physiology of performances; puberty; pre-planned agility; non-planned agility

1. Introduction

Vitamin D is a fat-soluble prohormone with the function of maintaining the concentrations of calcium and phosphate within the physiological ranges, which is crucial for bone mineralization, muscle contractility, and neurological conductivity [1]. A total of 80–90% of vitamin D is synthesized during exposure to sunlight ultraviolet B radiation, whereas 10–20% is obtained from food [2]. Many factors influence vitamin D status including age, season, latitude, nutrition, physical activity levels, and body-fat percentage [3]. The circulating form of vitamin D, 25(OH)D, is often used for determining its status in the human body, and is considered the most accurate indicator of cutaneous synthesis and nutritional intake [4,5].

During the last few decades, interest has increased in vitamin D research because it was discovered that almost every body tissue has a vitamin D receptor, meaning that this can directly and indirectly influence their functions [6]. Specifically, vitamin D affects the regulation of the differentiation, proliferation, and growth of cells; hormone production; and immune, nervous, and muscle systems [7]. Regarding these functions, vitamin D is considered to play a role in optimal sports performance since it is involved in muscle physiology as muscles express a high number of vitamin D receptors, affects the transport of phosphate and calcium across muscle cell membranes, modulates phospholipid metabolism, and induces the expression of several myogenic transcription factors and myotubular sizes, which together affect the contractile filaments [8–10].

Simultaneously, evidence points to suboptimal vitamin D status in the general population, including athletes, children, and adolescents [11–13]. Collectively, studies frequently evidence inadequate levels of 25(OH)D among athletes worldwide [13]. Athletes are more susceptible to being vitamin D deficient/insufficient compared with the general population, probably because of their increased enzymatic activity following exercise [14]. The problem is additionally accentuated in youth athletes, since adolescents have an increased risk of malnutrition due to an increased need for energy and nutrients required for proper growth and development [15,16].

Soccer (football) is one of the most popular sports, characterized by a combination of low- and high-intensity activities, alternating short periods of high intensity activity with long periods of low intensity [17,18]. Although the aerobic metabolic component prevails in the form of low- and medium-intensity running, its high-intensity (anaerobic) activities, such as sprints, jumps, stopping, changes of direction, and striking, are key determinants of game outcomes [19]. As these activities are crucial for soccer, studies have already investigated the association between the status of vitamin D and various performance capacities in this sport. A significant correlation was found between vitamin D levels and muscle functioning assessed with jumping, sprinting, leg press, and aerobic capacity tests in professional Greek soccer players [9]. Alimoradi et al., (2019) also recorded a positive correlation between higher vitamin D concentration and improvements in sprint and leg press tests in a group of Iranian soccer players supplemented with vitamin D in comparison with players who were not supplemented [20]. Conversely, Ksiazek et al., (2016) did not find an association between vitamin D levels and lower limb muscle strength in professional soccer players [21].

However, several studies conducted on younger soccer players have reported inconsistent results. Low vitamin D levels were associated with low muscle strength level, changes in direction, jumps, and sprints in 7- to 15-year-old children involved in soccer training [22].

A more recent study by Bezrati et al. (2020) recorded significant improvements in sprinting, change in direction, and running speed tests after vitamin D supplementation in youth soccer players aged 8–15 years who were vitamin deficit at the beginning of the research [23]. Bezuglov et al. (2019) also noted a weak association of 25(OH)D level with running speed and muscle strength among soccer players aged 13–18 years [24]. Moreover, after 8 weeks of vitamin D supplementation, Jastrzebska et al., (2016) and Skalaska et al. (2019) did not record any significant difference in sports performance between placebo and experimental groups of soccer players aged 16–18 years [25,26].

As even the smallest improvements in any aspect affecting sports performance can lead to improvements in sports results, it is important to further investigate the association between vitamin D and sports performance. However, as evidenced from our review of the literature, studies that examined the associations between vitamin D status and physical performance in youth soccer players reported inconsistent findings [22,24–26]. The authors of previous studies examined relatively narrow sets of physical performance (strength, jumping, and sprinting capacities), which are generally called tests of physical fitness. On the other hand, previous studies rarely observed any of the skill-based motor performance such as balance, sport-specific change of direction speed (soccer-CODS), and sport-specific reactive agility (soccer-AGIL), which are generally considered more important determinants of success in soccer [19,27]. Finally, as an important methodological issue it must be mentioned that previous studies involved players from different teams [22,24–26]. As a result, participants' physical performance could vary due to differences in their training regimes and training methodologies, irrespective of vitamin D status. Therefore, the aims of this research were: (i) to determine the status of vitamin D and (ii) to evaluate the associations of 25(OH)D levels with physical fitness and motor performance in youth soccer players that were members of the same team living in Southern Croatia (Mediterranean region) at the end of the winter period.

2. Materials and Methods

2.1. Participants

The 52 participants in this cross-sectional research were youth soccer players (15.98 ± 2.26 years old). All players were members of the same soccer team located in Split, Croatia, residing at the latitude of 43° N Prior to the testing procedures, all players were informed of the purpose of the research and provided informed consent (for participants under 18 years of age, informed consent was signed by the parent or legal guardian). Inclusion criteria were active soccer training for at least three years and presence at 80% of the training sessions during the last month. The exclusion criteria were illness or injury that could have reduced the intensity of training during the last month and presence of pain in any part of the body during the testing. The testing was conducted during February 2020. The study was approved by the Ethical Board of the University of Split, Faculty of Kinesiology, and was conducted according to the guidelines in the newest version of the Declaration of Helsinki (approval No.: 2181-205-02-05-14-001).

2.2. Variables and Testing

The variables included in this study were anthropometric measures (body mass and body height), vitamin D status (25(OH)D levels), fitness tests, and motor performance tests.

Anthropometric variables were measured using standardized equipment by an experienced technician. Body height was measured in cm (accurate to 0.5 cm), and body mass (BM) accurate to 0.1 kg. Body mass index (BMI) was calculated using the following equation: BMI = BM (kg)/BH (m^2).

The 25(OH)D levels were measured using the Elecsys vitamin D total assay (electrochemiluminescence binding assay (ECLIA)), and with a Cobas e601 analyzer (Roche Diagnostics International Ltd., Rotkreuz, Switzerland), using a competitive electrochemiluminescence binding technique. The vitamin D total assay employs a vitamin D-binding protein as the capture protein to bind vitamin D3(25-OH) and vitamin D2(25-OH). The detection range of the test is 7.5–175 nmol/L 25(OH)D, and the sensitivity of the assay is 5 nmol 25(OH)D/L. The intraclass CV ranges from 2.2% (at 174 nmol 25(OH)D/L), to 6.7% (at 165 nmol 25(OH)D/L), with the 5.0 nmol/L, 7.5 nmol/L and 12.5 nmol/L for the limit of blank, limit of detection, and limit of quantification, respectively. The blood samples were taken from the athletes prior to the morning exercise session and were analyzed at the laboratory of the University Hospital of Split, Croatia. For the purpose of this study the 25(OH)D values of >75 nmol/L, 51–75 nmol/L, and <50 nmol/L were used for the classification of vitamin D sufficiency, insufficiency, and deficiency, respectively [28]. In further statistical analyses (please see later text for details) players were divided into two groups according to the 25(OH)D levels: inadequate levels of 25(OH)D (vitamin D deficiency/insufficiency, <75 nmol/L), and adequate levels of 25(OH)D (vitamin D sufficiency, >75 nmol/L).

Physical fitness tests included sprint over 10 and 20 m distance, generic test of change of direction speed (20 Yards test), countermovement jump, and reactive strength index.

The 10 (S10M) and 20 m (S20M) sprint tests were assessed by photoelectronic timing gates (Powertimer, Newtest, Oulu, Finland). The timing gates were placed 10 and 20 m from the starting line. Players were instructed to run as fast as possible from the start line along the 20 m distance. The time was recorded at 10 and 20 m. Players performed three testing trials and the best result was further analyzed.

The 20 yards test (20Y) was performed on a 10-yard field. One cone was placed in the center, the second 5 yards to the left, and the third 5 yards to the right from the center cone. Players were laterally standing 50 cm from the central cone where timing gate (Powertimer, Newtest, Oulu, Finland) was placed. Players had to rotate 90° and run 5 yards to the left, turn, sprint 10 yards to the right, turn and sprint 5 yards back to the central cone. Players performed three trials, and the best result was used for further analysis. For the countermovement jump test (CMJ), players started from the standing position with hands placed on their hips. They performed a downward movement to approximately 90° knee flexion, followed by a maximum upward movement. The height of the jump was measured using Optojump (Microgate, Bolzano, Italy). Participants performed three testing trials, and the best result was further analyzed.

The reactive strength index (RSI) was calculated from the jumped height and the ground contact time while performing the drop jump with both legs. The RSI was measured with the Optojump system (Microgate, Bolzano, Italy). The test was performed three times, and the best result was included in the analysis.

Motor performance tests included tests of soccer-specific change of direction speed test (soccer-CODS), soccer-specific agility test (soccer-AGIL), and balance test of overall stability index (OSI).

The soccer-CODS, and soccer-AGIL were tested with specifically designed tests. An infrared (IR) sensor was used as input for time triggering, and LEDs placed in the cones were set as outputs. The cones were shaping a Y pattern. The first cone was placed at the start line, and in that cone, IR sensor was placed. The other two cones were placed 4.5 m diagonal from the starting cone. For both tests, players had to run from the start line, switching the IR sensor, which triggered one of the lighting cones, and when the timing began. They had to run to the lit cone, kick the ball placed in front of that cone and run back to the start line when the time was recorded. For soccer-AGIL, the players did not know which cone would be lit, whereas, for the soccer-CODS, they knew the testing scenario in advance. For soccer-AGIL, the players performed five trials; for soccer-CODS, they performed two trials. The best results were used in the analysis for both soccer-CODS and soccer-AGIL [29] (Figure 1).

The OSI was measured using the Biodex Balance System. The resistance level was set to number 6, with trials lasting 20 s. Participants performed three trials and the best result was further included in the study [30].

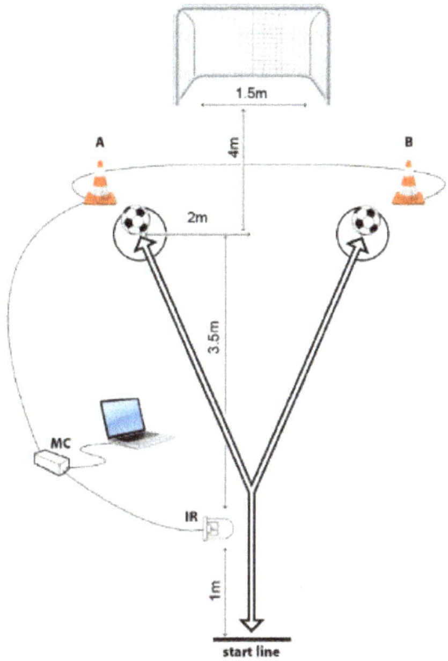

Figure 1. Soccer-specific change of direction speed and agility testing polygon.

2.3. Statistics

Variables were checked for normality of the distributions by the Kolmogorov–Smirnov test, and descriptive statistics included means and standard deviations. The test-retest reliability of the conditioning capacities was previously studied and reported in detail [29] and, therefore, in this study, all tests were checked for intratesting reliability by calculation of the intraclass coefficient (ICC) and coefficients of variation (CV).

To define the differences between the groups on the basis of vitamin D levels (vitamin D deficiency/insufficiency vs. vitamin D sufficiency), Student's *t*-test for independent samples was applied and further analyzed using the magnitude-based Cohen's effect size (ES) statistic with modified qualitative descriptors (ES ranges: <0.02 = trivial; 0.2–0.6 = small; >0.6–1.2 = moderate; >1.2–2.0 = large; and >2.0 very large differences) [31].

To identify the associations between 25(OH)D levels (measured values in nmol/L), and physical fitness and motor performance, Pearson's product moment correlation coefficients were calculated. These analyses were undertaken in two phases. In the first phase we calculated simple univariate correlations and regression coefficients between all pairs of variables. In the second phase, the significant correlations (regression coefficients) between 25(OH)D and physical fitness/motor performances that were evidenced in the first phase were additionally statistically controlled. For this purpose we included age and BMI as covariates in the correlation analysis, and statistically controlled the confounding effect of age in calculated correlations between vitamin D and physical fitness/motor performance variables. This was done since analyses showed significant correlation between age and vitamin D levels in the first phase, while there was a possibility of the confounding effects of age on association between 25(OH)D and fitness/performance. Pearson's product moment correlation coefficients (Pearson's R) of 0.1–0.29; 0.3–0.49; 0.5–1.0 (positive and negative values) represented low, moderate, and large correlation, respectively [32].

The type I error rate of 5% ($p < 0.05$) was set a priori and was considered statistically significant. StatSoft Statistica ver. 13.0 (Tulsa, OK, USA) was used for all analyses.

3. Results

Results of descriptive statistics for study variables and reliability parameters for physical fitness and motor performance tests are presented in Table 1. The intratesting-reliability of the applied tests of conditioning capacities ranged from appropriate values for OSI (ICC = 0.76, CV = 11%), to high reliability for S10 and S20M (ICC = 0.90 and 0.94, CV = 3% and 4%, respectively).

Table 1. Descriptive statistics and intra-testing reliability (ICC, CV) for the fitness tests and motor performance tests.

	Mean	Minimum	Maximum	Std.Dev.	ICC	CV
S10M (s)	1.77	1.53	2.06	0.11	0.90	0.03
S20M (s)	3.13	2.72	3.62	0.20	0.94	0.04
20Y (s)	4.86	4.25	5.78	0.36	0.89	0.06
CMJ (cm)	32.23	19.70	44.70	6.04	0.80	0.08
RSI (index)	1.11	0.44	1.79	0.29	0.78	0.08
Soccer-CODS (s)	2.59	2.24	3.29	0.23	0.80	0.08
Soccer-AGIL (s)	2.84	2.39	3.45	0.22	0.77	0.10
OSI (index)	1.63	0.60	6.40	1.03	0.76	0.11

Legend: S10M—sprint 10 m, S20M—sprint 20 m, 20Y—change of direction test over 20 yards, CMJ—countermovement vertical jump, RSI—reactive strength index, Soccer-CODS—soccer specific change of direction, Soccer-AGIL—soccer specific reactive agility test, OSI—overall stability index of balance.

Figure 2 presents the distribution of vitamin D sufficiency, insufficiency, and deficiency in youth players from southern Croatia, at the end of the winter period. In brief, 46.4% of players had sufficient levels of vitamin D (20(OH)D > 75 nmol/L), the vitamin D insufficiency (25(OH)D levels of 50–75 nmol/L) was evidenced in 44% of players, while 9.6% of studies players had vitamin D deficiency (25(OH)D < 50 nmol/L). The mean value for the 25(OH)D level was 79.03 nmol/L (standard deviation of 25.32), while median value was 73.55 nmol/L.

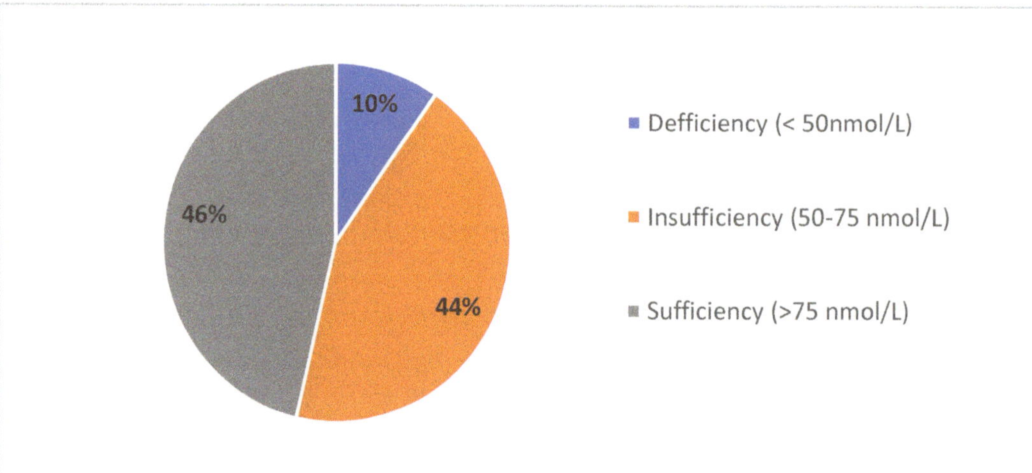

Figure 2. Vitamin D status (25(OH)D levels) in youth soccer players from southern Croatia.

Table 2 presents differences between groups of players on the basis of vitamin D status. Significant differences between groups were found in age (t-test = 2.2, p = 0.03), participants with sufficient vitamin D levels were older. Also, significant differences were found for S20M (t-test = 2.45, p = 0.02), and 20Y (t-test = 2.16, p = 0.04). In both cases, better results were achieved by participants who had sufficient vitamin D levels.

Table 2. Descriptive statistics and *t*-test differences between groups based on vitamin D status.

	Vitamin D Sufficiency (n = 23)		Vitamin D Deficiency/Insufficiency (n = 29)		*t*-Test	
	Mean	**Std.Dev.**	**Mean**	**Std.Dev.**	***t*-Value**	***p***
Age (years)	15.76	1.67	14.37	2.49	2.20	0.03
Body height (cm)	179.38	7.67	182.50	6.64	−1.06	0.30
Body mass (kg)	68.06	9.20	72.50	8.71	−1.22	0.23
BMI (kg/m^2)	21.09	1.94	21.71	1.77	−0.82	0.42
S10M (s)	1.74	0.11	1.80	0.11	−1.95	0.06
S20M (s)	3.06	0.18	3.19	0.20	−2.45	0.02
20Y (s)	4.74	0.27	4.95	0.40	−2.16	0.04
CMJ (cm)	33.67	6.26	31.09	5.71	1.55	0.13
RSI (index)	1.03	0.28	1.20	0.30	−2.00	0.04
Soccer-CODS (s)	2.52	0.20	2.63	0.25	−1.75	0.09
Soccer-AGIL (s)	2.80	0.15	2.88	0.25	−1.27	0.21
OSI (index)	1.94	1.43	1.39	0.45	1.99	0.05

Figure 3 presents ES differences between groups on studied variables according to their vitamin D status (sufficiency vs. insufficiency/deficiency). Moderate ES between groups based on 25(OH)D levels were evidenced for age (d = 0.65, 95%CI: 0.1–1.2), S20M (d = 0.68, 95%CI: 0.12–1.24), 20Y (d = 0.62, 95%CI: 0.06–1.17), while small ES were evidenced for all other variables.

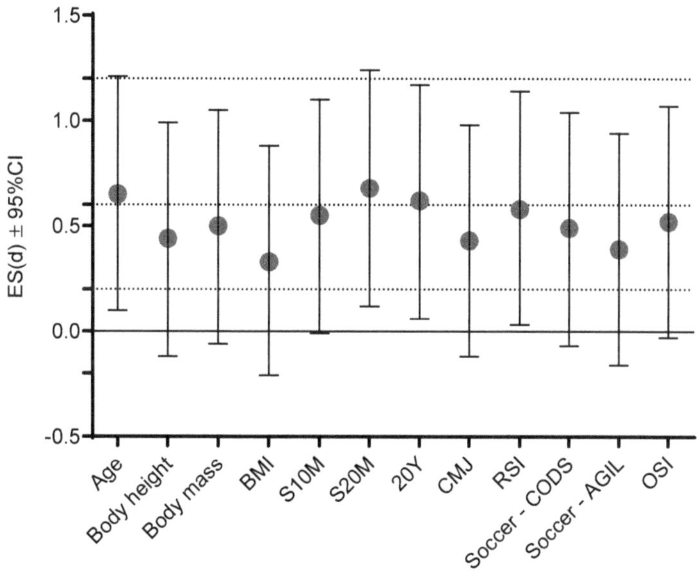

Figure 3. Effect size differences (Cohen's d) between groups based on vitamin D status (sufficiency vs. insufficiency/deficiency) dashed lines present ES ranges (<0.02 = trivial; 0.2–0.6 = small; >0.6–1.2 = moderate; >1.2–2.0 = large; and >2.0 very large differences).

Table 3 presents associations between study variables. Pearson's correlations between studied variables showed significant positive associations among most of the conditioning capacities. The negative correlation coefficients in some cases are results of opposite metrics of the variables (i.e., better achievement in sprinting is noted with a numerically lower result, while CMJ is noted with a numerically higher result), and practically highlight positive correlations between capacities. On the other hand, correlations between vitamin

D levels and conditioning capacities reached statistical significance ($p < 0.05$) for S20M (Pearson's R = 0.39, low positive correlation), 20Y (Pearson's R = 0.31, low positive correlation), CMJ (Pearson's R = 0.27, low positive correlation), and RSI (Pearson's R = 0.36, low positive correlation). Since vitamin D status was significantly ($p < 0.05$) correlated with participants' age (Pearson's R = 0.33, low positive correlation), the significant correlations between fitness variables (S20M, 20Y, CMJ, and RSI), and vitamin D status were additionally controlled for age as a covariate. Finally, significant ($p < 0.05$) partial correlations were confirmed for associations between vitamin D and (i) S20M (Pearson's R = −0.30, low negative correlation), (ii) 20Y (Pearson's R = −0.31, low negative correlation), and (iii) RSI (Pearson's R = 0.32, low positive correlation).

Table 3. Pearson's correlation coefficients between study variables (*** $p < 0.001$, ** $p < 0.01$, * $p < 0.05$).

	1	2	3	4	5	6	7	8	9	10	11	12
Age (1)	-											
Body height (2)	0.03	-										
Body mass (3)	0.38 **	0.74 ***	-									
BMI (4)	0.54 ***	0.23	0.80 ***	-								
25(OH)D (5)	0.33 *	−0.14	−0.14	−0.08	-							
S10M (6)	−0.39 **	0.14	0.08	−0.14	−0.26	-						
S20M (7)	−0.30 *	0.11	0.01	−0.20	−0.39 **	0.94 ***	-					
20Y (8)	−0.47 ***	−0.02	−0.19	−0.20	−0.31*	0.60 ***	0.77 ***	-				
CMJ (9)	0.32 *	−0.06	0.07	0.22	0.27 *	−0.65 ***	−0.73 ***	−0.57 ***	-			
RSI (10)	0.20	−0.15	−0.17	−0.06	0.36 **	−0.45 ***	−0.52 ***	−0.63 ***	0.45 ***	-		
Soccer-CODS (11)	−0.31 *	−0.06	−0.29 *	−0.18	−0.20	−0.59 ***	−0.66 ***	−0.46 ***	0.69 ***	0.41 **	-	
Soccer-AGIL (12)	0.17	0.15	0.25	−0.19	−0.14	0.48 ***	0.62 ***	0.67 ***	−0.52 ***	−0.45 ***	−0.37 **	-
OSI	−0.11	0.30 *	0.44 ***	0.37 **	−0.03	0.34 *	0.49 ***	0.50 ***	−0.32 *	−0.42 **	−0.32**	0.67 ***

4. Discussion

This study has several major findings. First, a deficiency of 25(OH)D was widespread among youth soccer players living in Southern Croatia. Second, low 25(OH)D levels were associated with lower results in the S20M, 20Y, and RSI (fitness tests). Last, there was no correlation between 25(OH)D levels and the results in motor performance tests.

4.1. Vitamin D Status in Youth Soccer Players

The mean value of serum 25(OH)D level of the studied players was 79.03 ± 25.32 nmol/L, with 54% of participants having low 25(OH)D levels (<75 nmol/L). These results are somewhat better than those with previously reported data in Croatia, neighboring countries, and countries with similar latitudes for samples of children and adolescents. For example, very recent report evidenced 38 ± 13 nmol/L in Italian, and 52 ± 14 nmol/L in Spanish children (42° N and 40° N, for Italian and Spanish samples, respectively) [33]. Furthermore, 72.3% Bosnian and Herzegovinian (B&H) adolescents (<18 years old) had low 25(OH)D levels (<75 nmol/L) [34], whereas 82.2% of Italian adolescents (10–21 years old) had hypovitaminosis D with a median serum 25(OH)D level of 50 nmol/L [35]. The somewhat better status of adolescents observed in our study can probably be attributed to the fact that our participants were athletes. Specifically, it is unlikely that adolescents with very low 25(OH)D levels will participate in systematic training (note that our participants had a minimum of 6 years of experience in soccer).

The results obtained are in line with several studies conducted on athletes worldwide. Specifically, a review study noted that 56% of athletes involved in different sports had inadequate 25(OH)D levels [13]. Similar results were noted for youth soccer players. A recent Russian study reported low 25(OH)D levels in 42.8% of youth soccer players [36] and 61.1% of youth Polish soccer players were found to have 25(OH)D concentration <50 nmol/L [26]. In addition to the general reasons for the lack of vitamin D (i.e., lack of vitamin D in regular nutrition, low bioavailability of the vitamin D) [15], another possible reason for low 25(OH)D levels could be the period or season in which the players were tested in our study (during February, which is the end of the winter). In brief, the skin synthesis of vitamin D is lower from October to March in regions far from the equator (i.e., above 35° north or south) [37]. As Croatia is located at 42°–46° N, the accumulation of vitamin D during the winter season is inadequate. In agreement with our findings, Morton et al. evidenced a significant decrease in vitamin D levels in professional English soccer players between the summer and winter seasons [38]. Some could argue that soccer is a sport practiced on an open field; therefore, vitamin D status could be better. However, the soccer players examined in this research held their practice and training sessions in late afternoon and evening (from 5:00 to 9:00 p.m.), when the sun has set during the winter; hence, it is not likely that the nature of the sport could have positively affected their vitamin D status during this period.

4.2. Vitamin D and Physical Fitness in Youth Soccer Players

We found association between serum 25(OH)D levels and results in the S20M, 20Y, and RSI, with better results in fitness tests among those players who had higher levels of 25(OH)D Meanwhile, the findings of previous studies that investigated the effects of 25(OH)D levels on sports performance are not consistent. Koundourakis et al., (2014) found a correlation of sprint and jumping test results with 25(OH)D levels in professional Greek soccer players, whereas several other studies did not [20,25]. Additionally, some studies concluded that soccer players with lower 25(OH)D levels have lower results in performance tests [39]. To explain the possible reasons for our findings (better performance in athletes with higher 25(OH)D levels), a short overview of the mechanisms of the influence of vitamin D on the muscle system and energy capacities of athletes is necessary. Specifically, authors of the study are of the opinion that mechanisms of potential influence of vitamin D on studied performances should be observed from two perspectives: (i) acute influence of higher vitamin D levels on performances, and (ii) chronic (long-term) influence of vitamin D levels on (development) of performance as a result of prolonged period of training. Therefore, the physiological background(s) of those two mechanisms will be presented accordingly.

First, it is important to highlight that vitamin D controls the expression of several proteins that are included in calcium signaling and phosphate-dependent metabolic processes, including ATP and creatine phosphate synthesis in the muscle cells [40]. Vitamin D regulates serum calcium concentrations which directly impacts muscle contraction [41]. Vitamin D increases the influx of calcium into the cytoplasm of muscle cells within minutes by activating two kinases, c-Src and PI3K, enabling calcium to bind to the troponin-tropomyosin complex resulting in exposure to active binding sites and allowing muscle contraction [42]. Increased calcium release and increased myosin movement across actin filaments may result in greater contractile muscle strength [43]. These processes could have all contributed to enhanced performance in tests that require a higher level of muscle excitation (i.e., sprint tests, generic CODS, and RSI). More precisely, all tests where 25(OH)D level was significantly correlated with achievement are dependent of the fast production of force [44]. Therefore, the better the muscular capacity to produce force, the better the sprinting, jumping, and generic-CODS performance. Therefore, there is a certain possibility that higher 25(OH) levels have a positive acute ergogenic impact on explosive performances, and consequently the established correlations should be observed taking into account the previously explained involvement of vitamin D in metabolic (energetic) processes in the human body. As a certain support to such a mechanism of (acute) influence we can note several studies that confirmed the short-term effects of vitamin D supplementation on improvement of explosive muscular capacities in athletes. Specifically, supplementing a high dose (5000–6000 IU per day) of vitamin D for 6–8 weeks among athletes and soccer players led to improved 5 and 10 m sprints, and vertical jumps, but also in aerobic capacity [45–47]. Additionally, young soccer players displayed improvements in vertical jumps, triple-hop jump, 10 and 30 m sprints, and shuttle run test after a one-time mega dose (200,000 IU) of vitamin D [23].

However (second), the authors of this study are more of the opinion that the mechanisms of vitamin D influence on fitness capacities should be observed from the perspective of prolonged, long-term effects of better vitamin D availability in athletes who have been involved in a regular training process. Specifically, vitamin D has been shown to play an active role in muscle maturation because, thanks to a vitamin D receptor (VDR)-mediated signal, myoblasts can differentiate into myocytes [48]. Activated VDR acts on cyclin-dependent kinases (serine threonine kinases) that actively participate in cell cycle regulation, stimulating muscle cells to proliferate and differentiate [49]. Vitamin D also regulates the expression of insulin-like growth factor-1, which has a well-recognized role in muscle hypertrophy and remodeling [50,51]. Vitamin D increases the size of myosin heavy chain type II-positive myotubes (e.g., fast-twitch muscle fibers), and increases the diameter and width of fast-twitch fibers [52]. This type of muscle fiber is a major determinant of the explosive type of human movement [53]. As a result, anaerobic activities of maximum intensity (e.g., jumping, running, accelerating) largely depend on the size of fast-twitch fibers. It is also important to note that vitamin D is considered to be involved in the production of testosterone as vitamin D metabolizing enzymes and receptors are expressed also in the Leydig cells where testosterone is produced [54]. It is known that the anabolic action of testosterone is stimulated by transcriptional genes regulation and amino acid uptake, leading to the synthesis of skeletal muscle proteins which later increased power and speed in athletes [45,55]. Supportively, several studies recorded that oral supplementation of vitamin D led to increased testosterone levels [45,56]. In our case it means that there is a certain possibility that players with higher 25(OH)D levels could have higher testosterone levels. It could allow them to undergo increased training load; more importantly, to achieve better supercompensation (supercompensation is a post training period during which the trained function/parameter has a higher performance capacity than it did prior to the training period); and subsequently to improve performance throughout prolonged period of time superiorly than their peers with lower 25(OH)D levels. However, since in this study we observed only 25(OH)D levels, we cannot currently draw a clear conclusion about the correlation between 25(OH)D levels and the studied capacities.

4.3. Vitamin D and Motor Performance in Youth Soccer Players

Our results indicated no significant correlation of 25(OH)D levels with motor performances (i.e., agility, balance). The most probable reason for the evidenced results should be found in specifics of soccer-specific performances studied herein. In most common words, these capacities are dependent on skill level, and in youth athletes are not strongly correlated to physical capacities. For example, tests of soccer-specific agility and change of direction speed are highly dependent on the ball-handling skills and perceptual abilities [27,29]. The same applies to balance, which is independent of most of the conditioning capacities, and actually depends on accurate control of body position over the surface [57]. As a result, the eventual positive effect of 25(OH)Don performance that largely depends on skill, and less on physical capacities, is logically limited.

One could argue that CMJ is a test of physical capacity (fitness) and, therefore, should be positively correlated with 25(OH)D. However, we must highlight that although the correlation between CMJ and 25(OH)D do not reach statistical significance when controlled for covariates, the Pearson's coefficient was actually similar to the correlations between other physical fitness tests (sprinting, jumping, and generic-CODS) and vitamin D. Additionally, CMJ is a relatively non-standard test in soccer players, especially if considering the CMJ-testing sequence used in our investigation (the hands remained on the hips during the test's execution), which probably resulted in a somewhat lower correlation between 25(OH)D level and CMJ simply because some players who were not highly familiar with the testing did not achieve their best results for CMJ.

4.4. Limitations and Strengths

The main limitation of this research comes from its cross-sectional design. Therefore, the results are not the final word of the problem as the causality cannot be determined. Second, we observed only 25(OH)D levels, and we observed no molecular data, which limited the possibility of the more accurate interpretation of the possible mechanisms of the influence of vitamin D on studied performances. Also, players were not supplemented with vitamin D and, therefore, we may not speak about the clear influence of vitamin D status on studied capacities. Furthermore, herein players were not separated according to playing positions since they were not specialized yet, meaning they played several positions in the game. Next, in this study we used 25(OH)D as an indicator of vitamin D status. Although it is globally accepted as the best marker of status of the vitamin D, other measures (i.e., free 25OHD, ratio of 24,25-dihydroxyvitamin D) could eventually be better indicators of vitamin D status, and this issue should be overviewed in future studies. Finally, in this investigation we have used definitions of sufficient/insufficient levels of 25(OH)D which is currently accepted in the country and the region where the study was commenced, but we cannot ignore the fact that there is an ongoing debate regarding the definition of vitamin D deficiency [58].

The main strength of this research is the inclusion of a wide range of performance capacities essential for soccer. Therefore, the results of this research are broadening the knowledge of associations of vitamin D status with numerous performance variables. Furthermore, tested players were from the same soccer club. Thus, the possible influence of other factors on performance is diminished. Therefore, we hope thatour results will improve knowledge in this field, and allow further investigations to be specifically focused on those variables/capacities on which certain influences of vitamin D could be expected.

5. Conclusions

Of the studied players, 44% had 25(OH)D insufficiency, which agrees with previous reports. Players were tested at the end of the winter season; although this may at least partially explain these results, the figure is still concerning and points to the necessity of further evaluating the reason for such results even in young athletes. However, only a negligible number of studied players had 25(OH)D deficiency (9.6%), which is a much lower prevalence of deficiency than that evidenced in age-matched youth from the same geographical region. Most probably, the very low 25(OH)D levels are a factor limiting successful participation in competitive sports (and, therefore, youth with very low 25(OH)D do not participate in sports), but it should be studied in more detail in the future.

It appears that 25(OH)D plays a greater role in tests where energetic capacity is essential than in tests where performance skill is crucial. Therefore, results from this research support the theory of the influence of 25(OH)D on the energetic capacity of athletes. Authors are of the opinion that this finding does not necessary imply an ergogenic effect of vitamin D on force production but should rather be observed as a long-term positive effect of appropriate vitamin D levels on supercompensation and consequently on positive training effects during a prolonged period of time (i.e., sports career), resulting in superior sprinting, jumping, and generic-CODS capacity.

However, 25(OH)D was not correlated with tests representing motor performance skill. Considering all the present findings and the results of previous studies, we concluded that 25(OH)D plays a supportive and not a crucial role in sports performance. Future studies should investigate the effects of 25(OH)D supplementation on the performance in various tests of physical capacities among young soccer players and in other sports.

Author Contributions: Data curation, J.K.; Formal analysis, B.G. and Z.K.; Investigation, J.M. and D.S.D.; Methodology, B.G., J.K., Z.K. and D.S.; Project administration, D.S.; Resources, D.S.D.; Validation, B.G. and D.J.-P.; Writing—original draft, B.G., J.K., D.J.-P., J.M. and D.S., Writing—revision, B.G., D.S., J.M., Z.K. All authors have read and agreed to the published version of the manuscript.

Funding: This research is partially financed by Croatian Science Foundation (grant numbers: IP-2018-01-8330; DOK-2018-09-1940).

Institutional Review Board Statement: The study was conducted according to the guidelines of the Declaration of Helsinki, and approved by the Institutional Ethics Committee) of University of Split, Faculty of Kinesiology, Split, Croatia (Protocol code: 2181-205-02-05-1418-0021, approved on 8 January 2018).

Informed Consent Statement: Informed consent was obtained from all subjects involved in the study.

Data Availability Statement: Data will be available upon reasonable request.

Acknowledgments: Biomedical Research Networking Center on Frailty and Healthy Aging (CIBER-FES) and FEDER funds from the European Union (CB16/10/00477); Support of the Croatian Science Foundation is gratefully acknowledged.

Conflicts of Interest: The authors declare no conflict of interest.

References

1. Anderson, P.; May, B.; Morris, H. Vitamin D Metabolism: New Concepts and Clinical Implications. *Clin. Biochem. Rev.* **2003**, *24*, 13.
2. Holick, M.F. Evolution, Biologic Functions, and Recommended Dietary Allowances for Vitamin, D. In *Vitamin D*; Springer: Berlin/Heidelberg, Germany, 1999; pp. 1–16.
3. Autier, P.; Boniol, M.; Pizot, C.; Mullie, P. Vitamin D Status and Ill Health: A Systematic Review. *Lancet Diabetes Endocrinol.* **2014**, *2*, 76–89. [CrossRef]
4. Lips, P. Vitamin D Physiology. *Prog. Biophys. Mol. Biol.* **2006**, *92*, 4–8. [CrossRef]
5. Wimalawansa, S.J. Vitamin D Deficiency: Effects on Oxidative Stress, Epigenetics, Gene Regulation, and Aging. *Biology* **2019**, *8*, 30. [CrossRef] [PubMed]
6. Zella, L.A.; Meyer, M.B.; Nerenz, R.D.; Lee, S.M.; Martowicz, M.L.; Pike, J.W. Multifunctional Enhancers Regulate Mouse and Human Vitamin D Receptor Gene Transcription. *Mol. Endocrinol.* **2010**, *24*, 128–147. [CrossRef]
7. Holick, M.F. Vitamin D: A Millenium Perspective. *J. Cell. Biochem.* **2003**, *88*, 296–307. [CrossRef] [PubMed]
8. Pojednic, R.M.; Ceglia, L. The Emerging Biomolecular Role of Vitamin D in Skeletal Muscle. *Exerc. Sport Sci. Rev.* **2014**, *42*, 76–81. [CrossRef]
9. Koundourakis, N.E.; Androulakis, N.E.; Malliaraki, N.; Margioris, A.N. Vitamin D and Exercise Performance in Professional Soccer Players. *PLoS ONE* **2014**, *9*, e101659. [CrossRef]
10. Knechtle, B.; Jastrzębski, Z.; Hill, L.; Nikolaidis, P.T. Vitamin D and Stress Fractures in Sport: Preventive and Therapeutic Measures-A Narrative Review. *Medicina* **2021**, *57*, 223. [CrossRef] [PubMed]
11. Bendik, I.; Friedel, A.; Roos, F.F.; Weber, P.; Eggersdorfer, M. Vitamin D: A Critical and Essential Micronutrient for Human Health. *Front. Physiol.* **2014**, *5*, 248. [CrossRef]
12. Cashman, K.D.; Dowling, K.G.; Škrabáková, Z.; Gonzalez-Gross, M.; Valtueña, J.; De Henauw, S.; Moreno, L.; Damsgaard, C.T.; Michaelsen, K.F.; Mølgaard, C.; et al. Vitamin D Deficiency in Europe: Pandemic? *Am. J. Clin. Nutr.* **2016**, *103*, 1033–1044. [CrossRef] [PubMed]
13. Farrokhyar, F.; Tabasinejad, R.; Dao, D.; Peterson, D.; Ayeni, O.R.; Hadioonzadeh, R.; Bhandari, M. Prevalence of Vitamin D Inadequacy in Athletes: A Systematic-Review and Meta-Analysis. *Sports Med.* **2015**, *45*, 365–378. [CrossRef]
14. Dahlquist, D.T.; Dieter, B.P.; Koehle, M.S. Plausible Ergogenic Effects of Vitamin D on Athletic Performance and Recovery. *J. Int. Soc. Sports Nutr.* **2015**, *12*, 33. [CrossRef]
15. González-Gross, M.; Valtueña, J.; Breidenassel, C.; Moreno, L.A.; Ferrari, M.; Kersting, M.; De Henauw, S.; Gottrand, F.; Azzini, E.; Widhalm, K.; et al. Vitamin D Status Among Adolescents in Europe: The Healthy Lifestyle in Europe by Nutrition in Adolescence Study. *Br. J. Nutr.* **2012**, *107*, 755–764. [CrossRef]
16. Wiciński, M.; Adamkiewicz, D.; Adamkiewicz, M.; Śniegocki, M.; Podhorecka, M.; Szychta, P.; Malinowski, B. Impact of Vitamin D on Physical Efficiency and Exercise Performance—A Review. *Nutrients* **2019**, *11*, 2826. [CrossRef] [PubMed]
17. Evangelos, B.; Lefteris, M.; Aristotelis, G.; Ioannis, G.; Natalia, K. Aerobic and Anaerobic Capacity of Professional Soccer Players in Annual Macrocycle. *J. Phys. Educ. Sport* **2016**, *16*, 527. [CrossRef]
18. Marin-Pagan, C.; Blazevich, A.J.; Chung, L.H.; Romero-Arenas, S.; Freitas, T.T.; Alcaraz, P.E. Acute Physiological Responses to High-Intensity Resistance Circuit Training vs. Traditional Strength Training in Soccer Players. *Biology* **2020**, *9*, 383. [CrossRef]
19. Reilly, T. *Science and Soccer*; Routledge: London, UK, 2003.
20. Alimoradi, K.; Nikooyeh, B.; Ravasi, A.A.; Zahedirad, M.; Shariatzadeh, N.; Kalayi, A.; Neyestani, T.R. Efficacy of Vitamin D Supplementation in Physical Performance of Iranian Elite Athletes. *Int. J. Prev. Med.* **2019**, *10*, 100. [CrossRef] [PubMed]
21. Książek, A.; Zagrodna, A.; Dziubek, W.; Pietraszewski, B.; Ochmann, B.; Słowińska–Lisowska, M. 25(OH)D_3 Levels Relative to Muscle Strength and Maximum Oxygen Uptake in Athletes. *J. Hum. Kinet.* **2016**, *50*, 71–77. [CrossRef]

22. Bezrati, I.; Hammami, R.; Ben Fradj, M.K.; Martone, D.; Padulo, J.; Feki, M.; Chaouachi, A.; Kaabachi, N. Association of Plasma 25-Hydroxyvitamin D with Physical Performance in Physically Active Children. *Appl. Physiol. Nutr. Metab.* **2016**, *41*, 1124–1128. [CrossRef]
23. Bezrati, I.; Ben Fradj, M.K.; Hammami, R.; Ouerghi, N.; Padulo, J.; Feki, M. A Single Mega Dose of Vitamin D(3) Improves Selected Physical Variables in Vitamin D-Deficient Young Amateur Soccer Players: A Randomized Controlled Trial. *Appl. Physiol. Nutr. Metab.* **2020**, *45*, 478–485. [CrossRef]
24. Bezuglov, E.; Tikhonova, A.; Zueva, A.; Khaitin, V.; Lyubushkina, A.; Achkasov, E.; Waśkiewicz, Z.; Gerasimuk, D.; Żebrowska, A.; Nikolaidis, P.T. The Dependence of Running Speed and Muscle Strength on the Serum Concentration of Vitamin D in Young Male Professional Football Players Residing in the Russian Federation. *Nutrients* **2019**, *11*, 1960. [CrossRef]
25. Jastrzębska, M.; Kaczmarczyk, M.; Jastrzębski, Z. Effect of Vitamin D Supplementation on Training Adaptation in Well-Trained Soccer Players. *J. Strength Cond. Res.* **2016**, *30*, 2648–2655. [CrossRef]
26. Skalska, M.; Nikolaidis, P.T.; Knechtle, B.; Rosemann, T.J.; Radzimiński, Ł.; Jastrzębska, J.; Kaczmarczyk, M.; Myśliwiec, A.; Dragos, P.; López-Sánchez, G.F. Vitamin D Supplementation and Physical Activity of Young Soccer Players During High-Intensity Training. *Nutrients* **2019**, *11*, 349. [CrossRef] [PubMed]
27. Pojskic, H.; Åslin, E.; Krolo, A.; Jukic, I.; Uljevic, O.; Spasic, M.; Sekulic, D. Importance of Reactive Agility and Change of Direction Speed in Differentiating Performance Levels in Junior Soccer Players: Reliability and Validity of Newly Developed Soccer-Specific Tests. *Front. Physiol.* **2018**, *9*, 506. [CrossRef]
28. Holick, M.F.; Binkley, N.C.; Bischoff-Ferrari, H.A.; Gordon, C.M.; Hanley, D.A.; Heaney, R.P.; Murad, M.H.; Weaver, C.M.; Endocrine, S. Evaluation, Treatment, and Prevention of Vitamin D Deficiency: An Endocrine Society Clinical Practice Guideline. *J. Clin. Endocrinol. Metab.* **2011**, *96*, 1911–1930. [CrossRef]
29. Krolo, A.; Gilic, B.; Foretic, N.; Pojskic, H.; Hammami, R.; Spasic, M.; Uljevic, O.; Versic, S.; Sekulic, D. Agility Testing in Youth Football (Soccer) Players; Evaluating Reliability, Validity, and Correlates of Newly Developed Testing Protocols. *Int. J. Environ. Res. Public Health* **2020**, *17*, 294. [CrossRef]
30. Sekulic, D.; Spasic, M.; Mirkov, D.; Cavar, M.; Sattler, T. Gender-Specific Influences of Balance, Speed, and Power on Agility Performance. *J. Strength Cond. Res.* **2013**, *27*, 802–811. [CrossRef]
31. Cohen, J. *Statistical Power Analysis for the Behavioral Sciences*; Academic Press: Cambridge, MA, USA, 2013.
32. Mukaka, M.M. Statistics Corner: A Guide to Appropriate Use of Correlation Coefficient in Medical Research. *Malawi Med. J.* **2012**, *24*, 69–71. [PubMed]
33. Wolters, M.; Intemann, T.; Russo, P.; Moreno, L.A.; Molnar, D.; Veidebaum, T.; Tornaritis, M.; De Henauw, S.; Eiben, G.; Ahrens, W.; et al. 25-Hydroxyvitamin D Reference Percentiles and The Role of Their Determinants Among European Children and Adolescents. *Eur. J. Clin. Nutr.* **2021**. [CrossRef] [PubMed]
34. Sokolovic, S.; Alimanovic-Alagic, R.; Dzananovic, L.; Cavaljuga, S.; Beslic, N.; Ferhatbegovic-Opankovic, E. Vitamin D Status in Bosnia and Herzegovina: The Cross-Sectional Epidemiological Analysis. *Osteoporos. Int.* **2017**, *28*, 1021–1025. [CrossRef]
35. Vierucci, F.; Del Pistoia, M.; Fanos, M.; Erba, P.; Saggese, G. Prevalence of Hypovitaminosis D and Predictors of Vitamin D Status in Italian Healthy Adolescents. *Ital. J. Pediatr.* **2014**, *40*, 54. [CrossRef]
36. Bezuglov, E.; Tikhonova, A.; Zueva, A.; Khaitin, V.; Waśkiewicz, Z.; Gerasimuk, D.; Żebrowska, A.; Rosemann, T.; Nikolaidis, P.; Knechtle, B. Prevalence and Treatment of Vitamin D Deficiency in Young Male Russian Soccer Players in Winter. *Nutrients* **2019**, *11*, 2405. [CrossRef]
37. Bolland, M.J.; Grey, A.B.; Ames, R.W.; Mason, B.H.; Horne, A.M.; Gamble, G.D.; Reid, I.R. The Effects of Seasonal Variation of 25-Hydroxyvitamin D and Fat Mass on a Diagnosis of Vitamin D Sufficiency. *Am. J. Clin. Nutr.* **2007**, *86*, 959–964. [CrossRef]
38. Morton, J.P.; Iqbal, Z.; Drust, B.; Burgess, D.; Close, G.L.; Brukner, P.D. Seasonal Variation in Vitamin D Status in Professional Soccer Players of the English Premier League. *Appl. Physiol. Nutr. Metab.* **2012**, *37*, 798–802. [CrossRef]
39. Hamilton, B.; Whiteley, R.; Farooq, A.; Chalabi, H. Vitamin D Concentration in 342 Professional Football Players and Association with Lower Limb Isokinetic Function. *J. Sci. Med. Sport* **2014**, *17*, 139–143. [CrossRef]
40. Berchtold, M.W.; Brinkmeier, H.; Muntener, M. Calcium Ion in Skeletal Muscle: Its Crucial Role for MuscleFunction, Plasticity, and Disease. *Physiol. Rev.* **2000**, *80*, 1215–1265. [CrossRef]
41. Hamilton, B. Vitamin D and Human Skeletal Muscle. *Scand. J. Med. Sci. Sports* **2010**, *20*, 182–190. [CrossRef] [PubMed]
42. Santillán, G.; Katz, S.; Vazquez, G.; Boland, R.L. TRPC3-Like Protein and Vitamin D Receptor Mediate 1alpha,25(OH)2D3-Induced SOC Influx in Muscle Cells. *Int. J. Biochem. Cell Biol.* **2004**, *36*, 1910–1918. [CrossRef] [PubMed]
43. Girgis, C.M.; Clifton-Bligh, R.J.; Hamrick, M.W.; Holick, M.F.; Gunton, J.E. The Roles of Vitamin D in Skeletal Muscle: Form, Function, and Metabolism. *Endocr. Rev.* **2013**, *34*, 33–83. [CrossRef] [PubMed]
44. Michailidis, Y.; Tabouris, A.; Metaxas, T. Effects of Plyometric and Directional Training on Physical Fitness Parameters in Youth Soccer Players. *Int. J. Sports Physiol. Perform.* **2019**, *14*, 392–398. [CrossRef]
45. Michalczyk, M.M.; Gołaś, A.; Maszczyk, A.; Kaczka, P.; Zając, A. Influence of Sunlight and Oral D(3) Supplementation on Serum 25(OH)D Concentration and Exercise Performance in Elite Soccer Players. *Nutrients* **2020**, *12*, 1311. [CrossRef]
46. Jastrzębska, M.; Kaczmarczyk, M.; Michalczyk, M.; Radzimiński, Ł.; Stępień, P.; Jastrzębska, J.; Wakuluk, D.; Suárez, A.D.; López Sánchez, G.F.; Cięszczyk, P.; et al. Can Supplementation of Vitamin D Improve Aerobic Capacity in Well Trained Youth Soccer Players? *J. Hum. Kinet.* **2018**, *61*, 63–72. [CrossRef] [PubMed]

47. Close, G.L.; Russell, J.; Cobley, J.N.; Owens, D.J.; Wilson, G.; Gregson, W.; Fraser, W.D.; Morton, J.P. Assessment of Vitamin D Concentration in Non-Supplemented Professional Athletes and Healthy Adults During the Winter Months in The UK: Implications for Skeletal Muscle Function. *J. Sports Sci.* **2013**, *31*, 344–353. [CrossRef] [PubMed]
48. Tanaka, M.; Kishimoto, K.N.; Okuno, H.; Saito, H.; Itoi, E. Vitamin D Receptor Gene Silencing Effects on Differentiation of Myogenic Cell Lines. *Muscle Nerve* **2014**, *49*, 700–708. [CrossRef]
49. Irazoqui, A.P.; Boland, R.L.; Buitrago, C.G. Actions of 1,25(OH)2-Vitamin D3 on the Cellular Cycle Depend on VDR and p38 MAPK in Skeletal Muscle Cells. *J. Mol. Endocrinol.* **2014**, *53*, 331–343. [CrossRef]
50. Ameri, P.; Giusti, A.; Boschetti, M.; Murialdo, G.; Minuto, F.; Ferone, D. Interactions between Vitamin D and IGF-I: From Physiology to Clinical Practice. *Clin. Endocrinol.* **2013**, *79*, 457–463. [CrossRef]
51. Kraemer, W.J.; Ratamess, N.A. Hormonal Responses and Adaptations to Resistance Exercise and Training. *Sports Med.* **2005**, *35*, 339–361. [CrossRef]
52. Agergaard, J.; Trøstrup, J.; Uth, J.; Iversen, J.V.; Boesen, A.; Andersen, J.L.; Schjerling, P.; Langberg, H. Does Vitamin-D Intake During Resistance Training Improve the Skeletal Muscle Hypertrophic and Strength Response in Young and Elderly men?—A Randomized Controlled Trial. *Nutr. Metab.* **2015**, *12*, 32. [CrossRef]
53. Deschenes, M.R.; Kraemer, W.J. Performance and Physiologic Adaptations to Resistance Training. *Am. J. Phys. Med. Rehabil.* **2002**, *81*, S3–S16. [CrossRef] [PubMed]
54. Nimptsch, K.; Platz, E.A.; Willett, W.C.; Giovannucci, E. Association Between Plasma 25-oh Vitamin D and Testosterone Levels in Men. *Clin. Endocrinol.* **2012**, *77*, 106–112. [CrossRef] [PubMed]
55. West, D.W.; Phillips, S.M. Anabolic Processes in Human Skeletal Muscle: Restoring The Identities of Growth Hormone and Testosterone. *Phys. Sports Med.* **2010**, *38*, 97–104. [CrossRef] [PubMed]
56. Pilz, S.; Frisch, S.; Koertke, H.; Kuhn, J.; Dreier, J.; Obermayer-Pietsch, B.; Wehr, E.; Zittermann, A. Effect of Vitamin D Supplementation on Testosterone Levels in Men. *Horm. Metab. Res.* **2011**, *43*, 223–225. [CrossRef] [PubMed]
57. Zech, A.; Hübscher, M.; Vogt, L.; Banzer, W.; Hänsel, F.; Pfeifer, K. Balance Training for Neuromuscular Control and Performance Enhancement: A Systematic Review. *J. Athl. Train.* **2010**, *45*, 392–403. [CrossRef]
58. Giustina, A.; Bouillon, R.; Binkley, N.; Sempos, C.; Adler, R.A.; Bollerslev, J.; Dawson-Hughes, B.; Ebeling, P.R.; Feldman, D.; Heijboer, A.; et al. Controversies in Vitamin D: A Statement From the Third International Conference. *JBMR Plus* **2020**, *4*, e10417. [CrossRef]

biology

MDPI

Article

Knowledge about Fibromyalgia in Fibromyalgia Patients and Its Relation to HRQoL and Physical Activity

María Mendoza-Muñoz [1], Jesús Morenas-Martín [2,*], Miguel Rodal [3], Judith García-Matador [1], Miguel Ángel García-Gordillo [4] and José Ignacio Calzada-Rodríguez [1]

1 Health, Economy, Motricity and Education Research Group (HEME), Faculty of Sport Sciences, University of Extremadura, 10003 Caceres, Spain; mamendozam@unex.es (M.M.-M.); judgm@hotmail.com (J.G.-M.); jocalzada@alumnos.unex.es (J.I.C.-R.)
2 Motor Control Research Group, Faculty of Sport Sciences, University of Extremadura, 10003 Caceres, Spain
3 BioErgon Research Group, Faculty of Sport Sciences, University of Extremadura, 10003 Caceres, Spain; mrodal@unex.es
4 Universidad Autónoma de Chile, Talca 3467987, Chile; miguel.garcia@uautonoma.cl
* Correspondence: jesusmorenas@unex.es

Simple Summary: Fibromyalgia (FM) affects 2.40% of the Spanish population and its most widespread treatment has been the combination of patient education, pain coping strategies and exercise, in this sense, with respect to patient education, few studies have tried to see the relationship between education in FM with an improvement in FM. Therefore, the aim of this study was to know the level of knowledge about FM among patients in Extremadura, to explore the relationship between knowledge of FM and health-related quality of life (HRQoL), and to analyze the relationship between knowledge of physical activity in FM and the practice of physical activity. For this purpose, 121 women with a mean age of 55.06 years were evaluated. It was found that 10% of these women had low knowledge of FM, 49% medium and 41% high. It was also found that the level of knowledge of physical activity was only weakly related to HRQOL and body pain. Thus, it was concluded that the level of knowledge about FM of patients in Extremadura was medium-high and that there was a weak relationship between knowledge about physical activity in FM and HRQOL. However, no relationship was found between knowledge of physical activity in FM and the practice of physical activity.

Abstract: *Introduction*: Fibromyalgia (FM) affects 2.40% of the Spanish population. The most widespread treatment has been the combination of patient education, pain coping strategies and exercise. With regard to patient education, there are few previous studies on the efficacy of relating FM education in isolation with an improvement in FM, although there are some studies that report that health education programs could modify the perception of quality of life and improve pain. *Objectives*: the aim was to find out the level of knowledge about FM among patients in Extremadura, to explore the relationship between knowledge of FM and Health-Related Quality of Life (HRQoL) and to analyze the relationship between knowledge of physical activity in FM and the practice of physical activity. *Methods*: A single-measure cross-sectional study was carried out with 121 women with a mean age of 55.06 (±9.93) years. The following questionnaires were used: Fibromyalgia Knowledge Questionnaire (FKQ); SF12v2 (Short-Form Health Survey); and EURO-QOL-5D-5L (EQ-5D-5L). *Results*: regarding the level of knowledge of the participants about FM, it was found that 10% had a low knowledge, 49% medium and 41% high. In relation to the associations between the level of knowledge and HRQoL, a weak correlation between EQ-5D-5L and the FKQ in the domain of physical activity (r = 0.243) were found. *Conclusions*: it can be concluded that the level of knowledge about FM of the patients from Extremadura was medium-high and that there is a direct weak relationship between knowledge about physical activity in FM and HRQoL. However, no association was found between knowledge of physical activity in FM and the practice of physical activity.

Keywords: HRQoL; knowledge; physical activity; rheumatic diseases; women

Citation: Mendoza-Muñoz, M.; Morenas-Martín, J.; Rodal, M.; García-Matador, J.; García-Gordillo, M.Á.; Calzada-Rodríguez, J.I. Knowledge about Fibromyalgia in Fibromyalgia Patients and Its Relation to HRQoL and Physical Activity. *Biology* **2021**, *10*, 673. https://doi.org/10.3390/biology10070673

Academic Editors: Badicu Georgian and Eugenia Murawska-Ciałowicz

Received: 3 June 2021
Accepted: 14 July 2021
Published: 16 July 2021

1. Introduction

Fibromyalgia (FM) was recognised as a disease in the Copenhagen Declaration by the World Health Organisation in 1992 [1]. It is the second most frequent condition among rheumatic diseases [2] and it can be considered as a form of non-articular rheumatism characterised by chronic diffuse musculoskeletal pain, together with the presence of multiple pressure sore spots [3], implying a great impact on the physical, psychological and social well-being of the patient [2].

FM affects 2.7% of the world population, being more prevalent in 50-year-old or older women [4]. In Spain, FM affects of 2.40% of the population [5], being associated with females aged between 40 and 59 years [6] who suffer from diffuse musculoskeletal pain, aches or stiffness associated with fatigue, anxiety, poor sleep, headaches, irritable bowel syndrome, subjective swelling in arthritic and periarticular areas, and numbness [7]. In this line, it has been shown that the factors that can reduce health and quality of life are widespread pain and tenderness, cognitive problems, non-recovery sleep, fatigue, depression, anxiety, poor physical fitness, stiffness and mobility or balance problems [8].

Currently, there is no cure for this pathology, but by applying different multidisciplinary treatments, improvements in quality of life at the physical, psychological and social levels can be achieved [9,10]. Therefore, treatment depends not only on the use of pharmacologic therapy, but also on the implementation of intervention programmes, often of a cognitive-behavioural nature and physical exercise [11–13].

Previous studies on the impact of FM on quality of life suggest that the most effective treatment would be a combination of patient education, pain coping strategies and aerobic exercise [14]. Specifically, related to patient education, previous studies on the efficacy of linking information/education on FM in isolation to a noticeable improvement in FM can be hardly found. Nevertheless, Koca, et al. [15] state that patients' knowledge about FM could contributes to the control of disease and other studies have reported that health education programmes could modify the perception of quality of life and improve pain relief, as well as decrease dependence on health services [16], or that education in pain physiology seems to improve health status and long-term endogenous pain inhibition [17]. However, in most studies, information/education is approached from a multidisciplinary point of view, i.e., it is associated with other types of treatment such as physical activity, with very beneficial results [18–20].

In relation to the content of information/education, at the societal level, the increased interest in FM has contributed to confusion with claims lacking scientific rigour, as many questions have been raised about people with FM and many media have tried to answer them. Information should be direct, objective and in accordance with the scientific knowledge that exists about FM, since improving knowledge about FM can be useful for creating a social bond between people affected by this disease and health professionals, family members, friends and/or work colleagues, thereby facilitating their adaptation to the difficulties that the disease causes in their daily lives.

Therefore, the aim was to find out the level of knowledge about FM among patients in Extremadura, as well as to explore the relationship between knowledge of FM and Health-Related Quality of Life (HRQoL) and, more specifically, to analyse the relationship between knowledge about physical activity in FM and the practice of physical activity.

2. Materials and Methods

2.1. Study Design

A single measurement cross-sectional study was conducted.

2.2. Ethical Approval

Ethical approval was provided by the Bioethics and Biosafety Committee of the University of Extremadura (approval number: 51/2013).

2.3. Sample Calculation

One hundred and thirteen participants were needed to reach a power of 80% to detect a difference of 0.30 between the null hypothesis correlation of 0.29 (very low or close to zero association) and the alternative hypothesis correlation of 0.60 (high association) [21]. The significance level was set at alpha equal to 0.05. Futhermore, about ten percent more were recruited to meet this estimate

2.4. Participants

The total sample consisted of 121 women with a mean age of 55.06 (±9.93) years. Participants were recruited along one month from different associations of patients affected by FM syndrome (Mérida, Don Benito, Valencia de Alcántara, Alburquerque, Toledo and Almansa) and the "Manuel Encinas" Health Centre in Cáceres by the professionals working there. The inclusion criteria followed were: to be of legal age, to have a diagnosis of FM by a rheumatologist according the American College of Rheumatology criteria [8], not to present any disease affecting the understanding of the test and to sign the informed consent form.

2.5. Procedure, Material and Measurement

In this study, several instruments were used to assess knowledge about both FM and HRQoL, as well as to obtain characterisation data on the participants (age, level of education, years of diagnosis of the disease, distance spent walking per week, hours spent doing physical activity per week, etc.). The questionnaires were sent to the participants both by e-mail and by post. These were completed along one week in the centres mentioned above, with the help of a trained professional who got in contact with the research team. Finally, this material was sent to the research team together with the participants' informed consent. The instruments used were:

Fibromyalgia Knowledge Questionnaire (FKQ) [22]. This questionnaire assesses the level of knowledge about FM. It is composed of 18 items, divided into four domains (general knowledge about the disease; knowledge about treatment, medication and possible side effects; knowledge about physical activity; and knowledge about day-to-day activities in relation to energy used or how best to save energy), with the total score of the questionnaire reaching 26 points (Cronbach's alpha > 0.69).

Short-Form Health Survey (SF12v2) [23]. It is a generic measure of HRQoL, which is an abbreviated form of the SF-36. This instrument can be self-administered or completed through an interview. It contains 12 items addressing eight domains (physical functioning, physical role, body pain, general health, vitality, social function, emotional role, and mental health) and two summary scores related to physical health and mental health (Cronbach's alpha > 0.70). With these dimensions, and by applying predetermined algorithms, the two summary scores are created: the physical health index (PHI) and the mental health index (MHI) [24].

EURO-QOL-5D-5L (EQ-5D-5L) [25]. This instrument was developed by the EuroQol Group with the aim of improving the previous version (EQ-5D-3L) [25]. The EQ-5D-5L describes the current health status of individuals across five dimensions (mobility, self-care, daily activities, pain/discomfort and anxiety/depression) (ICC: 0.69 to 0.94) [26]. Each of the dimensions has five possible responses with 1 being the best possible health status and 5 the worst. In addition, this questionnaire includes a visual analogue scale, which evaluates from 0 to 100 the health status of the person on the day the questionnaire is administered.

2.6. Statistical Analysis

The information collected in this study was tabulated in an anonymised database designed specifically for this study. IBM SPSS Statistics software (Version 21, IBM SPSS, Chicago, IL, USA) was used for statistical analysis.

Data were presented as mean and standard deviation (SD) and as percentage. After analysing the normality of each of the study variables using the Kolmogorov-Smirnov test and verifying that the data did not follow a normal distribution, Spearman's correlation coefficient was calculated between the SF12v2, FKQ and EQ-5D-5L questionnaires. The Bonferroni correction was applied based on the formula $\alpha^* = \alpha/n - 1$ [27], where α^* is the corrected value at which the null hypothesis should be rejected and n is the number of hypothesis pairs. Therefore, the alpha significance level was set at 0.001 for multiple comparisons between the FKQ and the SF12. To define the type of correlation between the variables, Cohen's ratio [28] was used: 0.10 to 0.29 small correlation; 0.30 to 0.59 moderate correlation; 0.6 to 0.79 high, and $\geq$0.8 excellent. To analyse the FKQ domains associated with a greater impact of the quality of life of FM (EQ 5D-5L), a multivariate linear regression model was designed, in which the dependent variable was the score on the EQ 5D-5L and the independent variables were those that were significant in the bivariate analysis. Statistical significance was defined for $p < 0.05$.

3. Results

The total number of participants was 121 with a mean age of 55.06 ($\pm$9.93). In this way, Table 1 shows the characterization of the participants.

Table 1. Characterization of study patients.

	N	Means (SD)
Age (years)	121	55.06 (9.93)
Years of accurate diagnosis	118	4.94 (0.65)
Years suffering from generalized pain	114	15.21 (10.24)
Number of members of the family unit	106	2.71 (1.17)
Monthly household income (Euros)	94	528.37 (862.50)
Number of trigger points (0–18)	85	7.07 (5.32)
Degree of pain at the time of the questionnaire (0–100)	119	68.06 (17.68)

Table 2 shows that most of the participants had primary education (47.8%), followed by secondary education (29.8%).

Table 2. Educational level of study patients.

	N (%)
Illiterate	1 (0.8)
No education but can read and write	9 (7.4)
Primary education	58 (47.9)
Secondary education	36 (29.8)
Graduate	14 (9.9)
Other	3 (2.5)

Table 3 shows the percentages of participants at each level according to the scores obtained in each domain of the FKQ and for the total of score. For this purpose, the following score ranges were established: less than 50% of the maximum possible score (low knowledge), from 50 to 75% (medium knowledge) and more than 75% (high knowledge). In this table, it can be seen that for all the domains and for the total of the questionnaire, the highest percentages are in the range 50–75%, followed by the range +75% regarding the general and physical activity domains, as well as the total of score.

The response percentages related to the total sample are shown in Table 4. These percentages are expressed for each response option within its corresponding item. It can be seen that there are several response options that are not correct, but with a high percentage of responses from the participants (item 1 option b = 43%; item 8 option d = 49% and option e = 63%; item 11 option b = 44%), as well as some correct responses with a low number of responses (item 17 option b = 13%).

Table 3. Percentage of participants at each level of knowledge according to the scores obtained for each domain of the FKQ.

	FKQ														
	General			Medication			Physical Activity			Energy			Total		
	Range	n	%	Range	n	%	Range	n	%	Range	n	%	Range	n	%
−50%	0–4	11	10	0–2	35	29	0–2	18	15	0–2	37	30	0–11	12	10
50–75%	5–7	62	51	3–4	66	55	3–4	67	55	3	71	59	12–17	59	49
+75%	8–9	48	39	5–6	20	16	5	36	30	4	13	11	18–23	50	41

Range: range of scores for the levels established according to each domain. FKQ: Fibromyalgia Knowledge Questionnaire.

Table 4. Percentage of responses for each possible option within each item in the FKQ.

Response Option		a	b	c	d	e	f
FKQ Domains	Item	(%)	(%)	(%)	(%)	(%)	(%)
General	1	13	43	56 *	7	58*	11
	2	5	77 *	15	3	94*	5
	3	3	0	0	97 *	1	
	4	75 *	7	12	13	62 *	10
	5	6	82 *	53*	2	31	
Medication	6	73 *	4	17	4	2	7
	7	2	41 *	17	13	28	
	8	4	50	3	49 *	63 *	12
	9	26	62 *	44 *	17	27	
Physical Activity	10	7	3	2	78 *	10	
	11	0	44	48 *	0	9	
	12	92 *	1	1	0	6	
	13	74 *	7	4	77 *	15	
Energy	14	4	0	93 *	2	2	
	15	2	95 *	83 *	3	5	
	16	75 *	1	15	0	14	
	17	11	13 *	2	23	53	
	18	6	3	0	10	72 *	10

* Correct answer for the corresponding item.

Table 5 shows the means in the different dimensions of the SF12v2 of the participants, compared to normal values for healthy women between 55–64 [29]. We can highlight that the values of the study sample are below normal values for healthy women between 55–64 in Spain, especially in the dimensions of physical function, body pain, general health vitality and physical health index [29].

Table 5. SF12v2 averages.

	Means (SD) (n = 121)	Means (SD) (n = 188)
Physical Function	30.55 (8.51)	45.4 (13.4)
Physical Rol	34.63 (7.68)	45.4 (13.8)
Body Pain	30.99 (9.07)	46.6 (6.5)
General Health	27.74 (8.48)	37.4 (11.0)
Vitality	37.93 (9.93)	52.0 (12.7)
Social Rol	36.70 (11.58)	47.6 (13.5)
Emotional Rol	37.04(10.69)	45.01 (14.6)
Mental Health	40.86 (5.63)	47.3 (14.2)
Mental Health Index	42.08 (8.21)	42.9 (13.1)
Physical Health Index	29.02 (8.37)	48.9 (14.1)

The association between the FKQ and the SF12v2 is shown in Table 6. It can be observed that there is not significative correlation between FKQ and the SF12v2.

Table 6. Association between FKQ and SF12v2.

	FKQ				
	General	**Medication**	**Physical Activity**	**Energy**	**Total**
Physical Function	0.074	0.019	0.161	−0.051	0.087
Physical Rol	−0.025	0.009	0.171	−0.021	0.015
Body Pain	0.023	0.051	0.203	0.057	0.095
General Health	0.109	−0.051	0.154	−0.008	0.058
Vitality	−0.033	−0.104	0.024	−0.155	−0.065
Social Rol	0.028	0.112	0.143	0.043	0.017
Emotional Rol	0.156	0.130	0.207	0.006	0.209
Mental Health	0.130	0.166	0.135	0.049	0.169
Mental Health Index	0.086	0.014	0.133	−0.009	0.138
Physical Health Index	−0.012	−0.050	0.105	−0.059	−0.008

FKQ: Fibromyalgia Knowledge Questionnaire.

Before performing the Bonferroni correction, a multivariate analysis was performed on those that were significant, and the results obtained were show in Table 7.

Table 7. Domain of FKQ that is associated with the body pain.

Body Pain (R^2: 0.067)	***B***	***t***	***p* Value**
Constant	12.60	1.566	0.120
FKQ Physical Activity	5.850	2.932	0.004

Table 8 shows the multivariate analysis revealed that the physical activity domain of FKQ was the only domain that were associated with a higher HQRoL.

Table 8. Domain of FKQ that is associated with a greater HRQoL.

EQ 5D-5L (R^2: 0.078)	***B***	***t***	***p* Value**
Constant	0.165	1.986	0.049
FKQ Physical Activity	0.066	3.180	0.002

Physical activity habits are shown in Table 9, where it is observed that with regard to the hours/week of practice, the highest percentages are found in "between 1 and 2 h" (31.4%) and "between 3 and 4 h" (24.4%). In relation to the distance walked daily, the highest percentages are found in "between 1 and 2 km" (46.3%) and "between 3 and 5 km" (21.5%).

Finally, Table 10 shows the association between knowledge measured by the FKQ and physical activity habits, showing that there is no correlation between the level of knowledge of the disease and the hours of physical exercise, nor in relation to the distance spent walking per week.

Table 9. Sports practice habits.

		N (%) 121 (100)
Hours (per week)	None	21 (17.4)
	Less than 1 h	13 (10.7)
	Less than 1 and 2 h	38 (31.4)
	Less than 3 and 4 h	30 (24.4)
	Less than 5 and 8 h	16 (13.2)
	Less than 9 and 14 h	2 (1.7)
	More than 21	1 (0.8)
Distance (daily)	Does not walk	19 (15.6)
	Less than 1 km	16 (13.)
	Between 1 and 2 km	56 (46.3)
	Between 3 and 5 km	26 (21.5)
	Between 6 and 9 km	2 (1.7)
	More than 9 km	2 (1.7)

Hours: How many hours per week do you exercise? Distance: approximately indicate how far you walk daily.

Table 10. Correlation between FKQ and physical exercise.

	FKQ				
	General	**Medication**	**Physical Activity**	**Energy**	**Total**
Hours	0.020	0.107	0.070	0.181	0.102
Distance	0.174	0.092	0.004	0.117	0.134

FKQ: Fibromyalgia Knowledge Questionnaire.

4. Discussion

To our knowledge, this is the first study on the assessment of knowledge of FM in women with FM in Extremadura. The main finding of the study was to determine that the level of knowledge about FM of patients in Extremadura was medium-high. As a secondary finding, we can highlight the existence of a weak direct correlation between knowledge about physical activity in FM and HRQoL, i.e., the greater the knowledge about physical activity, the greater the HRQoL. However, any association was found between knowledge of FM and HRQol evaluated by SF12v2. Furthermore, in accordance with the objectives of the study, no association was found between knowledge of physical activity in FM and the practice of physical activity.

Overall, the participants' level of knowledge about FM was medium (49%) and high (41%), with the dimensions medication and energy being the ones where the highest number of participants had low knowledge (29% and 30% respectively).

Specifically, for the general domain, in item 1, related to the causes of FM, it can be observed that there is a great diversity in the answers, with the correct options being chosen by only 56% and 58% of the participants, similar to the results reported by a similar study carried out by Moretti, et al. [30], which highlighted that 69% of their sample knew the cause of FM. In relation to the results obtained for this question (item 1), one of the most chosen options (43%) is that FM is due to physical trauma, so we can affirm that there is still a lack of knowledge about the cause among people with FM, in line with Alvarado Moreno and Oliva Arias [31], who highlight that the most deficient knowledge about FM is related to the origin and treatment.

It can be observed that the medication domain is one of those in which it was found the greatest lack of knowledge, since the responses are very diverse (items 7, 8 and 9), something which is supported by other studies in which this domain has shown the lowest score [31,32]. Specifically, in item 8 it can be observed how 50% of the participants chose the option of "regular exercise and anti-inflammatory drugs" as correct, which could be due to the lack of guidance from their doctors about the suitability of the different treatments or drugs, since as highlighted by several studies such as Blotman, et al. [33], 20% of the doctors participating in their study took nonsteroidal anti-inflammatory drugs as suitable medication. Furthermore, the study conducted by Kianmehr, et al. [34] stated that 53.2% of the general practitioners participating in the study had low or very low levels of knowledge about the treatment of FM, and more specifically 52.1% of them also marked the use of nonsteroidal antinflammatory agents. In this line and at a general level, Ortiz, et al. [35] tated that the percentage of physicians with knowledge about "Drugs with proven usefulness" was slightly higher than half of the respondents (59.3%).

In the domain of physical activity, the results show that in the question on exercise and body pain (item 10), a hit rate of 78% was found, in line with that reported by Moretti, et al. [30] (89% hit rate). However, Alvarado Moreno and Oliva Arias [31] found only a 29.2% accuracy rate, attributing this to the poor knowledge of physicians in their region about appropriate physical therapy and exercise-based treatments [35]. In relation to the importance of physical activity (item 11), we can find that there is a high response rate (44%) for "when the patient suffers pain, the best thing to do is to stay in bed" compared to the correct response "they should do physical exercise three times a week" (48%), far from what was reported by another study with 82% correct [30].

For the energy domain, in relation to item 17, we can say that there is a great lack of knowledge about protection in relation to energy, as 53% of the participants answered that they "did not know", with only 13% selecting the correct answer (carrying the bags on the forearm instead of in the hands), similar to what reported Moretti et al. [30], where 27% of the participants did not know the correct answer.

In relation to the participants' HRQoL, the scores on the different dimensions of the SF12v2 of the participants are well below the normal values for healthy women aged 55–64 [29], especially in the dimensions of physical function, body pain, general health, vitality and physical health index, which highlights the HRQoL deficiencies of FM patients.

Several studies claim that patients' knowledge of the disease contributes to disease control [15,36]. Most of them use disease education within a treatment programme or multidisciplinary programme, resulting in improvements in both the impact of FM and HRQoL [36,37]. Certainly the evaluation of an FM education programme in isolation has been little studied, so further studies would be relevant to determine the level of influence or weight that knowledge of the disease could have in such multidisciplinary programmes, as the results obtained in this study reveal that greater knowledge of the disease, in this case about physical activity, could be associated with better HRQoL.

Regarding the association between knowledge and physical activity practice, no association was found between the level of physical activity practice and the level of knowledge. In other words, it cannot be affirmed that those who do more physical activity have better knowledge of the disease, not even in the specific domain of physical activity.

This study has several limitations, including the fact that the sample was based on women only. In addition, descriptive data such as weight and height were not taken, so the anthropometric characteristics of the sample cannot be observed. Furthermore, it was not possible to establish cause-effect relationships due to the cross-sectional nature of the present study, for which experimental studies would be necessary.

This study shows that knowledge about FM in women with the pathology could influence some aspects of their HRQoL. Therefore, this possible improvements both at the level of the patient and the symptomatology, as well as at the level of the health service, if this would allow cost reduction, could be beneficial for patients as already being in other diseases such as diabetes [38–40] or arthritis [41,42], where there are many educational programmes aimed at different populations. In FM field this is still underdeveloped, so it would be very interesting to implement them. Furthermore, FM patients suffer a high lifetime rate of comorbid like migraine, irritable bowel syndrome, chronic fatigue syndrome, major depression, panic disorder, etc, therefore, it would be very interesting in future studies to know to what extent these comorbidities could influence both knowledge about the disease, as well as the relationship between knowledge and HRQOL.

5. Conclusions

Based on the results of the present study, it could be concluded that the level of knowledge about FM of patients in Extremadura was medium-high and there is a direct relationship between knowledge about physical activity in FM patients and their HRQoL. However, no association was found between knowledge of physical activity in FM patients and their practice of physical activity.

Author Contributions: Conceptualization, M.M.-M.; Data curation, M.R.; Formal analysis, J.M.-M. and M.R.; Investigation, M.M.-M.; Methodology, J.G.-M. and J.I.C.-R.; Project administration, J.M.-M. and M.Á.G.-G.; Resources, M.Á.G.-G.; Software, M.R.; Supervision, J.I.C.-R.; Validation, J.G.-M.; Visualization, J.G.-M.; Writing—original draft, M.M.-M. and J.I.C.-R.; Writing—review & editing, M.M.-M., J.M.-M., M.Á.G.-G. and J.I.C.-R. All authors have read and agreed to the published version of the manuscript.

Funding: This research received no external funding.

Institutional Review Board Statement: The study was conducted according to the guidelines of the Declaration of Helsinki and approved by the Bioethics and Biosafety Committee of the University of Extremadura (approval number: 51/2013).

Informed Consent Statement: Informed consent was obtained from all subjects involved in the study.

Data Availability Statement: The datasets used during the current study are available from the corresponding author on reasonable request.

Acknowledgments: The authors would like to acknowledge the associations and professionals who made it possible to carry out this study.

Conflicts of Interest: The authors declare no conflict of interest.

References

1. Csillag, C. Fibromyalgia: The Copenhagen declaration. *Lancet* **1992**, *340*, 663–664. [CrossRef]
2. Climent, C.d.O.; Ballester, M.C.; Cantus, D.S. Calidad de vida en pacientes con Fibromialgia. Revisión bibliográfica. *Revista Española Enfermería Salud Mental* **2019**, *50*, 12–22.
3. Álvarez Lario, B.; Alonso Valdivieso, J.; Alegre López, J.; Viejo Bañuelos, J.; Marañón Cabello, A. Síndrome de fibromialgia: Características clínicas de las pacientes españolas. *Rev. Esp. Reumatol.* **1996**, *23*, 76–82.
4. Queiroz, L.P. Worldwide epidemiology of fibromyalgia. *Curr. Pain Headache Rep.* **2013**, *17*, 356. [CrossRef] [PubMed]
5. Cabo-Meseguer, A.; Cerdá-Olmedo, G.; Trillo-Mata, J.L. Fibromialgia: Prevalencia, perfiles epidemiológicos y costes económicos. *Med. Clín.* **2017**, *149*, 441–448. [CrossRef]
6. Mas, A.; Carmona, L.; Valverde, M.; Ribas, B. Prevalence and impact of fibromyalgia on function and quality of life in individuals from the general population: Results from a nationwide study in Spain. *Clin. Exp. Rheumatol.* **2008**, *26*, 519.
7. López Espino, M.; Mingote Adán, J.C. Fibromialgia. *Clín. Salud* **2008**, *19*, 343–358.
8. Wolfe, F.; Clauw, D.J.; Fitzcharles, M.A.; Goldenberg, D.L.; Katz, R.S.; Mease, P.; Russell, A.S.; Russell, I.J.; Winfield, J.B.; Yunus, M.B. The American College of Rheumatology preliminary diagnostic criteria for fibromyalgia and measurement of symptom severity. *Arthritis Care Res.* **2010**, *62*, 600–610. [CrossRef]
9. Cruz, A.C.; Mata, X.T.i.; Gassol, A.A.i.; Gabaroi, D.C.; Vilarrasa, R.; Miyar, M.V.; Muñoz-Gómez, J. Eficacia del tratamiento multidisciplinario del dolor crónico incapacitante del aparato locomotor. *Med. Clín.* **2001**, *117*, 401–405.
10. Navarro, E.A. *Efectos de un Programa Multidisciplinar sobre el Estado de Salud y Procesos Psicológicos en Personas con Fibromialgia;* Universidad Miguel Hernández de Elche: Elche, Spain, 2017.
11. Gelman, S.; Lera, S.; Caballero, F.; López, M. Tratamiento multidisciplinario de la fibromialgia. Estudio piloto prospectivo controlado. *Rev. Esp. Reumatol.* **2002**, *29*, 323–329.
12. Pătru, S.; Pădureanu, R.; Dumitrescu, F.; Pădureanu, V.; Rădulescu, D.; Dragoi, D.; Matei, D. Influence of multidisciplinary therapeutic approach on fibromyalgia patients. *Exp. Ther. Med.* **2021**, *21*, 1–6. [CrossRef]
13. Masquelier, E.; D'haeyere, J. Physical activity in the treatment of fibromyalgia. *Jt. Bone Spine* **2021**, *88*, 105202. [CrossRef]
14. Carbonell-Baeza, A.; Aparicio, V.; Chillón, P.; Femia, P.; Delgado-Fernandez, M.; Ruiz, J. Effectiveness of multidisciplinary therapy on symptomatology and quality of life in women with fibromyalgia. *Clin. Exp. Rheumatol.-Incl Suppl.* **2011**, *29*, S97.
15. Koca, T.T.; Tugan, C.B.; Koçyiğit, B.F.; Nacitarhan, V. Fibromyalgia awareness in women aged between 18 and 75 years: A current view to fibromyalgia. *J. Public Health* **2019**, *27*, 491–497. [CrossRef]
16. Romero, E.B.; Moya, N.S.; Esteve, M.V.; Valer, S.V. Estudio de la calidad de vida en pacientes con fibromialgia: Impacto de un programa de educación sanitaria. *Atención Primaria* **2002**, *30*, 16–21. [CrossRef]

17. Van Oosterwijck, J.; Meeus, M.; Paul, L.; De Schryver, M.; Pascal, A.; Lambrecht, L.; Nijs, J. Pain physiology education improves health status and endogenous pain inhibition in fibromyalgia: A double-blind randomized controlled trial. *Clin. J. Pain* **2013**, *29*, 873–882. [CrossRef] [PubMed]
18. Rooks, D.S.; Gautam, S.; Romeling, M.; Cross, M.L.; Stratigakis, D.; Evans, B.; Goldenberg, D.L.; Iversen, M.D.; Katz, J.N. Group exercise, education, and combination self-management in women with fibromyalgia: A randomized trial. *Arch. Intern. Med.* **2007**, *167*, 2192–2200. [CrossRef]
19. Hassett, A.L.; Gevirtz, R.N. Nonpharmacologic treatment for fibromyalgia: Patient education, cognitive-behavioral therapy, relaxation techniques, and complementary and alternative medicine. *Rheum. Dis. Clin.* **2009**, *35*, 393–407. [CrossRef]
20. Novo, J.P.; Pereira, A.E.; García, A.R.; Martín, R.S.; Méndez, B.G. Guía para la rehabilitación de la fibromialgia. *Rev. Cuba. Reumatol.* **2015**, *17*, 147–156.
21. Zar, J.H. *Biostatistical Analysis*; Pearson Education India: Cranbury, NJ, USA, 1999.
22. Suda, A.L.; Jennings, F.; Bueno, V.C.; Natour, J. Development and validation of fibromyalgia knowledge questionnaire: Fkq. *Rheumatol. Int.* **2012**, *32*, 655–662. [CrossRef]
23. Ware, J.E., Jr.; Kosinski, M.; Keller, S.D. A 12-Item Short-Form Health Survey: Construction of scales and preliminary tests of reliability and validity. *Med. Care* **1996**, 220–233. [CrossRef]
24. Monteagudo Piqueras, O.; Hernando Arizaleta, L.; Palomar Rodríguez, J.A. Valores de referencia de la población diabética para la versión española del SF-12v2. *Gac. Sanit.* **2009**, *23*, 526–532. [CrossRef]
25. Van Hout, B.; Janssen, M.; Feng, Y.-S.; Kohlmann, T.; Busschbach, J.; Golicki, D.; Lloyd, A.; Scalone, L.; Kind, P.; Pickard, A.S. Interim scoring for the EQ-5D-5L: Mapping the EQ-5D-5L to EQ-5D-3L value sets. *Value Health* **2012**, *15*, 708–715. [CrossRef]
26. Merayo Alonso, L.A. *Eficacia Comparada de la Terapia Cognitiva-Conductual y la Gestión Personal en Pacientes con Fibromialgia*; Universidad de Sevilla: Sevilla, Spain, 2008.
27. McLaughlin, M.J.; Sainani, K.L. Bonferroni, Holm, and Hochberg corrections: Fun names, serious changes to p values. *PM R* **2014**, *6*, 544. [CrossRef]
28. Cohen, J. Statistical power analysis for the behavioral sciences: Jacob Cohen. *J. Am. Stat. Assoc.* **1988**, *84*, 19–74.
29. Monteagudo Piqueras, O.; Hernando Arizaleta, L.; Palomar Rodríguez, J.A. Normas poblacionales de referencia de la versión española del SF-12V2 para la Región de Murcia. *Gac. Sanit.* **2011**, *25*, 50–61. [CrossRef]
30. Moretti, F.A.; Heymann, R.E.; Marvulle, V.; Pollak, D.F.; Riera, R. Assessing knowledge on fibromyalgia among Internet users. *Rev. Bras. Reumatol.* **2011**, *51*, 7–19. [PubMed]
31. Alvarado Moreno, M.E.E.; Oliva Arias, E.J. *Conocimientos, Actitudes y Resiliencia Respecto a su Enfermedad en Pacientes con Fibromialgia de dos Hospitales de Lambayeque, 2017*; Universidad Católica Santo Toribio de Mogrovejo: Chiclayo, Peru, 2019.
32. Gomes, C.M.S.; Coutinho, G.C.; Miyamoto, S.T. Efeitos do programa de educação em pacientes com artrite reumatoide do Serviço de Reumatologia do Hospital Universitário Cassiano Antônio de Moraes (HUCAM)-projeto piloto. *Rev. Ter. Ocup. Univ. São Paulo* **2013**, *24*, 250–258.
33. Blotman, F.; Thomas, E.; Myon, E.; Andre, E.; Caubere, J.; Taieb, C. Awareness and knowledge of fibromyalgia among French rheumatologists and general practitioners. *Clin. Exp. Rheumatol.* **2005**, *23*, 697. [PubMed]
34. Kianmehr, N.; Haghighi, A.; Bidari, A.; Sharafian Ardekani, Y.; Karimi, M.A. Are general practitioners well informed about fibromyalgia? *Int. J. Rheum. Dis.* **2017**, *20*, 1917–1921. [CrossRef] [PubMed]
35. Ortiz, F.E.A.; de la Cruz, V.A.C.; Jiménez, F.E.L. Conocimientos de fibromialgia en médicos de atención primaria de la provincia de Chiclayo-Perú, 2016. *Reumatol. Clín.* **2017**, *13*, 326–330. [CrossRef] [PubMed]
36. Rodríguez, M.d.M.L. Education in fibromyalgia. A systematic review. In Proceedings of the Jornadas Internacionales de Investigación en Educación y Salud: Educación y Salud en una Sociedad Global, Almeria, Spain, 9–10 May 2014.
37. Oh, T.H.; Stueve, M.H.; Hoskin, T.L.; Luedtke, C.A.; Vincent, A.; Moder, K.G.; Thompson, J.M. Brief interdisciplinary treatment program for fibromyalgia: Six to twelve months outcome. *Am. J. Phys. Med. Rehabil.* **2010**, *89*, 115–124. [CrossRef] [PubMed]
38. Pérez Gutiérrez, C. *Educación para la Salud en la Mujer con Diabetes Gestacional*; Universidad de Salamanca: Salamanca, Spain, 2017.
39. Canché-Aguilar, D.L.; Zapata-Vázquez, R.E.; Rubio-Zapata, H.A.; Cámara-Vallejos, R.M. Efecto de una intervención educativa sobre el estilo de vida, el control glucémico y el conocimiento de la enfermedad, en personas con diabetes mellitus tipo 2, Bokobá, Yucatán. *Rev. Bioméd.* **2019**, *30*, 3–11. [CrossRef]
40. Modrego, B.A. *Programa de Salud Dirigido a Adolescentes con Diabetes Mellitus tipo I*; Universidad de Zaragoza: Zaragoza, Spain, 2019.
41. Esquivel García, J.A.; Gale Ojeda, J.V.; Rodríguez Florido, F.; Rodríguez Galindo, D.G.; Sánchez Ramos, K.J. *Evaluación de un Programa Educativo en salud para Pacientes con Artritis Reumatoide en Bogotá*; Faculdad De Ciencias Economicas Y Administrativas: Bogota, Columbia, 2018.
42. Lopez-Olivo, M.A.; Ingleshwar, A.; Volk, R.J.; Jibaja-Weiss, M.; Barbo, A.; Saag, K.; Leong, A.; Suarez-Almazor, M.E. Development and pilot testing of multimedia patient education tools for patients with knee osteoarthritis, osteoporosis, and rheumatoid arthritis. *Arthritis Care Res.* **2018**, *70*, 213–220. [CrossRef] [PubMed]

biology

MDPI

Article

Specific Bioelectrical Impedance Vector Analysis Identifies Body Fat Reduction after a Lifestyle Intervention in Former Elite Athletes

Francesco Campa [1,*], Catarina Nunes Matias [2,3], Catarina L. Nunes [4], Cristina P. Monteiro [5], Rubén Francisco [4], Filipe Jesus [4], Elisabetta Marini [6], Luís B. Sardinha [4], Paulo Martins [4], Cláudia Minderico [4] and Analiza M. Silva [4]

1 Department for Life Quality Studies, Università degli Studi di Bologna, 47921 Rimini, Italy
2 Bioperformance & Nutrition Research Unit, Bettery S.A., 2740-262 Lisbon, Portugal; catarina.matias@imetrico.com
3 CIDEFES—Universidade Lusófona, 1749-024 Lisboa, Portugal
4 Exercise and Health Laboratory, CIPER, Faculdade de Motricidade Humana, Universidade de Lisboa, 1499-002 Cruz Quebrada, Portugal; catnunes94@gmail.com (C.L.N.); ruben92francisco@gmail.com (R.F.); fastj96@gmail.com (F.J.); lbsardinha55@gmail.com (L.B.S.); pmartins@fmh.ulisboa.pt (P.M.); cminderico@gmail.com (C.M.); analiza.monica@gmail.com (A.M.S.)
5 Laboratory of Physiology and Biochemistry of Exercise, Faculdade de Motricidade Humana, Universidade de Lisboa, Estrada da Costa, 1499-002 Cruz Quebrada, Portugal; cmonteiro@fmh.ulisboa.pt
6 Department of Life and Environmental Sciences, Neuroscience and Anthropology Section, University of Cagliari, 09124 Cagliari, Italy; emarini@unica.it
* Correspondence: Francesco.campa3@unibo.it

Citation: Campa, F.; Matias, C.N.; Nunes, C.L.; Monteiro, C.P.; Francisco, R.; Jesus, F.; Marini, E.; Sardinha, L.B.; Martins, P.; Minderico, C.; et al. Specific Bioelectrical Impedance Vector Analysis Identifies Body Fat Reduction after a Lifestyle Intervention in Former Elite Athletes. *Biology* **2021**, *10*, 524. https://doi.org/10.3390/biology10060524

Academic Editors: Filipe Manuel Clemente, Badicu Georgian and Eugenia Murawska-Ciałowicz

Received: 21 May 2021
Accepted: 9 June 2021
Published: 12 June 2021

Simple Summary: The ability of specific bioelectrical impedance vector analysis (BIVA) to classify subjects according to the percentage of fat mass has been recognized in different cross-sectional studies, but no longitudinal designs have yet been applied. The results of this investigations showed that specific BIVA can be used as a practical solution for assessing body composition management in former overweight/obese athletes. In particular, reductions in bioelectrical vector length adjusted according to the specific BIVA procedure were found to be associated with reductions in percentage of fat mass.

Abstract: Background: specific bioelectrical impedance vector analysis (BIVA) has been proposed as an alternative bioimpedance method for evaluating body composition. This investigation aimed to verify the ability of specific BIVA in identifying changes in fat mass after a 16-week lifestyle program in former athletes. Methods: The 94 participants included in the Champ4life project (clinicaltrials.gov: NCT03031951) were randomized into intervention (n = 49) and control (n = 45) groups, from which 82 athletes completed the intervention (age 43.9 ± 9.2 y; body mass index 31.1 ± 4.6 kg/m^2). Fat mass was estimated by dual-energy X-ray absorptiometry. Bioelectric resistance, reactance, phase angle, and vector length were assessed by bioelectric impedance spectroscopy, and the BIVA procedure was applied. Results: A significant ($p < 0.05$) group x time interaction for fat mass, specific resistance, reactance, and vector length was found. Fat mass and vector length significantly ($p < 0.05$) decreased in the intervention group, while no change was measured in the control group. Considering the participants as a whole group, changes in vector length were associated with changes in fat mass percentage ($r^2 = 0.246$; $\beta = 0.33$; $p < 0.001$) even after adjusting for age, sex, and group ($R^2 = 0.373$; $\beta = 0.23$; $p = 0.002$). Conclusions: The specific BIVA approach is suitable to track fat mass changes during an intervention program aimed to reduce body fat in former athletes.

Keywords: body composition; BIVA; fat mass; weight loss

1. Introduction

Body composition analysis is fundamental to understand the effect of a diet and/or exercise intervention or to follow the progress of a disease [1,2]. As proposed by Wang et al. [3], it is possible to analyze body composition on the basis of five different levels: the sum of atoms, molecules, cells, tissues, or as different body segments. In sports-related populations, the most monitored parameters belong to the molecular level [1,4], in which body mass equals the sum of fat mass (FM) and fat-free mass (FFM) that can be additionally compartmentalized into total body water, minerals, and proteins.

At the molecular level, the four-compartment model is considered the "gold standard" for the assessment of FM [5]. Given that this model combines the use of several techniques, due to the assessment of bone mineral content by dual-energy X-ray absorptiometry (DXA); total body water by isotopes dilution, bioelectrical impedance analysis (BIA), or bioelectrical impedance spectroscopy (BIS); and body volume by air displacement plethysmography, it requires considerable time and high costs, as well as highly qualified personnel. In particular, BIA refers to the measurement technology that performs measurements at a single or multiple frequency. In addition, in order to clearly distinguish multi-frequency BIA from the analysis based on Cole plots or other models for fitting impedance data over the entire frequency range, the term BIS has been used to refer to the latter [4]. Among these techniques, DXA is the one that allows for the evaluation of the widest range of variables, factoring body mass according to bi-compartmental (FM and FFM) or three-compartmental (FM, lean soft tissue, and bone mineral content) models [6,7]; however, it remains a difficult technique to apply in sports populations due to the fact that the equipment is not portable and user-friendly.

BIS is currently a widely used technique for assessing body composition in several contexts [6,8–10]. However, the use of prediction equations limits the precision and accuracy of this method [9,11]. Therefore, it is preferable to conduct an alternative analysis that consists of the direct interpretation of the raw bioelectrical impedance parameters (resistance (R) and reactance (Xc)) [4]. One widely used approach to evaluate the raw bioelectrical impedance parameters is bioelectrical impedance vector analysis (BIVA) [4]. This technique consists of the standardization of the raw R and Xc for the height in meters of the subject and their combined interpretation within a R-Xc graph, where they are represented as a vector [12]. Recently, the effectiveness of BIVA has been shown in tracking changes in body fluids compared with a reference method (dilution techniques) [13,14]. Though the "classic BIVA" is considered the traditional method, a variant named "specific BIVA" has been proposed [6]. This approach standardizes R and Xc for the arm, waist, and calf circumferences, as well as for the height of the subject, thus contrasting the body volume effect and becoming informative about the relative quantities, such as FM%, and no-longer-absolute ones, such as TBW. The advantage of classic and specific BIVA, with respect to the only estimation of the body composition parameters, lies in the evaluation of total body water or FM% simultaneously with the phase angle, whose value is determined by the lateral position/displacement of the vector in the R-Xc graph [15]. The bioelectric phase angle is graphically represented as the angle between impedance and the x-axis, and it reflects the relationship between intracellular and extracellular fluids [15,16]. Furthermore, BIVA allows one to evaluate the vector position in comparison with population-specific tolerance ellipses [16].

The ability of specific BIVA to classify subjects according to FM% has been recognized in different cross-sectional studies, but no longitudinal designs have yet been applied [6,16,17]. This study used the results of the participants of the Champ4life project, a lifestyle intervention aimed to reduce weight in former athletes who were inactive and had overweight or obesity. Former athletes are considered an understudied group, and there is a lack of evidence regarding this topic. Therefore, this study aimed to compare FM% changes, assessed by DXA, with vector displacements measured with the specific BIVA approach after a 16-week intervention program aimed to reduce weight and body fat in inactive former elite athletes with overweight/obesity. Since specific BIVA has already been identified as a valid classification (e.g., higher or lower body fat content) tool without providing estimates of fat mass [6,16], our hypothesis was that vector displacements could reflect changes in body fat over the intervention period.

2. Materials and Methods

2.1. Participants and Study Design

This study was part of the 1-year Champ4Life lifestyle intervention targeting former athletes who had overweight/obesity and were inactive. Participants were divided in several sports, such as martial arts (25.6%), football (14.9%), athletics (mainly sprinters and middle- and long-distance track and field) (14.9%), dancing/gymnastics (10.6%), swimming (8.5%), volleyball (9.6%), handball (5.3%), rugby (3.2%), and others (7.4%). The baseline variables did not differ significantly between groups ($p > 0.05$), except for android fat ($p = 0.034$). For more details, the study protocol and the main results for this project can be found elsewhere [18]. Using a longitudinal design, two time points were considered: a baseline (0 months) and post-program (16 weeks). Ninety-four participants were included in the main study and computer-generated randomized to an intervention group ($n = 49$) or control group ($n = 45$). From those participants, 82 (control group $n = 41$; intervention group $n = 41$) completed the 16-week lifestyle intervention. Nevertheless, only 80 participants had all the data, and the intervention and control groups were therefore represented by $n = 40$ each. Details regarding recruitment, eligibility, and intervention or regarding participant characteristics and main results were stated elsewhere [18].

The study was approved by the Ethics Committee of the Faculty of Human Kinetics, University of Lisbon (Lisbon, Portugal) (CEFMH Approval Number: 16/2016) and was conducted in accordance with the declaration of Helsinki for human studies from the World Medical Association [19]. Additionally, the present study was registered at www.clinicaltrials.gov (clinicaltrials.gov ID: NCT03031951) prior to participants' recruitment.

2.2. Anthropometry

All measurements were performed on the same day (approximately 8 a.m.) after a 12-h fast. Furthermore, alcohol and stimulant beverage consumption were not allowed for at least 15 h prior to testing. Participants were weighted to the nearest 0.1 kg with a scale (Seca, Hamburg, Germany), and stature was measured to the nearest 0.1 cm with a stadiometer (Seca, Hamburg, Germany) using the standardized procedures described elsewhere [20]. BMI was calculated as body mass (kg) divided by height squared (m).

2.3. Dual-Energy X-ray Absorptiometry

DXA was performed according to the standard procedures recommended by the manufacturer and described elsewhere [21] on a Hologic Explorer-W, fan-beam densitometer (Hologic, Waltham, MA, USA) to obtain total whole-body FM and FFM.

2.4. Bioelectrical Impedance Spectroscopy

Raw data (resistance (R) and reactance (Xc)) were obtained using a bioelectrical impedance spectroscopy (BIS) analyzer (model 4200B, Xitron Technologies, San Diego, CA, USA) at a frequency of 50 kHz, as previously indicated by our group [22]. The device was calibrated using the standard control circuit supplied by the manufacturer that has a known impedance (Rz $\frac{1}{4}$ 380 Ohm 1% precision and Xc $\frac{1}{4}$ 47 Ohm 1% precision). The test-retest CV in 16 participants for R, Xc, and phase angle was 0.3%, 0.8%, and 0.9%, respectively. BIS was performed with patients lying supine with their limbs placed slightly away from their body after an overnight fast and having emptied their bladders. BIVA analysis was carried out using the specific BIVA approach, i.e., multiplying R and Xc by a correction factor (A/L), where A is the estimated cross-sectional area (or 0.45× arm area + 0.10× waist area + 0.45× calf area) and L is the length of the 'conductor' (1.1× height). The length of the vector (VL) was calculated as the hypotenuses of individual impedance values [15]. The phase angle was calculated as the arctangent of $Xc/R*180°/\pi$ [15]. In particular, by adjusting R and Xc according to the specific BIVA approach, the length of the vector became informative for the FM% (vector elongation = FM% increase; vector shortening = FM% reduction) and the lateral displacement of the vector responded to the changes in phase angle.

2.5. Statistical Analysis

Descriptive statistics were applied to characterize the sample. Linear mixed models included group and time as fixed effects, with baseline values and sex as covariates, to assess primary and secondary outcomes for the impact of treatment (control vs intervention group), time (baseline—0 months; post-intervention—4 months), and treatment-by-time interaction. The covariance matrix for repeated measures within subject over time was modelled as unstructured. Model residual distributions were examined graphically and by performing the Kolmogorov–Smirnov test, and no data transformations were necessary. The paired, one-sample Hotelling's t^2 test was performed to determine whether the changes in the mean group vectors (measured between the first and second time points) were significantly different from zero (null vector). The paired one-sample Hotelling's t^2 test is a multivariate extension of the Student's t test for paired data in comparison of mean difference vectors. Single and multiple regression analyses were performed to evaluate the associations between changes in specific VL and FM%. In addition, after the transformation of measured vector components into bivariate Z-scores (i.e., R/H and Xc/H minus the mean and divided by the standard deviation of R/H and Xc/H calculated in the reference population), bioelectrical variables were analyzed in relation to the distribution of the reference population [17]. To assess the effect size (ES) of the significant parametric test results ($p < 0.05$), the Hedge's ES and the Mahalanobis distance (D^2) were calculated. Threshold values were identified for ES < 0.5 as small, ES $\leq$ 0.8 as medium, and for ES $\geq$ 0.8 as large. Statistical significance for all analyses was defined as $p < 0.05$. SPSS v. 25.0 (IBM Inc., Chicago, IL, USA) was used for all statistical calculations.

3. Results

Eighty participants were included in this analysis. No significant differences ($p > 0.05$) in age or BMI were found between the two groups before the intervention period. A detailed description regarding the initial recruitment and the dropouts at each time point was presented elsewhere [18].

The mean values for intervention and control group before and after the intervention, divided by sex, are presented in Table 1. The intervention was effective at reducing weight and fat mass and increasing fat-free mass in the intervention group compared with the control group (interaction time x group $p < 0.001$). For bioimpedance variables, a significant ($p < 0.05$) reduction was observed for R, Xc, and VL in the intervention group. A significant time x group interaction was also found for R, Xc, and VL ($p < 0.05$). No significant ($p > 0.05$) time effect or group by time interaction for phase angle were observed.

At baseline, the mean vectors were found to be far from the athlete-specific tolerance ellipses in both groups due to a higher FM% of the participants compared to the athletic reference population [16], as shown in Figure 1 (left panels). After the intervention period, only the group that followed the lifestyle program showed a mean vector within the tolerance ellipses (Figure 1). The paired one-sample Hotelling's t^2 test showed a significant change in the mean vector between the first (PRE) and second measurements (POST) for the intervention group (T2 = 12.9; $p < 0.001$; D2 = 1.01), while no change in the control group was assessed (T2 = 0.2; $p = 0.920$; D2 = 0.10). Figure 1 shows the vector displacements in male and female participants for both groups. Vector displacements parallel to the major axis of tolerance ellipses indicated progressive changes in FM% (higher FM% with long vectors, out of the upper pole, and lower FM% with short vectors, out of the lower pole).

Table 1. Linear mixed models for the comparisons between the groups at baseline (PRE) and after 16 weeks (POST).

		Intervention (*n* 40)			Control (*n* 40)				
		Men (*n* 27)	Women (*n* 13)	Whole Sample	Men (*n* 26)	Women (*n* 14)	Whole Sample	ES †	Interaction *p*-Value
Weight (kg)	PRE	95.5 ± 14.9	83.4 ± 12.0	91.5 ± 15.7	94.6 ± 17.2	76.6 ± 11.6	88.3 ± 17.6	−0.053	<0.001
	POST	90.3 ± 17.1	78.6 ± 11.9	86.6 ± 16.6 *	95.0 ± 18.2	76.9 ± 11.1	88.7 ± 18.4		
BMI (kg/m^2)	PRE	31.7 ± 4.4	31.4 ± 3.5	31.6 ± 4.1	31.1 ± 5.4	29.4 ± 4.1	30.5 ± 5.0	−0.051	<0.001
	POST	30.2 ± 4.7	29.7 ± 3.3	30.0 ± 4.3 *	31.3 ± 5.7	29.4 ± 4.1	30.6 ± 5.2		
FM (kg)	PRE	28.0 ± 7.7	35.6 ± 6.5	30.4 ± 8.1	28.0 ± 11.3	30.7 ± 6.7	28.9 ± 9.9	0.06	<0.001
	POST	24.2 ± 7.8	31.9 ± 7.0	26.6 ± 8.3 *	28.6 ± 13.1	30.8 ± 6.5	29.4 ± 11.2		
FM (%)	PRE	29.1 ± 5.7	42.5 ± 3.1	33.5 ± 8.1	28.7 ± 6.4	39.8 ± 3.8	32.6 ± 7.7	0.058	<0.001
	POST	26.5 ± 5.4	40.2 ± 4.1	30.9 ± 8.2 *	29.0 ± 7.1	39.8 ± 3.8	32.8 ± 8.0		
FFM (kg)	PRE	67.5 ±10.9	47.8 ± 5.7	61.1 ± 13.3	66.6 ± 7.6	45.9 ± 5.9	59.4 ± 12.2	−0.093	0.054
	POST	66.1 ± 11.3	46.7 ± 5.2	60.0 ± 13.3 *	66.4 ± 7.4	46.1 ± 5.5	59.3 ± 11.9		
Rsp (Ω·cm)	PRE	436.7 ± 49.6	529.3 ± 40.0	466.8 ± 63.7	438.6 ± 54.1	501.2 ± 78.9	460.5 ± 69.7	0.12	0.001
	POST	397.6 ± 50.3	476.0 ± 71.1	423.1 ± 68.0 *	422.5 ± 57.5	489.7 ± 67.4	446.0 ± 68.4 *		
Xcsp (Ω·cm)	PRE	54.7 ± 7.1	58.7 ± 7.0	56.0 ± 7.2	53.9 ± 8.6	53.1 ± 6.8	53.6 ± 7.9	−0.105	0.009
	POST	49.9 ± 8.6	52.9 ± 9.2	50.8 ± 8.8 *	51.6 ± 7.5	51.4 ± 8.1	51.5 ± 7.6 *		
VLsp (Ω·cm)	PRE	440.0 ± 49.8	532.6 ± 40.0	470.1 ± 63.8	442.0 ± 54.4	504.6 ± 80.3	463.9 ± 70.4	0.123	0.001
	POST	400.8 ± 50.6	478.9 ± 71.5	426.2 ± 68.2 *	426.4 ± 58.8	492.4 ± 67.7	449.5 ± 69.0 *		
PhA (°)	PRE	7.2 ± 0.7	6.3 ± 0.7	6.9 ± 0.8	7.0 ± 0.8	6.1 ± 0.4	6.7 ± 0.8	−0.257	0.964
	POST	7.1 ± 0.9	6.3 ± 0.6	6.9 ± 0.9	7.0 ± 0.9	6.0 ± 0.6	6.7 ± 0.9		

Note: Data are expressed as mean and standard deviation.* = $p < 0.05$ vs PRE (whole sample); BMI = body mass index; FM = fat mass; FFM = fat-free mass; R_{sp} = specific resistance; Xc_{sp} = specific reactance; VL_{SP} = specific vector length; PhA = phase angle; ES = effect size; † the Hedges' *g* effect size was used.

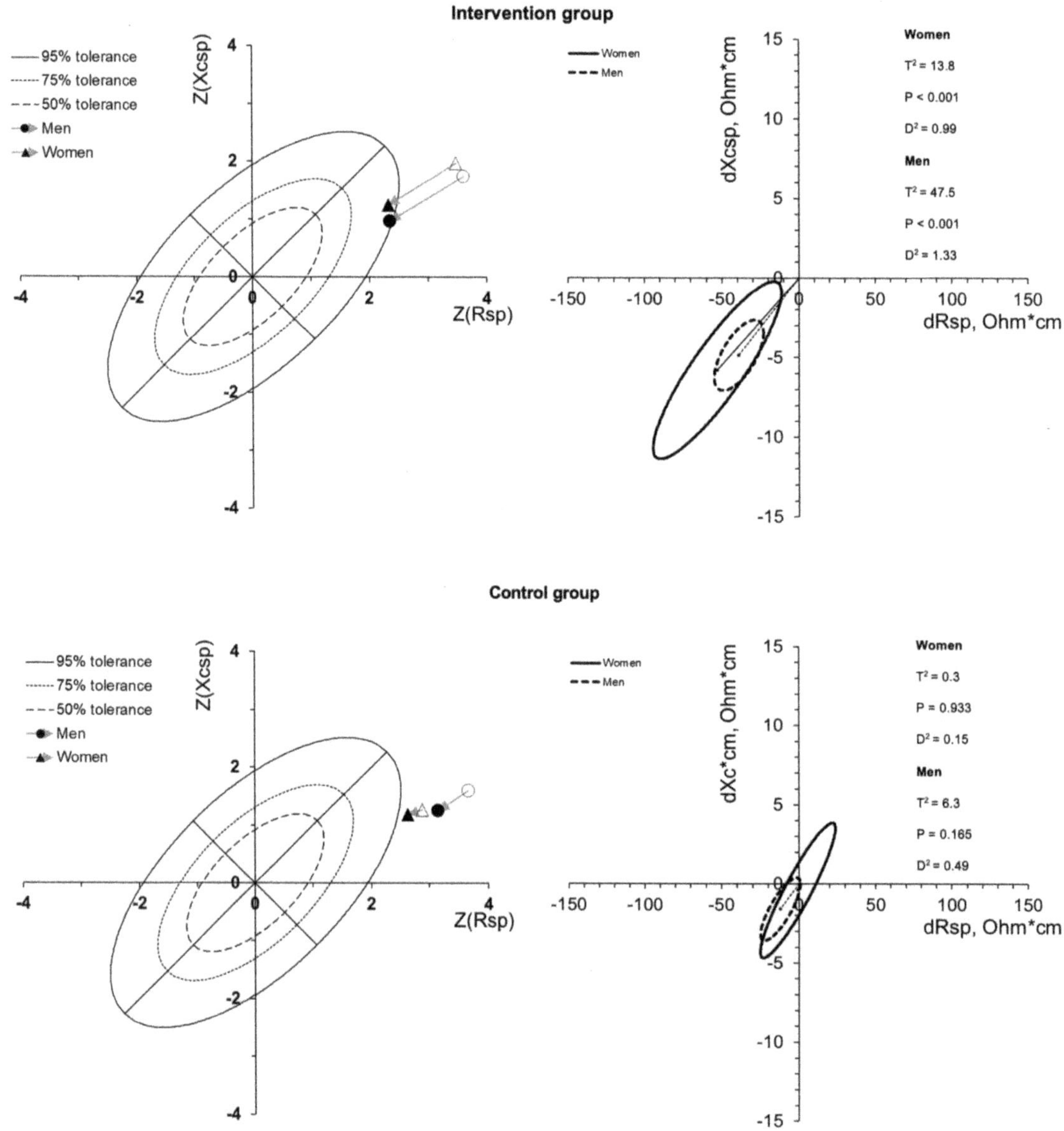

Figure 1. R-Xc z-score and paired graphs for the multivariate changes in bioelectrical parameters. In the left panels, bioimpedance data are plotted on the R-Xc z-score graph after the transformation of the impedance measurements from the athletes into bivariate z-scores (with respect to their reference population [16]). In the right panels, mean vector displacements with 95% confidence ellipses and results of the Hotelling's t^2 test are shown.

Considering the participants as a whole group, changes in specific VL were positively correlated with FM% changes (Table 2 and Figure 2), even when adjusted for age, sex, and group, as shown in Table 2. Therefore, reductions in specific VL corresponded to reductions in FM%.

Table 2. Linear regression analysis for independent variables and changes (Δ) in the percentage of fat mass considering all the participants as a whole group.

Independent Variable	R^2	SEE	β	95% CI	*p*-Value
Δ Specific vector length	0.246	2.36	0.33	0.020, 0.046	<0.001
Model 1	0.373	2.12	0.23	0.009, 0.037	0.002

Note: SEE: standard error of the estimate; β: unstandardized coefficients beta; CI: confidence interval. Model 1: adjusted for age and sex

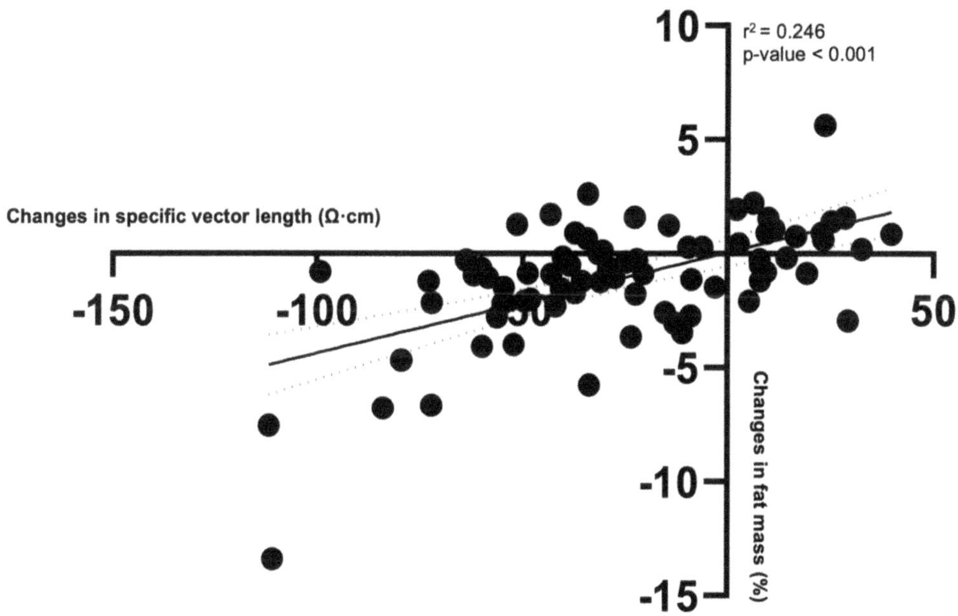

Figure 2. Scatter plot with changes in specific vector length and percentage of fat mass considering all participants as a whole group.

4. Discussion

This study aimed to explore the ability of specific BIVA to assess FM% changes over a 16-week intervention program. The aforementioned intervention had a positive effect on weight, BMI, and body composition. To the best of our knowledge, this investigation, with a longitudinal randomized study design, was the first study that investigated the ability of specific BIVA to identify FM% changes in former athletes. In particular, a vector-shortening reflecting the FM% reduction observed from baseline to the second assessment moment (post intervention program) was seen in the intervention group. On the contrary, no vector displacement was observed in the R-Xc graph in the control group, which also did not reduce FM% after the 16-week period. Furthermore, a direct association between changes in specific VL and FM% was shown.

The main difference between classic and specific BIVA consists of the interpretation of the vector displacements over the longitudinal/major axis of the tolerance ellipses included into the R-Xc graph [16,23]. In fact, in classic BIVA, vector elongations reflect reduction in body fluids and vector shortening is associated with fluid accumulations [14]. On the contrary, specific vector displacements represent changes in FM% and no longer in total body water [16,24]. In this regard and in accordance with this study, the positive associations between specific VL and FM%, both in athletes and in the general population, has already been highlighted [6,16,17,24]. However, while longitudinal studies have compared changes in classic BIVA patterns with reference methods [13,14], only cross-sectional studies have investigated the relationships between specific BIVA patterns and body composition parameters derived from DXA [6,16,17]. While changes in vector length were observed in the intervention group, the phase angle remained unchanged from baseline to the second measurement. In this regard, vectors falling (steady state) or migrating (dynamic state) parallel to the minor axis and above (left) or below (right) the major axis of tolerance ellipses indicate more or less phase angle, respectively (i.e., vectors with a comparable R value and a higher or lower Xc value, respectively) [4]. As with classic BIVA, lateral vector changes along the minor axis reflect changes in phase angle and thus in fluid distribution [23]. In fact, phase angle is a good predictor of the intra/extra cellular water ratio, so we could speculate that no change in the distribution of fluids among the compartments occurred as a result of this intervention. In contrast, the results of this study showed that a shortening of the specific VL is associated with a reduction in FM% after a weight reduction program in former athletes. Therefore, the use of specific BIVA could represent a valid alternative to laboratory techniques to indirectly evaluate fat mass variations after an intervention or a training program.

The analysis of body composition nowadays represents a challenge, especially in the sports context where laboratory methods such as DXA or dilution techniques cannot always be used. For this reason, evaluation techniques such as anthropometric analysis and bioimpedance analysis are commonly performed [4]. However, these techniques present some limitations such as the need for using prediction equations to estimate body composition parameters or, in the case of anthropometry, present between- and within-operator variability [25]. In contrast, BIVA represents a qualitative approach for assessing body composition that avoids the use of predictive equations commonly used by anthropometry and conventional BIA [4]. In particular, specific BIVA appears to be a suitable method for monitoring changes in FM% in athletes. In this regard, the analysis of body fat is crucial in both former athletes and ones still engaged in sports competitions. Indeed, fat mass is considered to be a non-functional mass, with increasing amounts mechanically and metabolically hiding sports performance and negatively affecting thermoregulation, physical functioning, and general health [26,27].

A strength of this study was the randomized design and the use of DXA as a reference method. However, this study was not without limitations. In fact, DXA is not considered the state-of-the-art method for evaluating FM, but it is employed in conjunction with air plethysmography and dilution techniques in the four-compartment model, which is currently considered the gold standard. Furthermore, the results of this study cannot be generalized, but they could be applicable to assessments performed at a single frequency of 50 kHz since differences in the analysis were highlighted between bioelectrical parameters measured using single and multifrequency devices [28–31].

5. Conclusions

BIVA seems to be a feasible field method to assess fat mass while avoiding the use of prediction equations. The specific BIVA approach represents a suitable method for evaluating FM% and its changes following a lifestyle intervention program. In particular, vector displacements towards the lower pole of an R-Xc graph were found to reflect decreases in FM% after an intervention program aimed to reduce body fat.

Author Contributions: Conceptualization, C.N.M., C.P.M., L.B.S., and A.M.S.; data curation, C.L.N., C.P.M., R.F., F.J., C.M., and A.M.S.; formal analysis, F.C., C.L.N., R.F., and F.J.; investigation, C.N.M., C.L.N., C.P.M., R.F., F.J., P.M., and A.M.S.; methodology, F.C., C.L.N., C.P.M., R.F., E.M., L.B.S., P.M., C.M., and A.M.S.; project administration, A.M.S.; supervision, L.B.S. and A.M.S.; visualization, L.B.S.; writing—original draft, F.C., C.L.N., C.P.M., E.M., and C.M.; writing—review and editing, F.C., C.N.M., C.L.N., C.P.M., R.F., E.M., L.B.S., P.M., C.M., and A.M.S. All authors have read and agreed to the published version of the manuscript.

Funding: This work was supported by the Instituto Português do Desporto e Juventude (Portuguese Institute of Sports and Youth). C.L.N. and R.F. were supported with a PhD scholarship from the

Portuguese Foundation for Science and Technology (SFRH/BD/143725/2019 and 2020.05397.BD, respectively).

Institutional Review Board Statement: The study was approved by the Ethics Committee of the Faculty of Human Kinetics, University of Lisbon (Lisbon, Portugal) (CEFMH Approval Number: 16/2016) and was conducted in accordance with the declaration of Helsinki for human studies from the World Medical Association [19]. Additionally, the present study was registered at www.clinicaltrials.gov (clinicaltrials.gov ID: NCT03031951) prior to participants' recruitment.

Informed Consent Statement: The participants provided a written informed consent.

Data Availability Statement: The data that support the findings of this study are available from the corresponding author upon reasonable request.

Conflicts of Interest: The authors declare no conflict of interest.

References

1. Silva, A.M. Structural and functional body components in athletic health and performance phenotypes. *Eur. J. Clin. Nutr.* **2019**, *73*, 215–224. [CrossRef] [PubMed]
2. Santanasto, A.J.; Goodpaster, B.H.; Kritchevsky, S.B.; Miljkovic, I.; Satterfield, S.; Schwartz, A.V.; Cummings, S.R.; Boudreau, R.M.; Harris, T.B.; Newman, A.B. Body Composition Remodeling and Mortality: The Health Aging and Body Composition Study. *J. Gerontol. A. Biol. Sci. Med. Sci.* **2017**, *72*, 513–519. [CrossRef] [PubMed]
3. Wang, Z.M.; Pierson, R.N.J.; Heymsfield, S.B. The five-level model: A new approach to organizing body-composition research. *Am. J. Clin. Nutr.* **1992**, *56*, 19–28. [CrossRef]
4. Lukaski, H.; Raymond-Pope, C.J. New Frontiers of Body Composition in Sport. *Int. J. Sports Med.* **2021**. [CrossRef]
5. Matias, C.N.; Campa, F.; Santos, D.A.; Lukaski, H.; Sardinha, L.B.; Silva, A.M. Fat-free Mass Bioelectrical Impedance Analysis Predictive Equation for Athletes using a 4-Compartment Model. *Int. J. Sports Med.* **2021**, *42*, 27–32. [CrossRef]
6. Buffa, R.; Saragat, B.; Cabras, S.; Rinaldi, A.C.; Marini, E. Accuracy of specific BIVA for the assessment of body composition in the United States population. *PLoS ONE* **2013**, *8*, e58533. [CrossRef] [PubMed]
7. Santos, D.A.; Dawson, J.A.; Matias, C.N.; Rocha, P.M.; Minderico, C.S.; Allison, D.B.; Sardinha, L.B.; Silva, A.M. Reference values for body composition and anthropometric measurements in athletes. *PLoS ONE* **2014**, *9*, e97846. [CrossRef] [PubMed]
8. Campa, F.; Matias, C.N.M.; Teixeira, F.J.; Reis, J.F.; Valamatos, M.J.; Toselli, S.; Monteiro, C.P. Leucine metabolites do not induce changes in phase angle, bioimpedance vector analysis patterns, and strength in resistance trained men. *Appl. Physiol. Nutr. Metab.* **2020**. [CrossRef]
9. Ward, L.C. Bioelectrical impedance analysis for body composition assessment: Reflections on accuracy, clinical utility, and standardisation. *Eur. J. Clin. Nutr.* **2019**, *73*, 194–199. [CrossRef]
10. Francisco, R.; Matias, C.N.; Santos, D.A.; Campa, F.; Minderico, C.S.; Rocha, P.; Heymsfield, S.B.; Lukaski, H.; Sardinha, L.B.; Silva, A.M. The Predictive Role of Raw Bioelectrical Impedance Parameters in Water Compartments and Fluid Distribution Assessed by Dilution Techniques in Athletes. *Int. J. Environ. Res. Public Health* **2020**, *17*, 759. [CrossRef]
11. Moon, J.R. Body composition in athletes and sports nutrition: An examination of the bioimpedance analysis technique. *Eur. J. Clin. Nutr.* **2013**, *67*, S54. [CrossRef] [PubMed]
12. Piccoli, A.; Rossi, B.; Pillon, L.; Bucciante, G. A new method for monitoring body fluid variation by bioimpedance analysis: The RXc graph. *Kidney Int.* **1994**, *46*, 534–539. [CrossRef] [PubMed]
13. Silva, A.M.; Nunes, C.L.; Matias, C.N.; Rocha, P.M.; Minderico, C.S.; Heymsfield, S.B.; Lukaski, H.; Sardinha, L.B. Usefulness of raw bioelectrical impedance parameters in tracking fluid shifts in judo athletes. *Eur. J. Sport Sci.* **2020**, *20*, 734–743. [CrossRef] [PubMed]
14. Campa, F.; Matias, C.N.; Marini, E.; Heymsfield, S.B.; Toselli, S.; Sardinha, L.B.; Silva, A.M. Identifying athlete body fluid changes during a competitive season with bioelectrical impedance vector analysis. *Int. J. Sports Physiol. Perform.* **2020**, *15*, 361–367. [CrossRef]
15. Khalil, S.F.; Mohktar, M.S.; Ibrahim, F. The theory and fundamentals of bioimpedance analysis in clinical status monitoring and diagnosis of diseases. *Sensors* **2014**, *14*, 10895–10928. [CrossRef]
16. Marini, E.; Campa, F.; Buffa, R.; Stagi, S.; Matias, C.N.; Toselli, S.; Sardinha, L.B.; Silva, A.M. Phase angle and bioelectrical impedance vector analysis in the evaluation of body composition in athletes. *Clin. Nutr.* **2020**, *39*, 447–454. [CrossRef]
17. Stagi, S.; Irurtia, A.; Rosales Rafel, J.; Cabras, S.; Buffa, R.; Carrasco-Marginet, M.; Castizo-Olier, J.; Marini, E. Segmental body composition estimated by specific BIVA and dual-energy X-ray absorptiometry. *Clin. Nutr.* **2021**, *40*, 1621–1627. [CrossRef]
18. Silva, A.M.; Nunes, C.L.; Matias, C.N.; Jesus, F.; Francisco, R.; Cardoso, M.; Santos, I.; Carraça, E.V.; Silva, M.N.; Sardinha, L.B.; et al. Champ4life Study Protocol: A One-Year Randomized Controlled Trial of a Lifestyle Intervention for Inactive Former Elite Athletes with Overweight/Obesity. *Nutrients* **2020**, *12*, 286. [CrossRef]
19. World Medical Association. World Medical Association Declaration of Helsinki: Ethical principles for medical research involving human subjects. *JAMA* **2013**, *310*, 2191–2194. [CrossRef]

20. Lohman, T.; Roche, A.F.; Martorell, R. *Anthropometric Standardization Reference Manual*; Human Kinetics Books: Champaign, IL, USA, 1988.
21. Santos, D.A.; Gobbo, L.A.; Matias, C.N.; Petroski, E.L.; Goncalves, E.M.; Cyrino, E.S.; Minderico, C.S.; Sardinha, L.B.; Silva, A.M. Body composition in taller individuals using DXA: A validation study for athletic and non-athletic populations. *J. Sports Sci.* **2013**, *31*, 405–413. [CrossRef]
22. Matias, C.N.; Santos, D.A.; Gonçalves, E.M.; Fields, D.A.; Sardinha, L.B.; Silva, A.M. Is bioelectrical impedance spectroscopy accurate in estimating total body water and its compartments in elite athletes? *Ann. Hum. Biol.* **2013**, *40*, 152–156. [CrossRef]
23. Campa, F.; Matias, C.; Gatterer, H.; Toselli, S.; Koury, J.C.; Andreoli, A.; Melchiorri, G.; Sardinha, L.B.; Silva, A.M. Classic Bioelectrical Impedance Vector Reference Values for Assessing Body Composition in Male and Female AthleteS. *Int. J. Environ. Res. Public Health* **2019**, *16*, 5066. [CrossRef]
24. Toselli, S.; Marini, E.; Maietta Latessa, P.; Benedetti, L.; Campa, F. Maturity Related Differences in Body Composition Assessed by Classic and Specific Bioimpedance Vector Analysis among Male Elite Youth Soccer Players. *Int. J. Environ. Res. Public Health* **2020**, *17*, 729. [CrossRef]
25. Kasper, A.M.; Langan-Evans, C.; Hudson, J.F.; Brownlee, T.E.; Harper, L.D.; Naughton, R.J.; Morton, J.P.; Close, G.L. Come Back Skinfolds, All Is Forgiven: A Narrative Review of the Efficacy of Common Body Composition Methods in Applied Sports Practice. *Nutrients* **2021**, *13*, 1075. [CrossRef]
26. Boileau, R.A.; Horswill, C.A. Body composition in sports: Measurement and applications for weight gain and loss. In *Exercise and Sport Science*; Garret, W.E., Jr., Kinkendall, D.T., Eds.; Lippincott Williams & Wilkins: Philadelphia, PA, USA, 2000; pp. 319–338.
27. Bosch, T.A.; Burruss, T.P.; Weir, N.L.; Fielding, K.A.; Engel, B.E.; Weston, T.D.; Dengel, D.R. Abdominal body composition differences in NFL football players. *J. Strength Cond. Res.* **2014**, *28*, 3313–3319. [CrossRef]
28. Silva, A.M.; Matias, C.N.; Nunes, C.L.; Santos, D.A.; Marini, E.; Lukaski, H.C.; Sardinha, L.B. Lack of agreement of in vivo raw bioimpedance measurements obtained from two single and multi-frequency bioelectrical impedance devices. *Eur. J. Clin. Nutr.* **2019**, *73*, 1077–1083. [CrossRef]
29. Dellinger, J.R.; Johnson, B.A.; Benavides, M.L.; Moore, M.L.; Stratton, M.T.; Harty, P.S.; Siedler, M.R.; Tinsley, G.M. Agreement of bioelectrical resistance, reactance, and phase angle values from supine and standing bioimpedance analyzers. *Physiol. Meas.* **2021**. [CrossRef]
30. Campa, F.; Toselli, S.; Mazzilli, M.; Gobbo, L.A.; Coratella, G. Assessment of Body Composition in Athletes: A Narrative Review of Available Methods with Special Reference to Quantitative and Qualitative Bioimpedance Analysis. *Nutrients* **2021**, *13*, 1620. [CrossRef]
31. Stratton, M.T.; Smith, R.W.; Harty, P.S.; Rodriguez, C.; Johnson, B.A.; Dellinger, J.R.; Williams, A.D.; White, S.J.; Benavides, M.L.; Tinsley, G.M. Longitudinal agreement of four bioimpedance analyzers for detecting changes in raw bioimpedance during purposeful weight gain with resistance training. *Eur. J. Clin. Nutr.* **2021**. [CrossRef]

MDPI
St. Alban-Anlage 66
4052 Basel
Switzerland
Tel. +41 61 683 77 34
Fax +41 61 302 89 18
www.mdpi.com

Biology Editorial Office
E-mail: biology@mdpi.com
www.mdpi.com/journal/biology